*Social behaviorism*

*THE DORSEY SERIES IN PSYCHOLOGY*

Editor
HOWARD F. HUNT
*Columbia University*

# Social behaviorism

**ARTHUR W. STAATS**
*University of Hawaii*

 1975

THE DORSEY PRESS *Homewood, Illinois* 60430
*Irwin-Dorsey International,* London, England WC2H 9NJ
*Irwin-Dorsey Limited,* Georgetown, Ontario L7G 4B3

*First Printing, January 1975*

ISBN 0-256-01537-6
Library of Congress Catalog No. 74–12931

*Printed in the United States of America*

To the first and second generation behavioral psychologists,
especially Ivan P. Pavlov, John B. Watson,
and Edward L. Thorndike,
and Clark L. Hull, B. F. Skinner, Edward C. Tolman,
and others referred to herein.

# *Preface*

I HAVE ATTEMPTED to formulate basic principles and an approach that could be generally extended to the various areas in psychology and in the social sciences—to child and adult behavior, normal and abnormal, individual, group, and cultural behavior—and to link with biological science as well. This effort has taken place over an extended period of years, beginning when I was a graduate student. It has led me to peruse literature in various areas of psychology and other disciplines. While my orientation was within behaviorism and learning theory my contact with these and other areas led me at an early time to the conclusion that a new generation of development was necessary before a behavioral approach could comprehensively deal with human behavior. Development was necessary in philosophy and at the basic theory level—one requirement being avoidance of partisanship for one or the other existing learning theory, since productive as well as unproductive elements abound in each. Development was also necessary in dealing with human behavior, to be able to include concepts, principles, and findings from non-behaviorally oriented approaches, and thus to break out of restraints imposed by the formulations of previous generations of behaviorists. I have called the resulting formulation *social behaviorism*, to indicate the debt to the past as well as the differences from it, and to indicate the bridge between the behavioral and non-behavioral.

The basic principles and the approach provide a way of looking at various areas of study in a manner that is integrative, enabling an increase in the range of areas that can be considered. I have attempted to show that this is a paradigmatic effort in Thomas S. Kuhn's sense. As Kuhn has recognized, a scientific paradigm is at first a framework which while interpreting many known events, has extensive implications and projections yet to be tested. In the present case a considerable literature is employed in presenting the paradigm. However, it is hoped that it has been adequately indicated that such a paradigm is a skeleton, a program, its body to be further fleshed in empirical and theoretical work in the progress of what Kuhn calls normal science. I have attempted to emphasize this characteristic by suggesting theoretical and empirical developments at various points that require elaboration and verification.

Following the goals outlined above has led to a product that is a general psychology of a systematic sort. I have written the book with its textbook uses in mind. Although the basic principles are dealt with in somewhat greater detail and depth than is usual in a general psychology book, the whole book is based upon those same principles. This produces, as the student progresses, a conceptual unification less demanding than an eclectic book that, while written to be simple, introduces new and sometimes competing concepts and principles in each area dealt with. Moreover, the systematic approach can lead the student to an integrated conception of human behavior that is not readily obtained from an eclectially oriented general psychology. That is, in addition to specific area treatments, the book *in toto* represents a conception or a model of man.

I hope the book will be useful in general behavior modification courses, and as a general psychology text, and in behavioral courses in specific areas such as personality or social psychology. I hope also that the book will provide an impetus toward the achievement of unified science in the study of man, and in this respect be of interest to the student of human behavior in various disciplines.

I am grateful to Reuben Baron, as well as to Howard Hunt and Norman Miller, for outstanding jobs of critiquing the manuscript and for their perspicacious and very helpful comments and suggestions. For help in preparation of the manuscript I wish to thank Frieda Helinger of the University of Hawaii Social Science Research Institute. Finally, to my wife Carolyn and our children Jennifer and Peter I express gratitude for their forbearance during the past several years when much of my free time was spent on this manuscript, and for the happy family atmosphere that gave me energy for this work.

*December 1974*                                              ARTHUR W. STAATS

# Contents

*gressive behaviors. Language-cognitive repertoires in aggression.* Imitation (modeling or observational learning): *Attention: Stimulus control of receptor-orienting behavior. Sensorimotor repertoires and imitation. Language as a mediator of modeling. Images and the mediation of imitation. Self-reinforcement mechanisms and social learning theory. Original behavior through imitation. Social learning and social behavioristic theories of modeling.* Instrumental personality repertoires. Instrumental, cognitive, and emotional personality repertoires: A tripartite but interactional approach.

*Part three*
*Functional behaviorism levels*

therapy: *Modeling and desensitization. Counterconditioning and aversion therapy*. The psychodynamic and conditioning conceptions in clinical psychology: *The conditioning paradigm*. The social behaviorism paradigm: *Three-function learning theory, A-R-D analysis and clinical treatment. Language-cognitive personality analysis and clinical treatment. Instrumental repertoire analysis and clinical treatment*. Additional elements of the paradigm and further rapprochements with traditional psychotherapy: *Language behavior therapy. Catharsis and A-R-D modification. Social interaction, transference and rapport. Insight, behavior analysis, the self-concept, and language behavior therapy*.

The elementary principle level: *Classical conditioning with children. Instrumental conditioning. Three -function learning. S-R mechanism learning*. The behavior modification level: *Sensorimotor skill development. Toilet training. Eating behavior and eating problems. Problem behaviors assumed to be organically caused. Disability and rehabilitation*. Additional studies of child behavior modification. The social behaviorism paradigm and child development: *Cumulative-hierarchical learning and child development. Cumulative-hierarchical learning and regularities and universals in child development. Social interaction learning principles. The personality level. Parents as trainers of their children*.

Behavior modification in education: *Producing adjustive behavioral skills. Early childhood educational behavior modification. Token-reinforcer system development. Token-reinforcer motivational systems and schools of the future. Behavioral counseling*. The personality repertoire level in education: *A theory of reading*. Conclusions.

The elemental behavioristic paradigm and psychological measurement. Behavior assessment: Foundations and status: *Behavior traits of personality. Abnormal psychology and assessment. Representative samples and behavior assessment. Behavior assessment*. The behavior assessment conceptual context. Social behaviorism and psychometrics: *Need for unity*. The A-R-D system and personality measurement: *The A-R-D properties of interest test items*. Intelligence (the language-cognitive system) and personality measurement: *Number-mathematical concepts and intelligence*. Instrumental behavior traits and verbal tests. Behavior analysis, psychometrics, and behavior modification.

*Part four*
*Biological science, social science, and humanistic levels*

Subjective and objective conceptions. Feelings and sentiments and thoughts. Awareness. The individual versus the general. Self-direction

# Part one

## Basic theory levels

# Models, theories, and paradigms

Science may be considered to constitute the systematic effort to gain knowledge about the events of the world. While that statement appears to be straightforward and reasonable, what constitutes knowledge is not always agreed upon. Many disagreements in beliefs occur because different standards of knowledge are basically involved. Comparison of different conceptions, and constructive synthesis of different conceptions, is difficult unless some set of standards can be agreed upon. This chapter will explore briefly some of the methods for seeking new knowledge as a preliminary step in the formulation of a hierarchical theory and paradigm of social behaviorism.

## What is theory?

The concerns, methods, facts, and technologies of the broad fields of physical, biological, and behavioral science are vastly complicated. Consequently, it would seem unlikely that general methodological principles could be abstracted that would be meaningful within the range from classic scientific theories down to common conceptions in our general language. The statements to be made in the next sections, however, have been formulated to attempt that type of generality, as the basis for the unification of the areas of knowledge to be treated herein.

## A THEORY INCLUDES A SYMBOLIC SYSTEM

One basic characteristic of all theories is their composition from some symbolic system. There are those who would separate scientific theories from prescientific theories on the basis of whether or not mathematics is basic to the theoretical structure—because scientific theory in the most advanced sciences depends so much on mathematics. Such a categorization is based upon superficial considerations, however. There are actually various symbolic systems. Mathematics is a symbolic system. Symbolic logic is another. Our common language is a symbolic system, and there are different languages that provide different symbolic systems. It should be understood that the various languages, including mathematics and symbolic logic, may constitute basic symbolic systems for theories.

A symbolic system consists of symbols and rules for relating the symbols. In a language the rules are set forth in a grammar. The rules of our number system are set forth in an algebra. Many statements can be made by following the rules of the symbolic system. For example, "The crepuscular lamb levitates tenderly" is a statement in a symbolic system of English. The statement

$$\frac{[a^{-2}\,(a^4 - 2a^3 + a^2)]^{-1}}{(1 - a)^{-2}} = 1$$

is a statement in the symbolic system of algebra. Both statements are true in the sense that they follow the rules of the symbolic system. Neither statement has any meaning regarding happenings in the world, however. They are merely statements in a language of symbols.

## LINKING SYMBOLIC STATEMENTS TO WORLDLY EVENTS

The importance of the language of algebra is that events of the world follow the same relationships as the rules of algebra do. When symbols from the algebra have been assigned to specific events, we can learn about how these events relate to one another. That is, by noting how the symbols relate to one another, we can find out how the worldly events relate to one another.

To illustrate, let us say that in the statement $a \times b = c$, the symbol $c$ stands for the distance traveled, $a$ stands for the rate of travel, and $b$ stands for the time of traveling. The algebraic statement given above now becomes a statement about the events of the world. If $a$ (rate) is 50 miles per hour, and $b$ (time) is 3 hours, we can predict that $c$ (distance) will be 150 miles. Following the rules of the algebra language, it should be equally true that $a$ (rate) $= c$ (distance)$/b$ (time). If we know that we have 150 miles to traverse and only three hours to travel, we can ascertain how fast we will have to drive. The algebraic statements constitute a useful "theory" about these aspects of the world. We can also make variations of these theoretical statements according to algebraic rules, that is time = distance/rate ($b = c/a$), and so on.

The same is true for natural languages. The symbols and rules of a language can be considered as a formal system—without worldly meaning. It should be noted, however, that many words of a language are firmly attached to particular worldly events, unlike the symbols of algebra, which are used in different ways depending on immediate interests. Nevertheless, there are words in a language—like pronouns—that can be varied in terms of the worldly events to which they are linked. Sentences employing such symbols will be empty of empirical meaning (will not be theoretical statements about the world) until the symbols are defined by worldly events. The following statement is an example. "If he pulls that, then it will die." The statement can be recognized as deriving from the formal system of the English language. In that sense it is true; it abides by the English rules of grammar. However, it cannot be seen to pertain to any worldly events, and in that sense it is meaningless—not either true or false, any more than the statement $a \times b = c$ is by itself empirically true or false.

When, however, certain of the words in the above statement are identified with certain worldly events, then the statement can be seen to pertain to those events. Let us say that the symbol *he* refers to a man, the symbol *that* refers to the trigger of a loaded gun, and the word *it* refers to a horse at which the gun is pointed. The statement of the formal system can now be seen to have meaning for the events of the empirical world to which it applies. That is, the statement can be true or false in a worldly (empirical) sense. It is when the symbols of a symbolic system have been defined by observable events that the statement has empirical truth value.

## THEORIES AS MODELS

It has been suggested that mathematics is a formal system of symbols with rules for relating those symbols (Stevens, 1951). Various statements may be derived from this formal system. The mathematical system, however, has its major significance in serving as a model for certain aspects of the world. It is when some of its symbols have been defined in terms of objects or events of the real world that its statements take on empirical meaning about those objects and events. If true statements regarding the world can then be generated, the model can be very useful, as indicated in the example involving rate, distance, and time.

The property of being *isomorphic* with events of the world is what characterizes a model. Various types of models are possible, of course. Models of airplanes are employed to study the aerodynamics of a full-scale airplane that is being planned. The model is studied because it performs in a manner that *parallels* the performance of the larger plane. There are many other mechanical models which make it possible for us to study the workings of the model and thereby obtain information about the workings of the actual events of the world in which we are really interested. In science a familiar case in point is the extensive use of models of atomic structures by chemists and physicists to stand for the minute events they cannot readily see. The model may be manipulated, and then expectations of how the events of the world might occur can be derived and checked.

The model may also consist entirely of symbols and be theoretical rather than mechanical. The same properties are important in either case. That is, the units of the model, as they are assigned to stand for events of the world, must behave with respect to one another as do the worldly events themselves. This is the primary characteristic of a theoretical model.

## FUNCTIONS OF THEORIES

A theory may describe a realm of events and thus prepare us to deal with that realm. For example, the men who described the rising and falling of the Nile River in ancient times and those who described the calendar produced symbolic statements that constituted models of certain aspects of the world. As models, their statements allowed people to respond in advance of the coming events. Such theory rests upon observation and description of the realm of events involved—the basic essential of any scientific theoretical endeavor. An important product of such theory is the prediction obtained.

Sometimes the events described are lawfully related to each other. When this is the case the theory can yield not only prediction but also control of the events involved. For example, when it had been observed that being bitten by a rabid dog was frequently followed by the fatal illness of rabies, the statement of this relationship constituted a theoretical model of a very simple type. The statement suggests that avoidance of the first event (being bitten) will preclude the occurrence of the second event (having rabies). The "theoretical model" suggests the manner in which the second event can be controlled. When a causative (explanatory) theoretical statement exists, and when the prior event (cause) can be manipulated, then the subsequent event (effect) can be controlled. Control is another important product of scientific theory.

## Theory characteristics

There are commonalities in the basic features of different theories. The word *theory* should not be restricted to the esoteric, the mathematical, or the highly technological laboratory. The differences in levels of theory—

from common language conceptions to the highly developed theories of the physical sciences—are differences not of kind but of degree. The differences occur along a number of dimensions, each of which can involve various levels of quality. It is relevant to discuss some of these dimensions here.

## PARSIMONY IN SYMBOLS

It has been suggested that theory is based in a symbolic system, which could be a natural language, a part of mathematics, or symbolic logic. The symbols of different symbolic systems vary. This, however, is largely arbitrary. It would not matter if the rules of algebra were stated in the Cyrillic script instead of the Latin script. As long as the rules themselves were the same, the particular symbols employed would be unimportant. The Spanish statement "*Veinte y cinco y diez y cuatro igual treinta y nueve*" is the same as "Twenty-five plus ten plus four equals thirty-nine." The symbols differ, but the rules relating the symbols are the same, and the two theoretical systems are equivalent.

It is significant to indicate the fact that the symbols themselves are not important, but rather how they are used. In the social sciences there are many cases where *different* terms will be considered differently although they actually are defined by the same things. One of the tasks of improving the theoretical system in such a case is the deletion of surplus terms in the system. It is also the case that the individual who seeks generality of theory across the boundaries of different disciplines may have to ignore the seeming differences that are engendered by the use of different symbols, looking rather to the things that define the symbols. As an example, it may turn out that the terms *utility* in economics, *value* in sociology, *cathexis* in psychoanalytic theory, and *reinforcement* in psychology may have uses (definitions) that suggest they should be considered the same. Since the present work is concerned with general theory, this matter of translating multiple terms from other systems and areas of study and reducing them into one term will occur repeatedly.

## RULES FOR LINKING SYMBOLS AND EVENTS

Important differences in the type of theory, as well as in the specific merits of the theory, can arise in the rules employed for linking the symbols to the events the theory is supposed to model (although the rules may not be explicit). An example of such a rule might be that a symbol would not be defined by a worldly event unless that event could be observed publicly, and that the symbol would only stand for (mean) those observations.

This rule, or at least a similar rule, has been suggested by the philosophy of science that has been called operationism. This explicit rule was advanced, as a matter of fact, because not all theories followed the practice, and because the rules that were followed sometimes resulted in ineffective or misleading theories. Certain primitive theories of nature may be used to indicate what can happen when the rule of linking symbols and events does not demand strict observation. In ancient times there was a theory of nature in which all objects were thought to have in them a living process (spirit, soul, or animism). Rocks were considered to have such an animism, as were trees, streams, mountains, the sun, moon, and so on. The terms suggesting spirits (or souls) inside these objects could not have been introduced according to the rule of operational definition. No one has ever observed an animism, spirit, or soul inside a rock, mountain, stream, the sun, or for that matter in a tree, fish, dog, or man.

The point is that it is important to differentiate the rules of definition that apply for any particular theory, for these rules give the theory different properties. There may be a theory that there is a god inside a volcanic mountain. The theory may further state that the god has certain characteristics: he is benign if his desires are followed, but when his desires are thwarted he becomes angry and the volcano erupts.

Let us say that there actually is a volcano god who does respond in this way. If the villagers who live at the foot of the mountain follow the god's desires, they prevent the eruptions they dread. The theory then has value in guiding them to actions they consider beneficial.

But if the rules of definition of terms have been incorrect, if there is no god in the volcano, the theory will be incorrect, and actions stemming from the theory may be irrelevant. Worse, if one of the things the villagers do is to throw human sacrifices into the volcano, the theory may be very antagonistic to their happiness.

What occurs in such theories as this is that individuals respond to the terms of the theory as if the terms were defined by actual events. However, if the terms have not been defined by actual events, then the statements that include such terms are likely not to parallel reality. A theory composed of such statements will yield poor prediction and control. It is for this reason that the rule of operational definition holds. When a theory maker follows the rule of operationionism strictly, he does not make the mistake of including terms in his theory that seem to refer to actual events but in fact do not.

In our common language, and in common languages in general, there are no statements for indicating the rule by which terms should be defined. Many terms are defined by observations. For example, most nouns —like chair, table, book, car, boy, and so on—are terms that are linked to easily observable objects. Many other terms, such as ego, soul, God, will, intelligence, talents, traits, minimal brain dysfunction, learning disability, and so on—are not so easily linked to observable objects or events.

OPERATIONISM, ELEMENTAL BEHAVIORISM, AND INTERVENING VARIABLES. It has been implied by example, at least, that theories that include terms not defined by observations are not isomorphic with reality and thus will poorly serve the functions of a good theory in guiding one's actions. Is it the case, then, that terms that purport to stand for the actual events of the world must be defined by direct observations of those events? That was the dictum of operationism in the field of physics (Bridgeman, 1928). John B. Watson, in psychology, proposed a similar approach in his behaviorism: "Let us limit ourselves to things that can be observed, and formulate laws concerning only those things. Now what can we observe? We can observe *behavior—what the organism does or says*" (Watson, 1930, p. 6). In the context of the time in which this was said, Watson's elemental behaviorism had a good deal to recommend it.

Concern with the appropriate rules to follow for linking terms in a theory to the events to which the terms were supposed to denote continued, however. Tolman (1936) suggested that there are events that are internal to the individual and which themselves cannot be observed but which can be specified by observations. He suggested that the characteristics of "mental processes" could become known by the manner in which they act as a "set of intermediating functional processes which interconnect between the initiating causes of behavior, on the one hand, and the final resulting behavior itself, on the other" (p. 2). An example of such intervening variables that Hull (1943) elaborates from Carnap (1935) is that of a man's inferred anger. Hull suggests that anger "lies between 1) the antecedent conditions of frustration and what not which precipitated the state, and 2) the observable consequences of the state; i.e., anger is an unobserved intervening variable" (p. 277).

For Hull, however, there was to be no other meaning to the intervening variable term than the meaning given by observing the antecedent situation and the consequent behavior (Spence, 1944). An intervening variable was considered to be only a logical term. It is suggested, however, that observations must be sought to characterize the inferred term, in addition to the observation of the stimulus situation and the overt behavior of the individual. Using Hull's illustration, the term *anger* may be considered to stand for an emotional response, an actual process, not an empty logical term. Observations to define the term must be sought. Thus, when the individual is subjected to a frustrating situation, does his heart beat increase in rate, does he perspire more, is there a change in the blood volume of various organs, are glandular secretions, such as adrenalin, increased, and so on? Observations such as these can help link the term *anger* to actual events.

It is suggested further that it is these and other *types* of observations that provide the empirical justification for terms in a theory of man that refer to events within the individual which are not directly observable. A philosophy of theory construction that suggests that terms may be introduced ·vhich do not refer to direct observables is seen to be appropriate. It is suggested, however, that such terms must be defined by various types of indirect observations, including physiological, anatomical, or other biological observations. As Chapter 15 will indicate, this aspect of the approach helps provide a basis for linking the study of behavior with biological science study.

INDIRECT OBSERVATION. It is not possible to provide any detail here on the very significant matters in the area being discussed. There are, however, several points that should be emphasized. For one thing, it is the case that there are various levels of the extent to which a term in a theory is based upon observations. A term like volcano god, as an example, is easily seen now to involve no direct observations. The events corresponding to the term are inferred. Peoples who accept such terms would argue, however, against a suggestion that there was no proof (observations) to support the term. They might recount cases where people had broken the volcano god's laws and this had been followed by an eruption. They might also point to similar occurrences with other gods and to other things as evidence: the fact that the sea provides fish proves there is a god of the sea, the fact that there are sometimes storms indicates the gods are angry, the fact that disease strikes some people and not others proves that there are those who are favored by the gods and those who are not, and so on.

Let us examine another term that is widely accepted in our common language theory, the term *intelligence*. No anatomical structure or physiological or biochemical process, or what have you, has ever been found to link with intelligence. Nevertheless, the term is very tenaciously held by almost everyone in our culture, including psychological scientists. Now, although there are no direct observations to link to the term, there are a number of indirect, or circumstantial, types of evidence. Naturalistic observations, for example, show that "intelligent" behavior runs in families. Moreover, children who demonstrate intelligent behavior at one age will generally do so at another, suggesting some internal personal quality. There is knowledge which shows that as species develop more complex nervous systems they are capable of more complex behaviors, and it is commonly observed that different species display different characteristics of behavior. The science of genetics has shown that biological inheritance can determine physical traits in animals, as well as some in man. The whole thrust of evolutionary fact is that genetic changes can occur that have marked effects upon the species.

Thus, there is a good deal of indirect or circumstantial evidence that leads many people to expect that individual differences in biological structure or physiological function are the underlying causes of many individual differences in man's behavior—including the quality of behavior we consider to be intelligent. In the case of the volcano god concept the indirect evidence was very loosely linked to the terms in the theory. The concept of intelligence is linked to a good deal of observation, but not of a direct type.

There are several central points here. There are concepts for which there are no direct observations. The indirect observations to which these concepts are linked can vary in quantity and quality. The rules of linking terms to observations thus may vary from theory to theory. Ordinarily these rules will be implicit rather than explicit. Frequently, a distinction is not made, even in science, between terms that are directly linked to observations and those that are not. While indirectly defined terms may be introduced into the theory productively, it should be understood that the variation in the quality and quantity of support for the term must be considered. The formal and informal statements of the theory must be weighted according to the quality of its terms. Statements derived from the volcano god theory must be weighted less than statements derived from a theory of intelligence. And the latter statements must be weighted less than statements from a theory in which the terms are all linked to direct observations, or more closely linked to indirect observations.

Mistakes have been made in the past where indirect (circumstantial) evidence has been the basis for a scientific concept (like the concept of mental faculties of phrenology) but where the theory does not prove to be productive. Where we continue to meet frustration in finding direct observations for a term in our theory there is the likelihood that there are none. This may be the case with the biological concept of intelligence. Although there are many indirect observations employed to support the term, there are questions involved, as well as a paucity of direct evidence, and a good deal of countering evidence (see Staats, 1971a; Layzer, 1974).

At any rate, the quantity and quality of evidence are important, as well as the strength of opposing observations. There is a continuum of observational justification for the terms we have in our theories. There should be a continuum of confidence that can be placed in the statements and decisions derived from those theories.

## COMPREHENSIVENESS

Theories may differ also in the number of events to which they pertain. Statements that pertain to only one relationship between events will not usually be called a theory. Moreover, theories that span a wide number of events that are significant will be considered more important than theories that span a lesser number. Classic theories of great import tend to include principles that extend to a large number of worldly events.

This is thus a characteristic upon which theories can be compared, especially when the theories are competitors within the same realm of events. It is not infrequent in the social and behavioral sciences for theories to be in competition. In such cases, one may ask whether the theories arise only in the realm of a limited sphere of events and a limited number of observations. One may also be concerned with the extent to which each theory in one realm of human behavior joins up productively with theories in other realms of behavior, to form a unified theoretical structure. The theory that arises in a limited sphere and does not link in a productive manner with theoretical statements in other related realms of events will not be as valuable as the theory that does so.

The possibility for a comprehensive theory in the study of human behavior is the concern of this book. At this point, however, it need only be indicated that one of the goals of theory construction in science is the development of theoretical statements that can incorporate a wide number of empirical relationships. It is thus appropriate to compare approaches in their ability to do this. There are other characteristics that differentiate theories. For example, the observations on which they are based may vary in detail, in precision of measurement, reliability, and so on. For present purposes, however, these need not be considered.

## CONCLUSIONS

It has been suggested that a theory is a symbolic system linked to worldly events. Further, a theory may be defined by its use. If symbolic statements are employed to understand natural events and to make decisions about them, or to predict and control such events, then the symbolic body is attempting to perform the functions of a theory. It is not necessary, as some have maintained, that developments such as the use of mathematics, refined measurement and observational technologies, and so on, be part of the theory. The use of these characteristics as criteria arbitrarily rules out of consideration many symbolic structures that serve the functions of theory. This is unfortunate, because by not considering such symbolic systems as theory, the systems are not subject to evaluation in the same way theories are.

It is productive to consider the functions as defining the theory. By so doing we can evaluate symbolic statements that are not ordinarily subjected to such scrutiny. Our common language contains many theories that apply to aspects of the physical and social world, in the sense that we make decisions and predictions about events and attempt to affect them through the use of statements from the language. In this sense the languages of religion, ethics, values, politics, and education, and so on constitute theories that apply to human behavior. These common language conceptions are theories that direct many people's decisions and actions, and it is worthwhile to recognize these symbolic systems as theories so they can be evaluated by the standards we have established for evaluating theories. It is suggested that we could markedly improve our commonsense theories by use of the criteria that have been outlined. At least, we could have a better realization, through this evaluation process, of the characteristics and justification of our commonsense theories.

## Differences in models, theories, and paradigms

Models, theories, and paradigms can be distinguished in several ways. A useful distinction can be made between the terms which in a sense places them on a continuum of generality, with a model being of least generality and a paradigm of greatest generality.

### MODELS AND THEORIES

Models are sometimes considered to be distinguished from theories in having mechanical or mathematical features. Sometimes they are considered, in contrast to theories, as not actually representing the realm of events to which they pertain but only constituting a "model" which shares characteristics with that realm of events. (But this should apply to all theoretical systems.) Sometimes models are thought of as being restricted to serving a descriptive purpose rather than including cause-and-effect statements as some theories do.

It is suggested, however, that models can have the various characteristics of theory. The major distinction is that models are not as general as theories. It is also suggested that a general theory may include subtheories, which could be called theoretical models. As will be seen, the present approach will outline a general theory of human behavior, but included in this general theory will be more restricted theoretical analyses. These more restricted theories will derive from the same principles as those in the general theory and will help compose it. As an example, a subtheory of human motivation will be described. This could be called a model of human motivation. The essential fact is that the subtheory of human motivation constitutes a part of the general theory. The human motivation theoretical model could serve as the basis for a very extensive theoretical and experimental series of explorations—and does, as recent research and theoretical developments indicate. It is less than the whole theory, however.

### PARADIGMS AND THEORIES

T. S. Kuhn (1962) has introduced a term that may also be considered in this context. In referring to such classics of science as Newton's *Principia* and Lavoisier's *Chemistry,* he states:

> I shall henceforth refer to . . . "paradigms," a term that relates closely to "normal science." By choosing it, I mean to suggest that some accepted examples of actual scientific practice—examples which include law, theory, application, and instrumentation together—provide models from which spring particular coherent traditions of scientific research. (p. 10)

The concept of the paradigm is important in recognizing that a general theory is much more than its formal statement, in various ways. The ongoing activity of a science includes, for example, implicit acceptance of what it is that is important to study. There is widespread agreement on the types of experimental methods to employ, the types of apparatus, the methods of analysis of research results, and so on. It should be added that an essential ingredient of a paradigm is a philosophy of science. The paradigm also has a philosophy of gaining knowledge (for example, empirical study versus revelation) and includes an experimental philosophy and a methodology for developing theory. These are in addition to formal theory statements and may be implicit, not directly stated.

A paradigm has its central importance in directing the observations made by those who accept the paradigm and in organizing what otherwise would be disparate investigatory activity.

> In the absence of a paradigm or some candidate for paradigm, all of the facts that could possibly pertain to the development of a given science are likely to seem equally relevant. As a result, early fact-gathering is a far more nearly random activity than the one that subsequent scientific development makes familiar. Furthermore, in the absence of a reason for seeking some particular form of more recondite information, early fact-gathering is usually restricted to the wealth of data that lie ready to hand. The resulting pool of facts contains those accessible to casual observations and experiment together with some of the more esoteric data retrievable from established crafts like medicine, calendar making and metallurgy. . . . it produces a morass. One somehow hesitates to call the literature that results scientific. (T. S. Kuhn, 1962, pp. 15–16)

This statement has much that is relevant for the social and behavioral sciences. That a common paradigm to guide and unify the study of man has not yet arisen is quite clear. The present book, in addition to treating various specific areas of study, will attempt to indicate how the principles involved constitute a general paradigm for the study of man.

## Social behaviorism: A hierarchical theory and paradigm

One characteristic of theory that was not described in preceding sections concerns the hierarchical structure that some classic theories possess. In advanced states of a science, higher-order (more general, abstract, or more elementary) statements or laws may be made that can be employed to generate the relationships of lower-order phenomena (lower-order laws or principles). Thus a simple, elegant, general set of theoretical statements can be developed that will account for a number of different, less general observations or empirical laws. In fact, hypotheses may be derived from the theoretical statements that suggest empirical laws and observations not yet discovered. This characteristic of some scientific theories has been in part described by Spence (1944):

> The physicist is able to isolate, experimentally, elementary situations, i.e., situations in which there are a limited number of variables, and thus finds it possible to infer or discover descriptive, lower-order laws. Theory comes into play for the physicist when he attempts to formulate more abstract principles which will bring these low-order laws into relationship with one another. Examples of such comprehensive theories are Newton's principle of gravitation and the kinetic theory of gases. The former provided a theoretical integration of such laws as Kepler's concerning planetary motions, Galileo's law of falling bodies, laws of the tides and so on. (pp. 47–48)

It might be added that from the elementary (higher-order) principles of gravitation have come derivations concerning how to correctly aim and guide space vehicles on their way to the moon or other astral bodies. These derivations are additional lower-order statements that illustrate the control (utility) properties of the higher-order theories.

To continue, this conception of hierarchical theory was followed by the traditional learning theorists. The laboratory experiments on the principles of learning (conditioning) were considered to be the lower-order laws. The learning theorist attempted to organize these laws around some higher-order, more abstract laws—such as Clark Hull's concept of habit strength, B. F. Skinner's concept of reflex reserve, and so on. Hull and Kenneth W. Spence were the most systematic in attempting the development of such a learning theory, but by 1960 Spence indicated the failure of this general theory strategy.

It is suggested that the lack of success was not because the philosophy of science was incorrect but because of the way in which it was applied. These learning theorists recognized the hierarchical characteristic of classic theory, but they considered the laws of learning as the lower-order laws, and they focused upon the logical derivation of a set of higher-order laws. This approach was in error, it is suggested, in too narrowly limiting the events the theory was to concern. The construction of classic hierarchical theory requires the present field of study be elaborated to include functional human behavior in all of its complexity, not just the events of the basic learning laboratory. Then it can be realized that the principles of learning are themselves the higher-order principles. It is true that these principles have been discovered in the laboratory, not through logical (mathematical) derivation, as Newton did. But the distinction is superficial. It is the position and function of the elementary laws of learning that must be the deciding factor.

The suggestion is that the empirical, elementary principles of learning must take a position at the apex of the theoretical structure. The task is to employ these higher-order learning laws to explain, deal with, and integrate the lower-order principles, concepts, and observations of human behavior as these occur in the study of the psychological and social sciences. This does not mean there is not a basic, general theory level in this study. The experimental facts of learning are diverse and complex, and in cases contradictory. The task of basic learning theory is the selection, elaboration, derivation, and systematization of a set of basic learning principles from the experimental facts that can organize and guide the basic work and serve as a foundation for the more general theory of human behavior.

## LEVELS OF THE HIERARCHICAL THEORY

The principles of learning are elementary statements of how the environment can determine behavior. The principles are relatively simple, and because they have been studied in the artificial simplicity of the laboratory, they have seemed to be of distant concern with human behavior. And, without further development, that is the case. It is not possible—as some elemental behaviorists have claimed—to account comprehensively for complex human behavior through the straightforward application of the elementary learning principles. Rather, the elementary principles have to be developed, in a lower-order theoretical formulation, into a theoretical structure that lends itself to consideration of human behavior. As following chapters will indicate, before many specific types of human behavior can be dealt with it is necessary to show how the simple principles can be combined to yield more complex possibilities. In addition, at the next level, it is necessary to formulate a personality theory level of the theory—a level that includes principles of human learning not derived from the basic animal laboratory. It is necessary at this level to analyze the various important repertoires of human behavior, man's "personality" repertoires. The way that a person will respond in a situation is a function not only of the elementary learning principles but also of the complex repertoires of behaviors (personality characteristics) he has learned. This concept and this level of the theory allow the heretofore separate domains of behaviorism and personality theory to be productively brought together. This makes it possible to treat topics that could not be approached with traditional behaviorism, for example, the self-concept, the human emotional-motivational system, values, purpose, self-determination and personal freedom, personality tests, and so on.

Chapters on behavioral humanism, behavioral psychotherapy, psychological measurement, education, language and personality, the social sciences, and so on, become possible with this elaborated approach.

In addition, the basic principles require other types of lower-order elaboration. The elementary learning principles, after all, were stipulated in the stark simplicity of the laboratory. That has meant dealing with one stimulus, usually, one response, one organism, and so on. However, most human behavior is social. More than one organism is in interaction. The basic principles must be elaborated to indicate what occurs in interactions where one person provides determinants for another's behavior, and vice versa. These lower-order elaborations enable the theory to deal with various types of social phenomena. It may be suggested that there are other human learning principles, not derived from the basic laboratory, that must be added to the elementary principles, as will be indicated. These elaborations allow consideration of yet other lower-order aspects of human behavior.

One other level of the theory may be mentioned here. The manner in which biological science can relate to the study of human behavior in psychology and the social sciences will be indicated in Chapter 15.

## THE EXPLANATORY NATURE OF THE THEORY

It is important to note that the theory that follows is of the explanatory type. This characteristic stems from the employment of elementary laws that are of a cause and effect variety. Each of the principles of learning that is to be described is of this nature. For example, one basic law is that when an organism makes a response and this response is followed by a rewarding circumstance, that behavior will occur more frequently in the future. In this example, the cause of the occurrence of the later behavior is the reward of the previous occurrence of that behavior. One can predict whether the later behavior will occur from knowledge of the earlier occurrence. One can also produce or prevent the later behavior by manipulating whether or not the behavior is rewarded on the earlier occasion. The principle is thus an explanatory, cause and effect law. The law can be demonstrated very publicly and reliably.

The explanatory nature of the elementary principles to be employed is important because it will determine whether the lower-order extensions to human behavior will be explanatory and yield prediction and control. When such elementary principles are extended to a realm of human behavior, the result is an explanatory theory. This type of theory, as noted above, makes it possible to predict behavior from the knowledge of the preceding conditions and to affect the behavior by affecting the conditions. In this sense, also, the theory to be outlined has the explanatory characteristic of classic theories.

These various points will be exemplified in the following chapters.

# 2

# *Basic learning and behavior theory*

CENTRAL to the social behaviorism paradigm is the concept that complex functional human behavior is learned. Man, above all other creatures, is distinguished by his extraordinary ability to learn.

Recognition of man's learning has existed since antiquity. Aristotle considered the child's mind to be a *tabula rasa*, an empty tablet to be written upon by his experiences. These conclusions were based upon naturalistic observations. Ordinarily it is difficult to isolate the elementary principles that explain the things in which we are interested when our only source of observations is the natural world. The complexity of the events makes it difficult to show the lawfulness and determinacy involved when countermanding conditions cannot be controlled.

The development of the scientific study of learning took the path that other natural sciences have taken. It became experimental in the sense of turning to the laboratory for greater control. Although the concern with learning stemmed from an interest in understanding human behavior, it can now be seen that it was necessary to turn to the artificial simplicity of the laboratory to isolate the elementary principles involved. In real life, one deals with a complex environment, a complex behavior, and a complex organism with a complex history. This can be simplified in the laboratory by dealing with samples of a simple stimulus, such as a pellet of food, a simple sound, and so on; a simple behavior, such as salivation, a bar press, and so on; and a simple organism with a controlled history, such as a rat, dog, and so on. By this means it has been possible to isolate and specify many of the elementary principles of learning—that is, the basic ways that environmental events can affect the organism's behavior.

While the philosophy that experience molds the mind dates from antiquity, the experimental study of learning principles is relatively recent, dating from around the turn of the present century. It is important to note that the learning (or behavioral) approach has a distinct character in terms of growth; although the general approach has suffered vicissitudes of acceptance in psychology, it has been continuous and additive, like the approaches of the established sciences. Since its self-conscious emergence with the work of Edward L. Thorndike and Ivan P. Pavlov, investigators concerned with the basic learning area have extended and detailed the principles of learning, making methodological as well as substantive contributions.

Superimposed upon this more continuous progress have been discrete advances in terms of specific learning theories. The first generation of learning theorists or behaviorists consisted of such men as Pavlov, Thorndike, and John B. Watson. During their time the facts of learning were just beginning to be discovered. By the 1930s the experimental facts were much more complex, and the dominant thrust began to be the organization of the facts into second-generation general learning theories. Prominent theorists in this generation were Clark L. Hull, B. F. Skinner, Edward C. Tolman, and Edwin R. Guthrie. These theories had much in common with the first-generation theories. For example, there is a good deal of continuity from Watson's *Behaviorism* (originally published in 1924) to Skinner's and Hull's later theories, with overlap in principles, concepts, and analyses.

Other second-generation learning theorists who were identified with either a Hullian or Skinnerian approach were Kenneth W. Spence, Neal E. Miller, O. Hobart Mowrer, Charles E. Osgood, Fred S. Keller, and William N. Schoenfeld, among others. They have elaborated learning concepts, conducted research, and extended learning principles to additional areas of behavior.

> [T]he traditional learning theories, especially those of Hull and Skinner, were very important advances in the progress of the science of learning. . . . In this era of [second-generation] learning theories the area became the model for scientific theory construction in psychology. A great deal of animal research was conducted which greatly expanded the findings concerning the basic principles. Many different behaviors were also subjected to experimentation. Moreover, great strides were made in experimental methodology, for both animal and human research. . . . The importance of this fund of knowledge should not be underestimated. These developments, and their systematization into general learning theories, it is suggested, will come to be ranked with the early achievements of the other natural sciences. (Staats, 1970a, p. 193)

This is not to say that the frameworks provided by the first- and second-generation learning theories constitute complete or unerring guides for the continued development of the approach. Each generation has retained elements from the past while introducing important novel developments. A major aspect of the second generation was its emphasis upon competing theories and the expectation that one or the other theory was correct. A great deal of research was conducted under the impetus of this competition, and the atmosphere produced militated against an integrational effort. The traditional learning theories thus abetted a separatism in psychology.

Additionally, however, as was indicated in the first chapter, the second-generation behaviorists tended to think that there were basic (higher-order) principles of the theory, to be found rationally. These higher-order principles were then to be applied to the lower-order facts of animal learning. Although the learning theories were applied to human behavior, the relationship of animal learning study and human-level study was not clearly indicated—especially the reciprocal nature of the two levels of study in the task of basic theory construction. The basic theories of the second generation were thus based primarily on findings from the animal laboratory.

It is suggested that there were specific weaknesses in the second-generation learning theories, some of which did not lead to what will be considered to be central lines of development. For example, Hull's emphasis upon mathematical statement of the higher-order laws of learning led to the development of the study of mathematical learning theory. Skinner's emphasis upon schedules of reinforcement led to a proliferation of studies of this type of variable (see Ferster and Skinner, 1957) that were beyond the range of practical or theoretical significance. Skinner's theory has also involved constraints on development at various levels, as will be indicated further on.

The second-generation learning theories were developed in the 1930s and 1940s, on the basis of the conceptual and empirical foundation that was then available. The empirical study of learning has continued to grow, as has the methodology of experimentation. It must be expected that a third generation of theory and theoretical methodology will also be needed.

The present author began the study of basic learning theory and its extensions to human behavior in the 1950s. The approach he formulated at this time was that the learning theories contained important elements, but changes and development were necessary. One important aspect of this approach was the consideration of the first- and second-generation theories as a common context, rather than as mutually exclusive. The theories were seen to contain principles and concepts that were susceptible to harmonious combination as well as those that could be discarded.

This development had to include an integration of classical and instrumental conditioning, the two major types of learning. Skinner treated these as distinctly separated, with classical conditioning playing a minor role in human behavior analysis. Hull did not distinguish classical and instrumental conditioning. In both cases the symbolic system did not reflect the interrelation of the two types of conditioning. At any rate, a selective and integrational effort appeared necessary as a first step in a third-generation behavioristic effort. (see Staats, 1956, 1957a, 1957b, 1961, 1963, 1968c).

In addition, it was felt that development of the basic learning theory required reference to the human level of study and to complex human behavior. Observations at this level had to have an influence on the basic formulation. Thus, as the present author's approach grew, the human level of study was given important consideration. The elaboration has continued to refer to both animal and human research. Additional elaboration of the learning theory at the basic level of animal and human research is still necessary (see Staats, 1970a, for a programmatic outline for this continued effort). The goal of this chapter, however, is to organize the elements of learning into a distilled learning theory to be employed as a foundation for dealing with the various areas and fields of human behavior, to be considered in the later chapters of the book. To begin, the experimental facts of classical and instrumental conditioning will be summarized.

## Classical conditioning

Philosophers in the British empiricist tradition, following the Aristotelian tradition, made systematic statements of the principles they felt were involved in the manner in which experience affected the contents of one's mind. These 18th-century philosophers were called associationists because a primary principle was that sensations become associated when they occur in contiguity with one another in the person's experience.

Some of these principles were later employed by Hermann Ebbinghaus in the latter part of the 19th century as the conceptual foundation for experimental work that showed that associated elements were learned and retained in a lawful manner (Boring, 1950). Ebbinghaus' interests were more in describing the higher processes of the mind than in establishing the elementary principles of learning, however.

It fell to Pavlov, a Russian physiologist, to begin the isolation of the principles of conditioning, employing animal subjects and precise laboratory controls. The use of animal subjects and physiological responses formed a basis for establishing the generality and biological nature of elementary learning principles.

Actually, the discovery of classical conditioning was in part serendipitous (that is, accidental), since Pavlov's main interest was in studying
digestive processes. As part of this study dogs were prepared by diverting
their salivary ducts through their cheeks so the drops of saliva could be
collected and measured. When food was placed in the animals' mouths
the salivary response was elicited, on an unlearned basis. The serendipitous finding was that the dogs sometimes salivated when food had not
yet been placed in their mouths, in an anticipatory manner. It was
noticed that a stimulus that had been presented in contiguity with the
food would later itself elicit the salivation.

This is the principle of classical conditioning which Pavlov began to
study systematically. In Pavlov's work a simple sample of the environment
was employed as the to-be-conditioned stimulus—a sound of a bell. The
stimulus that elicited the response was a bit of food stuff. The glandular
response under observation was indexed by the number of drops of
saliva. The bell, called the $C_S$ (conditioned stimulus), was presented and
followed by presentation of the food, called the $UC_S$ (unconditioned
stimulus). This was done for a number of conditioning trials. Then the
bell was presented alone, and the amount of salivation to the bell was
measured. The subjects employed, dogs, could be very reliably conditioned to make this internal physiological response to a new stimulus.
Classical conditioning is schematized in Figure 2.1.

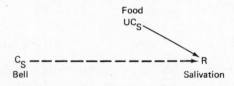

**FIGURE 2.1**

*Presentation of the food stimulus elicits the
salivary response. Once the bell has been
paired with food, the bell also comes to
elicit the response on its own.*

A witty but cynical colleague of the author once questioned the importance of the principle of classical conditioning by asking "Who cares
about slobbering dogs, anyway?" Certainly the scientist or professional
or man in the street who is interested in the wordly measure of human

behavior is justified in asking such a question from the standpoint of his interest. The answer will be given in various parts of the present book. It is the intent of the present chapter to present the basic principles of learning in their abstract form, without elaborating on the significance of the principles in human action. Later chapters deal with extensions that focus on the content of human behavior. It is relevant to indicate here, however, that a finding such as that of classical conditioning with dogs becomes relevant for understanding human behavior only when the principle has been extended on a theoretical, experimental, or applied basis to the human level.

An early example of such an extension can be seen in a study by Watson and Raynor (1920). They presented a white rabbit as the $CS$ to a child. When the child was looking at the rabbit, a loud noise was presented as the $UCS$. The noise elicited a crying response in the child. After a sufficient number of conditioning trials the white rabbit by itself came to elicit the crying response.

It should also be noted that many unconditioned stimuli exist in the world, and there are a number of types of responses involved. For example, various types of edible substances will elicit various responses in the organism. Such reactions may be indexed by the salivation that such food substances elicit when placed in the mouth, or by the change in heart rate, the change in the flow of gastric juices in the stomach, the change in blood flow to different internal organs, and so on, all of which are conditionable. Tactile stimulation of certain types elicits responses in the individual, especially tactile stimulation of the sexual (erogenous) areas. Evidence of such responses may be seen in the tumescence of erectile tissue, as in genital responses. Smells, sounds, and temperature also provide examples of internal, physiological response-eliciting stimuli.

These examples involve stimuli that elicit "positive" responses. There are also many stimuli that elicit "negative" responses. Organisms are constructed so that injurious environmental energies also act as unconditioned stimuli. A loud noise will elicit a change in the sweat gland activity of the skin. A painful mechanical, electrical, or temperature stimulus to the skin will elicit various responses, such as change in heart rate, change in blood volume to different organs, and so on. A bitter taste or malodorous scent will elicit negative responses. Stress on the muscles and tendons, as in sprains and wrenches, will activate pain receptors that serve as negative eliciting stimuli.

It can also be noted that most of the internal organs of the body may be involved in response to stimulation and thus be subject to classical conditioning. This includes the organs concerned with all the vital functions such as the heart; the respiratory, digestive, and reproductive systems; the liver, spleen, various glands, and so on. It can be said in a general way that it appears that the stimuli involved have a very direct biological relevance to the organism—being those that are necessary to either avoid or to gain conditions that are necessary for individual or species survival. These stimulus-response capacities of the organism appear to be built into the biological organism on the basis of biological evolution, contributing to the survival of the individual and the procreation of the species. The members of the human species are similar in respect to this responding. There may be differences in sensitivity, but all normally constructed humans appear to respond similarly on an unlearned basis to positive and negative stimuli—again probably because absence of these capabilities would not have led to biologically successful organisms that would duplicate their kind.

The suggestion is that all kinds of stimuli which the organism is sensitive to, but which do not elicit a response in him, can come to do so on the basis of conditioning. Any type of stimulus can come to be a response-eliciting stimulus. No rationality is necessarily involved. The only requirement is that the stimulus be paired with some stimulus that does elicit a response.

## COROLLARIES OF THE PRINCIPLE OF CLASSICAL CONDITIONING

Several subprinciples that are involved in classical conditioning or that further elaborate classical conditioning processes have been specified through research. These subprinciples are also important in making analyses of human behavior and in understanding the operation of conditioning in real life as well as in the laboratory.

HIGHER-ORDER CLASSICAL CONDITIONING. When a number of conditioning trials have occurred under the circumstances that are possible in the human's extended learning history, the $CS$ will come to very strongly elicit the response—exen though it is rarely ever again presented with the $UCS$. To all intents and purposes, the conditioning may become more or less permanent. In such a case, the $CS$ may serve the same function as a $UCS$.

To elaborate, if a new stimulus is paired some number of times with a *conditioned stimulus* that strongly elicits a response, the response will be conditioned to the new stimulus. The new stimulus thus becomes a *higher-order* conditioned stimulus. The person who has been conditioned to respond positively to smiles, as most of us have, will be conditioned positively to people who smile. The response to such people will have been acquired in this case on the basis of higher-order conditioning. Higher-order conditioning of responses can be produced in the animal laboratory (Brogden, 1939a; Murphy and Miller, 1957), although the effect is not so strong, ordinarily, as with first-order conditioning. Under appropriate conditions, however, the effects of higher-order stimulus pairing may be long lasting (Zimmerman, 1957), even on the animal research level. In human experience there are more general opportunities for such appropriate conditions, as will be indicated.

STIMULUS GENERALIZATION AND DISCRIMINATION. Stimuli in real life are rarely the same on each occurrence. Words are not pronounced the same, for example, from time to time or from person to person. Physical stimuli seen from different angles actually constitute different stimuli. Yet we respond to similar stimuli as if they were the same. This is adaptive; if we had to learn again the response to a stimulus each time it varied slightly, the effects of learning would be too specific to be of much help.

Organisms are so constructed, however, so that conditioning generalizes. A response conditioned to one stimulus will be elicited by a similar stimulus—in strength depending upon the extent of similarity. As an example, a boy who is conditioned to negative responses to his father will have such responses elicited by individuals who are similar to his father. This is a naturalistic example of stimulus generalization. Experimental description of the principle has also been made (Hovland, 1937a, 1937b; R. A. Littman, 1949).

There is another principle of conditioning, however, that applies in cases where the response to a similar stimulus is not adaptive. According to the principle of stimulus generalization, let us say that an organism has been conditioned to respond to a tone—as in Pavlov's procedures. On presentation of a tone of a different frequency, it will be found that the organism will also salivate. However, let us say that the $UCS$, the food stimulus, never follows the second tone, but the food stimulus continues to be paired with the original tone. In this case the organism will learn a discrimination. It will continue to salivate when the first tone is presented by itself but will salivate progressively less to the second tone. Such discrimination has been investigated with various responses, including the classically conditioned eye blink (Gynther, 1957).

Humans can learn different responses to stimuli that are quite similar. All that is required are differential conditioning circumstances. The human organism is thus constructed to be adaptive not only in behaving according to the principle of stimulus generalization when the consequences are the same but also in learning to discriminate when the consequences are different.

EXTINCTION AND COUNTERCONDITIONING. Included in the above description of discrimination was a principle that is basic to an understanding of learning. When the organism has been conditioned to respond to a stimulus, the learning will persist. Barring countermanding circumstances, if the stimulus is presented even at a much later period of time, it will still elicit the conditioned response (Marquis and Hilgard, 1936; Razran, 1939). It is adaptive that conditioning persists. But it is also adaptive that conditioning can be changed—for conditions in the world vary.

One way that conditioning can be changed is through the process called extinction. Even though a response has been conditioned strongly to a stimulus, if that stimulus is then presented a number of times more and is never paired with the $UCS$, the conditioning to the stimulus will progressively weaken (Pavlov, 1927; Hilgard and Marquis, 1935). That is what accounted for the discrimination in the example above; the response to the second tone weakened when that tone was never followed by the food stimulus.

As an example of extinction, a child may be conditioned to a negative response to the sight of a dog because the dog has jumped upon him to lick his face and has knocked him down. If the dog is trained not to jump on the child, and the child experiences seeing the dog repeatedly with no repetitions of the assault, the dog, as a stimulus, will gradually cease to elicit the internal response.

Conditioned responses may also be changed in another process called counterconditioning. In the above case, the child's negative response to the dog could also be changed by pairing the sight of the dog with some other $UCS$ that elicited a positive response in the child (Watson and Raynor, 1920; M. C. Jones, 1924). Thus, the dog could be introduced in the presence of the child and the child could be given a piece of candy. The candy would elicit a positive response which, when conditioned to the dog, would counter the negative conditioning. If the process is to be successful, it is necessary that the counterconditioning stimulus (the candy) elicit the new response more strongly than does the original conditioned stimulus (the dog). There are various ways for insuring this.

DEPRIVATION. An unconditioned stimulus will elicit a response to an extent depending upon the subject's deprivation of that stimulus. For example, food is an unconditioned stimulus for the salivary response when the subject has been deprived of food for a time. But when he has just eaten to satiation, a food stimulus elicits the salivary response, much less strongly, if at all. Thus, classical conditioning of a response, using food as the $UCS$, will depend upon the organism having had a period (not necessarily a long period) without food (Finch, 1938).

The same is true of sexual stimuli as well as other internal response-eliciting stimuli. The deprivation corollary to the principle of classical conditioning does not seem quite as relevant to negative response elicitation and conditioning (although the eliciting functions of a negative stimulus will decrease with repeated presentation). Viewed from the standpoint of adaptation, this makes sense. It is important that the organism is not repeatedly and continually occupied by a positive stimulus much longer than it takes to perform the consumatory function involved, such as eating or copulation. On the other hand, it is important that a harmful stimulus continue to have its effect, even after it has been presented a number of times.

ADDITIONAL COROLLARIES. A number of other subprinciples of classical conditioning have been specified through laboratory research. For example, the time interval between $CS$ and $UCS$ will affect the extent of conditioning, with about a half second being the most propitious interval (White and Schlossberg, 1952).

Another subprinciple that can be important in considering human behavior is that conditioning can occur even when the $UCS$ does not continuously follow the $CS$. While it produces a more intense conditioned response if the pairing is "continuous" on the various trials, the resistance to extinction of the conditioned response is actually greater if the pairing has been of an intermittent nature (Humphreys, 1939). Thus, if one wished to produce conditioning that would be very resistant to extinction, it would be wise to have continuous pairings at first and later have intermittent pairings. Life circumstances very commonly arrange just such conditions.

Finally, the number of classical conditioning trials may be a powerful variable in the extent of conditioning that occurs (Hovland, 1937b). Ordinarily, the more trials—up to some point—the stronger the conditioning will become, both in the intensity with which the conditioned stimulus elicits the response as well as in how resistant to extinction the conditioning is. The long and complex learning history of humans provides opportunities for conditioning far stronger than can be given an animal research subject in a brief experimental period.

## Instrumental conditioning and the principles of reinforcement

Pavlov reported his work on classical conditioning at the turn of the 20th century. In this same period an American psychologist, E. L. Thorndike, was performing research that had the same significance with respect to the other major type of learning, instrumental conditioning. Pavlov's work concerned the response of the salivary glands, one of a number of internal responses of glands and smooth muscles (like those in the blood vessels and the walls of the viscera). Thorndike's work concerned the responses of the skeletal musculature. These muscles move the body itself, and thus instrumental conditioning pertains to actions. Thorndike conducted a series of experiments with animals employing a puzzle box. The animal was placed in the box, or cage, with a piece of food outside the box. Inside the box was some type of lever or rope or bobbin which, if activated by the animal's movements, would release the door to the cage and enable the animal to exit from the cage and receive the food.

Thorndike kept records of the time necessary for the animal to get out of the cage on the successive trials and found that there was a learning curve. The animal took less and less time over a series of trials. Thorndike saw the importance of the consequences of the organism's response in the animal learning that response. He termed this principle the law of effect. If a response is followed by a pleasant state of affairs, the response is learned; if it is followed by an unpleasant state of affairs, the response is not learned.

The principle of instrumental conditioning can be stated more explicitly in the following form. If a motor (instrumental) response occurs in the presence of some stimulus or stimulus situation, and the response is followed by the occurrence of a certain type of stimulus which will be called a positive reinforcing stimulus, the response will tend to occur again in the presence of that stimulus condition. After a sufficient number of such instrumental conditioning trials, the stimulus condition by

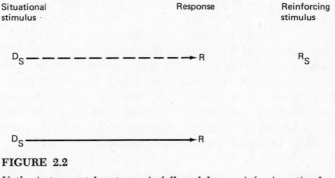

**FIGURE 2.2**

*If the instrumental response is followed by a reinforcing stimulus, then later the situation will become a directive stimulus and elicit the response, even when reinforcement is not presented.*

itself will tend to strongly elicit that instrumental response—even if the response is not followed by the reinforcing stimulus.

A reinforcing stimulus, symbolized as $^R$S, strengthens the tendency of the stimulus condition to elicit the instrumental response. The situational stimulus, or any part of it, that comes to elicit the response is called the directive stimulus, or $^D$S. Instrumental conditioning (or the principle of reinforcement) is schematized in Figure 2.2.

Several examples of instrumental conditioning will be mentioned. The T maze provides a good illustration (see Tolman, 1938). Let us say a rat is placed in the starting box of an enclosed runway that diverges into two paths at the far end, as indicated in Figure 2.3. Let us say that at the choice point the animal can see the alleys leading in each direction.

The numerals 1 in the figure indicate the presence of two small lights. A light may thus be turned on in either alley. The numerals 2 indicate where a pellet of food may be placed.

The animal is then given training in which, if he runs down the alley where the light is on, he will find a pellet of food. The light stimulus is randomly alternated, and the food is found only if the animal runs toward

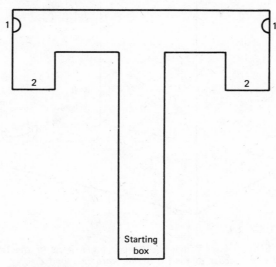

**FIGURE 2.3**

*The T maze. A pellet of food is placed at either point numbered 2 if the corresponding light, numbered 1, is turned on.*

the light stimulus. A choice of the other direction results in removal from the box without receipt of food. This system means that a response of approach to the light stimulus is followed by reinforcement.

In terms of the stimulus-response analysis, this situation involves a $^D$S (the light), a response (approaching the stimulus), and a reinforcing stimulus (the food pellet). Under such circumstances, the $^D$S will come to elicit the approach response, according to the principle of instrumental conditioning.

The same principle can be seen in another type of apparatus called an operant conditioning chamber or Skinner box, as shown in Figure 2.4 (see Skinner, 1938). Say that the organism is placed in a box that has a lever against one wall. In this apparatus, when a light is turned on above the lever, if the animal presses the lever a pellet of food will be delivered into the bin beneath the lever. When the light is off, the lever-pressing response is never reinforced, however. With sufficient learning trials the animal will come to make the response more frequently when the light is on and much less frequently when the light is off. Again, the light is the DS, the pellet of food the reinforcer, and pressing the lever is the response.

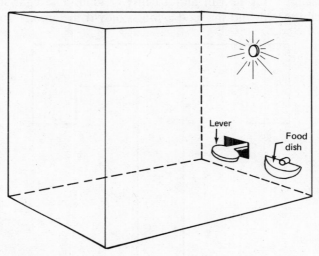

**FIGURE 2.4**

*The Skinner box. When the light is on, a press of the lever results in a pellet of food dropping into the food dish.*

In the present view instrumental conditioning principles are operative in various situations, including those described above. Skinner (1938; Holland and Skinner, 1961) changed the name of this type of conditioning to operant conditioning and based his principles only on data obtained in operant conditioning chambers. This apparatus, in which the animal is free to respond at will, has been considered to be different from apparatuses such as the T maze which involve discrete learning trials. It is suggested, however, that no difference in principle is involved.

Frequency, or rate, of response (as obtained in the Skinner box) has also sometimes been considered to be the unique datum (Skinner, 1938; Holland and Skinner, 1961). It is suggested, however, that there are other indications of the strength of conditioning besides rate of response, for example, whether or not the response occurs when the DS is presented, and how quickly. (Hull, 1943, also recognized multiple indices of learning.) When a child learns to read a letter, for example, rate of response may not be a meaningful criterion measure. There are also some human responses that occur only once—such as committing suicide or perhaps getting married. In such cases a measurement of response strength other than rate is necessary.

It is relevant to include also an example of instrumental conditioning with humans. Figure 2.5 shows the child learning apparatus in which various principles of learning may be studied, as well as the acquisition of various kinds of behavior (Staats, 1968b, 1973; Staats, Brewer, and Gross, 1970). The child sits facing the panel, with the tubes and the toys above them to his left. The experimenter sits in the other chair. The experimenter can present learning tasks to the child by placing 5″ × 8″ cards in the window slot at the top of the panel facing the child. Letter, word, or number stimuli, for example, may be presented. The child's response is reinforced by delivery of a marble down the chute in front of him. He can use the marble to obtain a trinket or small edible (raisin, peanut, or M&M) by placing the marble in the small hole above and to the right of the chute. Or he can place the marble in one of the four tubes below the toys and obtain the toy he has previously selected to work for when the tube is full.

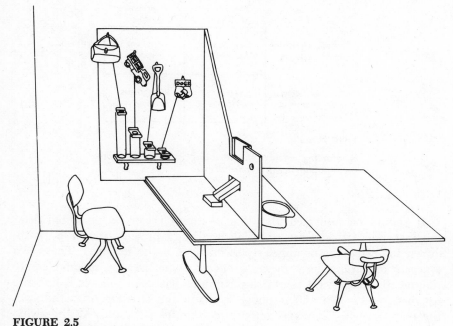

**FIGURE 2.5**

*The child learning apparatus. (From Staats 1968b).*

As an example of the study of learning principles using this apparatus, the effects on rate of response of different reinforcement schedules with children learning to read letters have been investigated (Staats, Finley, Minke, and Wolf, 1964). It was found that a child would respond more rapidly under partial reinforcement, in which he was reinforced for a varying proportion of his responses, than he would when he was reinforced for each response—as would be expected on the basis of animal learning studies. Discrete trial data and response rate data have been generated in this type of study (see Staats, 1968c). The results indicate the same principles are involved in each case.

## COROLLARIES OF THE PRINCIPLE OF
## INSTRUMENTAL CONDITIONING

The manner in which an instrumental response can be more strongly elicited by a stimulus has been described. A major aspect of instrumental conditioning, however, is that behavior to a stimulus can also be weakened. There are stimuli which make a response less likely to occur again. These have been called negative reinforcers, aversive stimuli, or, in common terms, punishments.

|                                         | Positive reinforcer     | Negative reinforcer     |
|-----------------------------------------|-------------------------|-------------------------|
| Stimulus presentation following response | Strengthens response    | Weakens response        |
| Stimulus withdrawal following response   | Weakens response        | Strengthens response    |

**TABLE 2.1**

*Reinforcing effects of presenting and withdrawing positive and negative reinforcers.*

It should be indicated also that a stimulus can be presented after the instrumental response or, if it is already present, it can be withdrawn following the organism's performance of the response. These two opposite operations have opposite effects upon the instrumental conditioning. If the positive reinforcing stimulus has been present before the response, and when the response is made the stimulus is withdrawn, the response will be weaker on future stimulus occasions. Withdrawing a positive reinforcing stimulus functions as a punishment. If the negative reinforcing stimulus has been present (such as an electrified grid for a rat in a Skinner box, or a person with a headache) and a response is followed by removal of the stimulus, the response will be strengthened. Withdrawing a negative reinforcing stimulus functions as a reward. The fourfold relationship between type of stimulus and effect on the response is shown in Table 2.1.

EXTINCTION, STIMULUS GENERALIZATION, AND DISCRIMINATION. It can be added that in a similar manner as in classical conditioning, extinction trials weaken the power of the directive stimulus to elicit an instrumental response. Extinction consists of not presenting the reinforcing stimulus following the response. If extinction trials are repeated in sufficient number, the strength of the stimulus to elicit the response will return to its preconditioning state (Rohrer, 1947). It is also true that an instrumental response conditioned to one stimulus will be elicited by a similar stimulus, a case of stimulus generalization. When a person says hello to someone mistaken as a friend, this is the operation of stimulus generalization. In general, if the response is reinforced to a particular stimulus, but not to similar stimuli, the organism will learn a discrimination—it will respond only to the specific stimulus.

COUNTERCONDITIONING. The extent to which a directive stimulus elicits an instrumental response may also be affected by an operation that is analogous to counterconditioning. That is, if the $^D$S is made to elicit another incompatible response more strongly than it elicits the first response conditioned to it, then presentation of the directive stimulus will elicit the stronger response. The second response will then supersede the first response (Hull, 1939). For example, the individual who learns touch typing more strongly than his original one-finger response will touch type when presented with a typewriter.

The fact is, however, that unlike as in classical conditioning, the stimulus does not lose its tendency to elicit the original response. If something weakens or prevents the occurrence of the second learned response, it will be observed that the first response will then be elicited. Moreover, positive reinforcers may be involved in conditioning both responses in instrumental counterconditioning.

HIGHER-ORDER INSTRUMENTAL CONDITIONING. Once a directive stimulus has been strongly conditioned to elicit a response, it may be employed to condition the response to a new stimulus—even without the use of additional reinforcing stimuli (Staats, 1968c). For example, the author tested this principle in the following way. His four-year-old daughter had learned to respond to the word *close* as a directive stimulus. The stimulus would elicit a closing instrumental response, as in responding to the verbal stimulus "Close the door." The author then told the child "Wug means close," several times. He then asked the child to "Wug the door," and the new stimulus elicited the instrumental response previously conditioned to the usual verb.

This principle indicates how it is that instrumental behavior can be learned without reinforcement (Staats, 1968c). Again, this principle of instrumental learning through contiguity has not been included in learning theories such as Hull's, Skinner's, and Miller and Dollard's. Higher-order instrumental conditioning will only be maintained, however, if it is sometimes subjected to reinforcing conditions. The same is true in classical conditioning. No conditioning will remain unless it is maintained in strength by at least occasional conditioning trials.

DEPRIVATION.  Deprivation also has an effect in instrumental conditioning. For example, a piece of food will not serve as an effective reinforcing stimulus in instrumental conditioning unless the organism has been deprived of food. This effect of deprivation on learning is shown clearly in an experiment by Hillman, Hunter, and Kimble (1953). Skinner (1938) has shown that higher rates of response occur as the animal is deprived of the reinforcer that is employed. The relationship of deprivation to extinction also been studied in a classic experiment by Perin (1942) which showed greater resistance to extinction under deprivation.

ADDITIONAL COROLLARIES.  The time between the response and the presentation of the reinforcing stimulus is important (Perin, 1943; Wolfe, 1934). With animals, a delay as brief as 30 seconds nullifies the conditioning. The time between the occurrence of the directive stimulus and the response is also important where an interval is involved.

Partial or intermittent reinforcement is also an important variable in instrumental conditioning (see Ferster and Skinner, 1957). It has been found that a response that is only sometimes reinforced, using certain schedules of reinforcement, will occur more frequently than if it is continuously reinforced. In fact, different schedules of reinforcement will have characteristic effects upon the rate of response. A response learned under partial reinforcement will also be more resistant to extinction than it will if continuously reinforced (McNamara and Wike, 1958), and this is true also of other irregularities, for example, delay of reward. In general, if one wishes to produce behavior that is strong in the sense of occurring at a rapid rate in the presence of a stimulus, and is resistant to extinction, the best training is to begin with a higher ratio of reinforced to nonreinforced occasions. Then gradually the ratio can be decreased.

Implicit in all conditioning is the occurrence of conditioning trials. A certain number of trials is necessary to produce conditioning. The number will vary according to different conditions.

## Three-function learning theory: The interrelationship of classical and instrumental conditioning

The present account has described classical and instrumental conditioning separately, following common contemporary practices. Actually, there have been different ways of considering the two types of conditioning procedures. The classic learning theories of Pavlov (1927), Hull (1943), Tolman (1932) and E. R. Guthrie (1935) attempted to consider the findings of both instrumental and classical conditioning in terms of one underlying principle or process—learning either through contiguity or through reinforcement.

Other theorists in animal learning, however, posited two separate learning processes to account for the different learning procedures involved in classical and instrumental conditioning (Konorski and Miller, 1937; Schlosberg, 1937; Skinner, 1935; Thorndike, 1932). In general, classical conditioning was seen as pertaining to involuntary (reflex) physiological responses of the smooth muscles and glands which were learned through contiguity. Instrumental conditioning was seen to depend upon reinforcement and to involve voluntary responses of the skeletal muscular system. Skinner (1938; Holland and Skinner, 1961) had a major influence in disseminating the separation of the two types of conditioning, including use of separate symbols to denote the types. These symbols did not indicate overlap or interaction between the types of conditioning. Skinner's theory and research stressed the importance of instrumental conditioning and paid little attention to classical conditioning (which he called respondent conditioning).

As has been indicated, Hull's general learning theory (1943) was a one-process theory. Nevertheless, it served as the foundation for what later developed into a type of two-process learning theory (Doob, 1947; N. E. Miller, 1948; Mowrer, 1947, 1954; Osgood, 1953; Rescorla and Solomon, 1967; Solomon and Wynne, 1954; Staats, 1956, 1968c, 1970a; Trapold and Overmeier, 1972). The data base for this approach in animal learning was formed in the context of avoidance learning. Thus, Kimble (1961) generally illustrates two-process theory by reference to an experiment by May (1948). Rats first learned to escape from shock in a box separated by a barrier (shuttle box). When the floor was electrified on one side, the animal could escape the painful stimulation by jumping the barrier to get to the other side. The animals learned the instrumental escape response through reinforcement (the removal of the shock). Following this training, they were given classical conditioning training with a buzzer as the $^{C}S$ and an electric shock as the $^{UC}S$. Later, when the animals were placed in the shuttle box and the buzzer alone was sounded, they would perform the escape response by jumping the barrier and moving to the other side. They had never learned the instrumental response to the buzzer. Two-process learning theory suggests that the animals learned a fear response to the buzzer in the classical conditioning, and they had learned to make the instrumental avoidance response when they experienced the fear response. When the buzzer sounded, the fear response was elicited, and this gave impetus to the performance of the instrumental response. Similar two-process learning demonstrations have been made with positive classically conditioned responses (for example, Estes, 1948). Solomon and his associates have conducted an impressive series of studies of two-process learning (Maier, Seligman, and Solomon, 1969; Rescorla and Solomon, 1967) as have Trapold and Overmeier (1972).

Two-process theory was also developed on the human level in the context of areas of human behavior, as will be indicated in later chapters. Some of these accounts have remained close to their foundation in Hullian theory (Doob, 1947; Osgood, 1953). In these theories, like the two-process learning theories based upon animal research, there was no systematic concern with the nature of the overlap between the two learning processes of classical and instrumental conditioning *in terms of the functions that a stimulus can have.*

These possibilities, however, were treated in a series of analyses (Staats, 1961, 1963, 1968c, 1968d, 1970a; Staats, Gross, Guay, and Carlson, 1973), which may be summarized here. It is suggested that there is overlap and interaction of various sorts between classical and instrumental conditioning. The same subprinciples (such as stimulus generalization) appear to hold, in the main, for both types of conditioning, as has been indicated—and it appears likely that there are central neural elements of a common nature that are involved in both types of conditioning and provide the basis for the similarity. It may be added that the interactions of the two types of processes may be conceptualized in terms of the three functions that stimuli can have for the behaving organism, and the manner in which these functions interact. The next several sections will describe these three functions and indicate their relationships, employing a symbolic convention that makes explicit the relationships involved.

## THE REINFORCING FUNCTION OF UNCONDITIONED STIMULI

Basic to consideration of the classical and instrumental conditioning overlap is the fact that most $^{UC}S$s have two functions. Most stimuli that function to elicit responses in classical conditioning will also function as reinforcers. This overlap may be obscured in traditional textbook treatments of learning. In one chapter food will be treated as an unconditioned stimulus ($^{UC}S$) that elicits the salivary response in classical conditioning. In another chapter food will be treated as a reinforcing stimulus ($^{R}S$), used to effect the learning of an instrumental response. But the relationships of these functions and of classical and instrumental conditioning are not usually indicated.

The fact is that food as a stimulus has more than one function, and this applies to most stimuli. A stimulus that elicits a response, like salivation, will serve also as a reinforcer. Thus, in actuality, except in cases to be discussed in a later section, every $^{UC}S$ is also an $^{R}S$. It may be convenient to label only the single function with which one is concerned in any particular case. But it should be recognized that such a stimulus is actually a $UC\text{-}R_S$—a stimulus with double functions. The fact is we think with the symbols we employ, and labeling only one function many times means that the other function will not be recognized.

## LEARNED REINFORCERS AND THE DUAL FUNCTION OF STIMULI

Statement of the dual functions of most unconditioned stimuli leads to reexamination of the simple conditioning situations already described. For example, in introducing classical conditioning it was stated that when a new stimulus was paired with an unconditioned stimulus that elicited a response, the new stimulus, as a conditioned stimulus, would come to elicit the response. The model should actually be elaborated to include *both* of the stimulus functions of the unconditioned stimulus.

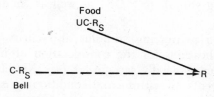

**FIGURE 2.6**

*A more detailed depiction of classical conditioning. Both functions of the $UC\text{-}R_S$, the response-eliciting and the reinforcing functions are transferred to the new stimulus, making it a $C\text{-}R_S$.*

As shown in Figure 2.6, the unconditioned stimulus should be labeled in terms of its two functions, as a $UC\text{-}R_S$. This immediately suggests that in the conditioning process both stimulus functions will be transferred to the new stimulus. The new stimulus, as it comes to elicit the internal glandular response, also becomes a conditioned reinforcing stimulus and is thus labeled as a $C\text{-}R_S$. This means that the $C\text{-}R_S$ could also serve as a reinforcing stimulus for the organism involved in the conditioning. Thus if, as in Pavlov's experiment, a bell is paired with a bit of food, the bell will come to elicit the salivary response. In addition, the bell should acquire rewarding properties. Any instrumental response that is followed by presentation of the bell should be strengthened. Zimmerman (1957) has conducted an experiment that supports this analysis. In this experiment a buzzer was presented in contiguity with food, in what may be seen as classical conditioning. Later, the buzzer alone was presented whenever the animal pressed a bar. The animal learned to press the bar many times, with only the buzzer as the reward. The buzzer was a $^{C\text{-}R}S$.

Note that the principles involved in this *learning* are those of classical conditioning. The principles involved in the *functioning* of these stimuli as rewards (or punishments) are those of instrumental conditioning.

## CONDITIONED STIMULI, REINFORCING STIMULI, AND DIRECTIVE STIMULI: MULTIPLE STIMULUS FUNCTIONS

Instrumental conditioning should be similarly elaborated. Traditionally, in describing instrumental conditioning, only the reinforcing function of the reinforcing stimulus and only the one function of the directive stimulus are described. As has been indicated, however, a reinforcing stimulus also elicits a set of internal physiological responses. This is true of unconditioned reinforcing stimuli ($^{UC\text{-}R}S$) as well as conditioned reinforcing stimuli ($^{C\text{-}R}S$).

The fact that a reinforcing stimulus also elicits such physiological responses has implications for the consideration of instrumental conditioning, as has been generally recognized (see Kimble, 1961). That is, the sequence of events in instrumental conditioning is that the directive stimulus is presented. The instrumental response then occurs, and this is followed by the reinforcing stimulus. This latter stimulus, however, elicits internal physiological responses which will be conditioned to the other stimuli that are present in the situation—that is, to the directive stimulus. This means that the directive stimulus in this process is also becoming a CS that will elicit the internal physiological responses. As a consequence of eliciting such physiological responses the directive stimulus will also acquire reinforcing properties for all instrumental responses and thus be an $^{R}S$. Figure 2.7 depicts the manner in which the directive stimulus comes to have three stimulus functions—conditioned stimulus, reinforcing stimulus, and directive stimulus functions—and should thus be labeled as a $^{C\text{-}R\text{-}D}S$. This more detailed analysis of instrumental conditioning is depicted in Figure 2.7.

Thus, it is suggested that the instrumental conditioning situation in which an instrumental response is brought under the control of a directive stimulus also provides this stimulus with two other functions. The stimulus will also become a conditioned stimulus as well as a reinforcing stimulus. While it is not intended to distinguish in detail the present learning theory from others, it may be indicated that the concept of the directive stimulus is different from Skinner's discriminative stimulus and Dollard and Miller's (1950) concept of cue. There is similarity in the several concepts in that in each case a stimulus is involved that comes to control an instrumental response. For both Skinner and Miller and

Dollard, however, this type of conditioning involves only reinforcement of the instrumental response.

For Skinner, also, the discriminative stimulus is not considered to be an inherent part of all instrumental conditioning but to occur only as a result of certain conditions of discrimination training. Nor is the discriminative stimulus considered to also become a conditioned stimulus.

The directive stimulus term is also different from the cue concept employed by Dollard and Miller (1950). Rather than just eliciting a rewarded instrumental response, in the three-functioning learning theory the directive stimulus comes to have the three stimulus functions. Moreover, as the next section will indicate, the three-function learning theory suggests that classical conditioning can give a stimulus directive control over instrumental behavior. Another difference resides in the fact that

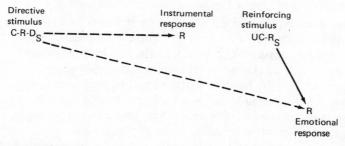

**FIGURE 2.7**

*A more detailed description of instrumental conditioning. The response is reinforced in the presence of a stimulus. This stimulus in the process becomes a directive stimulus, $^{D}S$; a conditioned stimulus, $^{C}S$, that will elicit an emotional response; and also a reinforcing stimulus, $^{R}S$, that can reinforce other instrumental responses.*

Miller and Dollard consider reinforcement to be necessary in instrumental conditioning. In Hull's theory, reinforcement is seen to involve drive reduction. Skinner's operant conditioning also depends on reinforcement. As has been indicated, the present position is that higher-order instrumental conditioning does not require reinforcement.

### COROLLARIES OF THREE-FUNCTION LEARNING THEORY

The corollaries described for classical and instrumental conditioning apply in the present context of social behaviorism. When classical conditioning is involved so are the relevant corollaries and the same is true when instrumental conditioning is involved. There are some additional considerations that should be mentioned, however.

CLASSICAL CONDITIONING AND REINFORCING AND DIRECTIVE STIMULUS VALUE. In the present view, classical conditioning plays a more central role than is given to it in Skinner's theory. In fact, classical conditioning can give a conditioned stimulus directive (instrumental) stimulus functions. This stipulation constitutes another difference from Skinner's use of the term *discriminative stimulus* or Miller and Dollard's use of the term *cue*.

An experimental example is relevant here. Trapold and Winokur (1967) first presented animals with classical conditioning in which a stimulus, let us say a tone, was paired with food, while another stimulus, let us say a clicker, was not. This would be expected to make the tone a $C_S$. Later, these animals were introduced to another type of training. In this procedure half of the animals were reinforced for an instrumental bar press when the tone was present and not reinforced for making the response when the clicker was present. For the other half of the animals this procedure was reversed. The animals that had learned to respond to the tone as a $C_S$ were found to learn to respond to the tone as a $D_S$ more rapidly than the other animals. These results can be seen to show that in such circumstances, a change in the conditioned stimulus value of a stimulus results in change in its directive stimulus value. Classical conditioning may thus be basic to instrumental response learning.

This type of experiment has been referred to as a "transfer of control" demonstration. Transfer of control can also involve changing the reinforcement value of a stimulus through prior classical conditioning— although this is not usually considered as transfer of control. That is, in the study of Zimmerman's referred to above, animals were first classically conditioned in a manner that would establish a buzzer as a $C_S$. Later, the buzzer was used as an $R_S$ in conditioning the animals to press a bar. This study can be interpreted to show that change in the conditioned stimulus value of a stimulus will change its powers of instrumental control by changing its reinforcement value.

DRIVE AND THREE-FUNCTION LEARNING. As has been indicated, experiments have shown that deprivation has effects on both classical and instrumental conditioning. Deprivation conditions have thus been included in the classic learning theories. Hull (1943), for example, included the term *drive* in his theory. Deprivation of food, for example, would increase the animal's level of drive. This would have the effect of energizing all the responses of the animal. Also, the drive state was considered to have cue properties that could elicit specific behaviors and thus guide the animal appropriately. Receipt of food was considered to reduce the drive, and it was the process of drive reduction that was reinforcing. Miller and Dollard (1941) had a similar view. Experiments have been conducted, it may be added, showing that animals can learn to make one response in the presence of one type of deprivation and another response when under another deprivation. Later additions to the theory by Hull (1952) and Spence (1956) included the concept of an anticipatory response (anticipatory of reinforcement) as an additional motivational mechanism that would energize and guide habits.

Skinner (1938), also employing the term *drive*, was primarily concerned with the fact that deprivation of a reinforcer would increase the reinforcing value of the stimulus. Two-process learning theorists (Mowrer, 1950; Rescorla and Solomon, 1967) have viewed the conditioned internal responses as motivational states. Thus, employing the experiment by May (1948) that has been described, the reason the animals avoid the buzzer, in this view, is because of the fear response that the buzzer elicits, the fear being considered to be a motivational state.

As an initial statement of the three-function learning theory position, it is suggested again that classical conditioning and instrumental conditioning are related. The primary effects of deprivation are on the $UC_S$ and $C_S$ stimulus functions, and hence on the $R_S$ and $D_S$ functions. That is, Finch's (1938) study showed that food-deprived dogs salivated increasingly to the $UC_S$ as well as to the $C_S$ to the extent they were deprived. This is seen to be the primary effect of deprivation. As a consequence of a stimulus becoming a stronger $UC_S$ or $C_S$ through deprivation, however, the stimulus will thereby be a stronger reinforcing stimulus and a stronger directive stimulus. These principles of the *two-process, three-stimulus-function* learning theory will be elaborated in Chapter 4 in describing the human emotional-motivational system.

Deprivation-satiation operations, it may be noted, may be seen to affect the *three functions* of stimuli and thus the manner in which the organism's behavior will be learned and maintained (Staats, 1968d; Staats and Warren, in press).

## The stimulus-response mechanisms

The goals of the discovery, isolation, and detailing of the higher-order (elementary) principles of conditioning, including the various subprinciples, have been central in the basic science of learning. In this task the aim is to establish the principles in the simplest, most controlled circumstances possible. That is the goal in any experimental science in establishing its elementary principles. Experimental precision in many cases requires that a single stimulus be manipulated and the effect upon a single response be observed. Thus, the elementary principles of learning are usually based upon the study of simple stimulus-response (S-R) mechanisms.

Behavior in life situations is rarely so simple. Human behavior in naturalistic circumstances, for example, is usually quite complicated. Ordinarily, the individual learns complex combinations of stimulus and response events. Most human acts involve several principles, as well as many stimuli, controlling many responses. For this reason, an important part of a learning theory in general, as well as one that is to be significant in the human realm, should include specification of the ways that the principles of learning can operate to produce more complex learned S-R mechanisms. Critics are quite correct when they say that human behavior cannot be adequately described in terms of single S-R events. To serve as a model of various human behaviors (or even complex behavior of lower organisms) the learning theory must outline some of the general ways that complex S-R mechanisms can be formed.

In certain cases the first- and second-generation learning theories referred to what may be seen as examples of complex stimulus-response mechanisms. And there are experiments that may be seen to involve S-R mechanisms, even though conducted with a different purpose in mind. Isolation, derivation, and systematic empirical description of the abstract S-R mechanisms, however, has not been seen as an explicit task of learning theory. Moreover, the place of the principles of S-R mechanisms in the basic learning theory requires explicit recognition (Staats, 1970a), especially in the context of human behavior.

The S-R mechanisms are not basic principles themselves; they are derived from the basic conditioning principles. The S-R mechanisms, on the other hand, must be seen as abstract principles independent of the particular behaviors involved in their study, for the S-R mechanisms when abstracted can be seen to apply to many different types of behavior. While this area cannot be considered in detail, a few examples of S-R mechanisms that function in human behavior will be outlined.

## THE RESPONSE SEQUENCE S-R MECHANISM

It is important to indicate that in general, as well as with respect to the present topic, a response ordinarily has stimulus properties. Every muscular response, for example, produces stimulus activation. That is, there are sensory nerves in the muscles and tendons that are activated by movement of the muscles and tendons. Also many internal responses have stimulus characteristics, as when we experience the contractions of an empty stomach. These internal stimuli produced by responses are like any other stimuli in the sense that they can have or acquire the various functions of stimuli.

Such stimuli have a special significance in that they can play a role in the learning of a very ubiquitous type of S-R mechanism, that of the response sequence or response chain. That is, any time a response occurs just following a preceding response, the second response will be conditioned to the stimulus of the first response. The result is that one response produces the stimulus that elicits the next response (see Hull, 1930; Skinner, 1938; Watson, 1930). Much of the importance of the study of serial learning with animals and with human verbal responses, it may be suggested, is simply in demonstrating this S-R mechanism. It is important to study the characteristics of the formation of sequences of responses and the manner in which such sequences relate to adjustive human behavioral repertoires. For example, the establishment of sequences of responses such as the example shown in Figure 2.8 appears to be essential in understanding aspects of language, problem solving and reasoning, mathematics, and so on.

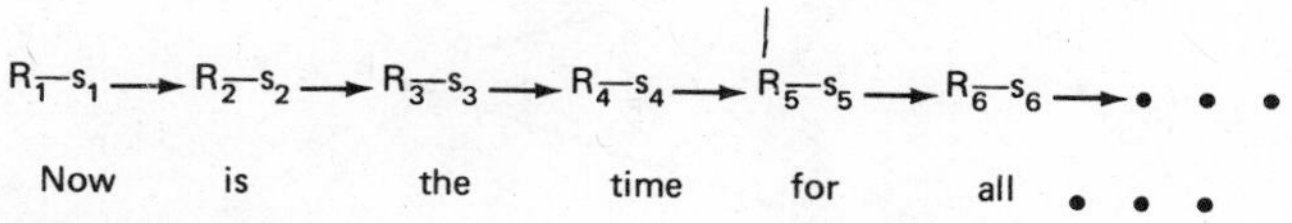

$$R_{\overline{1}}\text{-}s_1 \longrightarrow R_{\overline{2}}\text{-}s_2 \longrightarrow R_{\overline{3}}\text{-}s_3 \longrightarrow R_{\overline{4}}\text{-}s_4 \longrightarrow R_{\overline{5}}\text{-}s_5 \longrightarrow R_{\overline{6}}\text{-}s_6 \longrightarrow \bullet \quad \bullet \quad \bullet$$

Now        is        the        time        for        all    •   •   •

**FIGURE 2.8**

*A language response sequence. When the word responses in this familiar phrase ("Now is the time for all good men to come to the aid of the party") are repeated a number of times, each response comes to elicit the next response. After a sufficient number of repetitions (trials), the response sequence is well learned and will function as a unit. Most readers could finish on their own the sequence begun in the figure, because the sequence has already been learned.*

In this example and those to follow, the responses involved are instrumental (sensorimotor responses). Not all of the S-R mechanisms to be described would be relevant for the consideration of classically conditioned responses.

## THE MULTIPLE STIMULUS ELICITATION MECHANISM

There are several subtypes involved in the classification of multiple stimulus elicitation. One is depicted in Figure 2.9(a). In this mechanism one stimulus, $S_1$, has come to elicit a particular response. $S_2$ has also come to elicit that response, as has $S_3$. Each stimulus has a certain eliciting strength. When the three stimuli occur together there is a summation of strength, and the response will more strongly tend to be elicited (see also Hull, 1943, p. 206).

As another example, take the case where $S_1$ elicits one response and $S_2$ elicits another. When the two stimuli occur in close proximity, *both* responses will tend to be elicited. Figure 2.9(b) shows the two responses being elicited in sequence. This mechanism can result in the elicitation of a new *combination* of responses—a novel or creative act (Staats, 1963, 1968c). As an example, when a child sounds out a word he has never read before, according to individual syllable reading response units he has learned, he produces a novel response. After several repetitions the separate syllables will be combined into a unit sequence, and he will be able to read the whole word as a unit.

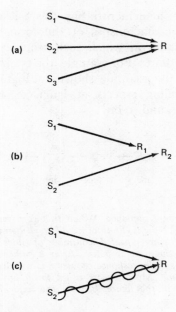

**FIGURE 2.9**
*Examples of multiple stimuli involved in response elicitation.*

There are also cases where the responses that tend to be elicited by a combination of stimuli are incompatible. In this case either one or the other will occur. Such mechanisms account for a good deal of uncertainty in human behavior in the naturalistic situation. For example, a response that has been observed to occur in the presence of a stimulus may not be made at a later time, when some other unrecognized stimulus interferes with the response. Thus, a situation may include stimuli for "not making" a response. For example, a good friend as a stimulus may have some tendency to elicit dirty-joke-telling behavior. However, when a maiden aunt is also present such behavior is unlikely to occur because she tends more strongly to elicit "no-dirty-joke-telling" behavior. This is depicted in part c of Figure 2.9, with inhibitory control shown by the second stimulus. (S. J. Weiss, 1972, has reviewed the literature on stimulus compounding and presented an analysis that covers both additive and suppressive summation.)

## THE MULTIPLE RESPONSE MECHANISM

The individual may also learn more than one response to the same stimulus. In fact, this is ordinary in human behavior. For example, a child in the presence of an ice cream vendor (or any other stimulus situation which involves something he wants) may have learned to make a polite request, to pout, to whine, to cry, to have a temper tantrum. Since these are incompatible responses, the strongest will occur first and when its strength extinguishes, the next strongest will occur, and so on (see Hull, 1939, for an example in the animal laboratory). Figure 2.10 depicts such a multiple response mechanism.

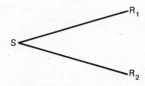

**FIGURE 2.10**

A *multiple response S-R mechanism, or response hierarchy.*

## Multiple stimulus, multiple response combinations

Combinations of the S-R mechanisms can occur that yield even more complex mechanisms. Hull (1943) outlined one type of complex mechanism called the habit-family hierarchy. There are many other types, especially in complex human repertoires like language (Staats, 1961). For example, a sequence of word responses may actually be a sequence in which the stimulus produced by one word response controls more than one following response. Thus, in our language culture, the stimulus of the word *give* will be followed in the child's experience by the personal pronouns *him, her,* and *me* and thus acquire tendencies to elicit each of those responses, among others. This will also be true of the word responses *throw* and *push.* A relatively simple sequence of these multiple S-R mechanisms which everyone in our language culture would have acquired through his language experience is shown in Figure 2.11.

It should be noted that in saying any one of the sentences in the above example the particular word sequence uttered would also be determined by the various other stimuli present that had elicitation power. Thus, the individual would be likely to say "Give him the ball" in a stimulus situation involving a child who had just taken a ball from another child, not one of the other sentences.

The fact is that in general people can and do learn fantastically complex sequences, arrays, and combinations of responses under complex stimulus-eliciting conditions. It must be realized that one of the requirements in developing the basic learning theory is the elaboration on a

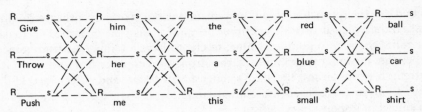

**FIGURE 2.11**

*At each point in the generation of a language response sequence (sentence), the stimulus produced by the particular word response has association tendencies to elicit several alternative responses.*

theoretical level of the abstract types of S-R mechanisms. Moreover, the basic task must also be one of showing in experimentation how such S-R mechanisms can be learned—and, once learned, how the mechanisms function in behavioral acts of all kinds. This area must include demonstrations of the different kinds of responses (for example, motor, verbal, and visceral) that can enter into S-R mechanisms. The basic principles of conditioning must also be studied in the context of S-R mechanisms. Finally, and most important for the present purposes, the S-R mechanisms constitute an important aspect of the theory for understanding human behavior and for dealing with human problems, as will be illustrated in later chapters.

## Additional elaborations

It has been suggested that the traditional learning theories and facts of learning require change at the basic level. Three-function learning theory constitutes such a change. Description of the S-R mechanisms is another area requiring development. But there are others also. Several additional examples will be given that are relevant in the present context.

### OVERT AND COVERT STIMULI AND RESPONSES

According to the doctrine that has been espoused by Watson—and presently Skinner—the subject matter of behaviorism concerns observable stimuli and observable responses. The fact is, however, that our common experiences tell us that much that is of importance in the study of human behavior occurs within the individual and is not open to direct observation. The individual thinks, plans, imagines, and feels, all on a covert level, as will be discussed in later chapters. This conflict between objectivism and the knowledge of what we experience has produced one of the sharpest schisms in psychology and the social sciences. Elemental behaviorists have sometimes ignored covert events, while humanistic psychologists have felt this to be the most important aspect of human nature.

The problem is, how can a behavioristic approach include the study of events that are internal to the individual? A few points on the basic level should be made concerning this issue. It has been suggested that it is possible to define concepts and establish principles based upon indirect observation. Generalizing this to the present case, if there are covert stimuli and responses, and conditioning principles, that cannot be specified through direct observation, perhaps it is possible to do so through indirect observation.

As an example, it has been suggested that there are various types of stimuli that are internal to the individual and can perform the functions of stimuli but are not externally observable. Since we have no way of getting into the behaving person, what justification is there for inferring internal sources of stimulation? For one thing, we can observe the nerve fibers in anatomical studies of cadavers. Anatomy tells us that there are various sources of internal stimulation. Moreover, the electrical action of sensory nerves can be monitored in the living specimen through the use of apparatus. It can be seen that when an internal stimulus is supposed to occur the nerve does indeed display the expected electrical activity. Conditioning methods may also be employed to indirectly observe an internal stimulus. Let us say, for example, that we wished to see if the internal stimuli in the muscles and tendons could serve the functions of a conditioned stimulus. It would be possible to have the subject perform a movement that would activate the involved muscle stimuli and follow this with some unconditioned stimulus that would elicit an observable response. After a sufficient number of trials, the subject could be instructed to perform the movement, and it would be expected that the internal stimuli would then elicit the conditioned response. Study of this type of evidence, which gives indirect specification to an internal stimulus, has been conducted. Razran (1961), for example, describes an experiment in which the response of blood vessel constriction was classically conditioned to the stimuli produced by a type of breathing response.

The same type of specification can be made of internal responses that cannot be directly observed. Hefferline (1963), for example, has shown that slight muscular movements which could not be directly observed, and which the individual is not aware of, can be conditioned according to the principles of instrumental conditioning. He employed electronic equipment to measure the electrical activity of the muscles moving the thumb so the response could be objectively observed. The experimenter provided reinforcement for slight movements of these muscles and increased their frequency, a case of instrumental conditioning.

One of the most important types of internal responding involves language or thinking. Much significant cognitive behavior occurs that cannot be directly observed. It will be assumed herein, for example, that language behavior takes place internal to the individual that has the same properties as the external language behaviors we can directly observe. We may ask what justification there is for this conception. An experiment performed by N. E. Miller (1935) demonstrates how "thinking" responses follow the same principles as overt language responses—that is, how a covert language response can come to elicit a conditioned response. In the study a sweat gland response was classically conditioned to the subject's saying of the letter *T*. This was done by pairing an electric shock as the unconditioned stimulus with the saying of the letter but not the saying of the word *four*. Later, the subject was presented with a series of dots appearing on a screen every so often. The subject was instructed to say the letter to himself when one dot appeared and to think the number when the next dot appeared—in an alternating manner. The subject's sweat gland response was measured. The times when the subject was thinking of the letter showed the conditioned response, which did not appear when the subject was thinking of the number. The subject did not even know that thinking of the letter elicited his sweat gland response.

This study indicates that internal responses do occur, even such "mental" responses as thinking. Moreover, the study shows that such responses have stimulus characteristics. That is, another response could be conditioned to the stimulus provided by the response of thinking of the letter. The internal response could be studied by "tagging" an observable response to it (Staats, 1963). This is not to say that internal stimuli and responses have been studied to the extent that is necessary. However, enough evidence is available to indicate that the same principles that hold for external stimuli and responses appear to be applicable to internal, covert behavioral processes. Realization of these possibilities establishes an important base for enabling a rapprochement of behavioristic and humanistic conceptions.

## CLASSICAL CONDITIONING AND IMAGES

Another elaboration not developed in the traditional learning theories involves a type of classical conditioning. As has been indicated, classical conditioning originally dealt with the salivary response, and later work showed that the principles applied to other glandular and smooth muscle responses. The autonomic nervous system is involved in such responses, whereas the central nervous system and its offshoots (the peripheral nervous system) are involved in instrumental responses. It has been suggested, however, that responses of the central and peripheral nervous systems are susceptible to classical conditioning, in the following manner.

The concept to be developed is that of the sensory response. To begin, it is suggested that the world—the environment—is made up of various physical energies. There are light waves, sound waves, mechanical energies, chemical energies, electrical energies, gravitational energies or forces, and so on. To some of these the human (among other organisms) is sensitive. This means that the human, via various specialized structures we call sense organs or sense receptors, can respond to certain environmental energies. He cannot respond to all of them. The physical energies to which the individual can respond have been called stimuli. It is on the basis of responding to these aspects of the environment that man comes to behave in a manner that is relevant to the events of the world—in the multifarious ways that he has at his disposal.

To continue, it is suggested that activation of the sense receptors, and the neural activity the receptors in turn activate, may be considered a response. Ordinarily, the response characteristic of sensory experience is not considered, only the stimulus characteristic. The activation of sense receptors is merely said to give rise to a sensation, which is what the individual experiences. For example, when a visual environmental stimulus occurs, the person sees it. He does not experience the response characteristics of the neural activation. However, it is suggested that a sensation is the stimulus produced by a sensory response. Moreover, it is important to consider both the response characteristics as well as the stimulus characteristics of the effects of environmental energies.

LEARNED (CLASSICALLY CONDITIONED) SENSORY RESPONSES. The reasons the concept of the sensory response has been introduced at this basic level are to stress the partial independence of the "sensation" from the environmental stimulus and to lay the basis for indicating how sensory responses, because they are responses, can be learned. The independence of sensation and external environmental stimulus is shown in various ways. As an example, the sensory response in a human may be activated by direct brain stimulation, in the absence of external environmental stimulation (Rasmussen and Penfield, 1947). The sensory response, it is suggested, is produced by activation of the appropriate neural elements. Activation can be by the external stimulus, but also by other means.

More generally, it is suggested that sensory responses may be "disconnected" from the environmental stimulus that originally elicited them through learning. The principle is as follows. When an environmental stimulus that elicits a sensory response is paired with another environmental stimulus that does not, the latter will come to elicit the sensory response, at least in part. The sensory response will be conditioned to the new stimulus and may be called a conditioned sensory response.

The principle involved is that of classical conditioning. The original environmental stimulus that elicits the sensory response may be considered to be an unconditioned stimulus, or $^{UC}S$. The new stimulus that comes to elicit the conditioned sensory response fulfills the role of a conditioned stimulus, or $^{CS}$. After a sufficient number of pairings, the conditioned stimulus wil come to elicit the conditioned sensory response even when the $^{UCS}$ is not presented. The individual may experience a sensation (an image) even when the sensory stimulus is not present (Staats, 1959, 1961).

There are various sources of evidence that support the principles of the classical conditioning of sensory responses (see Brogden, 1939, 1947; Ellson, 1941; Leuba, 1940; Phillips, 1958). Since this conception was employed in a systematic learning framework (Mowrer, 1960; Sheffield, 1961; Staats, 1959, 1961), it has been widely adopted (see Bandura, 1969; Paivio, 1971) and extended in various human research studies (Paivio, 1971; Staats, Staats, and Heard, 1961), as well as in animal experimentation (Friedman and Carlson, 1973; Logan and Wagner, 1965).

SENSATIONS, PERCEPTIONS, IMAGES, AND LEARNING. There are several reasons for this somewhat complex introduction to the concept of the sensory response. One is that the concept serves to integrate a number of areas that otherwise involve antagonistic conceptions. Examples will re-appear throughout the book. A few specific illustrations may be made here, however. A person may be said to have an image when he experiences some sensation in the absence of the stimulus that ordinarily gives rise to the sensation. We have images when we dream. We have images of loved ones who are absent, of places we have been and things we have experienced, and so on.

Such phenomena may be considered on a basic level as consisting of conditioned sensory responses. For example, a young man has a love affair at a time when a particular piece of music is popular. He hears the piece played various times when he is with his amour. Later in life he hears the song and experiences an image of his sweetheart of the earlier day. The song is the conditioned stimulus, the young lady the unconditioned stimulus who elicits in the young man the sensory responses. The sensory responses are conditioned to the song, which when later presented elicits the image (conditioned sensory responses). Much of our mental life involves such conditioned sensations, perceptions, and images. (See also Staats, 1963, 1968c, 1971a, and 1972). Consideration of the concepts of sensation, perception, and image in stimulus-response terms allows a rapprochement between the behavioral approach and the cognitive approach, as later discussions will indicate.

Finally, it should be noted that there is not complete agreement with this view of imagery, as a recent review by Pylyshyn (1973) has indicated. In this context it may be added that images are not directly observable and additional research is necessary before an unequivocal account of the principles involved in the acquisition and function of images can be provided.

# HUMAN LEARNING ELABORATIONS

In the development of learning theory, even at the basic level, a central concern is to elaborate the findings of the animal laboratory in the context of human subjects. Thus, it is important to show that the elementary principles generalize to the human level. Until this is done it cannot be concluded with surety that the principles are relevant.

As has been indicated, in the animal laboratory the concern is with isolating elementary principles; the specific behavior or the specific stimuli are not usually of focal interest and may only be matters of a practical nature. With human behavior there is interest not only in the elementary principles but also in content—in the analysis of the specific behaviors and stimuli involved.

It should be understood that although the principles on the animal and human levels may be the same, the specifics may differ. One must be prepared to conduct additional analyses and experimentation when generalizing principles across different types of subjects and different stimuli and responses. When the rigid, specific laboratory definition does not precisely fit the human example, this has been accepted as verification of the value of separating the areas of study, a position that is taken by some investigators.

For example, generalization of the animal laboratory concept of the reinforcing stimulus to consideration of such an event as graduation from college has been criticized (Mason and Bourne, 1965). There certainly are differences involved in the specific events involved in several ways. How can graduation from college be considered a reinforcing stimulus? It occurs only once in the individual's life, and it is not contingent upon significant instrumental behaviors, only those of preparing for the social event of graduation, which are already well learned. For these reasons, it could not be determined if college graduation strengthens an instrumental behavior—which is the basic laboratory definition of a reinforcing stimulus.

Rather than leading to the conclusion that laboratory and human principles are different, however, such questions should serve as the impetus for further development. Actually, the press of such discrepancies can lead to the further development of the learning theory concepts. Elaboration of the concept of reinforcement to the human level, for example, can indicate how reinforcing stimuli may be measured by other means than whether or not they can strengthen an instrumental response—as a later chapter will indicate.

The important point here, however, is that our philosophy should tell us that one of the tasks of creating a learning theory of human behavior involves the elaboration of the basic terms and principles on the human level. In this task superficial differences should not be taken to indicate differences in kind. We can expect changes and elaborations to be necessary. It is a mistake to think that the conditioning principles and definitions will translate without change from the animal laboratory to the human level, and without effort in considering the special conditions involved. This assumption has led many individuals to reject learning theory as an important orientation in the study of man.

The example that has been given involves a stimulus. The same types of changes and elaboration are necessary for the concept of response in going from the basic laboratory to the human level. For example, basic researchers have considered human behavior in terms of simple responses, atomistically analyzed. This is not an inherent quality, however. Naturalistic observation suggests that responses can differ in complexity and duration, as well as in type. It appears, also, that smaller unitary responses can be grouped into large configurations of responses, and then the learning principles will apply to the whole rather than to the unitary responses. For example, the ballet dancer may laboriously learn a complex movement composed of many submovements. But when the total movement has been well learned as a unit, it can be easily elicited as a whole by the choreographer in planning a new combination of various movements. Finally, the dancer will learn the whole ballet as a unit. Human behavior cannot be understood unless this concept is included—*that more elementary responses can be formed into larger and larger constellations through learning,* and that these larger constellations then have the properties of simple responses in further learning. One of the criticisms of a traditional learning theory (or behavioristic) approach is that it cannot deal with larger units of meaningfully integrated human behavior.

Thus, although traditionally there has been a separation between basic learning theory and human experimentation, the perspective should include both, with mutual enhancement of each level of study (Staats, 1970a). The following chapters will attempt to demonstrate this suggestion in various ways, including elaboration of the principles in the context of human behaviors.

# Part two

## Personality and social interaction levels

# Personality: An effect
# and a cause

A VERY BRIEF characterization of the traditional concept of personality may be given by way of introduction to consideration of personality as an effect and a cause. In simplified form, it can be suggested that the traditional concept of personality refers to an internal process or structure within the individual that determines how he behaves.

Those who have studied personality have tried to infer the nature of the internal personality processes. According to this view, if the nature of the personality can be stated, then human behavior can be understood. If the personality structure or process can be ascertained for the individual, it should be possible to understand, predict, and deal more effectively with the individual and his problems. The determinants of personality are sometimes considered of less direct importance, for if one can directly tap the personality events by tests, prediction and control of behavior are possible without stipulating the determinants. The strategy seems especially suitable if the determinants are considered to be unavailable for modification in any case.

There are various personality concepts that have been introduced by theorists in the study of human behavior, some of them with precursors in antiquity. For example, Hippocrates proposed that human behavior was the result of varying mixtures of four underlying bodily states, which he termed "humors." If the bodily humors were mixed in good proportion, the individual's behavior would be in good balance. An excess of one humor or the other would lead to extremes in behavior.

Another common concept of long standing is that of the personality trait. Traditionally, traits are considered by personality theorists to be internal processes, structures, characteristics, or personal qualities (Allport, 1937, 1966). Like other personality concepts, a trait is usually considered to be a cause of overt behavior (Allport, 1966; Cattell, 1950). Personality traits are considered to be bases for individual differences in behavior among people. They are also considered to be enduring—to be characteristic early in life and to show continuity throughout the life span. Personality traits are also considered to underly the characteristic behaviors that people display across different situations. For example, some individuals will be assumed to have a "dependent" personality because they behave in a dependent manner. Other examples of traits would type individuals in terms of aggressiveness, sociability, honesty, selfishness, extroversion, activity, intelligence, bravery, emotional stability, independence, and so on.

Another personality concept that has been employed is that of the self or self-concept Mead (1934). The "self" has been considered to be the individual's self-perceptions, attitudes, and feelings—an inferred process that determines the individual's behavior (Hall and Lindzey, 1957). Some theorists have considered the self to be central in a theory of personality. For example, Rogers (1951) suggests that as a product of interaction with the individual's interpersonal environment part of his perceptual field becomes established as the self. The person's concept of the self is heavily invested with values that are imparted, largely from experience in the home. According to this view the major motivation of the individual is to maintain and enhance the self-concept. In striving to maintain and enhance the self-concept he has acquired, the individual may have to reject, deny, or avoid certain experiences, thoughts, and situations. This may make the individual behave less effectively than he would if his self-concept had been formed under more benign conditions, so that he could incorporate and benefit from his life experiences more fully.

Probably the most widely known and employed personality theory has been the psychoanalytic theory of Sigmund Freud and its many offshoots. For Freud there were three primary personality structures or processes that interacted in determining the individual's overt behavior. The *id* is the name Freud gave to the individual's biological, instinctual energies. It was said to operate according to the pleasure principle—a striving to reduce states of tension, for example, those arising from deprivation of some kind. Dreaming, hallucinating, and other behaviors of this type, and other impulsive, pleasure-oriented actions, were seen to be produced by the id.

The *ego* was seen to be that part of the personality that is developed for commerce with reality. It was thought to determine various behaviors that enable the individual to adjust to the physical and social worlds, such as learning, thinking, problem solving, and so on.

In this theoretical context, the *superego* was considered to be that part of the ego developed to represent cultural values, ideals, morals, and ethics. It was thought to be acquired by the child through the efforts of the parents and other authority figures. The superego became "the representative of all moral restrictions, the advocate of the impulse toward perfection. In short, it is as much as we have been able to appreciate psychologically of what people call the 'higher things in life' " (Freud, 1949, p. 94).

The generality of the concept of personality in the psychoanalytic view can be seen in the breadth of behaviors that could be explained by reference to personality concepts. Dreams, defense mechanisms, humor, memory and memory loss, response to tests, and all manner of normal and abnormal behavior were considered to be determined by inner personality structures and motivations.

## Personality and elemental behaviorism

John B. Watson set the foundation for considerations of personality by behaviorists, as he did in various other things. In his words, "*Personality is the sum of activities that can be discovered by actual observation of behavior over a long enough time to give reliable information. In other words, personality is but the end product of our habit systems*" (1930, p. 274). In this view personality is considered as behavior, the complex behavior that the individual has learned. This general framework considers personality as a dependent variable—as an effect. The independent variables, or causes, of personality are seen to be the laws of learning and the particular conditions of learning that the individual has experienced.

Later behaviorists followed a similar view. For example, Keller and Schoenfeld (1950) state: "While all human beings obey the same laws of behavior, each individual ends up with a unique behavioral equipment that defines his 'personality' " (p. 366). Another example of this view is the following:

> It may be concluded that the subject matter of interest in the study of "personality" is the behavior of human beings. Although many terms have been used to describe this behavior—and often these terms direct one's interest away from behavior and toward inferred internal processes and entities—the actual problems are those concerning the way complex human behavior is acquired and maintained. (Staats, 1963, p. 284)

Lundin (1961) discusses traits as behaviors that have some "common descriptive characteristic" (p. 16). He states, "Traits are, then, not the causes of behavior but merely descriptive terms applied to a general class of responses which appear to have something in common" (p. 16).

The elemental behavioristic approach has included related methodological principles, as illustrated in the criticism of traditional personality concepts. The concepts of personality theorists are considered to be in-

ferred from observations of the behaviors of people. When some commonality of behavior is observed, it is given a name. Then this name is
thought to indicate an internal personality process that was the basic
cause of the behavior. Personality theories in this sense were considered
to be circular in nature, suggesting that the personality processes were
independently observed, when actually only the behaviors were being
observed. Moreover, personality theories such as psychodynamic theories
are considered to direct more attention to inferred internal personality
processes and less attention to human behavior and the conditions under
which behavior is acquired and maintained (Lundin, 1961; Mischel, 1968;
Skinner, 1959; Staats, 1963; Ullmann and Krasner, 1969).

As later chapters will indicate, the behaviorist focus upon behavior and
the learning conditions that give rise to behavior provides a conceptual
scheme for a very pragmatic concern with the problems of human behavior. The principles of learning, moreover, provide a means for dealing
directly with various problems of human behavior. However, the behavioristic rejection of personality theory also implies a rejection of concern
with individual differences, with the concepts and observations of personality theory, and with the concepts and methods for measuring personality. The behavioristic position has come to be atomistic, in the sense
of being concerned with relatively specific behaviors and environmental
circumstances, and in its focus on present rather than past conditions of
learning. Mischel (1971) notes:

> Psychodynamic theorizing assumes a set of basic personality motives
> and dispositions that endure, although their overt response forms may
> change. . . . Such a Freudian-derived model is widely shared by many
> personality psychologists.
>
> In the opinion of social learning theorists, however, this model is
> inappropriate and has led to some tragic mistakes in clinical treatment
> and diagnosis for 50 years (e.g., Bandura, 1969; Mischel, 1968, 1969;
> Peterson, 1968). They suggest that seemingly diverse behaviors do not
> necessarily reflect a uniform underlying motivational pattern. Instead
> they view behaviors as relatively discrete and controlled by relatively
> independent causes and maintaining conditions [pp. 75–76]. . . .
> [P]sychodynamic theories look for the motivational roots of personality
> in childhood. In contrast, . . . behavior analyses seek the current causes
> of the person's behavior [p. 77]. . . . Influenced heavily by Skinner,
> most . . . behavior approaches emphasize what an organism *does* rather
> than make inferences about the attributes it has (Mischel, 1968). . . .
> [A]nalyses of human problems involve descriptions of the covariation
> between environmental conditions and what the person does, but they
> avoid inferences about the meaning of the behavior as a sign of some
> trait or underlying motive [p. 83].

In recent years Mischel has become a primary spokesman for this
elemental behavioristic interpretation. This interpretation has been criticized (Bowers, 1972; Alkers, 1972) for suggesting that "people show no

consistencies, that individual differences are unimportant, and that situations are the main determinants of behavior" (Mischel, 1973, p. 254).

> According to the traditional trait paradigm, traits are the generalized dispositions in the person that render many stimuli functionally equivalent and that cause the individual to behave consistently across many situations (Allport, 1937). The present view, in contrast, construes the individual as generating diverse behaviors in response to diverse conditions. . . . (Mischel, 1973, p. 264)
>
> [E]mpirically established behavioral consistencies, however, do not seem large enough to warrant the belief in very broadly generalized personality traits. The evidence . . . seems to be discrepant with the enduring conviction that people show marked behavioral generality. (Mischel, 1968, p. 43)
>
> Thus, while the traditional personality paradigm views traits as the intrapsychic *causes* of behavioral consistency, the present position sees them as the *summary terms* (labels, codes, organizing constructs) applied to observed behavior. (Mischel, 1973, p. 264)

It can generally be said that the elemental behaviorist position has not provided a framework for utilizing the observations and concepts of personality theory. The behavioristic approach has been critical of the methods, concepts, and findings of personality theory. This has been true in the other direction, also. Personality theorists have considered behaviorism to be lacking as a conceptual framework for considering central concerns of human behavior. The behavioristic and personality theory schism has been one of the primary issues in the separatism that characterizes psychology.

## Behavior as a cause: Some foundations for a social behavioristic view of personality

It is suggested that this separation is disadvantageous. The principles of learning are not synonymous with only one approach, such as the elemental behaviorism that has been described. The principles of learning can be utilized in a framework that is closer to the interests of personality theorists. Several lines of thought that can be joined in a rapprochement between behavioristic and personality theory will be outlined. These involve mediation theory, self-reinforcement theory, the self-concept, the principle of cumulative-hierarchical learning, and the concept of the basic behavior repertoire.

### MEDIATION THEORY

While the elemental behavioristic position has suggested a focus on a functional analysis of the environmental stimulus conditions causing a specific overt behavior, there have actually been elements that could provide a basis for a broader conception. Watson (see 1930), for example, had schematized sequences of behavior in which the stimulus produced by a preceding response would elicit a following response. Hull (1930) further developed this concept in the discussion of mechanisms that could be considered as illustrative of purpose and knowledge. These concepts can be interpreted in the present context as a basis for considering that processes internal to the individual can affect the nature of overt behavior —there may be additional determinants for behavior other than the external environment.

The concept of overt behavior mediated by covert behaviors of the individual has been elaborated a great deal in considering various aspects of language (Mowrer, 1960b; Osgood, 1953; Staats, 1961, 1968c). Moreover, there is research showing that overt responses (Shipley, 1933) and covert responses (Hefferline, 1963) can mediate other responses, as well as many studies that show the importance of mediating responses in the context of significant human behaviors.

## SELF-REINFORCEMENT

It has been suggested by Mowrer (1960a) that the basis for learning consists of the stimuli produced by the individual's response coming to have reinforcement value. The concept is that an instrumental response produces internal stimuli, as has already been described. When the response is reinforced, according to this account, the stimuli produced by the response will, through classical conditioning, come to be conditioned reinforcers. Thus, the individual will perform the response more frequently in the future, because his response produces a stimulus that is reinforcing.

This conception has some weaknesses as a basic theory of learning (see Staats, 1968c, pp. 450–451). However, the concept can be elaborated to indicate how the individual can provide *self-reinforcement* when he responds. If the individual responds in a way that produces a stimulus that has reinforcement value, he has provided himself with a reinforcing state of affairs (Staats, 1963, pp. 95–98). A number of studies have been conducted that demonstrate the phenomenon of self-reinforcement (Aronfreed, 1968; Kanfer, 1966; Kanfer and Duerfeldt, 1967; Kanfer and Marston, 1963), as is described further in Chapter 6. Again, this suggests that behavioral events internal to the individual may influence the nature of the individual's behavior.

## THE SELF-CONCEPT

As will be indicated in more detail in the next chapter, the self and the self-concept are concepts that have been employed in the field of personality theory in dealing with significant aspects of human activity. This term will only be mentioned here in the respect that its behavioral definition played a role in the development of the present concept of personality. That is, a beginning in the behavioristic-personality rapprochement was made through developing a behavioral analysis of the self-concept indicating how learning principles could be employed in dealing with a topic important to personality theory (Staats, 1963, pp. 261–266). For example, the behavioral analysis included an outline of the manner in which the individual learns to describe himself—a type of learned behavior. At this point the analysis was in concert with the elemental behavioristic view that personality is behavior. In addition, however, it was indicated the the individual's self-concept (self description or self-labeling) would affect the way the individual would behave in various situations, as well as the way other people would behave toward the individual.

The manner in which the self-concept has the causal properties usually given to personality processes was accounted for within this framework. In this manner, the behavioral account began to be imbued with characteristics that were much like those of traditional personality theory (Mead, 1934; Rogers, 1951; Snygg and Combs, 1949). Other behavioral accounts have included the concept of the self also (Aronfreed, 1968; Bandura, 1969; D. J. Bem, 1967; Kanfer and Phillips, 1970; Krasner and Ullmann, 1973; Mischel, 1968, 1971). However, the full implications of the treatment of complex constellations of behavior as the causes of other behaviors, and of reactions to the person by others, in terms of providing a basis for a behaviorally oriented personality theory were not elaborated. Thus, the behavioral analysis of the self-concept was employed in a specific rather than in a general way.

At any rate, mediation theory, self-reinforcement principles, and the beginning behavioral analyses of personality concepts provided some of the basic elements for a social behavioristic concept of personality. However, additional principles and additional conceptual development were needed before a concept of personality could be developed, within the behavioral tradition, that could serve as a basis for rapprochement with traditional personality conceptions. Some of the principles will be outlined in the next two sections before returning to the concept of personality as an effect and as a cause.

## Cumulative-hierarchical learning and the concept of the basic behavioral repertoire

A prominent characteristic of elemental behaviorism as it has been displayed in the field of clinical treatment has been that of specificity. This characteristic grew out of the laboratory science background of the approach, in which analysis into elementary events made possible the isolation of the basic learning principles and the stipulation of the lawfulness of the principles. The characteristic can also be seen as an antidote to the lack of specification of some concepts that is true of personality approaches.

One of the foundations of a behavioral approach to clinical psychology, for example, involved criticism of the then dominant psychoanalytic approach. In critizing the approach, one of the characteristics that was rejected was the concern with the early experience of the child and with the individual's life history, topics of central interest in psychoanalytic theory. Eysenck (1960a), for example, in the context of providing productive analyses, established the ahistorical characteristic that suggests we treat the individual as he is: "All treatment of neurotic disorders is concerned with habits existing at *present*; their historical development is largely irrelevant" (p. 11). B. F. Skinner's philosophy of the functional analysis of behavior problems—that of finding the reinforcement variables that are affecting a behavior—also gave less importance to the person's historical development.

> The experimental laboratory design and the objective of isolating functional relations place restrictions on certain kinds of questions that the investigator may want to raise concerning abnormal behavior—for example, about the parent's role in the etiology of childhood schizophrenia. Answers to such questions, though often intriguing, entail so much confounding that they are meaningless in a functional analysis of abnormal behavior. (Lovaas, 1966, pp. 111–112)

> The focus of . . . behavior theory is on what the person is doing in the "here and now" rather than on reconstruction of his psychic history. (Mischel, 1971, p. 86)

It may be added that not all personality theories place the same stress upon the experiential history of the individual as psychoanalytic theory does. For example, the field theorist Kurt Lewin stated as a primary assumption "Any behavior or any other change in psychological field depends only upon the psychological field *at that time*" (1943, p. 294). Although Lewin stated that this did not mean that field theory was ahistorical in nature, those who utilized field theory for clinical purposes and personality theories were not concerned with exploring the experiential history of the individual (Rogers, 1951; Snygg and Combs, 1949).

To continue, however, a learning conception of human behavior can actually be very coincident with psychoanalytic conceptions in terms of interest in the historical experiences of the person. Moreover, a primary development of the learning conception involves considering the complexity and the length of human learning. When the complex repertoires of human behavior are subjected to analysis in terms of the constellations of skills involved, it becomes evident that a long time and many learning trials are involved in the acquisition process, as well as various learning principles and stimulus-response mechanisms, and that the repertoires are composed of various subrepertoires of various kinds.

It is natural that the basic theories of learning have not considered learning in this sense, since they are concerned with simplification and the elementary principles. However, a theory of human behavior must be concerned with the length of the individual's learning history. This is one of the primary and unique characteristics of man. Human life presents possibilities for long-term, cumulative, skill acquisitions of various types. The importance of learning can only be seen by stepping out of the laboratory and considering such cases of human development.

The principles of cumulative-hierarchical learning are necessary in indicating that much of human behavior is acquired in learning processes of great duration and complexity. The important aspects of human personality involve extended series where the acquisition of one skill enables the individual to acquire another skill, or an elaboration of the first skill, and this then enables the next learning level to be attained. Such concepts and principles must be added to the elementary principles of conditioning to provide a basic framework for the consideration of human behavior.

## The basic behavioral repertoires (personality systems)

The conception that is focal to the present approach is that there must be a personality level to a theory of human behavior. In this sense the approach is in agreement with traditional personality theory. However, it is suggested that personality does not consist of unspecified internal mental structures or processes. The present account of personality involves the concept of the basic behavioral repertoire, which suggests that there are repertoires of behavior that are learned. These repertoires are complex constellations of skills that are learned only over a long period, on the basis of the elementary learning principles and the cumulative-hierarchical learning principles described in the preceding section. These basic behavioral repertoires constitute the compendium of skills by which the individual adjusts. Moreover, acquisition of the incipient basic behavioral repertoire in childhood provides the foundation for the acquisition of additional elements in the repertoire. As the child adds additional elements the repertoire will enlarge, advance, and become more complex.

It is because these repertoires provide a foundation for additional learning that the term *basic behavioral repertoires* is used (Staats, 1968c, 1971a). It was stated in introducing the cumulative-hierarchical learning principles that there are skills, once learned, that provide a basis for learning additional skills. The basic behavioral repertoires are constellations of complex skills which are evoked by many situations but also have the quality of providing the basis for additional learning.

As an illustration, let us take the case of a daughter of attentive parents who play with the child a great deal and teach her very cute social behaviors. Let us say the child consequently learns early to approach people, to be affectionate, and to play in ways that are mutually rewarding. Let us say that the child is also very physically attractive and that the parents are social and have many relatives and friends. Thus, the child learns these incipient social behaviors so they are relevant to people of various ages. Even at an early age the child could be described as friendly, outgoing, social, attractive, and so on. This would ordinarily be considered to be a trait of personality in the child.

In addition, however, the incipient basic behavioral repertoire of the child would produce a social environment that would affect her further learning. That is, because the child has had positive experiences with various people she will approach them. And, because she does this, and because when they interact she is rewarding to others, she will have a positive effect upon them. They will thus approach her in various ways and provide her with positive learning experiences. She will have more than the usual opportunities to interact with others and to learn social skills.

A child who has a continuing history like this will acquire a different basic behavioral repertoire in the areas of social motivation and social skills than will a child who lacks this history. In contrast, let us take the case of the child who has parents who love the child but who do not have the same types of skills as the parents in the above example. Let us say the parents are not very social, but rather spend their time in solitary pursuits, reading, listening to music, and the like. For them, playing with a young child is not rewarding. They go to the child when the child needs care, which many times means when the child is crying. Let us say that it is customary for the parents to leave the child alone as long as he is playing quietly and is not unhappy, but that they will go to him when he cries or indicates unhappiness. The child in this situation will learn the behavioral "skills" that bring the parents' attention, that is, the behaviors indicating unhappiness. However, these are not skills that will attract other people. A child who cries a good deal, whines, and performs in other unpleasant ways will not be sought out for interaction. Without such interaction, moreover, the child will not acquire either the social or behavioral skills for maintaining pleasant, fun types of activities with others.

This child would not likely be described as friendly, happy, outgoing, fun, and so on. His inferred personality trait in this area would be considered quite the opposite. Moreover, his basic behavioral repertoire (trait) in this area would establish conditions that would result in further development of this nonsocial basic behavioral repertoire. That is, this child would not have the same responses to people and would thus not approach them in the same manner as the first child. Moreover, the behavior of this child would not be as rewarding to others and they would not seek him out or respond positively to him. In both cases the basic behavioral repertoire that is acquired, even at an early age, would set the stage for later behavioral development.

It is necessary for the theory of personality to include analyses of the basic behavioral repertoires that are involved, as the next several chapters will indicate. Many important concepts and observations for such analyses already exist within the personality theories. However, some traditional personality categories have not been analyzed in terms of the basic behavioral repertoires of which they are composed. For example, a person might be said to have a strong ego, or to be a dependent personality, or to have a negative self-concept. Each of these personality concepts is actually composed of various basic behavioral repertoires. If the person's behavior that led to the inference of a strong ego was examined, it would be seen that some of the behaviors were language behaviors, some social behaviors, and some of them could best be understood in terms of the individual's emotional-motivational system. The same is true of other personality terms.

The various personality theories provide classifications that to some extent overlap and are redundant. For example, the term *ego* in psychoanalytic theory includes reference to some of the behavioral repertoires by which the individual copes with the world (Cameron, 1963). But these same behaviors are thought by other personality theorists to index intelligence (Wechsler, 1944), traits of various kinds (Guilford, 1959), the self-concept (Hall and Lindzey, 1957), cognitive styles (Gordon, 1963), and so on.

The goal here is to describe the general orientation. Several later chapters will exemplify the manner in which specific analysis of basic behavioral repertoires can be made. It is pertinent at this juncture to characterize additional behavioral interaction and to consider it as related to principles involved in the present conception.

## *Behavior-behavior and behavior-environment interplay: Principles of a behavioral interaction approach*

Perhaps the primary divergence between personality theory and elemental behaviorism has been in the attribution of causal status to personality. Behaviorism, in rejecting concepts of personality and in considering personality as behavior, placed personality in the role of an effect, not a cause. Behaviorism did not accept the method of hypothesizing internal mental structures and processes to account for observable behavior.

But personality theorists of varying persuasions could not be content with a psychology that restricted the causes of behavior only to the external events of the environment. They were sure that their phenomenological experiences indicated that there were internal happenings that contributed to the determination of their own behavior, and presumably to the behavior of others. These two views have long appeared to be antagonistic; at least they have been widely treated in this manner.

It is suggested, however, that the door to rapprochement was opened by recognition of behaviors that, while learned themselves, could serve as determinants of other behaviors—especially in the analysis of the self-concept. This avenue was expanded by the concept of the learned basic behavioral repertoire that could be considered as an independent variable (cause) as well as a dependent variable (effect).

> [T]he concept of intelligence . . . *may be considered to refer to a wide sample of the basic behavioral skills the child (or adult) has acquired which are important to the acquisition of further skilled behaviors.* Looked upon in this manner certain enigmas concerning intelligence and intelligence tests can be clarified.
>
> To elaborate, behavioral psychology has at one time attempted to remove some of the ambiguities in the use of the term intelligence by stating that intelligence was what intelligence tests measured. This is in agreement [with the behaviorist position] that intelligence refers to behavioral differences but does not explain them. However, this statement is incomplete, for intelligence test results also index an independent variable for, or a general determinant of, human behavior. That is, for example, people can be selected who score high on an intelligence test, along with a low scoring group. As an independent variable in an experiment, the intelligence test differences will produce effects. That is, the first group will learn many things more rapidly, will read more rapidly, will make higher grades in school, and so on.
>
> Intelligence seems to have the status of a determinant of human behavior. This is not a paradox, however, nor is it inconsistent with the present analysis. Intelligence may be considered as a dependent variable (an effect) and an independent variable (a cause) within the same analysis. When the concept of the basic behavioral repertoire is related to the concept of intelligence, the seeming paradox is resolved. The child through learning experiences acquires a basic behavioral repertoire (intelligence). This basic behavioral repertoire, and its measurement, is a dependent variable. If we wish to account for it we must look to the conditions and principles of learning.
>
> However, we are interested in this basic behavioral repertoire in the first place because we can observe that children who have the constituents of such a repertoire do well in their life tasks and problems. They do better in school and so on. (Staats, 1968c, p. 389)

In the next chapter a section will deal explicitly with the behavioral skills that constitute intelligence, indicating how these skills enable the child to perform better in various situations, including in school. This analysis of intelligence and related basic behavioral repertoires was further elaborated and generalized (Staats, 1971a), and the implications of the conception for establishing a rapprochement between behavioristic views and personality theory was indicated. A central part of this approach to personality involved indication of the several ways that the individual's personality repertoires could act as causes, both for himself and for other people, as the following brief statement indicates.

> In this level of theory construction the general principles of behavior-behavior interaction, and behavior-environment interaction, in their long-term, cumulative, hierarchical and cyclical relationships, also must be elaborated. For example, the personality repertoires play their role in a (1) directly determining future behaviors, (2) making it possible to respond and learn in new situations, and (3) altering the individual's social and physical environment in other ways that affect his development. . . .
>
> [W]hat the individual is at the beginning depends upon what has happened to him. But what he *is* then determines what will happen to him, and therefore what he will further become. The individual's behavior is determined. But his behavior determines what he becomes. The interaction even at an early point becomes . . . involved, and continues in . . . a complex, cyclical and reoccurring way. (Staats, 1971a, pp. 335–336)

The following sections will outline the types of behavioral interaction principles that have been suggested, including behavior-environment interactions. Two types of behavior-behavior interaction will be described first. Sometimes the interaction is direct—where one behavior itself produces the stimulus circumstance that in turn affects a following response. There are also additional cases of behavior-behavior interaction where the person's behavior has certain causal effects upon some later behavior, but the causation is not direct or immediate.

## DIRECT BEHAVIOR-BEHAVIOR INTERACTIONS

There are many ways in which one behavior of the individual can be the determinant of another behavior (Staats, 1971a). Chapter 5, for example, will indicate ways in which the individual's language repertoires will elicit other behaviors of the individual. As an example, it commonly occurs that if a person goes through one type of reasoning sequence with respect to a problem, his overt behavior will be of one type. On the other hand, if he goes through another type of reasoning with respect to the same problem, his overt behavior will be different. To illustrate, one individual who sees someone tending to avoid him will label the other's behavior as shyness, and this verbal response will elicit overt behaviors to make the person feel more at ease. Another individual, in response to the same social behavior, may say that the other person does not like him, and this will elicit overt behaviors of an avoidant or aversive kind.

There are various ways that a behavior can produce a stimulus that elicits another response. In some cases the stimulus is inherent in the response. Thus, any muscular movement produces kinesthetic cues, as has been indicated. Speech responses produce other types of stimuli—namely the auditory stimuli of the sounds. Thus, although speech must be considered in its role as a response which is learned, speech is also a stimulus that has multifarious functions in eliciting behaviors in the speaker as well as in other people. It has also been suggested that there are sensory responses or images. While such events have response characteristics and can be learned like other responses, the images also have stimulus characteristics. One can imagine various scenes, which will then elicit the emotional responses that the actual scenes would elicit. A man who responds by imagery of a nude woman will thereby produce a stimulus that elicits emotional responses that correspond to the sexual arousal he would experience in the presence of the actual stimulus object.

The central point is that one behavior may serve as the cause of another behavior. Certain conditions of learning will have been responsible for the first behavior. But the first behavior itself is also a cause of the second behavior. The first behavior is a dependent variable, an effect, and an independent variable, a cause, at the same time.

### INDIRECT BEHAVIOR-BEHAVIOR INTERACTIONS

The above examples involved single responses whose stimuli directly elicited other responses. The causal behavior, however, may actually be a repertoire rather than a single response. And the causal relationship may not be as close as direct elicitation. An example has already been given in suggesting that "intelligence behaviors" enable the child to do well in many later learning situations, such as in school. The intelligence skills may not be causative in terms of eliciting other behaviors, but they can be causal in terms of helping determine what later behaviors the individual learns. The one basic behavioral repertoire will thus act as a determinant for the acquisition of other general behaviors, as is depicted in Figure 3.1.

The figure shows that certain learning conditions produce a basic behavioral repertoire according to the principles of learning. This basic behavioral repertoire then constitutes an independent variable at a later time. Along with the conditions of learning in the later situation, the basic behavioral repertoire may be considered, in part, to determine what the individual's actions are in that situation (Staats, 1971a).

It should be noted that some basic behavioral repertoires are large systems that play a pervasive role in determining the types of experience the individual will have, and thus the types of behavior he will acquire. There-

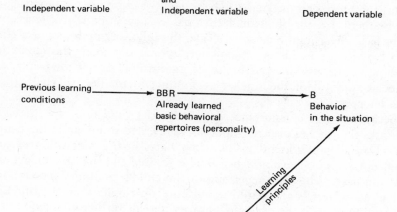

**FIGURE 3.1**

*Learning conditions constitute a determinant of the basic behavioral reper-toires (personality), and hence the way the individual behaves in some later situation. But the stimuli of that later situation also have a direct effect upon the behavior displayed in that situation.*

fore, a basic personality repertoire may determine what later behavioral acts the individual will demonstrate—not by directly eliciting them, but by producing circumstances that will result in whether or not the other behaviors are acquired or maintained.

## BEHAVIOR-ENVIRONMENT-BEHAVIOR INTERACTIONS

As has been suggested, there are also important cases of behavior inter-action that involve the individual's behavior affecting his social and physical environments, the environment then being an important determi-nant of the individual's further personality development (Staats, 1971a). Since the environment itself is a determinant of the individual's behavior, when he affects the environment he in turn helps determine what his own behavior will be.

This occurs ubiquitously and commonly. An example was given in describing the concept of the basic behavioral repertoire. The example involved the child who, because she has learned attractive social behaviors, creates a social environment for herself that results in further learning of attractive social skills.

As was suggested in this example, more than one basic behavioral repertoire may be involved in behavior-environment-behavior interactions. Thus, the formation of one basic behavioral repertoire may "produce" an environment for the individual that has a very important effect upon other basic behavioral repertoires. Take the following example, of a child who has deficits in his emotional-motivational system in the sense that learning academic skills (achievement motivation) is not rewarding, nor is social approval of the teacher or winning in intellectual competition with his peers.

Because of this circumstance, let us say, the child does not attend to the tasks the teacher presents, does not study, and so on, and after a time has advanced much less than his schoolmates in the acquisition of such intellectual basic behavioral repertoires as reading and the like. This means that the basic behavioral repertoire of the child's emotional-motivational system has adversely affected his intellectual development in the situation.

Continuing with the example, however, reveals further interactions. Let us say that at some point the child's intellectual deficits become evident. The child with a reading deficit (or any other evidence of underachievement) is likely to meet derisive social responses when he is young and social rejection when he is older. Moreover, he will not receive the same social rewards as other children of more fortunate circumstances. All of these are aversive conditions. Such presentation of aversive stimuli, advertent or inadvertent, by successful children and by the child's teachers would be expected to produce additional effects. That is, the child would be expected to learn negative attitudes toward other schoolmates, his teachers, his books and school materials, his topics of study, any type of formal learning situation, and the school in general. This will then produce additional effects of a negative nature. Such attitudes would induce in the child undesirable instrumental behaviors in a direct behavior-behavior type of causation. In a case the author has in mind, there was much evidence of this in that the child baited teachers and students, fought in school, disrupted class, and the like. When the child grew older he was apprehended for vandalizing a school. Such a child will in general also attempt to escape from the school in various ways, such as daydreaming and absenteeism, which will further affect the child's intellectual basic behavioral repertoires (Staats and Butterfield, 1965).

Such behavior-environment-behavior interactions occur generally in personality development. One can see, even in the present few examples, how futile it would be to attempt to account for human behavior only on the basis of simple stimulus-response principles of learning and simple stimulus-response analyses of behavior, although these elementary principles are centrally involved.

To continue, however, recent accounts in the behavioral literature have begun to accept the principle of behavior-environment interaction (even while, in the next example, holding to some of the behavioral tenets with respect to personality tests), in a manner very congruent with the behavioral interaction conception (Staats, 1971a).

> The mutual interaction between person and conditions (so easily forgotten when one searches for generalized traits on paper-and-pencil tests) cannot be overlooked when behavior is studied in the interpersonal contexts in which it is evoked, maintained, and modified. . . .
>
> Generally, changes in behavior toward others tend to be followed by reciprocal changes in the behavior of those others. . . . The person continuously influences the "situations" of his life as well as being affected by them in a mutual, organic two-way interaction. (Mischel, 1973, p. 278)

The example employed to demonstrate the principle also refers to the area of intellectual repertoires that has already been described.

> The relevance of cognitive-intellective competencies for personality seems evident in light of the important, persistent contributions of indices of intellignce to the obtained networks of personality correlations (Campbell and Fiske, 1959; Mischel, 1968). In spite of extensive efforts to minimize or "partial out" the role of intelligence in personality studies, for example, cognitive competencies (as tested by "mental age" and IQ tests) tend to be among the very best predictors of later social and interpersonal adjustment. . . . Presumably, brighter, more competent people experience more interpersonal success and better work achievements and hence become more positively assessed by themselves and by others on the evaluative "good-bad" dimension which is so ubiquitous in trait ratings. . . . (Mischel, 1973, p. 267)

Krasner and Ullmann (1973) refer to the behavior-environment interaction as behavioral influence, without developing the concept of personality:

> We prefer behavioral influence as an alternative concept because it emphasizes that the person acts within a physical and social environment. We wish to emphasize that . . . the necessary interaction is between the person and his social situation. (p. 21).
>
> One person's response frequently serves as a stimulus for a second person, and the second person's response provides further stimulation for the first person. (p. 143)

An example given of behavior-environment interaction concerns a girl engaging in intercourse outside of marriage, the original behavior. "If her boyfriend publicizes her new generosity, her environment may change considerably; and she may interact with others in a far different manner than she had previously, not because of the act of intercourse, but because of the recategorization following it" (Krasner and Ullmann, 1973, p. 162). That is, her behavior would affect her social environment in ways 'that would in turn have an effect upon her.

## Continuity, generality, and uniqueness in personality

As has been indicated, traditional behaviorism does not accept the concept of personality in the manner that it has been used in the field of personality theory, as well as in common usage. Behaviorism has not dealt with personality as a causal condition. Moreover, behavioristic accounts have minimized or rejected some of the characteristics of personality that have usually been accorded, for example, the continuity, the generality, and the uniqueness of personality. The present section will summarize these three principles as a prelude to later discussions.

### CONTINUITY AND PERSONALITY

One of the considerations that has led to the inference of internal personality traits that determine external behavior has been that of continuity. It can be observed that the young child will display characteristics that will continue throughout his life. The child may be observed to be aggressive and outgoing, for example, and this may continue, in elaborated form, as he grows older. The intelligent child tends to become the intelligent adult. The child who early develops sensorimotor athletic skills of a high level will tend to display commensurate skills at a later age.

Such observations have lain at the base of beliefs that there must be some internal structure or process, native to the individual, that is the determinant of the external behavior. This internal structure or process is thought to grow and mature, and result in the growth and maturity of the child's behavior—in a manner that parallels the continuity of the internal structure or process. In the realm of abnormal behavior, such continuity has led to the inference of long-developing, internal disease processes.

It is important that such observations are not simply ignored. The concept of basic personality repertoires, the principles of cumulative-hierarchical learning, and the principles of behavior interaction provide a means for accounting for such continuities. That is, basic behavioral repertoires, once acquired in incipient form, may set other events that will produce future learning of a similar nature (see Staats, Gross, Guay, and Carlson, 1973). Examples have already been given of this process, and later sections will amplify this topic.

There are, of course, cases where conditions can change and result in changes in development of the basic behavioral repertoires. For example, the child may move from a small community and school, where he is considered to be bright and is treated accordingly, to a new community and school where the social responses to him are quite different. Psychotherapy may also aim for personality change. With such change, a change in behavioral development may be noted. Detailed observations are necessary in any particular case to show continuities and discontinuities of personality development.

## GENERALITY AND PERSONALITY

As has been indicated, there is an issue between the traditional view of personality as general, across situations, and the elemental behaviorist view that behavior is largely situationally specific (Mischel, 1968; 1971). There are abundant naturalistic observations to support the traditional personality view. For example, the socially aggressive person tends to behave this way in various situations. Even when he meets new stimulus situations (new people, and so on) he is more outgoing than the person who generally behaves in a more socially reticent way.

On the other hand, as Mischel emphasizes, there are studies in which generality has not been found.[1] For example, LaPiere (1934), in a classic study, found little correspondence between attitude questionnaires in terms of the food and lodgings that would be offered Chinese guests and actual behavior toward these guests. That is, a large percentage "said" they would not give service to Chinese, but a well-dressed Chinese couple, accompanied by a Caucasian, was almost never refused service. Another classic study (Hartshorne and May, 1928; Hartshorne, May, and Shuttleworth, 1930) found children behave inconsistently with respect to lying, stealing, and cheating across different situations, whereas their expression of moral attitudes was consistent. This has been interpreted to mean that a personality concept such as that of the superego is incorrect in suggesting generality and, hence, is itself suspect.

The fact is, naturalistic observations provide a convincing type of evidence that there is considerable generality of behavior across situations. However, there is also convincing evidence of the experimental type that has been described in which current situational cues appear to determine behavior. A conceptual framework is needed that can encompass both types of evidence which, without further analysis, appear to be antagonistic.

It is suggested that the framework can be summarized in the schematization presented in Figure 3.1 above. There are situational cues that have a heavy influence upon the individual's behavior. These are causal, as is depicted under the label *present stimulus situation*. But the individual's past history of learning has produced or failed to produce certain personality repertoires, which also constitute independent variables.

---

[1] Mischel (1973) has recently begun to modify his approach toward recognition of a behavioral interaction view. The rejection of the generality of personality across situations continues to be central to the social learning approach, however.

It should be understood that these two sources of behavior determination can work in conjunction or in antagonistic or disjunctive ways. For example, in reference to the LaPiere study, the individual may have learned a negative attitude toward Chinese people. This may be part of a general personality repertoire, as later discussions will indicate. Whenever a Chinese person is present, the covert negative attitude will be elicited. However, what the individual does in terms of overt behavior will also be a function of the other source of behavioral determination, the situational stimuli. In a voting booth, where the individual is isolated and situational influence is weak, he may vote against a person with a Chinese name. Similarly, he may vote to exclude a Chinese applicant from admission to a university where his vote is secret and there are no countering situational stimuli affecting his behavior. On the other hand, in the presence of a disapproving audience—an important part of a current situation—the individual may give an affirmative vote, in a manner at odds with the influence of his attitude.

In the latter case, the situational stimuli may have overridden the individual's personality disposition. The determining source in other cases may be the personality disposition. For example, there are widespread examples of people who act in accordance with their attitudes, values, and other personality repertoires in the face of situational cues that would lead them to behave in a different manner. Sometimes this is maladjustive for the individual. His personality repertoires may lead him to behaviors that result in harmful or undesirable circumstances. The individual may, in fact, indicate that he realizes the unwanted consequences that will occur but can nevertheless behave in no other manner. It may be added that in cases where the individual has such strong personality repertoires, in terms of determining his overt behavior in the face of situational cues that should have restrained the behavior, he may be considered as abnormal.

Both the traditional and the elemental behavioristic views are correct, but not solely correct. The major point is that the dual sources of determination of human behavior should be recognized. The task then becomes one of isolating and specifying what these sources of causation are, to be able to understand human behavior and to deal with it.

## UNIQUENESS AND INDIVIDUAL DIFFERENCE IN PERSONALITY

A closely related topic concerns the uniqueness of personality, a concept traditionally referred to, and the importance of the measurement of individual differences in personality. It is commonly observed that in many cases several individuals placed in the same situation—where the principles and circumstances of conditioning are the same for all—will nevertheless behave differently. Traditionally this is accounted for by suggesting that each individual has a unique personality and thus will behave differently in response to the same circumstances.

The thrust of basic learning behavioristic research is the discovery of *general* principles that apply to all. Contemporary behavior modification has reflected this characteristic. For example, behavior modification work aims to develop procedures that are effective for everyone. It is true that treatment procedures can be created to which many persons will respond similarly, by employing generally effective conditions. Token reinforcer procedures and systematic desensitization constitute such procedures, as will be indicated later.

But success in producing such generally applicable procedures should not obscure the facts of uniqueness and individual difference as these are important to other problems. Whether or not the individual will respond adjustively to any situation will usually depend upon his personality repertoire development. It is for this reason that personality measurement is important. There are cases where different situational circumstances are appropriate for different personality repertoires and a means of assessing the repertoires would be very valuable. It is necessary to provide a conceptual framework for rapprochement between behaviorism and the measurement of personality. The concepts of the basic personality repertoires, along with specific analyses of these repertoires, and the principles of behavioral interaction provide a framework for this rapprochement.

## Behavior analysis and human learning

In the preceding chapter it was indicated that smaller response units may become combined into larger response units. This can be considered a separate principle of learning that has considerable significance at the human level. The principle does not derive from the animal laboratory, again because the behavior involved at this level is unimportant. Whether or not an animal's behavior under study is composed of several subbehaviors is not of central significance.

At the human level, however, it is important to explicate the principle that subbehaviors can combine into a larger behavior which will function as a unit in further learning. It is also important to be interested in the constituents of which a particular behavior—or a task of learning—is composed. Topics of interest are involved at the human level that are not crucial in the animal laboratory.

As an example, a response, even in a simple learning experiment, many times consists of a complex of responses, not just one. Pressing a bar, to illustrate, is a complex of responses. Since the complex as a unit will follow the principles of learning, analysis of the components of the complex is not important. Analysis, however, may be essential on the human level. That is, the course of human learning may be influenced by this fact, and lack of analysis may hide the processes involved. Take the child's learning to write the letters of the alphabet. Several different skills are actually involved for the beginning learner: (1) attending to the relevant letter stimuli, (2) scrutinizing them in terms of their individual characteristics, (3) holding a pencil, (4) making lines at will of the type desired, (5) applying skills of copying a new stimulus, and finally (6) the learning of a specific new letter.

If skills 1 through 5 have been learned, the learning of a new letter will take place easily, with but a trial or two. The learning process in this case is different from the process where skills 1 through 5 must also be learned, as well as the new letter. For this reason the learning task for a beginning learning child is quite a different process from the learning of an adult or an accomplished child.

The important point is that understanding human learning may involve a behavior (or task) analysis. It may not be possible to see reliably the principles of learning involved unless the behavior is stipulated in enough detail. The point to stress is that an abstract concept of personality is not sufficient. *It is necessary to have observations and analyses of the specific personality repertoires.* Social learning accounts, even those beginning to recognize some need for a concept of personality, have not provided such observations and analyses. It should be understood that *a behavior analysis in terms of stimuli and responses and learning principles constitutes a theory.* As will be indicated in later chapters, such theories can provide the basis for research on clinical findings.

## Conclusions and overview

Elemental behavioristic approaches have rested upon the elementary principles of conditioning. The present approach of social behaviorism also includes a statement of the S-R mechanisms by which more complex skills can be acquired, and it adds human learning principles to those that derive from the animal laboratory. In addition, the principles of behavioral interaction are applied in the present approach to specify how functional human learning involves the acquisition of repertoires of behavior that show interaction within themselves and between repertoires, as well as interactions with the environment. It has been suggested that these repertoires are the individual's personality. The conception allows a rapprochement between behavioristic principles and personality theory.

The personality theory level of the approach will be further specified in describing the personality repertoires of the emotional-motivational system (Chapter 4) and the language or cognitive system (Chapter 5), as well as sensorimotor (instrumental) skills, including imitation (Chapter 6). The framework will be further expanded in introducing principles of social interaction in Chapter 7. Finally, Chapter 8 will utilize these concepts and principles to outline a behavioral approach to the consideration of abnormal behavior.

Part three of the book will deal with the elaboration of the social behaviorism paradigm into the problem areas concerned with the study of man. Chapter 9 will indicate some of the progress in extending behavior principles to clinical psychology and how this can be elaborated within social behaviorism. Chapter 10 will be concerned with the area of developmental psychology, and Chapter 11 will deal with educational psychology. The concepts and principles of personality provide a foundation for the consideration of personality measurement in Chapter 12.

Part four will elaborate on the behavioristic-humanistic rapprochement and indicate how the social sciences and the biological sciences are related to the paradigm. In Part five, an epilogue, Chapter 16 further characterizes the social behaviorism paradigm and indicates its generality.

# Motivation theory and the emotional-motivational (A-R-D) personality system

THE CONCEPTS OF emotion and emotional states have figured prominently in explanations of human behavior for ages. Our common language includes a myriad of terms that refer to emotions. The individual is said to be glad, happy, joyful, in love, delighted, pleased, titillated, or blissful, all of which are terms used to denote positive emotional states. The individual may also be spoken of as angry, upset, malevolent or hateful, enraged, irritated, annoyed, contemptuous, unhappy, fearful, grief-stricken, bereft, hurt, tearful, embarrassed, and so on. These terms refer to negative emotional states.

The terms noted above also refer to positive or negative emotional states of varying degrees of intensity. Being pleased or satisfied is not as positively emotional in character as being joyful or delighted. On the negative side the same difference in intensity is seen between being embarrassed and being mortified, in disliking and hating, in being annoyed and being enraged, in being unhappy and in being grief-stricken.

Moreover, there is a time dimension involved in distinguishing some of the common terms for emotional states. Being in love suggests an emotional state of longer duration than being infatuated or being sexually aroused. Hating implies a more permanently established emotional response than does annoyance.

There are also terms that refer to emotional states that are characteristic of the individual and part of his continuing personality. For example, a person may be described as emotionally immature, emotionally labile, or emotionally dependent. He may also be considered to have a fixated emotion of some type, such as a phobia or fetish, that is a significant part of his personality.

## The concept of emotion

The established and commonly accepted view of emotions is that they represent mental states aroused by objects, events, or ideas. The emotional state in this view is thought to play a causative role when it in turn elicits or brings about overt behaviors that are appropriate to the emotion. The state of the individual's emotions is considered to be central to his personality, central in determining what the individual does and how he differs from others.

William James (1884) proposed a different relationship between the internal emotional state and the expressive response. He suggested that the stimulating event led to some response expression. In turn, it was the expression that produced the emotion, or the awareness of the emotion. In this view, different expressions would lead to different experienced emotions. Later work began to investigate the physiological basis of emotions. In 1927 Cannon (1929) failed to find different physiological measurements that were distinctive for the various emotions.

These views, although of historical vintage, have later-day counterparts. For example, there has been a good deal of research attempting to establish the physiological correlates of experienced emotions. And, as indicated in Chapter 2 in the context of animal learning principles (and as will be elaborated), emotion is still considered as a mediating state that occurs between the arousing stimulus and the overt instrumental behavior. On the other hand, there are contemporary views that have a similarity to James in considering emotion to be an epiphenomenon that parallels overt responding but does not act as a determinant of that responding (McGinnies and Ferster, 1971). This view derives from Skinner's approach, which places instrumental conditioning principles in a central position. While classical conditioning is recognized, the conditioning principles tend to be given a secondary and noncausal role in explaining human behavior. Accounts from this approach are apt to consider such concepts as emotion unnecessary (D. J. Bem, 1968; Holland and Skinner, 1961). Traditional conceptions of personality, as has been suggested, give a more central and causal role to emotions. This interest has prompted research along various lines, some of which can be noted.

### PHYSIOLOGICAL CORRELATES OF EMOTION

If emotional response expressions are based upon physiological response states, then it might be expected that there would be correlations between physiological emotional states and other indices of emotion. Cannon's work did not support the idea that different physiological states would be found for different emotions. Cannon (1929) found (1) that when the visceral functions thought to be the seat of emotions were surgically separated from the central nervous system, this did not alter emotional behavior, (2) the same visceral changes could occur over different emotional states or in nonemotional states, (3) visceral changes were too slow to be the cause of emotional states, and (4) induction of visceral changes artificially did not produce the emotions. Schachter (1970) has suggested that such results indicate a lack of identity between physiological changes in the viscera and emotions.

It is also true that various measures made on the same person do not produce consistency. In response to an emotion-inducing stimulus, thus, there may not be homogeneous indications of physiological emotional responding. The individual's heart rate, sweat gland activity, muscular tension, brain waves, and so on, may not show the same type of change (Lacey, Bateman, and Van Lehn, 1953).

The evidence is not all negative, however, in terms of the relationships between emotion-inducing stimuli and physiological measures of emotion. A study by Ax (1953) employed a measure of the amount of the hormones epinephrine and norepinephrine in the blood. These hormones are secreted by the adrenal glands, and both function to increase blood pressure. Epinephrine raises blood pressure by affecting the heart; norepinephrine raises blood pressure by constricting the blood vessels. Ax found that subjects presented with anger-inducing situations responded with an increased secretion of norepinephrine. Conversely, subjects made to experience fear showed an increase in epinephrine secretion.

There are other data of a supportive nature also. For example, Geer (1965) selected subjects who indicated either high or low fear of spiders. The subjects were then shown pictures of spiders and snakes, and their physiological emotional response was measured by means of the GSR (a measure of sweat gland activity). Those who reported fear of spiders responded more to the pictures of spiders than the other subjects, and this difference in response was not shown with the pictures of the snakes.

## EMOTION AS A DETERMINING STATE

The original conception of emotion was of a mental state that determined overt behaviors. Some of these overt behaviors were expressive. In addition, a frequently employed index of an emotional state has described the striving of the individual toward or away from the stimulus situation eliciting the emotional reaction.

> [E]motions also serve in the regulation of behavior as *incentives,* or anticipated goal "objects." (Dember and Jenkins, 1970, p. 581)

> [F]ear can serve as a drive. . . . Other emotional reactions have drive properties also. Anger, which is often aroused by a frustrating situation, . . . can energize behavior and encourage animals to learn responses that will reduce the anger. (Kendler, 1968, p. 270)

> [E]motion becomes a *felt tendency toward anything appraised as good, and away from anything appraised as bad.* This definition allows us to specify how emotion is related to action: if nothing interferes, the felt tendency will lead to action. (Arnold, 1970, p. 176)

However, it may be noted that experiments to demonstrate the mediation of instrumental behavior by emotional states have not been uniformly successful when the usual physiological measures are employed. On the one hand studies, such as that by Gantt and Dykman (1957), have found that as dogs learned an avoidance response to shock upon presentation of stimulus, that stimulus came to elicit a conditioned heart rate response, one indicator of emotion. However, when Wynne and Solomon (1955) further tested the possibility that emotional responses such as conditioned heart rate and blood pressure were actually mediating the instrumental avoidance response by removing the sympathetic nervous system involved in such emotional responses, it was found that the avoidance response was impaired but not completely prevented. Such results may be interpreted as negative evidence. Or, the impaired avoidance responding may be seen to indicate that the emotional response contributed to mediation. (See Rescorla and Solomon, 1967, for other relevant studies.) It may be added that Solomon and his associates (see Maier, Seligman, and Solomon, 1969) have conducted a number of studies showing that the process that mediates overt instrumental responding in the transfer of control type of experiment follows the principles of classical conditioning. This suggests that the mediating process is an emotional response. These findings agree with others conducted with humans (see Staats, 1959, 1968d), some of which will be referred to later.

## THE PERIPHERAL VERSUS CENTRAL DISTINCTION IN EMOTIONS

In describing classical conditioning in Chapter 2, the examples of peripheral emotional response given—such as heart rate, blood pressure, and gland activity—may be considered as types of peripheral emotional responses. It is suggested that a distinction between peripheral emotional responses and central emotional responses is necessary to understand certain important features of emotional responding.

To elaborate, a food stimulus elicits a salivary response and other digestive responses, while a sexual stimulus elicits responses in other organs. Both food and sex stimulation are considered as positive emotion-eliciting stimuli, but what provides the basis for considering both of these emotional stimuli in the same class? One justification lies in the fact that the two stimuli have certain common functions in affecting how the individual behaves, in terms of approaching the stimuli, for example. However, there is another type of justification. The peripheral responses described are only part of the emotional response the organism makes to a stimulus. In addition there are central nervous system emotional responses, and various lines of research indicate that there are brain mechanisms that are involved in emotional responding.

It is thus suggested that a positive emotional stimulus will have a specific effect in eliciting certain peripheral emotional responses. In addition, however, it will elicit an "emotional" response in the brain which will be like that elicited by another positive emotional stimulus. This central emotional response is general to the various positive emotional stimuli. It is suggested that it is for this reason that all positive emotional stimuli can have the same functions for the behavior of the organism— functions yet to be described in full. This same peripheral-central distinction can be made for negative emotional responses. In support of this possibility, LoLordo (1967) has shown that a loud noise and an electric shock have the same function in eliciting a negative emotional response. When an animal has learned to avoid a stimulus paired with loud noise, he will avoid a stimulus paired with shock (and vice versa), with no additional training. It may be added that various investigators in this area of study have suggested a central rather than peripheral interpretation (Bitterman and Schoel, 1970; Bolles and Grossen, 1969; Rescorla and Solomon, 1967; Shapiro, Mugg, and Ewold, 1971).

The equivocal nature of the evidence in the study of emotions pertains to the study of peripheral measures of emotion. As has been indicated, these measures do not always coincide for the same individual or across individuals. Moreover, it does not appear that the functions of an emotional stimulus are obviated in entirety by excising the organs of peripheral emotional response. Understanding of how visceral responses are involved in emotions and in the functions of emotional stimuli requires further investigation. Whether there are different patterns of peripheral responses involved in the different emotions remains to be shown. Perhaps the major differences in various emotional states are those of intensity, generality (in terms of the number of peripheral emotional response organs involved), and duration. At any rate, as employed herein, the concept of the emotional response will be assumed to have central components as well as peripheral indices.

## SITUATIONAL CUES AND EMOTIONAL STATES

The basic position elaborated in the context of emotions is that there are emotional responses, peripheral and central, that serve to affect the way the individual behaves. As has been noted in part, however, it has been suggested that emotional states are undifferentiated and that the nature of the situational stimuli and the labels the individual gives to his emotional states determine what he feels and how he behaves (Schacter, 1970).

These conclusions were derived from experiments in which subjects were given the drug epinephrine, which results in sympathetic nervous system arousal. The result is the elicitation of visceral responses that are like those elicited by emotional stimuli. Following the drug administration some subjects were placed in a situation in which a confederate of the experimenters acted either euphoric or angry, in the latter case in response to an insulting questionnaire also given to the subject. The subjects who experienced the confederate who was euphoric responded in a similar manner. The subjects who experienced the insulting questionnaire and the angry confederate showed signs of anger (Schacter and Singer, 1962). Subjects who were told the visceral responses they would experience, or subjects who were not given the drug, did not respond either to the euphoric or anger-producing situations.

It is suggested that Schacter and Singer's experiments illustrate that situational cues are important, as these investigators concluded. A smiling person elicits a positive response in others, ordinarily. A person behaving in an angry manner ordinarily elicits a negative emotional response in others, because of this past learning. More will be said on this principle later. It may be added that the way the individual has learned to label his emotional responses would also be expected to be a central determinant of the overt behavior the emotional response will mediate. For example, the individual who experiences sexual emotional responses to another person and who labels the responses as sinful will respond differentially than the individual who does not label the responses or who labels the responses as love. The labels themselves, as will be indicated more fully further on, can themselves elicit emotional responses and in this and other ways serve as determinants of the individual's behavior. Situational cues are also important sources of overt behavioral control.

It is not necessary, however, to exclude internal emotional states of a peripheral or central kind as determinants of external behavior. Whether drugs such as epinephrine, although producing emotional response effects, duplicate the patternings of peripheral emotional responses that actual emotional stimuli produce is still an open question. Whether or not central involvement is the same when epinephrine is employed as in actual emotional stimulation is also open to question.

Schachter and Singer's experiments and conclusions can thus be viewed as an addition to the study of emotions, in indicating additionally how learning and cognition (language) may be important. The findings do not obviate the need for consideration of learned emotional responses and the manner in which such responses can affect the individual's behavior.

## Motivation, drive, and arousal

The concept of motivation—or related terms such as drive, instinct, need, purposes, incentives, goals, and so on—is prominent in treating the topic of behavior. The commonsense view of motivation is to consider a motive to be a state that organizes, energizes, and directs behavior. Commonsense terms for motivation include ambitions, desires, and wants, on the one hand, and anxieties, fears, and so on, on the other. The person who is described as having an ambition to become a doctor is seen to have as a result a source of energy that organizes and directs his behavior.

It turns out that there is much in common between the terms *emotion* and *motivation*, at least in terms of the types of events that are referred to in each case. Thus, there are studies that attempt to provide a physiological basis for the study of motivation, as well as behavioral studies.

### AROUSAL AND MOTIVATION

One of the aspects of the concept of motivation that has been widely held is the characteristic of energizing the individual. Freud, for example, considered there were instinctual sources of energy within the personality, that is, instinctual drives. Hull (1943), as has been indicated, posited that drive had a general energizing effect.

An area of study in physiological psychology has gained some interest value because the investigations appear to fit into the concept of motivation as an energizer. The area may be labeled as the study of arousal and activation. At any rate, as in the study of emotion, there has been an attempt to relate physiological mechanisms to the concept of drive.

For example, Malmo (1959) proposed a neuropsychological theory of activation which suggests that there is a continuum of neural excitation. At its low level the continuum refers to sleep states, and it ranges to states of high excitement. "According to this theory there is an optimal amount of this background [neural excitation] activity for best supporting the organized [neural] activity. Below and above this optimal amount . . . the efficiency of the organized activity is relatively impaired" (Malmo and Belanger, 1967, p. 288).

Measures of neural excitement or activation have involved such responses as the heart rate, respiration, sweat gland activity, muscle tension, and so on, as well as the EEG, or brain waves. (It has been suggested, however, that different forms of arousal are involved in different measures; Lacy, 1967.) As an example, the brain continuously emits weak electrical activity. The characteristics of this electrical activity vary, in relationship to the state of activity displayed by the individual. When the individual is in a deep sleep his brain impulses have a slow wave quality. As he increases his behavioral activation level through being less deeply asleep, through being drowsy, to a relaxed state, and finally to a state of excitement, the waves change in a characteristic manner.

The reticular activating system in the brain has been considered to be the most likely mechanism involved in arousal (Hebb, 1955; Lindsley, 1951). In a study with monkeys, stimulation of this part of the brain with a mild electric shock, prior to engaging in a problem task, resulted in better problem solution than no stimulation (Fuster, 1958). Speed of reaction of the electrically stimulated monkeys was also greater. The results thus support the idea of a relationship between physiological activation and the concept of drive.

Berlyne (1967) has also reviewed a number of studies of arousal and learning. He has attempted to indicate the relationship of arousal, and its increase and decrease, to positive and negative reinforcement and to positive and negative affect. The physiological level of study, it may be suggested, should ultimately provide evidence which is coordinated with the behavioral level of study and with the principles of the learning theory.

## ANIMAL BEHAVIORAL THEORIES OF MOTIVATION

Bindra (1968) has attempted to relate both the physiological evidence and the behavioral evidence in an analysis of motivation. In his view there "are two types of motivational effects (a) drive-induced, those arising directly from the physiological consequences of drive manipulations; and (b) incentive-motivational, those produced by the stimuli that are associated with reinforcement" (p. 17). The former process is said to be responsible for increasing the level of general activity; the latter, while it can have this effect also, is said to be responsible for determining the occurrence of *particular* instrumental responses. This effect is thought not to be the result of the conditioning of specific instrumental responses, but "such facilitation arises from the creation of general central states that may be neither response-specific or drive-specific" (p. 17).

Another related conceptual development is that of P. T. Young (1967). He has suggested a theory of incentive motivation that is called "affective arousal." In this view the attractiveness of goal objects is a function of their pleasantness or unpleasantness—their hedonic value. Through experience with these objects, on the basis of sensory feedback, hedonic information is processed through a central mechanism. There it interacts with "cerebral dispositions" such as habits, expectancies, and so on, to influence external behavior. Young includes an energizing role for motivation.

It has also been suggested, however, that "whatever can be said in the language of motivation can be said as well in the language of reinforcement" (Bolles, 1967, p. 439). It is quite true that there is great overlap in the various terms that have been applied to emotional-motivational phenomenon (Staats, 1968d); additional examples of this will be related herein. However, other terms besides reinforcement and reinforcement principles appear to be necessary to formulate a comprehensive theory of motivation.

To continue, however, these various theories of emotion-motivation have much in common with what has been called the A-R-D theory of human motivation (Staats, 1968d). The letters naming the theory stand for the emotional (or, as Chapter 7 will indicate, the attitudinal), rein-forcer, and directive characteristics that some stimuli have for the indi-vidual. The principles involved correspond to the three functions of stimuli ($C$-$R$-$D_S$) described in Chapter 2. The manner in which these abstract principles, based on animal learning, may be elaborated and extended in developing a basic motivational theory will be described in the following sections.

## Three-function learning theory and the emotion-motivation mechanism

There has been a proliferation of concepts and theories in the study of emotion and motivation. Some of this proliferation has been due to the complexity of the events involved, but to some extent it is caused by duplication, overlap, and superfluity. This can be seen by noting how material dealing with emotions overlaps greatly with other material that refers to motivation. The same findings may be referred to (for example, physiological measures such as heart rate and sweat gland activity) to in-dex both emotion and motivation.

The same is true in the area of theorizing. There is actually a great deal of commonality among the various theories. Not all refer to all of the events involved in the field, and there are differences in subsidiary con-siderations. However, there is much overlap, as some of the examples that have been given indicate.

It can also be suggested that some of the differences between ap-proaches rest upon the fact that different aspects of the characteristics of emotion-motivation may be considered. In any event, at this point a frame-work will be developed within which the several aspects of emotion and motivation may be considered.

The principles involved are those of classical and instrumental condi-tioning. In Chapter 2 it was indicated that there are stimuli that elicit internal responses of the glands, viscera, and blood vessels. These stimuli may be considered to be emotional stimuli, and the responses may be considered to be peripheral emotional responses. Following the principles of Chapter 2, a stimulus that elicits an emotional response is said to have two additional stimulus functions: by virtue of eliciting an emotional response, without learning being involved, the stimulus will also have the function of a reinforcing stimulus. Thus, any stimulus that elicits a positive emotional response could also be employed as a reinforcing stimulus —any instrumental response the emotional stimulus followed would be strengthened. In addition, an emotional stimulus will also have the properties of a directive stimulus. Emotion-eliciting stimuli will directly elicit appropriate instrumental responses. The reason that emotional stimuli have this property is through learning, a process that needs further stipulation.

To elaborate, it has been stated that in the instrumental conditioning situation the reinforcing stimulus, besides functioning as a reinforcer, also elicits an emotional response which is conditioned to the directive stimulus (See Figure 2.9, Chapter 2). Thus, as the conditioning process progresses, in addition to the overt motor response being conditioned to the directive stimulus, the *emotional* response is also conditioned to the $D_S$ (Perkins, 1968; Staats, 1963, 1968d). In the previously described instrumental conditioning situations involving food as a reinforcer, whenever the light that is the directive stimulus for pressing the bar comes on, it elicits not only the instrumental bar press response but the conditioned emotional response as well. If physiological measurements were taken it would be seen that the animal would begin to salivate when the light goes on (Wenzel, 1961). In addition, the light would elicit other emotional responses, such as a change in heart rate. This may be depicted as shown in Figure 4.1 (a).

It should be noted that as this experience continues the organism is being reinforced for the instrumental bar press *in the presence of the emotional stimulus*. This stimulus should thus come to be a directive stimulus itself for the instrumental response. This result is depicted in Figure 4.1 (b).

The concept of the S-R mechanism, introduced in Chapter 2, refers to the learning of complex constellations of stimulus-response events. The concept of the S-R mechanism applies in the context of discussing emotion and motivation. The principle that the organism learns instrumental responses under the control of the stimuli produced by his own emotional responses provides the basis for the formation of a very central type of S-R mechanism.

To elaborate, the above account suggests that in any instrumental conditioning situation, the instrumental response is conditioned to the emotional response that the directive stimulus elicits. The reason this is of major importance is because the organism learns a large number of instrumental responses that are elicited by positive emotional responses, and a large number elicited by negative emotional responses. The organism inevitably will have a vast number of conditioning experiences in its history. For example, the rat sees a piece of food, a stimulus that elicits

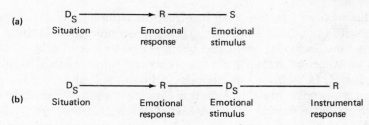

**FIGURE 4.1**

*Part a indicates that in the process of instrumental conditioning an emotional response is conditioned to the situational stimulus. The emotional response produces its emotional stimulus component. Part b shows that the instrumental response becomes associated with the emotional stimulus. After this is well learned, any situation that elicits the emotional response, with its stimulus characteristics, will thereby also elicit the instrumental response.*

a positive emotional response in it. It approaches the food stimulus and this instrumental behavior is reinforced, in the presence of his positive emotional response. The organism has been reinforced for approaching a stimulus that elicits a positive emotional response. It learns generally to approach stimuli that elicit positive emotional responses. As another example, the animal senses a female animal in heat, which elicits a positive emotional response. The animal approaches the female animal, mounts the female, and these approach behaviors are reinforced. Thus, another set of approach instrumental behaviors is learned to a stimulus that elicits a positive emotional response. In sum, the animal learns a whole class of different approach behaviors, elicited by the stimulus characteristics of a positive emotional response. This mechanism is depicted in Figure 4.2.

Once the organism has learned this mechanism, anything which increases the tendency of a stimulus situation ($^DS$) to elicit a positive emotional response in the organism will increase the tendency of that situation to elicit one of the "striving for," or approach, responses that the organism already has in its repertoire. Thus, classical conditioning procedures that condition an emotional response to a stimulus will also increase the power of that stimulus to elicit overt instrumental behaviors. In this way classical and instrumental conditioning are inextricably interrelated. The results of recent experiments may be interpreted by this S-R mechanism. For example, as has been indicated in Chapter 2, Trapold and Winokur (1967) have shown that prior classical conditioning which establishes a stimulus as a $^CS$ will later enhance the $^DS$ value of the stimulus. The stimulus will elicit overt approach behavior to a greater extent than will a neutral stimulus. While the transfer of control experiments have not in-

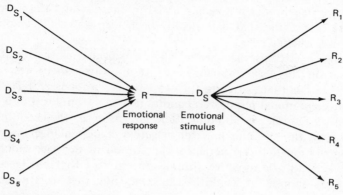

**FIGURE 4.2**

*Various instrumental approach responses are reinforced in the presence of various situational stimuli that elicit a positive emotional response. The positive emotional response (or its stimulus property) through this process comes to elicit a large number of instrumental approach responses.*

dicated this, this phenomenon depends on the prior establishment of the type of response sequence depicted in Figure 4.1 above. Thus, in the Trapold and Winokur study, the animals were first trained to press the bar for food, before the experiment commenced—thus establishing an association between the positive emotional response elicited by the food and the bar press response. As a consequence, any stimulus that elicited a positive emotional response would also elicit the instrumental bar press response. It is suggested that without this prior training, and the formation of the sequence, the transfer of control would not occur. This analysis could serve as an experimental hypothesis.

This example has been presented here to indicate that the elaboration of the three-function (A-R-D) learning theory can be employed to interpret basic animal learning experimentation, as well as findings at the human level, and to suggest new research as well. Another example at the basic learning level can be seen in the phenomenon called "auto-shaping." Brown and Jenkins (1968), using pigeons as subjects, paired a lighted disc with the delivery of food grains which the pigeons consumed by pecking. Later it was found that the pigeons would perform the instrumental response of pecking the lighted disc when it was presented. The instrumental response to the disc had not been reinforced (conditioned). In terms of the three-function A-R-D principles, it is suggested that pairing of the lighted disc and grain resulted in classically conditioning a positive emotional response to the disc. This conditioned emotional response (or its stimuli) was then elicited by the disc and the instrumental response of pecking that had already been learned to the emotional response was elicited.

Analogous circumstances would also be expected in the negative case. However, in this case "striving away from" (avoidance) responses are involved rather than "striving for" (approach) responses. That is, when a negative emotional response is conditioned to a stimulus it also becomes a negative reinforcing stimulus. The organism is then reinforced for escape from the stimulus *and the negative* emotional response (and its stimuli). Through this experience, the stimuli of negative emotional responses come to be directive stimuli (incentives) that elicit escape and avoidance responses. It would be expected, thus, that negative emotional classical conditioning to a stimulus would make that stimulus a $^{D}S$ that would tend to elicit any one of a large class of "striving away from" behaviors.

## ADDITIONAL SPECIFICATIONS

As has been indicated, there is considerable overlap between the concepts of emotion and motivation. In both cases conceptual formulations have usually involved some of the aspects of the A-R-D principles. Thus, as an example, definitions of both emotion and motivation have referred to the incentive (attractive) nature of some stimuli. In terms of study, however, there is perhaps a tendency for studies of emotion to be more concerned with the A (attitudinal or emotional) function of stimuli—and with specifying the emotional responses involved, the experience of emotion, and so on. Studies of motivation, on the other hand, have a tendency to be concerned with the directive, or incentive, function. There is a tendency in both cases to neglect the fact that emotional-motivational stimuli have a general reinforcing function.

It is additionally suggested that, although the A-R-D functions of emotional-motivational stimuli are related, so that as the A value is increased the R and the D values also increase, the base of the relationships between all the functions is not the same. That is, it was suggested in an abstract way that the emotional and reinforcing functions of stimuli are related on a biological basis. A stimulus that elicits an emotional response in the organism will also have reinforcing value, because of the organism's biological characteristics. Thus, an unconditioned stimulus, like food, has an emotion-eliciting function for the organism on an unlearned basis. It also will have a reinforcing function on an unlearned basis. As was indicated, the elicitation of an emotional response is seen as the basis for reinforcement.

Typically, theories of incentive motivation, which indicate that the organism approaches positive stimuli and avoids negative stimuli, do not explicate the learning (the S-R mechanism), involved. As has been indicated, however, the directive function of emotional stimuli is learned. Emotional stimuli may elicit overt responses, like facial expressions and startle responses, on an unlearned basis. Various studies have attempted to see if emotions reliably give rise to characteristic facial expressions and if these expressions arouse appropriate feelings in others (Feleky, 1922; Schlossberg, 1952). However, it is suggested that instrumental responses that either approach or avoid emotional stimuli have been learned.

This may seem strange in view of our general feeling that certain stimuli just naturally attract or repel us. For example, it seems like reflex action to withdraw from a painful (emotional) stimulus. Infants, however, appear only to respond only in a general manner to painful stimuli, and naturalistic observations suggest that children must learn even simple escape responses. This is supported by research with dogs raised in restricted environments (Melzack and Scott, 1957). In any event, it would seem that complex instrumental skills must be learned, and in this sense at least the directive functions of emotional stimuli are learned.

One additional point can be inserted here. As has been suggested, whether emotional responses are peripheral or central is a matter of some concern. The A-R-D principles help provide support for the interpretation that at least part of the basis of emotional-motivational stimuli is in the central nervous system. That is, it has been said that various stimuli can elicit an emotional response. For example, food is such a stimulus, as is tactile stimulation of the sexual organs. Since each stimulus elicits different peripheral emotional responses—for example, salivation versus tumescence—what is the basis for considering both to be examples of positive emotional stimuli?

One answer refers to the A-R-D functions. Food stimuli will display the three A-R-D functions, and so will sexual stimuli. This similarity in general functions suggests that there is an organ that is common to both emotional stimuli. As will be indicated in a later chapter, brain research suggests that there are centers in the brain that are involved in positive emotional responses and centers involved in negative emotional responses. In suggesting central mechanisms for emotional-motivational phenomena, A-R-D theory (Staats, 1968d) and various theories in this area (for example, Bindra, 1968; Young, 1967) are coincident.

## Deprivation and motivation

Thus far the account has treated the concepts of emotion and motivation as very similar, perhaps with some differences in emphasis. However, the study of motivation has included a focus of study that is characteristic. It has long been recognized that there is a cyclicity of behavior. The organism does not eat all the time, copulate continuously, and so on. Although food stimuli and sexual stimuli are among those that elicit positive emotional responses, these stimuli are not always effective as emotion elicitors, elicitors of instrumental approach responses, or as reinforcing stimuli.

> Some . . . activities appear rhythmically when an animal is observed under constant conditions; others appear sporadically. Eating, for example, is periodic; where there is an adequate supply of food the periodicity is approximately constant. Similarly, the drinking of water, the eliminative processes, female sexual behavior [and male sexual behavior], sleeping, breathing (respiratory cycle), all have their characteristic periodicities. (Young, 1936, p. 49)

Such observations of cyclicity in behavior, depending upon the state of deprivation for the substance that the behavior is instrumental in obtaining, introduce a dimension that is characteristic of the term *motivation*. It may be noted that the physiological mechanisms involved in deprivation are not the same across different emotion-eliciting stimuli. Deprivation of food, for example, produces internal tissue needs that unless fulfilled will result in damage to the biological organism. The same is true of water deprivation. Deprivation of sexual contact, on the other hand, appears not to involve such physiological mechanisms. The basis of the deprivation-satiation effects appears, in the case of sex, to be based upon the sex hormone concentration in the blood. The behavioral effects are the same, in any event.

As has been indicated, the traditional approach to motivation includes an energizing principle. The present approach emphasizes other principles instead. Thus, however, it is suggested that the cyclicity in behavior that is related to deprivation-satiation conditions depends upon the manner in which the strength of the *three* A-R-D functions are affected. In Chapter 2 it was indicated that food deprivation increased the extent to which a conditioned stimulus would elicit an emotional response (Finch, 1938). This should have the effect of increasing the reinforcing value of the stimulus as well as the directive value of the stimulus. It has been shown, for example, that the reinforcing value of food is increased by deprivation (Hillman, Hunter, and Kimble, 1953).

It is suggested that the behavioral effects relating deprivation to strength of behavior depend upon one or more of the A-R-D stimulus functions. Thus, for example, reinforcement appears to be involved in the learning that occurred in the following study with rats.

> Males were trained to traverse a straight runway and enter a goal box containing a receptive female. After one copulation, the male was returned to the starting point and given another trial in the alley. The day's testing concluded with the trial on which ejaculation occurred [the reinforcement].

> There was a positive relationship between the male rat's running speed and the intensity of his sexual reactions to the receptive female in the goal box. Males that had been running rapidly and mating promptly were castrated. The operation produced a gradual loss of sexual ability. . . . When the consummatory responses [reinforcement] began to weaken and disappear, the time spent traversing the alley increased. When copulatory responses were abolished, many males remained in the start box and failed to enter the alley. (Beach, 1956, p. 12)

Actually, the performance described above depended upon the directive function as well as the reinforcing function. That is, running down the runway involves the cues of the straight runway coming to elicit a conditioned emotional response that in turn elicits the running. The directive (or incentive) function of the female rat as an emotion-eliciting stimulus can be seen even more clearly in a study conducted by Seward and Seward (1940). These investigators placed a receptive female rat on one side of a hurdle and a male rat on the other side. They found that male rats would cross the hurdle to get to a receptive female more readily when they had been sexually deprived for several days than if they had recently copulated. The receptive female served as the A-R-D stimulus to "attract" the male animals.

If the animal studies are relevant to human behavior, it would be expected that laboratory studies with human subjects would also show the effects of deprivation operations upon the A-R-D value of stimuli.

An integrated series of studies has been conducted to test these possibilities. To begin, it was necessary to have a conditioned stimulus that was effective for people in eliciting an emotional response. The basis for the conditioned stimulus had to be an unconditioned stimulus that was susceptible to deprivation. It was reasoned that words that name foods fit the requirements for such a conditioned stimulus. That is, food is an unconditioned stimulus that elicits emotional responses (for example, salivation). Food words occur systematically in continguity with food. Not on each occasion, to be sure, but this is not necessary for conditioning to take place.

Since there are so many conditioning trials in the individual's life, it would be expected that food words would constitute a strong conditioned stimulus. Moreover, as Finch (1938) showed with dogs, deprivation of food should increase the extent to which the conditioned stimuli—food words—should elicit the emotional responses conditioned to them. The first step in demonstrating the effects of deprivation upon human motivation, measured by the strength of the A-R-D functions, was to see whether the physiological response of salivation was affected by deprivation. This was done by having one group of subjects skip their breakfast and lunch, the latter being just prior to the experiment. The other group of subjects ate breakfast and lunch, the latter just before experimental participation.

In the experiment a group of words was presented to the subjects, one at a time. Just before each word presentation a roll of cotton was placed just under the subject's tongue, where the salivary gland ducts empty saliva into the mouth. The word was presented and the cotton roll was left in for 20 seconds. The cotton was weighed very precisely on an electronic scale before and after it was placed in the subject's mouth. The increase in weight in the second measurement was due to the amount of saliva absorbed.

Half of the words presented to the subjects were food words—like *pancakes, roast beef, bacon, french fries,* and so on. The other half were nonfood words—like *square, advice, edge,* and so on. It was found that the food words as conditioned stimuli elicited a stronger salivary response than did the nonfood words. Thus, even in a relatively satiated condition, the food words may have some $^{C}S$ value. More importantly in the present context of motivation, however, the subjects who were deprived of food salivated significantly more to the food words than did the subjects who were not deprived of food (Staats and Hammond, 1972).

Another experiment was conducted to see if food words, since they elicit the emotional response of salivation, would serve to condition an emotional response to another stimulus in a higher-order conditioning procedure. Again, subjects were either deprived of food or had recently eaten—the hypothesis being that the higher-order conditioning would be stronger for deprived than for satiated subjects. This expectation was supported (Staats, Minke, Martin, and Higa, 1972).

If the A-R-D analysis of motivation holds, it would also be expected that deprivation should affect the strength with which food words would serve as reinforcing stimuli. In this area there are studies that show that deprivation will increase the efficacy of stimuli that serve a reinforcing function for people. For example, Gewirtz and Baer (1958) conducted an experiment in which children were reinforced for making one of two instrumental responses. The reinforcing stimulus consisted of approving verbal responses. The children were divided into three groups who experienced differing degrees of social deprivation and satiation. The deprived subjects played alone in a room by themselves for 20 minutes prior to the experimental situation. Another group participated directly after a school class. A third group, the least deprived subjects, spent the pre-experimental 20 minutes drawing and cutting out designs while the experimenter talked to them and approved of their work. It would be expected that the social approval would be an increasingly effective social reinforcer the more the children had been deprived of social stimulation and approval. This is what occurred, in a simple instrumental conditioning task. "Thus, a reinforcer appearing to be typical of those involved in children's social drives appears responsive to deprivation and satiation operations of a similar order as those controlling the effectiveness of reinforcers of a number of the primary appetitive drives" (Gewirtz and Baer, 1958, p. 172).

A number of other studies have supported this finding (see Eisenberger, 1970). However, to tie the principle involved firmly with the basic research on motivation, the principle that deprivation increases the reinforcement strength of an emotion-eliciting stimulus should be tested with stimuli known to elicit a physiological emotional response. The experiment employing food words described above suggests the use of the same stimuli to test the effect of deprivation on reinforcement value.

A study has been completed in the present author's laboratory to test this effect (Harms, 1973). Adult subjects were introduced to an instrumental conditioning task where they could make either a left-hand or a right-hand response on each trial. They were told the response was employed only to insure they were doing the same thing when they were presented with each one of a list of words to which their sweat gland response was supposedly being taken. The words were actually being presented as reinforcing stimuli. When the subjects made a left-hand response they were presented with a food word. When the right-hand response was made subjects were presented with a nonfood word. Half the subjects were deprived of food, and the other half had more recently eaten in the manner already described. It was found that the deprived subjects made the left-hand response more frequently than the less deprived subjects.

Finally, the A-R-D theory would suggest that deprivation would increase the extent to which an emotion-eliciting stimulus would serve as a directive stimulus. It has been shown that food words will elicit a physiological response and also have reinforcing value. On this basis it would be expected that food words would have directive stimulus value, tending to elicit approach responses. Further, the theory would suggest that food words would have stronger directive value for subjects deprived of food than for subjects not so deprived, since these conditions affect the strength with which the words elicit the emotional response.

This was tested employing subjects either deprived or not deprived of food, again using the previously described methods. In the experimental task, similar to the one developed by Solarz (1960), the subjects were presented on each trial with a word that was mounted on a stage. When the word was exposed by the shutter apparatus, the subject had to either pull the word toward himself or push it away. The time between the exposure of the word and the movement was precisely timed by electronic devices. The words were either food words or nonfood words. Half of the subjects had to learn to pull the food words toward themselves and to push the nonfood words away. The other half of the subjects had to learn to do the reverse.

The results showed that all subjects could learn to pull the food words toward themselves more quickly than they could push them away. This agreed with the finding that food words generally elicit a positive emotional response and thus elicit an approach response, not an avoidant response. Perhaps food words would have some A-R-D value even with food-satiated subjects. More centrally, however, it was found that deprived subjects could learn to pull the words toward themselves significantly faster than another group of deprived subjects could learn to push the words away. This difference in the strength of approach and avoidance responses to food words was not evident with the nondeprived subjects (Staats and Warren, in press).

It is thus suggested that the A-R-D emotional-motivational theory has support in basic human research. It appears that there are stimuli for humans that elicit an emotional response and that will as a consequence have both reinforcing and directive value. All three of these properties of the stimulus will vary, depending upon conditions of deprivation. As the individual is deprived of the primary stimulus that elicits the emotional response, the ability of the relevant conditioned stimulus to have the A-R-D properties will be increased. It would seem that deprivation of the conditioned stimulus may also have the same effects. That is, studies like that of Gewirtz and Baer (1958) deprived the child of social approval stimulation—which must be considered to be composed of learned A-R-D stimuli—and found an effect upon the reinforcing value of social approval. Extrapolation from this finding would suggest that deprivation-satiation conditions have an effect upon learned as well as unlearned emotional-motivational stimuli. Further research is needed to explore these possibilities and the three-function motivation principles.

## The human emotional-motivational system

Theorists concerned with human personality inevitably must consider the individual's emotional-motivational characteristics, because these are such important aspects of his make-up. A concern with this area was central in Freud's psychoanalytic theory, for example. Freud thought that basic instincts or drives resulted from biological tensions that pressed for release or satisfaction. He considered such things as hunger, thirst, and sex tensions as constructive drives, which he called erotic (or libidinal) in the general sense he employed. Later, Freud added the "death instinct," the source of hate, as a class of drives in contrast to the erotic motivation.

For Freud the organic needs underlay the individual's development of other motives. That is, libidinal energy could become invested in new objects. This was called cathexis, but the principles by which objects gained or changed their libidinal cathexis were not precisely stated. Freud's central concern was with the manner in which individuals came to be neurotic, or to demonstrate maladjusted behavior. A primary principle was that an obstacle to satisfaction of basic needs, because of societal restrictions incorporated in the superego, would result in the buildup of undesirable tension, a constituent of neurosis. However, frustration of a basic drive could be overcome by the ego, through providing a satisfaction of the drive in the form of neurotic symptoms or in a sociably acceptable and useful way. Freud (1935) notes that:

> People fall ill of a neurosis when the possibility of satisfaction for the libido is removed from them—they fall ill in consequence of a "frustration," as I called it . . . their symptoms are actually substitutes for the missing satisfaction. . . . In general, there are very many ways by which it is possible to endure lack of libidinal satisfaction without falling ill. . . . One amongst these processes [involves] . . . the adoption of a new aim—which new aim, though genetically related to the first, can no longer be regarded as sexual, but must be called social in character. We call this process *sublimation*. (pp. 301–303)

Thus, sublimation can be seen to indicate one way that the individual's basic motivation can be transformed into a social motivation. An example might be the woman who, frustrated in her maternal drive, becomes a nursery school teacher, or the artist who satisfies his sexual drive not directly but through painting.

As another example of the importance of motivation in personality theory, the work of Murray (1938) can be cited as an attempt to arrive at a catalog of human motivations. According to Murray there are two classes of needs, the viscerogenic, or organic, and the psychogenic needs. The latter are derived from the former. Some of the 28 psychogenic needs are: achievement, affiliation, aggression, dominance, exhibition, independence, nurturance, succorance, and so on. As the following quotation indicates, Murray tied the process of acquiring psychogenic needs into the principle of reinforcement, although not in a close explanatory way.

> What factors determine the establishment of a need as a ready reaction system of personality? This is an important problem to which only vague and uncertain answers can be given. . . . [T]he strength of some needs may be attributed to intense or frequent gratifications (reinforcements), some of which rest on specific abilities. . . . Some needs may become established because of their success in furthering other more elementary needs. The gratification or frustration of a need is, of course, largely up to the parents, since they are free to reward or punish any form of behavior. . . . Certain cultures and sub-cultures to which an individual is exposed may be characterized by the predominance of certain needs. . . . There are still other factors, no doubt, that work to determine what needs become dominant. (Murray, 1938, pp. 128–29)

Maslow (1943) has proposed a sequential theory of motivational development. According to Maslow there are physiological needs that are primary to man. "Undoubtedly these physiological needs are the most important of all needs . . . , in the human being who is missing everything in life in an extreme fashion, it is most likely that the major motivation would be the physiological needs rather than any others" (Maslow, 1943, p. 372). The physiological needs thus provide the lowest level of the motivation theory.

The theory is sequential or hierarchical in the sense that several levels of motivation are proposed, each successive level being higher than the preceding one. The levels, however, are not conceived to be related in that one would grow out of a preceding level or levels or would in some other way be based upon preceding motivational development. Maslow simply posited that when one level of need was satisfied the next level of motivation would then become functional. "If the physiological needs are relatively well gratified, there then emerges a new set of needs, which we may categorize roughly as the safety needs" (Maslow, 1943, p. 374). "If both the physiological and the safety needs are fairly well gratified, then there will emerge the love and affection and belongingness needs. . . ." (Maslow, 1943, p. 379). In addition, the next higher level of motivation was seen to involve the esteem needs, "a need or desire for a stable, firmly based, (usually) high evaluation of themselves, for self-respect, or self-esteem, and the esteem of others" (Maslow, 1943, p. 382). Finally, at the top of the hierarchy of motivations, the individual would experience the need for self-actualization.

> Even if all these needs are satisfied, we may still often (always?) expect that a new discontent and restlessness will soon develop. That is, it will appear unless the individual is doing what he is fitted for, what his talents allow him to do. A musician must make music, an artist must paint. . . . What a man *can* be, he *must* be. (Maslow, 1943, pp. 383–384)

As this quotation indicates, this theory does not specify the circumstances or the principles by which the higher needs would arise. There is an implicit assumption that man is structured in that manner. The determining factor for the sequential development of the motivation levels is the satisfaction of the preceding level of motivation. Again, this indicates the need for an explicit statement of the principles involved in the development of the human emotional-motivational system.

## LEARNING THE A-R-D SYSTEM

The important principle for discussing the *formation* of the individual's A-R-D system is that of classical conditioning, it is suggested. It has been noted that there are stimuli that naturally have functions as unconditioned stimuli ($UCS$). Food, water, air, sexual stimulation, warmth, and so on elicit upon presentation positive "emotional" responses (when the organism has been deprived of them). On the other hand, intense tactile, auditory, visual, and chemical stimuli elicit negative "emotional" responses. When these various stimuli are paired with neutral stimuli, eliciting their particular emotional response, the emotional response is classically conditioned to the neutral stimulus. When a stimulus has acquired this emotional quality, it can be transferred to new stimuli in the process of higher-order conditioning.

Thus, it is suggested that the individual's A-R-D system is founded upon the stimuli that originally elicit emotional responses in him on an unlearned basis, and it is elaborated by extensive first-order and higher-order classical conditioning. The individual's conditioning history in this respect is infinitely complex and extends over his lifetime, providing ample opportunity for a fantastically large number of conditioning trials and all the uniqueness we see in the human emotional-motivational system. The various subprinciples of classical conditioning would be expected to be involved. It should be noted that the conditions exist for producing much more varied and much more durable conditioning with humans than that which occurs in animal studies. The period of the individual's learning is a lifetime rather than a brief experiment, and this provides opportunity for an immense number of conditioning trials. A meal itself is extended over a considerable time and involves multiple conditioning opportunities. During the meal the individual may experience family members, friends, music, TV characters, political ideas, and so on, all associated with the food that elicits a positive emotional response.

Moreover, intermittent conditioning is normally involved in ways that make the learned A-R-D stimulus resistant to extinction, even when it is rarely paired with a primary stimulus that elicits an emotional response. Thus, for example, a smile may be considered to be an emotion-eliciting stimulus that gains this property in the beginning because the individual has received primary positive emotion-eliciting stimuli while being smiled at. This pairing is not continuous—sometimes the child is smiled at when no other positive emotional stimulus occurs. This intermittent conditioning, along with the vast number of conditioning trials involved, would be expected to make a smile a very enduring positive emotional-motivational stimulus, as seems to occur.

It should also be noted that through classical conditioning the individual learns a great multitude of different emotional-motivational stimuli, not just a few different types, as some of the personality theories suggest. Thus, there is an exceedingly large number of emotional stimuli for the individual. Various types of A-R-D stimuli will be the topic of later discussions.

Culture, class, and group effects. It has been suggested that the individual's A-R-D system is developed in an infinite number of conditioning trials that produce a unique characteristic. It should be noted, however, that there are similarities in the A-R-D systems that different individuals will have, because they are subjected to similar experiences. Thus, for example, different members of a family will have different conditioning experiences, and to that extent different stimuli will come to have emotional-motivational value for the members. It would also be expected, however, that the individual would also share certain conditioning experiences with the other family members and to that extent would acquire an A-R-D system that was similar to theirs.

It should also be noted that the group constitutes the learning environment for the child. This is the case with the formation of the individual's A-R-D system, as with other areas of personality. Ordinarily, a child raised in a primary group, for example, will acquire an A-R-D system that is similar to that of the group's common A-R-D system. The group will present conditioning experiences to the child that are similar to those that have already conditioned the members of the group.

*The family.* As an example of the manner in which the family can affect the emotional-motivational system of the child, take the case of the athletic father. He will have high A-R-D value for participating in sporting activities, in contrast to sedentary activities. Because he is skilled in this area and "enjoys" these activities, the father will be better able to provide experience that will yield the same effects in his child, including influencing the formation of the child's A-R-D system.

*Other primary groups.* While the family as a social institution has an early and profound effect upon the learning of the individual's A-R-D system, other groups and social institutions also will have an influence. The effects may be very similar and overlap. Many of the A-R-D values the individual learns in the middle-class home will be similarly produced in other social institutions, such as the Boy Scouts, the schools, the Junior Chamber of Commerce, peer groups, and so on.

> The peer group exists for the sake of "sociability." But behind this innocuous interest is a powerful force for conformity. Like any other socializing agency, the peer group represents a system of rewards and punishments, of approval and disapproval. It rewards the skills of sociability. It rejects the personality that disrupts the flow of good feeling and hinders smooth personal relations. . . . Peer groups transmit the skills and values of sociability. (Broom and Selznick, 1963, p. 112)

Such primary groups as the family and peer associates have an important effect upon the formation of the A-R-D system of the individual, as well as on the development of other aspects of his personality. These groups have their effect, it should be noted, as representatives of the social class or culture. The characteristics of the larger groupings of social class and culture also require description, for this description will also describe the characteristics of the individuals who compose the grouping.

*Social class.* One of the types of needs that have been described by

Murray is that of *achievement*, which is characterized as follows: "To accomplish something difficult. To master, manipulate, or organize physical objects, human beings, or ideas. To do this as rapidly and as independently as possible. To overcome obstacles and attain a high standard. To excel oneself. To rival and surpass others. To increase self-regard by the successful exercise of talent" (Murray, 1938, p. 164). This account includes descriptions of behavior, but the emotional-motivational characteristics of stimuli are also implicitly described in the definition. It is suggested, thus, that the stimuli involved are A-R-D stimuli, that these are learned, and that the child's experience, in good part depending on his social class and cultural groups, will determine this learning.

> Consider the stimuli associated with attaining a high standard. Certainly in a naive organism, overcoming obstacles and doing something difficult is not itself originally reinforcing. Effortful behavior produces stimulation that is aversive (Azrin, 1961; Hull, 1943). Without some change in these aversive features, it is to be expected that an organism would escape from hard work, that is, the cessation of work would act as negative reinforcement (Azrin, 1961). The same might be true of working as rapidly as possible. Since rapid responding is not itself reinforcing, . . . the behavior would not be expected to occur.
>
> The same is true of accomplishment, attaining high standards, excelling oneself, rivaling and surpassing others, and so on. These consequences are not by themselves positive [A-R-D stimuli]. Prior to the appropriate training, there is no reason to suppose that surpassing others, matching standards, and so on, will be [A-R-D stimuli]. Such events only become [A-R-D stimuli] because they have in the past been paired with positive [emotional-motivational stimuli]. Children have to be trained to find "winning" [to be a positive emotional event], to hold "standards," and so forth. Anyone who has attempted to instruct small children in group games finds immediately that those aspects of "competitive" events that are [positive A-R-D stimuli] for adults may be quite neutral for the children. (Staats, 1963, pp. 293–294)

There are large differences in the extent to which these aspects of the A-R-D system are learned by children. Sociological descriptions of the different ways that social classes treat their children indicate the conditioning involved. For example, a study by Rosen (1956) describes the differences in "aspirations" (A-R-D stimuli) among high school boys from different social classes. Rosen found that "members of the middle class tend to have considerably higher need-achievement scores than individuals in the lower social strata" (p. 206). His descriptions of the training practices of middle- and lower-class parents indicate the emotional-motivational conditioning involved.

> In the pre-school period the tendency for middle-class parents to make early demands upon their children is reflected in such practices as early toilet training and the intense concern with cleanliness [another

need described by Murray]. As the child grows he is frequently urged and encouraged to demonstrate his developing maturity (e.g., early walking, talking, and self care). Signs of precocity are signals for intense parental pride and often lavish rewards. It is precisely this . . . which provides . . . a most fertile environment for the growth of the achievement motive.

When the child starts his formal schooling, the achievement oriented demands and values of his parents tend to be focused on the school situation. From the beginning of his school career the middle-class child is more likely than his lower-class counterpart to have standards of excellence in scholastic behavior set for him by his parents. In fact, the relatively higher position which scholastic attainment has in the middle-class than in the lower-class value system means that more frequently for the middle than for the lower-class child parental demands and expectations, as well as rewards and punishments, will center around school performance. (Rosen, 1956, p. 211)

The lower class child does not have the advantage of this type of experience with respect to "achievement" events in the family. Moreover, he faces other drawbacks in learning achievement motives.

[I]f a group is discriminated against so that the individuals from the group cannot hold high positions which demand considerable educational background, then it is not likely that the individuals in that group will come to value educational achievement, that is, find educational achievement reinforcing. Specifically, if the skilled verbal and motor behaviors we call knowledge are not allowed to be paired with strong positive [A-R-D stimuli] (money, social approval, material possessions, and so on) then the consequence of these skills—achievement—will not become [A-R-D stimuli]. Minority group members who face such discrimination would not be expected to acquire the necessary behaviors for achievement, *nor would they even acquire the* [A-R-D systems] *which ordinarily would produce that type of behavior.* In common sense terms they would not even "desire" or "want" to achieve. They would appear unmotivated, lazy, disinterested, and so on. (Staats, 1963, pp. 296–97)

This account went on to show that there was support for the contention that this was the case in the United States. The 1950 census of population indicated that blacks in very small numbers occupied higher paying, more prestigeful occupations, and they were represented in much larger proportions than might be expected in lower, menial, less prestigeful occupations. Blacks, having no positive emotional-motivational experience with respect to achievement stimuli, should not be expected to have the necessary achievement motivation. This argument was supported by reference to statistics that indicated blacks as a group demonstrated poorer educational motivation in the United States.

This was written in 1963. It should be noted that since that time there is general recognition of the need for the members of minority groups to have access to prestigeful, rewarding jobs, not only for the individuals who obtain the jobs, but also for the development of the emotional-motivational systems of the group. It may also be noted that this conception, in contrast to approaches that explain blacks' relatively poor educational achievement in terms of inheritance (Jensen, 1969), suggests the deficit is one of learning—and suggests as well some of the conditions that are necessary to correct the problem. It should be added that this topic involves other ethnic groups also. A recent study employed a test of achievement motivation with preschool children from 10 different ethnic groups and found the following differences:

> [T]he three middle-class samples—Mormons, Catholics, and Jews—had higher mean total scores than the lower-class samples. Among the lower-class groups, the Negro-Urban, White-Rural, and Puerto Rican samples emerged with substantially higher mean scores than the remaining groups. The Mexican-American, Oriental (West Coast), American Indian, and Hawaii samples had the lowest average total scores. (Adkins, Payne, and Ballif, 1972, p. 564)

There are other differences in experiences that would tend to produce differences in A-R-D systems across social classes. For example, Maccoby and Gibbs (1954) found that lower-class mothers, in contrast to middle-class mothers, were more severe in their training procedures and employed more physical punishment, deprivation of privileges, and ridicule. The result should be a difference in the A-R-D systems of the children involved, as well as in other behaviors such as physical aggression.

*Cultural effects.* There are many interesting descriptions of cultural groups in anthropology that attest to the differences that have developed in A-R-D systems. Several examples pertinent to the emotional-motivational characteristic of surpassing others and winning will be given, since they are relevant to Murray's achievement need.

> The Kwakiutl are a people of great wealth and they consider it honorable to amass a fortune. But it is not hoarding they are interested in. Wealth, such as blankets, boxes, and copper plates, is used in a game of rising in rank, or validating honorific titles and privileges. Upon the occasion of taking a name a man distributes a considerable quantity of blankets among the men . . . in the presence of the entire community. The recipients are obligated to accept the property and must be prepared to repay it at the end of the year with 100 percent interest. Such men probably have property out at interest, which they call in at the end of the year to meet their payments. Should a man be unable to repay he is "flattened" and falls in social status. The victor, on the other hand, rises another rung in the social ladder. With each successful potlatch a man accumulates more renown as well as more property with which to conduct even greater potlatches. With prestige the driving motive in

> Kwakiutl society and with the basic intent of the potlatch the crushing of a rival, these property bouts take on a fiercely competitive tone. (Goldman, 1937a, p. 188)

A recent enlightening historical account of the ancient Greeks reveals a similar emotional-motivational system in a culture that has contributed heavily to our Western European culture. R. J. Littman (1974) describes the competitiveness of the Greeks and then suggests the following:

> Much of the competition, however, was non-productive. The Greeks had a shame culture, rather than a guilt culture. In the former one's sense of worth is entirely determined by the opinion of others, while in the latter an internalized set of standards control behavior. As in many shame cultures, the Greeks regarded any kind of defeat as disgraceful, regardless of circumstances. Yet no victory was possible unless someone else lost. The glory of winning accrued to the victor from the lost glory of the defeated. The loser could not leave the contest with as much prestige as he entered. Contributing to this contentiousness was the Greek obsession with fame, honour, and achievement. (p. 14)

The Zuni people, unlike the Kwakiutl people, do not ascribe such positive value to winning, nor is having a great deal more than someone else valued, according to anthropological observations.

> Strikingly characteristic of all social relations in Zuni is the relative lack of emphasis upon wealth. Property does not figure in marriage. Individuals do not compete for a fixed supply, and in terms of the prestige an individual may achieve, property in itself is not the determining factor. This does not at all imply that the Zuni are unmindful of the blessings of material comfort or that they are completely disinterested in the accumulation of wealth. But they do frown upon any undue interest in material possession, upon acquisitiveness, covetousness, stinginess, or sharp practice in economic transactions. If a material object has value, it has that only as a means toward a specific utilization end. But hoarding—the piling up of goods far beyond what is necessary for a comfortable existence—is practically unknown. Wealth circulates freely, and property rights are neither clearly defined nor strictly enforced. For one thing, material effects are never valued as a means to power and are only indirectly a source of prestige. (Goldman, 1937b, pp. 326–27)

It would be expected that a child raised in one of these cultures would acquire an A-R-D system that was coincident with that of the culture, under usual conditions.

DISJUNCTIVE LEARNING. This is not to say that learning always progresses in the direction desired by the social agent who deals with the child or individual. In the above example of the father who is an athlete, it should be realized that the experiences the father provides the son may not produce in the son an A-R-D system that is conjunctive with his own. Thus, while an athlete himself, and while having positive emotional responses to athletic activities, the father may have a negative emotional response to playing with someone who is inept—as the child is bound to be. Moreover, the father may have techniques of child training that are aversive. He may admonish the child, deride him, or be overly demanding. The child may thus be conditioned to a very negative emotional response to athletic endeavors, an aspect of his A-R-D system that is disjunctive with that of the father.

There may also be a disjunction between the characteristics of other groups and the A-R-D learning the child acquires. The child may not acquire the A-R-D system that is characteristic of various of his primary groups, or of his social class or cultural group. The ways in which the A-R-D system of the group is transmitted to the child through learning should be a topic of systematic study. This should include the manner in which both conjunctive and disjunctive individual and group systems are formed, and the effects their formation has upon the individual's adjustment.

## STRUCTURE OF THE A-R-D SYSTEM

As has been indicated, Maslow (1943) posited a hierarchical structure for the human motivation system. Implicit in this conception has been a genetic base for the structure, with the sequential development of the major motives dependent to some extent upon maturation.

It is suggested, however, that there is a twofold categorization of A-R-D stimuli. There are those that meet organic deficiencies and those that do not. This is similar to Murray's separation of viscerogenic and psychogenic needs. The latter can be considered to be learned in the manner already described. It would seem, however, that there is also a hierarchical order in the value of various A-R-D stimuli within the system. This would be expected to depend upon such factors as the strength of the conditioning that has been involved, the strength of the physiological emotional-motivational stimuli with which the A-R-D stimuli have been paired, and so on, as well as upon the conditions of deprivation for the individual (which will be discussed in a later section). The fact that some motives appear before others can also be a function of the opportunity for learning. Thus, for example, the individual may have to spend long years of study before acquiring a profession that provides him with financial conditions that allow him to engage in self-actualization activities. Moreover, the value of some A-R-D stmuli is only acquired after a long learning history, and motives based upon this learning will only appear at a later time. Thus, self-actualization motives in science must be preceded by a long course of learning in which values for scientific activities are gained.

For example, it is not unusual for undergraduate students in college to be surprised to find that a professor works hard for the reward of publishing an article when no pay or other obvious reinforcement is [involved]. Unless the "approval" of a select group of interested colleagues has become . . . positive . . . , it is difficult to see how one would "work hard and long" for such consequences (Staats, 1963, p. 294).

At any rate, it is important to describe the conditions of learning the A-R-D system and the relationships that exist among the aspects of that system. Although there may not be good proof of a sequential advancement to successively higher states of motivation, there appears to be some structure to the A-R-D system. Murray suggested 28 psychogenic needs. It would be expected that there would be relative strengths of the A-R-D stimuli represented by these categories. The task of describing the classes of A-R-D stimuli and measuring them, as will be indicated in a later chapter, is important to understanding and dealing with human behavior.

## The A-R-D system as a personality trait

In discussing the human emotional-motivational system thus far, the emphasis has been on the manner in which the system is developed and on the characteristics of the system. It can be added that the development and nature of the individual's emotional-motivational system is important in and of itself. Whether a particular stimulus situation will elicit a positive or negative emotional response in the individual—whether he will be happy or sad—will in many instances depend upon the emotional motivational system he has learned. An individual who had learned negative emotional responses to many more stimuli than another person had could be expected to be less happy than the other person, even when being faced with the same conditions of life. In essence, the individual's A-R-D system helps determine what he experiences in life.

Explication of the manner in which the A-R-D system is learned deals with this topic as a dependent variable, an effect. However, when it is suggested that the individual's emotional-motivational system helps determine what he experiences in life, the system begins to take on the characteristics of a cause, an independent variable. This quality becomes clear when it is suggested in addition that the individual's A-R-D system helps determine how he learns, whether he learns, how he behaves, and so on, across many different situations. It is because the individual's emotional-motivational system has these characteristics of a cause that the system may be considered a personality trait in the traditional sense.

The present discussion will indicate to some extent the principles by which the individual's emotional-motivational system plays the role of a personality trait. The individual's emotional-motivational system will play a role in the individual's further emotional learning, in the instrumental behaviors he learns, and in the stimuli that elicit approach (striving for) or avoidance (striving away from or against) behaviors.

## A-R-D SYSTEM DETERMINATION OF EMOTIONAL LEARNING

It has been said that the stimuli that come to have emotional-motivational properties for the individual gain those properties through classical conditioning. It should be understood, however, that the nature of this learning, once it has occurred, will help determine the individual's further emotional learning.

Let us take classical music as a stimulus that has emotional-motivational properties for some people. It is apparent that the response is learned. Different individuals and different groups have different emotional values for different types of music. For one individual classical music has positive A-R-D properties, while for another this auditory stimulus is neutral, or negative. Let us say that a young man, in hopes of providing a positive experience for a new girl friend and in this way enhancing her feeling for him, takes her to a symphony concert. If her A-R-D system is such that the music elicits a positive emotional response which will be paired with his presence, he will have accomplished his purpose. If the music elicits a neutral or negative emotional response, he will have defeated it. From her standpoint, whether or not she learns a positive emotional response to the boy will to an extent be determined by her own A-R-D system.

This is only one example. In life there are many situations that will result in different emotional conditioning for people because they come to the situations with different A-R-D systems. Experimental results relevant to this will be presented in Chapter 12, in the context of the measurement of interests.

## A-R-D DETERMINATION OF INSTRUMENTAL LEARNING

The principle of reinforcement states that following a response by a reinforcing stimulus will increase the strength of the response in that situation. When this is studied in the laboratory, a reinforcer is employed that will be effective. In human life, however, whether or not the stimulus that occurs in the role of a reinforcer actually has reinforcing properties may be crucial.

As an example, it has been suggested that there are social class differences in the extent to which achievement stimuli—those of doing well in some task, of getting good grades, of social approval, and so on— have positive A-R-D properties. It has been suggested, for example, that middle-class children come to attach much more positive emotional value to achievement stimuli than do lower-class children. With these givens it is not difficult to see how child learning (for example, classroom learning and adjustment) may be affected by the child's A-R-D system. The approval of the teacher and other students and the product of one's own developing skill are among the most important sources of reinforcement for student behaviors in the traditional classroom. Let us say that two children with differing A-R-D systems are placed in the classroom. For one child, child A, the teacher's approval and the child's own achievements are reinforcing; for the other, child B, these stimuli are not reinforcing. Let us say that both children receive the same treatment in class. Whenever they pay attention to materials the teacher presents and respond in the manner directed they receive the teacher's approval, and their instrumental behaviors produce stimuli that evidence their skill (achievements). Under such a circumstance, child A's attentional and working behaviors will be maintained in good strength, and as a result he will continue to develop new skills. Child B's behavior, on the other hand, will not be maintained. His attentional and working behaviors will wane, and other competitive behaviors that are strengthened by stimuli that *are* effective reinforcers will become relatively dominant.

Child A will be seen as interested, motivated, hard-working, and *bright*. Ultimately, he will also measure as very able and bright on class, achievement, and intelligence tests. Child B will be seen as disinterested and dull, and possibly as a behavior problem if problem behaviors are reinforced. Later he will also be measured this way, and this evidence may be used to support the contention that the child's behavioral failure is due to some personal defect.

Various examples from life could be cited. The boy for whom being rough and combative and winning over other boys has reinforcement value will be likely to develop athletic skills in contact sports such as football. The boy for whom reading elicits positive emotional response will be likely to learn reading skills much more easily than a boy for whom reading is less positive, for the first boy's reading behavior will be reinforced more than the second's. The girl for whom signs of personal beauty have positive emotional value will be likely to learn cosmetic skills. The boy for whom strenuous activity has negative emotional value will be likely to learn sedentary behavior—which, incidentally, will have an effect upon his physical appearance and function.

These are simple statements. It must be realized, however, that the individual's A-R-D system has its effects over the individual's lifetime. To the extent that this aspect of the individual's personality has continuity—and, as later discussions will indicate, the A-R-D system, by affecting the individual's behavior, helps insure continuity of development —the individual's emotional-motivational system will be involved in cumulative-hierarchical learning. Thus, very complex social behavior repertoires, work repertoires, recreational repertoires, intellectual repertoires —all other personality repertoires, actually—will be affected by the nature of the individual's A-R-D system. It is important to realize that the emotional-motivational system is a pervasive personality trait. Although the individual's emotional-motivational system is learned, it is a significant determinant of what the individual becomes in many ways. The A-R-D system has its effects early in life and continues to help determine what the individual experiences, how he behaves, and what he becomes throughout life.

## A-R-D DETERMINATION OF ELICITED INSTRUMENTAL BEHAVIOR

Freud's concept of drive suggested that an instinct has an aim and a particular behavior for attaining the object of the aim. Newcomb (1950) and Klineberg (1954), as other examples, discuss motives as including states of drive as well as directing behavior toward some goal. Basic learning theories have not clearly stated the relationship between reinforcement value and directive (incentive or goal) value (Holland and Skinner, 1961; Mowrer, 1960a), nor have the personality theories done so in an explanatory way.

The manner in which the emotional and reinforcement value of stimuli becomes related to their instrumental value was described in terms of animal behavior in an earlier section. A similar process occurs with the child. Let us say, to begin, that when a child sees a food stimulus, if he crawls toward the stimulus this response is followed by obtaining the stimulus, which is a reinforcement. This process fulfills the requirements for making a stimulus a directive stimulus; that is, a stimulus in the presence of which a response is reinforced will come to control the response.

This brief analysis must be expanded in human behavior in several directions. First, the child ordinarily will learn a large class of "striving" behaviors that will come under the directive control of such reinforcing stimuli, for example, crawling, walking, or running toward; climbing over and around obstacles; reaching and grabbing for; fighting and struggling for; asking, begging, and crying for; working, wheedling, arguing, or flattering for, being ingratiating or being respectful for, as well as competing for in various ways.

In addition, the child's class of striving behaviors will come under the directive control of a large number of different A-R-D stimuli on the basis of his experience with those, or similar, stimuli. Thus, in the child's conditioning history a wide variety of stimulus objects that are reinforcers will come to control responses that result in obtaining those objects. Moreover, as has been indicated, the emotional response elicited by such A-R-D stimuli will come to control the class of striving behaviors. It is important to indicate here that all of the stimuli that elicit a positive emotional response in the individual will have this directive function, once the mechanism has been well learned. The mechanism has been depicted in Figure 4.3, in an abstract way. The reader may fill in the figure with specific examples. Thus, each $^DS$ (which is actually an A-R-D stimulus, since it has those three functions), could be considered to stand for some positive case—perhaps, using a child as the subject, an ice cream store, a toy, a piece of candy, a TV commercial, a pet dog, and so on. Each response, on the other hand, could be considered to stand for some approach response—perhaps reaching for, walking toward, listening to, looking at, working for, begging for, crying for, helping behavior, affectionate behavior, and so on. The $^DSs$ each elicit the positive emotional response, and the stimuli of the emotional response tend to elicit the

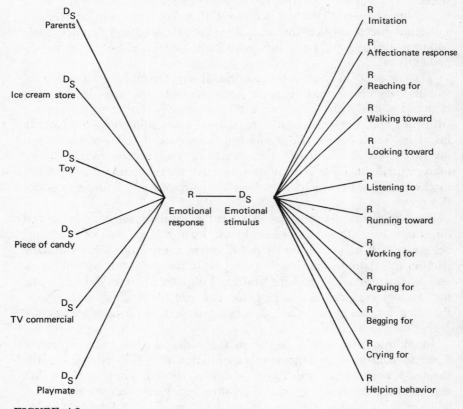

**FIGURE 4.3**

*The positive emotional-motivational (A-R-D) mechanism. All stimuli that elicit a positive emotional response will by virtue of this tend to elicit a large class of approach, or "striving for," instrumental responses.*

hierarchy of instrumental responses. Which particular instrumental response would occur would depend upon the various other stimuli in the situation. Thus, the child might ask for a piece of candy in the presence of another person, but reach for it himself if alone.

At any rate, any stimulus that is in the individual's emotional-motivational system, or that comes to elicit an emotional response, will by virtue of eliciting the emotional response also tend to elicit all of the approach instrumental responses the individual has already learned. This mechanism has been diagrammed with only a few stimulus members and only a few response members. It should be realized that there is an exceedingly large class of positive emotional stimuli, and the individual will ordinarily have a repertoire of instrumental behaviors that is very large.

It should be noted that in contrast to the laboratory situation, it is usually the directive control of instrumental behaviors that social and behavioral science investigators use to index motivational stimuli. The social psychologist, clinician, sociologist, or anthropologist, for example, ordinarily observes what people strive for (verbally or otherwise) when he studies motivation. This is why he has so generally introduced the concept of goal-directed behavior. The social theorist, or clinical theorist, does not see whether the receipt of a stimulus elicits an emotional response, which is usually not readily observed. Moreover, he does not see whether it strengthens *future* occurrences of an instrumental behavior.

A major difference between the present three-function learning theory and Skinner's operant conditioning approach may be made explicit in this context. The Skinnerian concept of the reinforcing stimulus is included in the more general concept of the A-R-D stimulus. Moreover, the things people are reinforced by are those things they strive for— which refers to directive value. It is because of this multi-function quality that many theorists have been unsatisfied with the simple concept of reinforcement, and have needed a goal directing (purposive) principle in considering human behavior.

## Conflict: Competing A-R-D control

The present discussion has considered the manner in which emotional-motivational stimuli can elicit approach or avoidant behaviors from the individual. The straightforward presentation of the principles may suggest a strict determination of behavior that is uncharacteristic of much that occurs in life. It is therefore necessary to indicate that the events that affect the individual's behavior are not simple. This suggestion has been made in several previous examples. In Chapter 3 it was suggested that the person's basic behavioral repertoires help determine what the individual does, but so does the present stimulus situation in which the individual finds himself. This topic of interacting sources of determination of the individual's behavior may be discussed here in the context of considering emotional-motivational principles.

The fact is that the situation in which the individual finds himself may involve more than one A-R-D stimulus that has directive influence over his behavior. Sometimes the multiple effects are conjunctive in that they both tend to elicit the same type of approach or avoidance behavior. Thus, a student may have strong avoidant responses to attending a class because the subject matter elicits a negative emotional response in him. In addition, the room involved may elicit a negative emotional response because it is too warm.

Naturalistic evidence suggests, however, that there are important circumstances that involve conflict for the individual. Freud recognized this. Lewin (1935) suggested that conflict results when there are events in the individual's environment that have a positive valence and attract him, as well as events that have a negative valence and repel him. N. E. Miller (see 1959) elaborated the principles of conflict and conducted a series of studies to demonstrate the principles experimentally. For example, he suggested that there are goal gradients: the nearer the goal (positive A-R-D stimulus), the stronger would be the organism's tendency to approach it. The converse was posited in the negative direction. Studies by J. S. Brown (1948) demonstrated these principles, employing rats as subjects in harnesses that could measure the strength of pull of the animals as their proximity to the positive or negative A-R-D stimulus was varied.

More centrally, however, Miller outlined several types of conflict situations and tested the expectations. The first to be described has been called *avoidance-avoidance* conflict. For example, let us say that the life situation of a woman involves two very strong negative A-R-D stimuli that may be labeled *prostitution* and *destitution*. No source of funds, destitution, elicits a negative emotional response, as does the act of prostitution. The woman who faces these choices and has this A-R-D system will be in conflict. Miller (1944) has shown that rats in such a situation, with negative A-R-D stimuli at either end of a chamber, will go a bit in one direction and a bit in the other, remaining in the center. It would be expected, moreover, that such a conflict situation would elicit a negative emotional response until the individual is able in some manner to remove himself from confrontation with the negative stimuli.

*Approach-avoidance* refers to conflict where a stimulus situation contains positive emotion-eliciting properties as well as negative emotion-eliciting properties. Miller (1944) showed that rats that had been alternately shocked and fed in one end of a runway would approach the goal (A-R-D) stimulus. That is, at first they would be under the control of the positive emotional response elicited by the stimulus. However, when they got close to the stimulus, they would stop. Miller suggested that the gradient for the negative stimulus event was steeper than for the positive. At the point where the two gradients crossed the animal would be equally pulled toward the goal stimulus and repelled by it.

Such a conflict would tend to imprison the individual in certain situations where he is drawn in but unable to go the whole way to satisfaction. Individuals with fears (negative emotional responses) to sexual stimuli may find themselves in this situation. They will approach the sexual object, but at a certain point the negative emotional responses will equal the strength of the positive emotional responses, and they will be unable to perform the instrumental behaviors in a manner that leads to satisfaction.

A study by J. J. Conger (1951) suggests the way that alcohol can have an effect in lessening conflict. Rats were given training of the type described above where they would not approach the end of the runway where food was located. Conger found that when injected with alcohol the animal would go to the food. The effect that alcohol has in loosening human "inhibitions" and "restraints" can be inferred from this study.

Sometimes the conflict situation involves two positive A-R-D stimuli, but approach toward one involves losing the other. It will be remembered that withdrawal of a positive reinforcing (A-R-D) stimulus constitutes a punishment. This situation thus has been described as *double approach-avoidance* conflict, and it can also produce drawn-out vacillation in making a decisive act.

There are many examples of this type of conflict in everyday life, as there are for the other types of conflict. As one example, a person may have the opportunity of markedly improving his professional position but be required to move away from a geographic location that he loves. As another illustration, women, at least in previous times, could opt for a career, but at the risk of losing the opportunity for marriage and family. This has been the occasion for much feminine conflict.

## Motivational changes in the A-R-D system

It has been suggested that the A-R-D system is learned. This type of learning continues throughout life. Thus, the A-R-D system may undergo change in important areas, as well as elaboration and development. The manner in which classical conditioning operations can change aspects of the system, affecting social behavior of various kinds, will be described in Chapter 7. These principles also underlie important methods in clinical psychology, as will be indicated in Chapter 9.

In addition to these learning methods for producing change, the emotional-motivational system can be changed by conditions of deprivation and satiation. This has been recognized on the basis of naturalistic observations. For example, as has been noted, Maslow (1954) suggested that there is a motivational hierarchy. In stating his conception Maslow actually included a statement of deprivation-satiation, although not in the succinct manner of the laboratory-established principle. That is, according to Maslow progress in developing the higher motives depends upon satisfaction of the lower motives. The following statement is an example: "[W]hen a need has been satisfied for a long time, this need may be underevaluated. People who have never experienced chronic hunger, are apt to underestimate its effects and to look upon food as a rather unimportant thing" (Maslow, 1948, p. 99). Dollard and Miller (1950) have also referred to the manner in which achievements that are rewarding, and are rewarded, at one age for a child will not be so rewarded at a later age. They also indicated that attainment of a reward in a graded series enhances the value of the next reward in the series.

The effects of deprivation-satiation conditions on the A-R-D system, and thus on behavior, can be elaborated by reference to the basic principles. To begin, it has been suggested that the A-R-D elements have a hierarchical relationship. Some emotional-motivational stimuli have more strength than others. It may be added that there appear to be classes of A-R-D stimuli, such as food (and food-associated) stimuli, sexual stimuli, recreational stimuli, work stimuli, and so on, which may have certain commonalities and coherence as a class. It is suggested that this structure provides one basis for use of the term *system* in reference to the individual's emotional-motivational characteristics, in the sense that the elements of a system have a mode of interaction.

At any moment in time the various stimuli in the individual's system have *relative* intensity. Relative, as well as absolute, strength has an important effect upon the individual's or group's behavior. Let us say, for example, that two individuals have a stimulus in each of their A-R-D systems that has precisely the same reinforcing value, reinforcer A. Let us also say, however, that for one individual there are no stronger reinforcers in his system, while for the other individual there is another stimulus, reinforcer B, which is an even stronger reinforcer. Now let the individuals be placed in a situation in which both reinforcers are available, but one reinforcer is presented contingent upon one instrumental behavior, and the other reinforcer is presented contingent upon an incompatible behavior. Under these circumstances the two individuals will develop different behaviors. The individual with the reinforcer system in which reinforcer B is the most "dominant" reinforcer will develop most dominantly the behavior that is followed by that reinforcer. The other individual, with the system in which reinforcer A is relatively stronger, will develop predominantly the other behavior. It is thus suggested that it is not only the absolute value of reinforcers that determines individual and group differences in behavior, but also the relative values of the various reinforcers in the A-R-D system.

This specification provides a basis for stipulating more clearly the effects of deprivation-satiation on the A-R-D system and on behaviors dependent upon the system. As has been indicated, deprivation enhances the value of a positive A-R-D stimulus, and satiation lessens its value. In both cases the operation can also change the *relative* dominance of A-R-D stimuli in the system. When the condition of deprivation or satiation is returned to the starting point, however, the A-R-D stimuli will regain their former relative positions. The changes produced directly follow the deprivation-satiation variations.

The manner in which groups and individuals can vary in behavior because different stimuli are effective in their A-R-D systems has been discussed. A few examples will be given here in which the relative (hierarchical) ordering of the A-R-D system is affected by deprivation-satiation variations, producing variations in behavior. As one example, stimulus variation is rare for the housewife with several small children, stuck at home all day. Her husband (a salesman) may, on the other hand, be satiated with diversion. Because of his work situation, he may find spending quiet evenings at home more attractive while she prefers going out. As another example, deprivation of sex stimulation would be expected to increase the A-R-D value of this class of stimuli, at the expense of other emotional stimuli in the individual's system. Under this type of deprivation other emotional stimuli will be relatively weak, and the behavior maintained by those stimuli will weaken.

The examples that have been given thus far have concerned only the change in the absolute strength of an A-R-D stimulus as a function of deprivation-satiation operations. It should be noted that there is an effect upon the relative strength of A-R-D stimuli as well. Let us say that there is for an individual a large class of food-type A-R-D stimuli, as well as a large class of sex-type A-R-D stimuli. The strength of these two classes of stimuli with respect to each other will be in part a function of the conditions of deprivation-satiation that pertain. Let us take the case of the married soldier who returns home after some absence. His home will offer both sexual and alimentary stimuli, both emotionally positive. Which class of stimuli will elicit the strongest emotional response in him will depend to a large extent upon his conditions of deprivation-satiation, and which class of stimuli most attract him will thus depend on these conditions. The principles involved are relevant to many types of human behavior. This last example indicates that deprivation-satiation operations affect the strengths of A-R-D stimuli in ways that are important to understanding conflict. Where approach-approach conflict is equal, the balance will be changed as there is a difference in deprivation conditions which makes one A-R-D stimulus stronger. The same is true of the other types of conflict. For example, deprivation may affect the balance of the avoidance-avoidance conflict. In the illustration employed, as the woman confronted with the destitution-prostitution conflict is deprived of food, shelter, and so on, the increased aversiveness of destitution will push her to prostitution.

It has been mentioned that there are subsystems within the major emotional-motivational system. Thus, there appear to be sexual A-R-D stimuli, food A-R-D stimuli, and so on. It may be added that when one is deprived of a class of stimuli, the A-R-D values of all the stimuli in the class are increased. For example, if the strongest food stimulus in the system's hierarchy is not available, then the next strongest would be the dominant A-R-D stimulus. The theory would also suggest that a relatively weak member of a class of A-R-D stimuli could be raised into a position of relative dominance in the total system through deprivation of stronger members of its class. A man deprived of potable fluids might be more affected by some brackish water than by a usually stronger A-R-D stimulus in some other class, such as sexual stimuli, food stimuli, social approval stimuli, and so on.

It can be indicated that deprivation may also affect the A-R-D system, and thus behavior, in another way that does not emerge from consideration of laboratory learning principles. For example, when a prisoner is deprived of contact with the opposite sex, the general class of sex reinforcers will increase in relative strength. As a consequence, reinforcers of lesser value in the class, but which are more accessible, such as homosexual contact, will be relatively stronger and may instrumentally condition behavior that would be unlikely to occur without the deprivation. It should be emphasized that each instance of a homosexual act constitutes a whole series of classical conditioning trials, since a sexual act extends over a considerable length of time. The homosexual conditioning experience would thus also be expected to increase the sexual emotional value of the class of social stimulus involved (members of the same sex), thus further altering the structure of the A-R-D system on a more permanent basis.

Changes in the A-R-D values of a stimuli can also occur through biological variations, in conjunction with deprivation. For example, adolescence markedly changes the A-R-D system and thus the individual's behavior. The adolescent experiences an increase in the secretion of sex hormones. This has the effect of raising the strength of the class of sex stimuli in the individual's A-R-D system—whatever these stimuli are. As a consequence, it will be seen that sexual stimuli elicit emotional responses more strongly in the adolescent than they formerly did. Sexual stimuli will have more reinforcing value and the adolescent will learn new behaviors—such as grooming himself, dating, and so on—under the action of such reinforcers. Moreover, sexual stimuli will have stronger directive power than formerly was the case. The adolescent will be attracted to things that include sexual stimuli more strongly than he has been in the past. In terms of relative strength, other things that formerly controlled his attention and participation—reading, family affairs, hobbies, and so on—will no longer compete favorably with sex-related stimuli.

In concluding this chapter, it can be suggested that the generality of the principles of the A-R-D system extends beyond the considerations dealt with in the present discussion. Chapter 7 will elaborate the principles into consideration of the manner in which A-R-D principles are important for the study of interaction between individuals and between individuals and groups. Chapter 8 will elaborate the concept of the emotional-motivational system as a personality trait into the realm of abnormal psychology. Chapter 12 on psychological measurement will prominently treat the measurement of the A-R-D system, and Chapter 14 will show how each of the social sciences is intimately concerned with events that can be seen to centrally involve A-R-D principles. In this progression the hierarchical structure of the theory can be seen, beginning with the basic level of the three-function learning theory and advancing to the social science levels. This hierarchical feature of classic theory can be seen also in the other areas to be treated, including the next personality system to be considered.

# Language theory and the language-cognitive personality system

A DISTINCTION is frequently made between what is called cognition and what is called language which considers these realms to be of a different order. Benjamin Whorf, an early noted linguist, suggested that language determines cognition (1956). The prominent theorist Jean Piaget (1970; Piaget and Inhelder, 1964) has considered child development in general in terms of the development of cognitive structures. In this approach language development also is viewed as depending upon cognitive (or intelligence) development. Noam Chomsky (1965, 1967) has proposed that language development depends upon the development of the individual's reason. This account has been considered to be nativistic in the sense that the determinants of language development are seen to be inherited cognitive structures.

A number of recent theories of language development have also accepted cognitive development as basic and as the determinant of language development. Slobin (1973), although originally tending toward the Chomskian position (Slobin, 1971), has generally proposed that the child learns those aspects of language that are within the scope of his cognitive development. The child's gradually developing cognitive complexity is said to provide the basis for his more complex language learning. E. V. Clark (1974) and Bloom (1974) have similar views of the relationship between cognitive development and language development.

John Watson established the original behaviorist position concerning language. "Language as we ordinarily understand it, in spite of its complexities, is in the beginning a very simple type of behavior" (Watson, 1930, p. 225). "In language we have something . . . to start on, *namely, the unlearned vocal sounds the infant makes at birth and afterwards*" (Watson, 1930, p. 226). Watson described an instrumental conditioning method by which he "established a conditioned vocal response" in a 6½-month-old infant and noted that: "After conditioned word responses have become partly established, phrase and sentence habits begin to form. Naturally single word conditioning does not stop. All types of word, phrase and sentence habits thus develop simultaneously" (Watson, 1930, p. 228).

Skinner (1957) elaborated this behavioristic account systematically and productively, based upon his more advanced learning theory. He suggested additional analysis of the types of stimuli that can come to control speech responses through instrumental conditioning, for example, aspects of the environment, printed words, and conditions of deprivation.

Mowrer (1954, 1960) also extended learning principles to the consideration of language. In an early paper (1952) he suggested that secondary reinforcement is involved in learning to imitate. Mowrer hypothesized that the parents' voices become conditioned reinforcing stimuli because these sounds are paired with positive unconditioned stimuli in their caretaking activities. He also further considered certain aspects of communication in terms of classical conditioning principles.

Osgood (1953) suggested that words could come to elicit covert responses. He proposed that such responses can be conditioned to words according to instrumental conditioning principles. These implicit responses were said to constitute the meaning of words, and Osgood and his associates focused their work in language on the measurement of meaning (Osgood, Suci, and Tannenbaum, 1957).

The latter two treatments of language were based upon Hull's (1943) learning theory. In the same manner that Skinner's, Mowrer's, and Osgood's theories of language were based upon preceding behavioristic or learning analyses, Staats (1961, 1963, 1968c, 1971a, 1971b) employed the previous behavioral formulations to develop a third-generation language theory. It was based upon a learning theory that combined both instrumental and classical conditioning principles and included experimental work that tested some of the major principles involved in the theory. Prior behavioral accounts involved little experimentation.

Two of the traditions in the study of language thus have been described. The first is a cognitive approach that has derivations from linguistics. The second is behavioristic. It should be indicated that in the 1950s there was an attempt to effect an interdisciplinary approach to language. This led to an exchange of papers among those who were involved as active researchers, but little genuine interaction (Cofer and Musgrave, 1963; Osgood and Sebeok, 1954).

This situation was soon changed, however. In 1959 Chomsky published a very critical review of Skinner's approach. Others influenced by Chomsky began to extend this criticism to the several learning theories of language (Ervin-Tripp, 1971; Fodor, 1966; McNeill, 1971). The two orientations that have been described then became quite separated.

It is suggested that the issue in this area has much in common with the schism that has existed between personality theory and behaviorism. Cognitive theorists in language have been dissatisfied with the description of language simply as learned behavior. A need has been felt for a conception that recognizes the causative aspects of language (or cognition) and that in so doing provides some basis for rapprochement between the approaches. The present account, while behavioristic in nature, will attempt to illustrate how language is a cause as well as an effect—that is, how language has personality trait (cognitive) characteristics. It has been suggested that cognitive development and language development concern the same thing (Staats, 1968c, 1971a, 1974). At least, it is proposed, there is a great overlap between the two. What is referred to as cognitive activity in many cases turns out, on specific analysis, to be functioning repertoires of language skills. This will be made more clear in later sections. It should be stated at this point, however, that many of the mechanisms in which cognitive theorists are interested can be isolated in the study of language development and language function.

## Language acquisition: The formation of cognitive repertoires

A first step in the development of a conception of language is to consider what it is composed of and how it is acquired. As with other aspects of human behavior, it is characteristic of the traditional conceptions of language, communication, or meaning to be concerned with one or two aspects of the phenomenon and to assume some unitary causal process. Thus, for example, some psycholinguists have been concerned primarily with the grammatical nature of language and have inferred an innate mental mechanism to account for this (Chomsky, 1965, 1967). It has been suggested that the restriction to single principles and only certain aspects of language has been true also of learning theory approaches as well (Staats, 1971b).

One long-standing categorization of language phenomena recognizes two divisions, receptive and productive language. This categorization usually assumes that there is a process involved in language comprehension (reception) that is different from that involved in language production. A continuing issue has concerned whether comprehension must precede production (Chapman and Miller, 1973; Fraser, Bellugi, and Brown, 1963; Ingram, 1974; McCarthy, 1954; Shipley, Smith, and Gleitman, 1969). The categorization of receptive and productive language suggests that there is a unity in principle involved in each category. It has been suggested, however, that this assumption, and the categorization system itself, are misleading (Staats, 1968a). Analysis of the language acts involved indicates, in actuality, that variegated aspects of language are involved within the categories, as well as different principles of learning. Moreover, not all language phenomena are covered by the two categories or can be segregated to them. The language act, as an example, may involve both receptive and productive language. When language is considered in a more analytic way, the issue concerning the relative precedence of comprehension and production disappears (Staats, 1968a).

The present approach is a pluralistic one. It suggests that language is actually composed of a number of repertoires. They may be considered separately, and the child can be trained to acquire the repertoires separately, at least in some cases. However, the repertoires are intertwined to yield the individual's functional language. None of the repertoires by itself will give the individual a functional language. Some of the language-cognitive repertoires will be described briefly in the next sections. These repertoires will be grouped into receptive, productive, and receptive-productive categories. These terms are simply descriptive, however, and suggest nothing about explanatory mechanisms. The first category includes aspects of language in which language is the stimulus and the concern is with the individual's nonlanguage response to these stimuli. The second category refers to cases where the individual produces a language stimulus, and the third category involves cases where language is the stimulus and it elicits a language response.

## LANGUAGE STIMULI ELICITING NONLANGUAGE BEHAVIOR: RECEPTIVE LANGUAGE

There are several types of language repertoires that fit into the category of receptive language, some of which will be described. It should be noted that the order of discussion of the repertoires does not follow the order in which they are learned. Although the repertoires are described independently, for conceptual reasons, interaction can be considered to occur between the repertoires in both their acquisition and function.

THE VERBAL-MOTOR REPERTOIRE. Language stimuli are crucially important in human behavior in part because they extensively elicit other already learned instrumental behaviors. People tell us things, and we respond in appropriate ways. This is so ubiquitous that only a few examples cannot demonstrate the importance of this repertoire. Consider, however, the number of times a day that one responds to the verbal stimuli provided by someone else. The wife who takes a shopping list to the store, reads an item, and goes looking for the item demonstrates the principle. The child is controlled by verbal stimuli when he responds appropriately to "Take your shower now," "Turn the television down," "Put on your tennis shoes," and so on.

Actually, to have a functional language we must learn a vast repertoire of motor responses under the control of single words, phrases, sentences, and longer language pieces (Staats, 1963, 1968c, 1971a). There are hundreds of verbs in our language that have come to control, under appropriate stimulus circumstances, specific motor responses. Examples are push, pull, walk, run, hop, give, take, hand, pass, climb, reach, lift, close, come, go, look, see, examine, listen, feel, smell, follow, watch, swing, raise, and so on.

The individual also must learn a vast repertoire in which he comes to respond to nouns. The response in this case is not specific in terms of the instrumental response involved, as in the case where different verbs must control different responses. For example, the noun stimulus in the request, "Close the door, please," serves only to determine which particular object the response will be made to. The verb determines the specific response that is elicited. Since there is a very large class of verbs and also one of nouns, and many ways of combining them, an infinite number of actions may be elicited by language, many of which are novel.

The principle of learning involved is simple—that of instrumental conditioning. When a response is followed by a reinforcing stimulus when a particular stimulus is present, the response will come to be elicited by the stimulus. The learning can be demonstrated with a preverbal child. An example is the author's experience with his daughter when she was six months old. Kneeling several feet away from her, he said "Come to Daddy," and jingled a ring of keys held before her. The keys had already been seen to be reinforcing for the child. Under this circumstance the child crawled forward and obtained the keys. Thus, in the presence of the verbal stimulus the child's approach response was reinforced. The training was conducted with various rewarding objects and actions (a hug, and so on), and after some trials the verbal stimulus alone reliably controlled the behavior.

The author has employed verbal-motor training, based upon these principles, to develop a more complete repertoire in his own children. The procedures have also been extended to treating the language deficits of an emotionally disturbed child who had no functional language. This treatment involved the use of a token reinforcer apparatus (Staats, 1968b) and was designed to train the child to respond appropriately to a number of verbs. The principles have also been employed in several other studies (Staats, 1968c; Staats, Brewer, and Gross, 1970).

In addition, although not focusing on the formation or function of the verbal-motor repertoire as such, there are a number of studies that involve this type of learning, as Luria (1969) has reviewed. The focus of these studies has been upon the manner in which the developing child begins to employ his own speech to control his impulsive behavior. However, the principles involved are the same, and they can be seen to support the more general analysis. For example, several Russian investigators have been interested in the fact that young children do not profit from instructions not to perform an action, whereas older children do. A child of two to two and one half will respond appropriately when told to squeeze a rubber ball when the light is red, but he will not refrain from squeezing the ball, even when told to do so, when the light is green (Yakovleva, 1958). The child could be gradually trained to respond to the commands to "Don't squeeze any more" by having him remove his hand from the ball immediately after making the squeezing response, until the response to the verbal stimulus was well learned. Miller, Shelton, and Flavell (1970) have failed to confirm this type of finding. However, there are other studies that support the result (Bem, 1967; Birch, 1966; Lovaas, 1964). Lovaas, for example, trained children to say the word *fast* to one stimulus light and the word *slow* to another light. The children had also learned to press a lever for reinforcement. It was found that their rate of lever responding increased when they said the word *fast* and decreased when they said *slow*. Meichenbaum and Goodman (1971) found further evidence of verbal-motor control employing a similar type of experimental situation.

Again, it should be noted that there has been insufficient recognition of the conception that there is a very large class of verbs in a language, each of which has to be learned as a controlling stimulus for motor actions. This learning begins very early. However, the child has to learn a very large repertoire, and the learning requires many, many trials and a great period of time. It appears that at first the child learns only to respond to other's speech stimuli. Later, he comes to respond to the stimuli that he himself produces. It is this learning that accounts for the advancement in the self-regulatory aspects of language that Luria has noted. The experimental efforts that have been conducted so far have focused on Luria's developmental hypothesis. It is suggested that a focus must be directed upon the original learning of the verbal-motor repertoire (which has not been developed in this framework), the number of elements in the repertoire, and the functions of the repertoire, as has been indicated (Staats, 1963, 1968c, 1971a).

To continue, however, the "Come to Daddy" example employed above indicates that in establishing the first elements of the child's verbal-motor repertoire, the learning is long term and tortuous in nature. The training is basically like that conducted with a lower animal—one has to laboriously arrange for the response to occur in the presence of the stimulus, and then provide reinforcement.

However, the child's learning does not remain on this level. After the child has learned other basic behavioral repertoires, the learning becomes much easier and progresses more rapidly. For example, when the child has learned to follow in motor action what the parent indicates he should do, the verbal-motor learning of the child may progress more rapidly. When the parent can demonstrate an action, and motion to the child to repeat the action, then the parent no longer has to go through the extensive task of originally training the child to make the responses— as was the case with the infant learning to crawl to his daddy. As an example, the parent may say "Push . . . push," at the same time he pushes something and thereby get the child to push it also. He may say "Stick out your tongue," and stick his out and in doing so get the child to perform the action also. These training processes assume the child has already acquired imitation skills, another basic behavioral repertoire to be discussed in the next chapter.

After the child has learned a partial verbal-motor repertoire, the further learning of the child may progress even more rapidly—which is another important function of this basic behavioral repertoire. To elaborate, the principle of higher-order instrumental conditioning (Staats, 1968c) states that if a stimulus has come to elicit a motor response and this stimulus is paired with a new one, the new stimulus will also come to elicit the motor response. This principle was illustrated in Chapter 2 in the example where the child learned a closing response to the word *wug*. Actually, a response may not be required at the time of the training. A college student may read in a dictionary "Osculate is to kiss" and merely repeat it to himself. Later, when instructed by a suitably learned coed to "Osculate me," the verbal stimuli may elicit an appropriate response—in this case with a good source of reinforcement.

The studies of the Russian investigators actually support this analysis. That is, they have employed a procedure in which an instruction to "squeeze" a ball is paired with another stimulus (a light), and the new stimulus comes to control the motor response. The same principle is involved in the higher-order instrumental conditioning described above, only the new stimulus is another word. It is interesting that the principle of higher-order conditioning (Staats, 1968c, p. 93) suggests that reinforcement is not necessary for the process to occur. Social learning theory has incorporated a similar principle that there is no need for reinforcement in associative learning (Bandura, 1969). Luria (1969) suggests also that "human conditioning can occur without unconditioned reinforcement" (p. 152) in reference to similar processes.

As is the case with most human repertoires, higher-level (more "abstract") verbal-motor units may be composed from earlier learned (more elementary) units. Thus, the child told to "Judge the matter well" may indicate that he does not know what that "means." At this point the adult may say, "Look into the matter thoroughly. Get everyone's points of view. Do not take either side. Then decide who is right." Through experiences of this type words that will control complex sequences of behaviors can be learned. The child learns to respond to the word *judge* as a higher unit, based upon already acquired responses to other words. In this context it may be mentioned that Paivio (1971) has suggested a dichotomous categorization between concrete words and abstract words. He has found that these two characteristics of words influence the manner in which they are learned and remembered. However, he has not indicated what it is that accounts for the difference between concrete and abstract words. It may be suggested that one variation for verbs, at any rate, is the extent to which they involve specific versus more abstract verbal-motor units. This possibility could easily be tested, and should be, as should other principles in the formulation.

To conclude, this section on receptive language can only hint at the extent, complexity, and abstractness of the language learning in this basic language repertoire. The importance of the repertoire, as well, can only be suggested by saying that it is involved in every type of human activity and learning—in everyday life, in science, in the arts, and so on. A later section will indicate how the individual's self-produced language stimuli are involved in problem solving and reasoning.

THE VERBAL-EMOTIONAL REPERTOIRE. In the previous chapter brief mention was made of the manner in which words can be paired with other stimuli that elicit emotional responses. In the process the word involved will also come to be a conditioned stimulus that elicits an emotional response. The principle is that of classical conditioning.

There is abundant laboratory evidence that an emotional response elicited by a stimulus will be conditioned to another stimulus through contiguous occurrence. The conditioning occurs with lower animals and with man—as the early experiment of Watson and Raynor (1920) showed. Better controlled studies have since been conducted. It has been shown in a number of studies that emotional responses can be conditioned to words (see, for example, Staats, Staats, and Crawford, 1962; Zanna, Kiesler, and Pilkonis, 1970).

With humans, moreover, it appears that this is a very central process that occurs innumerable times and that produces a large repertoire of words that elicit emotional responses. The conditioning can be of either a positive or negative sort; a very large class of words results in each case. Thus, for example, there are many nouns, adjectives, adverbs, and verbs that are emotion-eliciting stimuli.

The experience that yields such large repertoires of emotional meaning (or attitude) words begins very early. When the parent says "No, bad!" to his child as he applies some aversive stimulus, the parent is conditioning his child to respond with a negative emotional response to the words. The parent who asks his child "Good? Is that good?" as he feeds the child a desirable food is conditioning his child to respond with a positive emotional response to the word *good*.

The fact is that very important functions of language reside in the emotional qualities of words. Simple emotional elicitation through words, as in literature and poetry reading, is one example. Another great power of language is that words can transfer their emotional eliciting power to other words and to other nonverbal stimuli, a case of higher-order classical conditioning. If emotion-eliciting words are paired with another stimulus, this stimulus will also come to elicit an emotional response. A number of experiments have verified this finding employing the present author's experimental method for demonstrating this conditioning (see, for example, Early, 1968; Hekmat, 1972, 1973; Staats 1968c), although there are different explanations of the findings (see Page, 1969; Staats, 1969).

When one realizes the ubiquity of words that elicit emotional responses in language usage, the importance of emotional learning through language can be seen. For example, if a verbal reference to sexual contact is paired in the child's experience with words like evil, dirty, immoral, sinful, slut, disease, brute, and so on, verbal sexual stimuli will come to elicit a negative emotional response. When verbal stimuli elicit emotional responses, so will the objects and events that elicit those verbal responses. Considering the potential conditioning trials that are available in the individual's life, intense emotional responses can easily be accounted for on the basis of language conditioning.

It can be added that emotional responses can be learned to very abstract stimuli through language. Thus, a child may learn a strong emotional response to the words *God, liberty, justice,* and so on. It is not necessary that the word denote an actual object for the individual to learn a strong emotional response to it. All that is necessary for a word to elicit an emotional response is that emotional stimuli be paired with it. It is interesting in this context to note that emotional words could fit into either of Paivio's concrete or abstract categories.

The fact is that words play an exceedingly important role in establishing for the individual (or the group) what objects, events, people, activities, ideologies, values, and so on will elicit positive or negative emotional responses. Language is thus one of the primary ways that the A-R-D system is formed and changed. When this process occurs in a political realm, it is apt to be referred to as propaganda. When the process occurs in promoting a product, it is apt to be called advertising. When a mass medium is involved, the process will be referred to as communication.

When the process occurs in psychological treatment, it will be referred to as psychotherapy. In the school the process is included in teaching. In all of these cases an important means of changing the individual's (or group's) emotional response to certain aspects of life is the use of words that constitute a basic behavioral repertoire for the individual. Later discussions will elaborate these functions of language.

The examples employed have been those in which there is a common group response to the word. Thus, for the most part all people in our language culture will have a positive emotional response to *joy, fun, intelligently,* and *candy.* It should be noted, however, that there are wide individual differences in *what* words will elicit emotional responses. There are also group differences in this aspect of personality.

*Reinforcing function of emotional words.* The discussion in this chapter is divided into the present section on language acquisition and a later section on language function. However, the several functions of emotional words can be mentioned here. For example, the manner in which emotion-eliciting words may transfer this quality to other stimuli has already been mentioned. In addition, as the preceding chapter indicated, a stimulus that elicits an emotional response will also be a potential reinforcing stimulus. This is true of words as for other stimuli.

This conclusion is supported by experimental evidence. Finley and Staats (1967) first showed that words like holiday, food, swim, famous, and dollar, when presented contingent upon a motor response, with 12-year-old children, served to strengthen that response. On the other hand, with another group of children, when such words as spoil, ugly, starving, lost, and fat were presented following the response, the response decreased in frequency—indicating these words had a negative reinforcing value. This evidence has been corroborated by additional studies. For example, Pihl and Greenspoon (1969) used a procedure in which different groups of subjects received varying amounts of money while responding to a nonsense syllable. This should have made the syllable a conditioned stimulus for an emotional response, and hence a reinforcing stimulus, to the varying degrees. It was found that the nonsense syllable in this procedure did acquire reinforcing value for instrumental learning in the degree that was expected.

It is suggested that one of the powerful functions of language resides in the way it can serve to strengthen or weaken behaviors. In our everyday interactions with other people, we mold their behavior repeatedly with the reinforcing words we present, contingent upon their actions. And they do the same to us. For example, the father praises the son for working so conscientiously on his chores, and the boy's working behavior is strengthened. Because words are not in many cases explicitly rewarding or punishing stimuli, as praise is, and because the principle is obscure, the effects of such reinforcement upon our behavior to a large extent escapes our notice. We are thus unaware of many of these circumstances that determine how we behave and what we become.

*Directive function of emotional words.* The A-R-D theory indicates that a stimulus that elicits an emotional response, and thus has reinforcing value, will also tend to elicit a class of instrumental responses—"striving for" behavior if the emotional response is positive, and "striving against" if the emotional response is negative. Emotional words thus also constitute verbal motor units controlling approach or avoidance classes of behavior.

It was suggested in the previous chapter that food words had directive properties (as well as reinforcing properties). There are experimental results that support the contention that the very large class of positive emotional words has directive value for approach responses, and the class of negative emotional words has directive value for escape and avoidance responses. For example, Solarz (1960) had subjects move a lever either toward themselves or away from themselves each time he exposed a word to them in an apparatus. It was found that they moved the lever toward themselves with a shorter interval when the word was one that elicited a positive emotional response. They moved the lever away from themselves more readily when the word elicited a negative emotional response.

It is suggested that emotional words have an important function in determining our behavior in life circumstances, as will be more fully described in later discussions. One example may be given here. Let us say that the advertising for a movie employs words that have a sexual connotation. For some individuals the words will elicit a positive emotional response, for others a negative emotional response. It is suggested that the individual's language characteristics will help determine whether he is drawn to approach or avoid attending the movie. In general, the emotional values of words affect our behavior as directive stimuli in every area of human behavior.

THE WORD-IMAGE AND IMAGE-WORD REPERTOIRE. In Chapter 2 it was indicated that some stimuli in the world will elicit sensory responses in the individual. These sensory responses could be conditioned to other stimuli that were present. Conditioned sensory responses, usually reduced in vividness or completeness, are commonly called images.

It has been suggested that some words are systematically paired with sensory stimuli and come to elicit images on the basis of classical conditioning (Staats, 1959, 1963, 1968c, 1971a, 1971b). Several experiments, by Ellson (1941), Leuba (1940), and Phillips (1958), were cited to support this conception. Mowrer (1960) and Sheffield (1961) have also employed this analysis in learning formulations.

The author has elaborated this conception (1968c) to indicate that the individual learns a large repertoire of words that elicit conditioned sensory responses. Partial aspects, or general attributes, of stimuli (like color, shape, and so on) are also conditioned as images to words and serve as general concepts. For example, the individual learns appropriate conditioned sensory responses to the words *white, furry, barking, small, soft, sharp*, and so on. Furthermore, when combinations of such words are presented to the individual, complex combinations of images may be elicited. The individual who has learned the appropriate imaginal word repertoire may be told about a dog that is *small, white, furry, soft*, and *barks sharply*—and experience a composite image. These word "concepts" can be infinitely combined, giving infinite possibilities for sensory experience solely on the basis of language, without requiring primary experience.

It should be noted that combinations of words, because of the composite image they elicit, can have emotional (A-R-D) value. The words *dark, stomach, against, triangle, her, white, hair* individually would not elicit emotional responses. Placed in a passage like "the dark triangle of hair below her white stomach" they may, by eliciting a composite image, elicit a sexual response. Very significant cognitive and emotional activities may be explained on the basis of conditioned sensory responses, especially those that are elicited by words.

It has been suggested that words can elicit images; the reverse also holds (Staats, 1959, 1968c). Images, as responses with stimulus characteristics, as has been indicated, can come to elicit instrumental behaviors, including verbal labeling responses. The individual learns to label his internal conditioning sensory responses (images) in the process of learning the labeling repertoire when external stimuli are present. It is through the "image-word" learning that the individual is able to report his daydreams, dreams, and other imaginings.

## LANGUAGE RESPONSES ELICITED BY NONLANGUAGE STIMULI: PRODUCTIVE LANGUAGE

The term *receptive language* has been used loosely to refer to the fact that people respond to language stimuli. Even the brief learning analysis of the preceding section indicated that there are actually various ways that people learn to respond to language stimuli. Moreover, both classical conditioning and instrumental conditioning are important in establishing these stimulus functions of language. Thus, the category of receptive language (or comprehension) in no sense implies a unitary, basic, or explanatory process, it is suggested.

To continue, however, it is not only important that humans can learn to respond to language stimuli. We also must learn to make language responses ourselves, and in doing so to produce language stimuli for others (and ourselves as well). The actual learning of a speech repertoire is a long, drawn-out affair that will not be described here (see Staats,

1968c, 1971a). The motor skills involved in making speech responses are complex, and the "training" involved is not very direct. Thus, it is no surprise that it ordinarily takes a year or more of experience before the child comes to make his first speech response reliably.

As a later section will indicate, the child must later learn to be able to imitate all of the speech sounds made in his language community. The verbal imitation repertoire is basic to the child's accelerated acquisition of new words. In learning a new word, the child only has to imitate what someone else has said, rather than going through the laborious process of original instrumental conditioning of a new skill. It may be added that the learning of a verbal imitation repertoire—which includes all of the sound elements (phonemes) in a language—is no mean task, as will be indicated in the next chapter. In this section, however, imitation skills will simply be assumed, since the subject matter here will concern the very large repertoire the child must learn in labeling—in making word responses to specific stimulus objects, events, activities, and so on.

THE LABELING REPERTOIRE. The principle involved in the acquisition of the labeling repertoire is that of instrumental discrimination learning. For example, if the child happens to say "Da-da" in the presence of the father and the latter responds by saying "Dada, yes dada," with some affectionate interchange that reinforces the child, the child will thereby receive a conditioning trial for making the response in the presence of the father. The father as a stimulus will then tend to elicit the response. Learning words in this manner is a long, drawn-out affair.

When the child can imitate what the parent will say, the stage is set for rapid learning of a labeling repertoire. The parent may see the child looking at a toy dog out of the child's reach. The parent may then say "Dog, dog," and, when the child has performed a gross imitation, hand the toy dog to the child—which will serve as the reinforcer. Such conditioning trials, or those with the necessary elements, will ordinarily occur by the hundreds in the child's acquisition of the labeling repertoire. It should be noted that for the usual parent these training trials will be informal and not something he will be aware of, so will not realize his role in the child's learning (as is also the case with the other repertoires). Furthermore, after a time the child will not need instruction to imitate words and may not even say the word aloud in the learning situation. For example, when the father says "Hand me the plyers, please" and points to the tool, the child will learn a new labeling response. The child will also ordinarily learn to ask the name for items, a process, again, that will in its sporadic and informal occurrences go unnoticed (Staats, 1971a). Later, the child will acquire labeling responses solely through reading. An object or event will be described, and the label for the stimulus will be given. Ultimately, through these various experiences the individual will acquire a labeling repertoire that numbers into the thousands.

The author utilized, and tested, these principles and procedures in training a basic labeling repertoire in his own children. In addition, he has employed the procedures in a teaching laboratory where graduate students were trained to work with preschool children in language learning tasks. In this procedure the child was presented with a series of picture stimuli on cards. The child had, through his previous experience, already learned labeling responses to some of the pictures. When a picture occurred for which the child had no labeling response, he was given the name. When he repeated the label while looking at the picture stimulus, he received a reinforcer (in the apparatus that is depicted in Figure 2.5 in Chapter 2). This training was conducted for all of the unknown pictures until the child had learned the complete repertoire of labeling responses. The principles involved, as well as the conditioning procedures, are so well stipulated that this type of learning can be demonstrated in a standard manner. Yet it is a central type of language learning that is not generally understood.

COMPLEX LABELING. *Grammatical classes.* The examples given so far have involved nominals (nouns), which are usually the very first words that are learned under stimulus control. This is probably the case because objects are discrete stimuli, easily isolated, and because such important stimuli as the parents, food items, toys, utensils, and so on are objects. Nelson (1973), employing a sample of 18 children, has reported that of the first ten words learned, 65 percent were nominals, 16 percent were action words (verbs or words like *up*), 8 percent were modifiers, and 6 percent were function words (for example, *what?* or *that?*) or personal-social words (for example, *ouch*, which will be described later as self-labeling). The most common words among the first ten learned in Nelson's sample were *Mommy* (83 percent of children), *Daddy* (72 percent), *dog* (61 percent), *hi* (56 percent), and *ball* (44 percent).

As the child's labeling learning elaborates, it includes more difficult elements. At first the labels are learned to specific objects, like the parent's names. Later, the child must learn to label classes of objects which have certain commonalities but which also differ (Staats, 1959). Adjective learning also involves a similar increase in difficulty of the stimulus involved. For example, a color is only part of a stimulus. Moreover, the color to which the adjective label must be learned will be a part of various objects. The process of this type of learning has been described, as has early adverb learning, verb learning, and so on (Staats, 1971a, pp. 132–138). Much additional research should be conducted to explore this important type of learning systematically.

*Word endings and a linguistic-learning rapprochement.* In the introduction to this chapter it was indicated that there has been a schism in the study of language between behaviorists and cognitive theorists. This has prevented interdisciplinary cooperation in the study of various aspects of language, and a rapprochement has been proposed (Staats, 1974). One of the necessary steps towards rapprochement is the utilization of the concepts and principles of the one approach by the other to form a combination. It has been suggested by example that this can be done (Staats, 1963, 1968c, 1971b) and that the structure that results can provide the impetus for further research.

To begin, it can be suggested that linguists have made many observations of language. Their understanding of the grammatical features of language is especially elegant in comparison to usual observations. Many instances of grammatical usage have been systematically described that are not understood by native speakers of the language, even though their language abides by the grammatical rules.

Habitually, linguists do not make observations of the types of events that might explain (cause) grammatical language (Staats, 1971b). Their observations of some of the features of language are systematic, but the manner in which the human comes to perform such language behaviors is not studied in the science of linguistics. It is for this reason that a wedding of a learning theory of language and the linguistic description of important aspects of language has such significant possibilities. It is suggested that a learning theory of language that is oriented toward the problem can provide the vehicle for *explaining* linguistic phenomena. This possibility can be exemplified here.

In 1961 Brown and Fraser, in a paper for an interdisciplinary conference, presented some linguistic observations as an example of grammatical rules (see Brown and Fraser, 1963).

> The rule in English is: a word ending in a voiceless consonant forms its plural with the voiceless sibilant . . . as in *cats, cakes,* and *lips;* a word ending in either a vowel or a voiced consonant forms its plural with the voiced sibilant . . . as in *dogs, crows,* and *ribs;* a word ending in the singular with either /s/ or /z/ forms its plural with /z/ plus an interpolated neutral vowel as in *classes* and *poses.* We all follow these rules and know at once that a new word like *bazooka* will have, as its plural bazooka/-z/, even though most speakers of English will never know the rule in explicit form. (Brown and Fraser, 1963, pp. 5–7)

Berko (1958) had developed a method for studying the manner in which children evidence correct inflections. A child was shown a small animal and told: "This is a *wug.* Now there are two of them. There are two ———." She found progressive improvement in skill in this type of grammatical speech in first-, second-, and third-grade children and concluded that the abstract mental processes called grammatical rules develop with age.

Thus, this is a case where there has been systematic linguistic description of a type of language behavior. Rules describing the behavior can be explicitly formulated. Developmental research indicates that skill in this grammatical behavior increases with age in children. The next step in the quest for knowledge is an explanation of the phenomenon. The nativistic position might suggest that biological maturation of cognitive structure is involved, which would then provide an impetus for the search for such biological mechanisms. Another avenue of possibility is that the child *learns* the grammatical rules. The method to be used in this case, which it is suggested has general applicability, is to make an explicit analysis of the "grammatical rule" in terms of the stimuli and responses that are involved. Such an analysis should then serve as a theoretical body for the projection of research to test the possibilities. The following learning analysis was made concerning the sibilant inflections.

It is . . . suggested that grammatical word endings depend upon the formation of the appropriate response associations. . . . Actually, in this case the S-R account is rather simple. It is suggested that the stimuli which come to control the voiceless sibilant /-s/ vocal response are, for example, the plural stimulus object and the labeling response which ends in the voiceless consonant. After the child has had many, many trials where he is reinforced for the sibilant /-s/ following the voiceless consonant, this stimulus would elicit the appropriate response. After the appropriate associations have been formed between the ending of a word and the "plural" response, the appropriate ending would be expected to occur even when a novel word was introduced. (Staats, 1963, pp. 177–178)

This analysis has been tested in a series of studies by Guess and Sailor and associates (Guess, 1969; Guess, Sailor, Rutherford, and Baer, 1968; Sailor, 1971; Sailor and Tamar, 1972). Guess, Sailor, Rutherford, and Baer, using as subjects retarded children without the relevant language development, reinforced the correct imitation of singular and plural verbalizations to objects presented singly or in pairs. "A generative productive plural usage resulted, the subject correctly labelling new objects in the plural without further initial direct training on each newly presented pair of items" (Sailor, 1971, p. 305). The correct usage was also reversed by additional training and then returned to its correct form.

In addition, however, it was found in analyzing the errors the children made that the voiceless sibilant response, the /-s/ sound, occurred in cases when a voiced sibilant, the /-z/ sound, should have been produced. In theorizing on this finding Sailor was not sure whether it was due to the manner in which the training had been conducted or to some inherent quality of speaking itself.

[I]t was suggested by Cofer (1963) that those words requiring the /-s/ ending encourage the expression of this sound because the final position of the speech mechanisms in producing the word permits continuity of air flow in the plural formation. The same holds true for continuity in the case of voiced plural endings.

Staats (1968c), on the other hand, suggested a hypothesis based on a pure stimulus control-reinforcement position. According to Staats, a child in the process of acquiring the plural morpheme is differentially reinforced by the environment over many trials in the presence of plural stimuli for emissions of the correct plural response. After a sufficient number of trials, a discriminated response class (rule) will generalize to novel stimuli, and the child will use plurals correctly. (Sailor, 1971, p. 306)

To test these two theoretical positions, Sailor again employed retarded subjects who had not learned to make plural endings to words. The subjects were trained solely on one or the other of the plural allomorphs (either the /-s/ plural allomorph or the /-z/ plural ending). In this training single or paired objects served as the stimuli. After this learning was complete with words that ended in either unvoiced or voiced sibilants, objects were presented to the children that were of the other type. When the child was trained with only unvoiced, /-s/, sibilant endings, he generalized this plural ending to objects whose labels customarily take a voiced plural ending. The same inappropriate speech occurred when the original training was with voiced pluralization, which was generalized to objects eliciting labels that usually take an unvoiced plural ending.

Data from both subjects lent clear support to the expectation that productive allomorphs of the plural morphological class can be taught, using reinforcement procedures, to a retarded child such that he will generalize from specific allomorphic response class to the entire morphological class regardless of appropriateness. This finding lends support to the general notion that "rules" of grammar may be acquired through differential reinforcement in the presence of verbal models as suggested by Skinner (1957) and Staats (1968c). (Sailor, 1971, p. 309.)

Various other specific learning analyses of linguistic observations have been made (Staats, 1971b). It has been suggested that these can be employed as hypotheses that can serve as the basis for additional experiments.

At any rate, it is suggested that the principles that have been proposed for learning a repertoire of verbal labeling responses also applies, in an elaborated and analytic way, to complex linguistic observations. Such analyses can lead to productive research which explains the formation of grammatical language usages. Moreover, this type of theoretical research endeavor can lead to a rapprochement between linguistic and behavioristic conceptions.

*Multiword labeling.* Even the child's early single word labeling repertoire will be adjustive for him. When he can say "Outside," for example, he may provide an auditory stimulus that can direct the parent's behavior appropriately. However, the child will also begin making combinations of words. This will occur in part with no specific training, solely because the objects and events for which he has learned labels occur in combination. The child, for example, may say "Milk, all gone," without ever having heard it before. He has simply learned to label milk and to label the stimulus circumstance of something having disappeared that was recently there.

The child will also learn multiword labels of various kinds. Since this will also involve the establishment of word associations, the topic will be treated in later sections. It will only be added here that labeling responses of varying complexity and abstraction are learned, and frequently these are the basis for special skills, as in the artist labeling colors, or the anatomist labeling body parts. Many times the child's labeling repertoire also is considered as evidence of his cognitive maturity. If the child can label something as longer (involving a relational stimulus) or bigger, or equal, he will be considered to have a more mature mental development than if he cannot. As will be indicated in a later chapter, however, it is suggested that such indications of mental maturity—central in the formulations of many child developmentalists—are learned very straightforwardly, as has been experimentally demonstrated (Staats, Brewer, and Gross, 1970).

SOCIAL AND SELF-LABELING. The individual also learns to label the actions of others, since such actions also constitute a type of stimulus. In addition to learning to label instrumental actions such as walking, reaching, talking, and so on, we learn to label more complex combinations of subtle stimuli such as someone being angry, sophisticated, insulting, boring, acting suspicious, entertaining, and so on.

We also learn to label our own behavior as a stimulus. The child will receive experience in which his behavior is labeled by his parents as troublesome, argumentative, demanding, courageous, babyish, affectionate, kind, and so on. He will thereby learn a set of labels for his own behavior. He will also perceive the responses of others to his behavior and in labeling those responses come to label himself as congenial, aloof, entertaining, intelligent, attractive, or what have you. The child will also ordinarily receive experience that will train him to label internal stimuli, such as hunger pangs, a dry throat, pains, emotions, the stimuli produced by motor movements, and so on.

## LANGUAGE STIMULI ELICITING LANGUAGE RESPONSES: RECEPTIVE AND PRODUCTIVE LANGUAGE

The breakdown of the traditional categorization of language into receptive and productive can be seen in the cases of language in which both are involved. Many important language responses occur to language stimuli. These represent cases where there is receptive and productive language in one act. Verbal imitation, an important language repertoire, is one example of the receptive-productive act.

THE VERBAL IMITATION REPERTOIRE. The topic of imitation will be dealt with in greater detail in the next two chapters. It may be suggested here, however, that the skills of speech imitation are complex and difficult to learn for the child, but once learned they form an essential repertoire for the learning of additional speech responses in the individual's language repertoire.

As a simple example, if the child can make the speech responses *foot* and *ball* when asked to do so, he is capable of learning a new labeling response very easily. Let us say the child sees a football and says to the parent, "What that?" and the parent responds, "Football . . . can you say football?" When the child imitates the word—one that he has never heard or said before—while he is looking at the football, he has received a learning trial in acquiring a new labeling response.

This type of learning can be demonstrated very clearly with preschool children in the apparatus depicted in Figure 2.5, as has been shown in the author's teaching laboratory referred to previously. One of the types of learning graduate students in the laboratory conducted with preschool child subjects was to create a new, complex word for the children according to the principles described above. In the demonstration the goal was to train the child to say *Drosophilus Melanogaster*. The children had already acquired an imitative speech repertoire in which they could say each of the syllables in the words. The procedure for each trainer consisted of presenting the words to the child syllable by syllable. For each correct response the child was reinforced. A new syllable was not added to those already learned until the preceding ones had been well learned as an integrated combination. It is edifying to see how standardly children learn such language-cognitive units as new labels, according to the elementary principles of instrumental conditioning.

Moreover, the demonstration indicates the manner in which cumulative hierarchical learning takes place. That is, the child is able to learn the complex word label relatively readily because he has already acquired a basic behavioral repertoire of vocal imitation. If one had to train a child without the basic behavioral repertoire to a complex labeling response such as the one described above, the learning task would be seen to be of surpassing difficulty. The vocal imitation repertoire is important because it enables the child to learn new words very easily. It is in the learning phases of language, thus, that the vocal imitation repertoire is so important. Later in life its functions are not so prominent.

It can be added that cognitively oriented investigators of language, following Chomsky's nativistic view, for some time ignored or rejected the importance of vocal imitation in language acquisition (Ervin, 1964). This view suggested that the speech of parents was too different from that of the child for imitation to be involved. This, however, did not take into account the fact that what adults say to children cannot be indicated by studying the language employed by adults in speaking to each other. Parents speak to children in a different way than they do to each other. It was suggested in a 1967 symposium paper that "When speaking to the young child, the adult's speech is ordinarily appropriate to the listener's skill development. Thus, with preverbal infants and young children, the adult will emit many more one-word utterances, or utterances that are very simple" (Staats, 1971b, p. 136). Reanalysis of records made of mother-child interactions showed this to be the case. The grammar of speech addressed to children is simple (Drach, 1969), and speech to children is repetitive (Kobashigawa, 1969). Moreover, mothers increase grammatical complexity with increasing age of the child (Pfuderer, 1969).

VERBAL SEQUENCES (WORD ASSOCIATIONS) AND GRAMMAR. The strong linguistic emphasis upon the study of the grammatical aspects of language for a time led to neglect of the importance of the child's learning of single words. It should be indicated, however, that this orientation did produce important observations. For example, Brown and Fraser (1963) systematically observed and described the natural speech habits of children between the ages of 24 and 36 months of age. There was an increase in the number of word responses included in each separate sentence (complete sequence of verbal responses) with an increase in the age of the children. These investigators also noted that the speech of the young children was systematically abbreviated, and the extent of the abbreviation was related to the number of word responses the children produced in their average utterance.

Brown and Fraser further studied the abbreviation effect by having the children imitate sentences produced by the experimenter. It was found that with increasing age the match made by the child included more of the individual words presented by the experimenter. In addition, when words were not included they tended to be less essential words:

> . . . words that occur in intermediate positions in the sentence, words that are not reference-making [labeling] forms, words that belong to such small-sized grammatical categories as the articles, modal auxiliaries, and inflections; words that are relatively predictable from context and so carry little information, and words that receive the weaker stresses in ordinary English pronunciation. (Brown and Fraser, 1963, pp. 37–38)

These characteristics led Brown and Fraser to describe the speech of the children as "telegraphic English." They attributed the determinants of telegraphic speech to immaturity in cognitive development that resulted in a decrease in "memory span." It was suggested, however, that rather

than internal maturational processes, the phenomena that had been observed were due to learning (Staats, 1963). It was also suggested that
the child first acquires single word responses. The parent at the beginning
stages speaks to the child with single words. When the child has acquired
single words, the parent will provide training wherein he attaches another word to an already known one. Having learned to say "ball," the
child may be given training in saying "red ball." When this has been
learned he will be asked to say "Give red ball," in the appropriate situation, then "Give the red ball," "Give me the red ball," and so on in an
elaborating training.

This suggestion was supported on the basis of later evidence. Brown
and Bellugi (1964) made records of child-parent speech and found many
instances of what they called "expansions." The mother would listen to
what the child had said and then respond by repeating the child's utterance with the addition of other words. These expansions occurred
about 30 percent of the time, which indicates the important part played
by the parent in providing learning conditions for the child. More recently, Slobin (1968), Kemp (1972), and Kemp and Dale (1973) have
also agreed that new aspects of the child's developing language may be
added through imitating utterances more complex than his own.

In the learning account of the development of multiple word sequences
(Staats, 1963, 1968c, 1971b), it has been suggested that associations of
words are formed. Thus, the fact that children can repeat, or remember,
increasingly long sequences of words is seen, at least in part, to be a
function of the fact that as they have more experience with language they
form associations between words where the first one as a stimulus tends
to elicit the next one as a stimulus.

The manner in which humans can learn sequences of verbal responses
has been the subject of innumerable studies (McGeoch and Irion, 1952;
Underwood and Schultz, 1960), in the investigation of various types of
human characteristics. It is suggested, however, that the manner in which
word associations are established and function in actual language must be
studied. The case of forms of sibilant plural endings already described is
an example of word associations that function to produce grammatical
speech. The pluralization rules were seen to be based upon the learning
of an association between the voiced consonant word ending and the
voiced sibilant response, and so on.

Other grammatical phenomena can be explained, or explained in part,
through an understanding of word association formation and function.
Another example exists in the child's learning of verb endings. It was
suggested in a 1967 symposium paper that a learning theory of language
was challenged by the fact that the child first comes to use irregular
verbs correctly, but later adds regular endings to irregular verbs in an
incorrect manner. To elaborate, the child first learns verbs that are ir-

regular in their past tense (for example, from *eat, sit, drink,* to *ate, sat,* and *drank*). The regular verbs do not change but only add *-ed* to the root of the verb (*walked, moved,* and so on). The finding that has seemed paradoxical in terms of learning (see Slobin, 1971) is that children who have been using the correct past tense forms of the irregular verbs begin making errors. The child may say "I ated," "I dranked," and the like. A learning analysis of this phenomenon suggests, however, that this is due to the associations established between saying a verb and the *-ed* response. The developmental history is that the child first learns the irregular past tense verbs as whole words. The irregular verbs are the most common, those that the child learns first. Later, he has experience with a larger number of regular verbs in which past tense cues come to elicit the *-ed* response. The past tense cues thus elicit the *-ed* response even when an irregular verb has been involved until, with further learning, the discrimination becomes fixed (Staats, 1971b). Palermo and Eberhart (1968) have attempted to test this hypothesis, with positive results. However, as Slobin (1971) has indicated, there is still equivocation in interpreting the results, and the study should be conducted with young children or with retarded children to obtain better experimental control.

Other grammatical phenomena have been analyzed in terms of word associations (Staats, 1963, 1968c, 1971b). Experimental studies are needed to begin to explore this fertile field of investigation. Other types of S-R mechanism are involved besides word associations. For example, in the verb endings example above, the past-tense cues of the situation must also be thought to elicit the *-ed* response, not just the verb itself. This should be tested experimentally. Thus, the suggestion that word associations are involved in grammatical speech does not mean that word associations are the only determinants (see McNeill, 1971; Staats, 1971b). However, word associations must be studied, along with other principles and mechanisms.

Finally, it should be noted that the very complex language experience of the individual produces very complex word association mechanisms. When the child has experience with hearing or saying "pretty car," "pretty dog," "pretty picture," and so on, he is learning a single stimulus, multiple response S-R mechanism where the word *pretty* is the single stimulus tending to elicit multiple responses (see Figure 2.10). When the child experiences "Give him," "Throw him," "Push him," and so on, he will learn the type of S-R mechanism depicted in Figure 2.9(a). Figure 2.11 depicts combinations of these two types of mechanisms in which each word response (with its stimulus characteristics) tends to elicit several alternative responses and where several different responses tend to elicit the same alternatives. What the individual says, and the order in which he says these things, will be in part determined by the nature of such word associations but also by other stimulus controls over speech (such as the external environment).

The very complex nature of the individual's language experience in speaking, reading, and in life experience of a nonverbal sort (Staats, 1968b) produces fantastically complex word association mechanisms. It should be expected that these are characteristic of the individual, depending upon his experience, and can be expected to play a role in all of his language-cognitive experience.

## Language: The organization of the basic repertoires

In exploring the various approaches to the study of language it can be seen that, faced with a phenomenon of great complexity, there is a tendency to simplify the problem by considering only one aspect of language. These partial treatments may then be interpreted as full theories of language. Thus, while the justifiable fractional treatment may be very productive in its realm, it has drawbacks when it is considered as a theory of language. As an example, Osgood (1953) presented a very productive analysis of the instrumental conditioning of implicit responses to words. He further studied methods for measuring some of the implicit responses involved in which the meanings of many words were rated by subjects on such scales as kind-cruel, relaxed-tense, big-small. It was found that there were three types of meaning that are common to many words: evaluative meaning, potency meaning, and activity meaning. This analysis of implicit responses and the measurements of the three types of meaning became for many investigators a general theory of meaning, or even of language.

Consideration of language, however, reveals that there are various aspects that are not included in this analysis. Other learning principles are also involved. Moreover, analysis suggests that there are various types of meaning (Staats, 1968c). Actually, most of the repertoires that have been referred to herein, as well as others, constitute types of meaning. A word may be spoken of as meaningful because it elicits an emotional response (which may be called evaluative). A word may also be meaningful because it elicits a conditioned sensory response or because it has word associations—for example, the meaning of articles like *the* may be considered to be their word associations that help control the order of words emitted in a sentence. A word may be considered meaningful when it functions as a reinforcing stimulus in strengthening or weakening some instrumental behavior or because it functions as a directive stimulus. Verbs are meaningful in this sense when they elicit specific motor responses. Emotional words can be meaningful when they elicit approach or avoidance behavior. These are but examples. It is suggested that verbal meaning must be considered to be plural, in both the principles and the specific characteristics involved (Staats, 1968c).

As another example, it should be indicated that grammatical classes of words, as described by linguists, cut across more than one language repertoire, as described in the learning analysis. A verb, for example, may be a labeling response under the control of the stimulus of someone's action. But the verb itself may perform in the role of a stimulus that elicits an individual's response. This is true of other word classes such as nouns. And all of the grammatical classes of words are involved in different word association patterns, in addition to their membership in such language repertoires as labeling, imitation, verbal-motor, and so on. Description of the different "psychological" classes that words can belong to requires systematic description—perhaps of no less importance than the grammatical classification of words.

It is suggested that a pluralistic approach is necessary in studying language itself, as well as in more specific topics such as meaning, communication, grammar, and so on. Various learning principles are involved in language acquisition, and language has various functions. It is the acquisition of the various language repertoires in the necessary complexity, and with the necessary intertwining with other repertoires, that constitutes a functional language competence. None of the repertoires in isolation is sufficient, and some of them are functionless by themselves—for example, imitation.

It is also the case that language phenomena are important to every area of human behavior. Reference to language will thus occur in different parts of the present book, although it will not be possible to do this in every case where an analysis of language would be an important contribution. The next section will begin by outlining some of the functions that language has for the individual.

## Language function: The explanatory (personality) characteristic of cognition

Typically the approach to the study of language has been in isolating some of its features and, in the case of behavioristic accounts, suggesting the manner in which those features have been learned. This, however, does not deal with the explanatory aspects of language, how language is involved in cognitive (mental) activities in ways that influence the individual's behavior and that distinguish him from other people. Language, or cognition, it must be stressed, has its major significance in its *function*. It is because language or cognition enables the individual to give and receive communications, to solve problems, reason, plan, hypothesize, be intelligent, and so on, that this aspect of human behavior has its major importance. One of the reasons cognitive theorists have been dissatisfied with behaviorism, it is suggested, is because learning theories of language have tended to avoid these topics. In this area, again, behaviorism has treated language only as a dependent variable. Cognitive theory in language, as in other areas, has been interested in the personality trait value of human behavior. It is suggested that this is a necessary interest. The language repertoires the individual has learned may be considered to constitute a personality trait in the sense that they contribute to the determination of how he will behave in many ways, how he will learn, what he will experience, and what he will further become.

Several examples of cognitive activity will be dealt with in the following sections. Analyses of some of the language repertoires that have been described will be employed in this discussion.

## INTELLIGENCE

The concept of intelligence can be seen as one in which the traditional approach has resulted in insoluble problems. The common concept of intelligence is that it is an internal quality of the individual that affects his behavior in many ways and many situations. A high quality of intelligence will enable the individual to learn rapidly, to reason well, to solve problems well, and the like, so the conception goes. Individuals with exceedingly high intelligence, or talent, will be capable of performing exceedingly skilled behaviors in science, art, literature, the professions, and so on.

As with other personality traits, behaviorism has generally not found the concepts to be useful. There has been no radical behavioristic treatment of intelligence, or the field of individual differences in learning. This view, however, does not square with the cognitive theorists' interests, or with the evidence that intelligence is a *cause* of human behavior.

Based upon the idea that there is an enduring personality trait of intelligence, test instruments have been constructed to measure the internal quality. Studies that have used intelligence tests have shown wide individual differences in performance. Moreover, studies show that children who do well on IQ tests tend to do better in school and the like—an indication of causal characteristics. The same has been found with groups. For example, blacks in the United States do not do as well as whites on intelligence tests (or in school).

The interpretation of the causes of intelligence differences has been the source of continuing controversy. It has been widely accepted that individual differences in the internal quality of intelligence are to a large extent determined by inherited (genetic) characteristics. This interpretation leads logically to the conclusion that group and racial differences in achievement are due to genetic causes (Jensen, 1969). There are many socially significant implications of this view. Some people in the field have shied away from these social implications, while still accepting the concept of intelligence, including the genetic interpretation of individual differences (Kleinfeld, 1973).

These conceptions of intelligence have guided research. Data have been collected to show that familial background is related to IQ test performance, for example. These studies have assumed success in controlling or taking account of any possible effect of environmental determination. On the other side, studies have been conducted to show (1) that environmental stimulation produces higher IQ scores, and (2) that such conditions as education of parents, social class, number of books in the home, and so on, are related to IQ scores.

There has been an impasse in these antagonistic interpretations for many years. The issue becomes central from time to time, currently with the genetic claims of Arthur Jensen (1969) and William Shockley (1971), whose views have come to popular attention. But the issue is never resolved, and it is suggested that this failure arises from the error of the paradigm. Both the genetic and the environmentalist orientations have accepted the basic concept of intelligence as an internal quality and have recognized the rationale for its measurement. The concept of intelligence is that of a general internal quality, and the intelligence items employed to measure this quality are unimportant in and of themselves. As a consequence, after these many years of concern with individual differences in intelligence there still is no generally accepted definition of what intelligence is. Without a specification of the dependent variable, it follows that the independent variables will be poorly specified. Gross environmental conditions do not indicate how intelligence skills could be learned. Moreover, because there has been no specification of the environmental conditions that affect intelligence, and no specification of the principles involved, the studies that attempt to show that intelligence is inherited have lessened credibility. The basic assumption of such studies is that they have held environmental conditions constant or taken account of them. This does not appear likely when the conditions affecting intelligence have not been known. It has been concluded that environmental artifacts in studies of family-intelligence relationships have not been controlled or appropriately considered (Staats, 1971a).

It has been suggested, in an effort to resolve the issues, that intelligence is actually composed of a number of basic behavioral repertoires. Rather than considering the items on intelligence tests to be inconsequential— only important as they index an internal quality—the items are considered to be samples of these various repertoires. The particular item has general significance because it constitutes a *sample*. If the child responds correctly to the sample, it indicates that he has a larger universe of such behavioral skills than if he misses the item. Since intelligence tests have been composed to predict school success, this means that the skills measured by the test are ones that enable the child to be successful in school (Staats,

1971a). (This does not mean that intelligence test construction has led to the inclusion of all the skills that account for the child's success.)

It is suggested, further, that it is necessary to specify the repertoires that compose intelligence. The manner in which these repertoires are learned also must be specified, as must the manner in which these repertoires determine the individual's adjustment—how he learns, how he reasons, how he solves problems and plans, and so on. It may not be productive to continue general studies in which environmental and biological variables are gross and the measure of intelligence involved is of an undetermined set of behavioral skills. Specific explication is necessary for explanatory theory. The notion of general mental quality should be discarded and replaced with the specification of the basic behavioral repertoires involved in the personality trait called intelligence, as indicated below.

### INTELLIGENCE ANALYSES

*The labeling repertoire and intelligence.* A conception of intelligence in terms of learned basic behavioral repertoires has been outlined (Staats, 1971a). A few examples will be given here as illustrations. To begin, it may be suggested that each of the language repertoires described in the preceding section is an important constituent of intelligence. Let us take as an illustration the labeling repertoire. The child learns to name his parents and familiar objects of various kinds, events, and actions, and so on, as has been described. The relationship of this aspect of the language repertoire and intelligence may be seen by referring to standard intelligence tests. The Stanford-Binet (see Terman and Merrill, 1937), for example, includes various items that measure the labeling speech repertoire. At the two-year level, to illustrate, the items that ask the child to point at various objects such as a button, a cup, a spoon, and so on, actually involve this labeling repertoire. Only if the child has learned the appropriate verbal label of the object can he select it from among the several presented. In addition, however, the young child is shown pictures of common objects and is asked in each case "What's this?" or "What do you call it?" (Terman and Merrill, 1937, p. 77). The child must then name the picture correctly to make points toward his intelligence score.

Even at early ages, it should be noted, the labeling repertoire is important to the child's intelligence in more situations than just the IQ test. From the standpoint of the child or someone dealing with him, when the child has acquired even a rudimentary single word speech repertoire he is able to obtain objects and events he wants more readily and less effortfully than can the preverbal child. A child who lacks such speech responses must obtain things in a more primitive manner, by tugging at the parent, gesticulating, and so on. Moreover, as has been indicated, the child's labeling repertoire is one basis for his verbal-motor repertoire, and thus his ability to follow directions—a most important intelligence reper-toire, as will be indicated.

*The verbal-motor repertoire and intelligence.* It can readily be seen that the verbal-motor repertoire of language is a central aspect of intelli-gence. Actually, every item on an intelligence test, or any intellective test, involves following instructions—that is, the child's behaviors must be under the explicit control of verbal instructions. As one example, the first item on the Stanford-Binet for children includes a board with insets for a circle, square, and triangle. The examiner presents the following verbal stimuli to the young child, "Watch what I do." The examiner then removes the blocks and places each before the appropriate recess in front of the child. Then the examiner presents the verbal stimuli, "Now put them back into their holes" (Terman and Merrill, 1937, p. 75). When this intelligence item is considered specifically the essential stimulus and response components can be seen. The first verbal stimulus, "Watch what I do," must control the child's attentional behaviors. The second verbal stimulus must control the child's motor behaviors of attempting to place the pieces back in the holes in the board. The verb *put* must control a type of motor response. The pronoun *them* must control a response to the pieces involved. The word *holes* must control a response to the holes in the board. What this means is that if the child is to succeed he must have acquired some pretty complex verbal-motor units in his basic behavioral repertoire. If he has not learned these units he will be judged less intelli-gent.

And such a child will be less intelligent in a more general way. Even by the time the child is only a few years old, many of his social inter-actions will be based upon this repertoire. A child will be instructed to do many things in a preschool class, for example. If he does not look at what he is directed to, if he does not make the appropriate verbal and motor responses as directed, he will not learn as other children do. And he will again be considered stupid and deficient. His own social interactions with other children will be largely inhibited if he cannot respond to speech as directed. Even at preschool ages most children's interactions involve mutually following directions. When the child is told "Let's play house, you be Daddy," and he does not respond appropriately, he will be left out of the game. After several such interactions, other children will generally leave him out and he will be a social isolate, with the restrictions on learning that this imposes.

*The verbal imitation repertoire and intelligence.* One additional ex-ample, that of vocal imitation, will be given. At the three-year level of the Stanford-Binet there is an item in which the child is instructed to repeat two numbers pronounced by the examiner. The examiner states, "Listen; say 2." "Now, say 4–7" (Terman and Merrill, 1937, p. 79), and so on.

At the five-year level another item asks the child to imitate a sentence ten words in length. It should come as no surprise that a test of the child's proficiency in this task would be correlated with how well the child would perform in other situations. As has been indicated, the same skills are requisite for learning in those other situations. Other cognitive skills such as number concepts and so on are also learned on the basis of the verbal imitation repertoire (Staats, Brewer, and Gross, 1970).

LANGUAGE AND INTELLIGENCE, AND RESEARCH ON INTELLIGENCE. In the introduction to this chapter it was indicated that there has been an interest in the relationship between cognitive (intelligence) development and language. It has been rather widely accepted that language development depends upon or is heavily influenced by the child's cognitive development—considering the latter as an internal mental quality. A number of studies have employed this conception. For example, as Cromer (1974) has indicated, there are studies that have related intelligence development to language development, employing retardates as subjects. In one study Graham and Graham (1971) correlated a linguistic analysis of conversational performance with mental age, showing the two to be related. This was interpreted as evidence that the internal process of intelligence influences the rate of language development. The same interpretation was made of the data in Lackner (1967), which similarly showed that the sentence length of children increased with mental age, and the sentence length of retarded children of a certain mental age was the same as the sentence length of normal children of that mental age.

The present suggestion is that intelligence and language measurements are correlated because they are in good part measurements of the same thing. It was suggested in Chapter 3 that the same behaviors are employed to define different personality traits. This is an example. It is suggested that the language repertoires constitute parts of various personality concepts and studies of human behavior. The same is true of the other basic personality repertoires, such as the emotional-motivational system, and the instrumental behaviors yet to be described.

The present conception of intelligence in specifying how the language-intelligence repertoires can function in the individual's adjustment is a behavioral interaction conception. It also provides a foundation for conducting explanatory research considering, for example, such questions as: Do children really learn the types of skills represented by intelligence test items according to the straightforward principles of conditioning? Can a naive young child be trained to skills that allow him to succeed on an intelligence item, and on a number of items? Can the various item skills be produced in the young child through the learning conditions? Much research should be conducted in this area. Such research can provide a basis for procedures to affect intelligence development.

In addition, of course, it is important to begin the study of how such behavioral skills as those on intelligence tests aid the child's learning in school. It is as important to specify the explanatory functions of intelligence skills as it is to indicate how the skills are acquired.

INDIVIDUAL DIFFERENCES IN LANGUAGE-INTELLIGENCE LEARNING. Theories of language customarily are concerned with general principles rather than individual differences, a focal interest in the study of personality. In considering language as a personality trait, impetus is provided for a concern with individual differences, however. There are very wide differences in the learning conditions that children receive, for example, in the quality and extent of experience parents provide for the child within the confines of taking good care of him. As a mother dresses her child, for example, she may label everything she puts on him, as well as the movements she has the child make—such as raising his hands above his head in putting on a sweater. Another mother may perform the same acts but provide no verbal training for the child. This constitutes an important difference. A father may give a piece of bread to a young gesticulating child without verbal interaction, while another says the name of the food and has the child attempt to repeat the word. In the informal interactions of parent and child there are innumerable opportunities for language-intelligence training, even in the child's early years.

> It should be realized that parents will vary in the amount and complexity of their own language usage, and in the extent to which they interact verbally with the child. Ordinarily, a parent who considers that the child develops mostly through biological maturation will be less systematically concerned with ensuring that the child has extensive informal training circumstances—at least in comparison to the parent who accepts that a *child is what he learns.* Moreover, the other children and adults who interact verbally with the child will affect his language learning—and there is great variation in these factors also. Thus, children will receive different degrees of learning experience in these areas as well as in all the other areas of learning. (Staats, 1971a, p. 160)

The study of individual differences in training children in their language-intelligence repertoires needs a concentrated effort. This type of study, which would provide part of the foundation needed for a full understanding of individual differences, has only incipient beginnings. Nelson (1973) has reported data that are interesting in this context. In her study of 18 children, beginning when they were only 10 to 15 months of age, she found it is possible to see two different types of language development very early. Some children appear to learn a large proportion of labeling responses, in contrast to other children who learn what Nelson describes as personal, social, and function words.

> Analysis of the records of individual children revealed some striking differences. . . . In terms of the function of language, one child seems to be learning to talk about things and the other about self and other people; one is learning an object language, one a social interaction lan-

guage. . . . This division by functional type of language was found to be related to another difference that had been noted in the sample, namely, the early learning of speech units of two words or more. . . . The number of phrases produced by the [labeling] group ranged from 0 to 5 (mean 2.4), while those of the [personal-social] group ranged from 6 to 18 (12.6). (Nelson, 1973, pp. 21–22)

Limber (1971) reports data that can be similarly considered. Neither of these studies made observations of the childrens' learning conditions that produced the language-intelligence differences. There is evidence, however, of this type. Bernstein (1960) originally reported that mothers in different social classes had different language interactions with their children. So has Deutsch (1965). Olim (1970) investigated language differences of black mothers from three socioeconomic levels and their relationship to the cognitive development of their four-year-old children. The mothers' language styles differed and could be seen as a predictor of their children's cognitive performance.

In terms of manipulative studies, Levenstein (1970) compared the intelligence test scores of three groups of preschool children before and after the mothers of one group were given training designed to stimulate verbal interaction between mother and child. The training, and presumably the consequent mother-child verbal interaction, lasted seven months. The results showed that the children in this group showed significant gains in intelligence scores, in comparison to the two non-treated groups. Another relevant study was conducted by Whitehurst, Novak, and Zorn (1972) with a three-year-old child who had a language comprehension score of a four year old but practically no speech. They had the mother vary in her interaction with the child in terms of amount of conversation and amount of prompting of the child to repeat words she said in learning labels, and so on. The child's speech production was greatest when the mother employed a high amount of both conversation and language prompting.

It is suggested that the various types of language training that have been outlined herein, and elsewhere in more complete form (Staats, 1971a), could be employed to enhance the child's language-intelligence learning in a more comprehensive manner. These possibilities remain to be empirically tested in a systematic series of studies. In any event the present conception represents a general theory of intelligence and intelligence development which includes directions for conducting both basic research and clinically oriented studies.

## LANGUAGE AS A THEORY: REASONING AND OTHER COGNITIVE ACTIVITIES

It has been suggested that the individual's language repertoires constitute a theory for him (Staats, 1968c). His language behaviors may be seen to perform the same functions for him that the scientist's theory perform. The individual's language repertoires largely embody the knowledge he has. His language sequences enable him to predict that which has not yet happened, and his language helps him do things to affect the natural events of the world. The individual's language repertoires provide the basis for personal and social decisions of all kinds. Thus the individual's language in action constitutes what is usually considered as cognition—his reasoning, problem solving, planning, purpose, hypothesizing, and so on. The repertoirial elements of which language is composed should be considered in this light: recognition of these cognitive functions of language helps establish a rapprochement between behaviorism and cognitive theory.

As has been indicated, there is a traditional tendency to consider cognition, such as problem solving or reasoning (Maier, 1930), as made up of unitary processes. However, the acts that in everyday life would be referred to as examples of problem solving or reasoning appear to be extremely varied and to involve various principles and various basic behavioral repertoires. It is suggested that the *various* language repertoires may be involved. However, for present purposes, a more simplified account will be presented. Frequently, a reasoning or problem-solving act will include (1) the manner in which the stimuli involved are labeled, (2) verbal response sequences made to the labels, (3) and some final overt act elicited by the individual's verbal processes. Although all cognitive acts by no means are composed of these elements, they can be referred to in indicating some of the cognitive functions of combinations of language repertoires.

To illustrate the operation of these several repertoires in reasoning and problem solving, let us take the case of a mechanic at work on a car that has an engine problem. He will begin by observing the operation of the engine and labeling its defective functioning or obtain a description of the symptoms from the owner. The engine may be missing, backfiring, have poor acceleration, cut out for brief intervals, die when idling, or have a knock or ping. Depending upon the labels the mechanic applies to the cues, different reasoning response sequences (word associations) will be elicited. The fact that an engine cuts out for brief periods might be a puzzle for the mechanic, unless he has learned the labels that the car has a fuel injection system governed by an electronic mechanism, as well as the instructions concerning electronic failures that produce such symptoms. Moreover, in solving the problem, verbal response sequences that indicate to him where the part failure is located, what tools he will need, and the actions he must take will be central components of the problem-solving process. Finally, the individual's verbal-motor repertoire will be relevant. His self-instructions are not functional unless they can elicit the instrumental actions of selecting tools, going to the part of the car involved, seeking out the offending part, and so on. As has been indicated, repertoires of responses are involved in the three components identified in this example, and as will be indicated variation in reasoning and problem solving can occur in any component.

LABELING, PERCEPTION, AND THOUGHT DIFFERENCES. There are individual and group differences in labeling that will determine reasoning. For example, Brown and Lenneberg (1954) showed that the perception of colors depended upon whether or not, and in what strength, the subjects had previously learned to label colors. In the procedure the subject was shown four colors. After the four had been removed the subject was shown a chart with 120 colors and asked to point to the four he had seen. The stronger the verbal label the subject had learned to the color, as judged by cultural norms, the more surely could he perceive it among the many others. Katz and Seavey (1973) have more recently demonstrated the effect of labeling on perception of skin color.

Successful reasoning or problem solving many times involves selective perception of the essential cues among a number of stimuli. It is suggested that the individual's labeling repertoire will help determine the perceptual process, and it is interesting to interpret a very influential conception in the present terms. The linguist Benjamin Whorf suggested that different languages lead to differences in thought (1956). For example, Eskimos have 11 labels that differentiate types of snow, labels that the English language does not contain. On the basis of the present analysis, it would be expected that these labeling differences could affect reasoning and problem-solving processes involving snow.

*Emotional value and labeling.* To continue, it can be added that the emotional value of the labels will also affect the process of reasoning and problem solving. Lazarus and McCleary (1951) showed that nonsense syllables that had been paired with electric shock, coming to elicit a negative emotional response, required longer visual exposures to be perceived than did other nonsense syllables. As has been indicated, attention to and perception of a stimulus may be the first step in reasoning.

It should also be noted that the emotional value of the *label* made to a stimulus has the same significance. For example, DiVesta and Stover (1962) conditioned children to respond with a positive or negative emotional response to nonsense syllables. Then the children learned to label objects with the nonsense syllables, and finally the children could select which objects they wished to take home. The children selected the object that had a nonsense syllable label that had positive emotional value. The emotional value of the label of an object would be expected to influence the manner in which it was selected for attention in a reasoning situation. Moreover, it would influence the reasoning process by helping determine the types of verbal response sequence subsequently elicited. For example, a physician who had a negative emotional response to the label *asthma* would be more likely to reason that a patient he has labeled with this affliction should be treated by another doctor.

It can also be added that labels that are awry may result in reasoning processes that are awry. The physician who labels (diagnoses) a patient incorrectly will reason incorrectly and treat the patient incorrectly. Incorrect labeling in everyday life may also produce inappropriate (abnormal) behavior, as will be indicated in Chapter 8.

WORD ASSOCIATION SEQUENCES AND REASONING. As has been indicated, the initial labeling of the stimuli in a problem situation can elicit a sequence of additional word responses. The process of reasoning many times also depends upon the characteristics of the individual's word association repertoire and of the particular word sequences the labeling elicits in him. An experiment by Judson, Cofer, and Gelfand (1956) can be used as an example. The problem for the subjects was to tie two strings together that were too far apart to be reached by hand (the Maier two-string problem). The solution required that one of the objects available in the situation had to be tied to one string so that it could be swung as a pendulum, bringing it close enough to the subject to reach. Prior to being presented with the problem some of the subjects first acquired the word response sequence *rope-swing-pendulum* embedded in a word association list they had to learn. The first word, *rope*, may be considered the label; the other two constitute the elicited word association sequence, which in turn could elicit the problem-solving behaviors. The other subjects had to learn the same words embedded in the list, but not in sequence. The results showed that the subjects who had learned the relevant labeling and word association sequence solved the problem better.

Word association sequences that mirror the characteristics of the events involved can aid in problem solution and decision making. The converse may be true when the word association sequences are awry. Specialized training is many times aimed at producing useful sequences. The mechanic, for example, learns the word association sequence *On this car replacing the main bearings begins with removing the pan.* The electrical worker learns *Amperage is equal to the voltage divided by resistance.* The Jivaro Indians of South America learn that *Illness is the result of evil spirits put into the person by an enemy;* the early Christians learned that *Illness stems from the will of God.* Each of these may be considered to be a word association sequence, but the sequences have differing degrees of isomorphism with events of the world. Since language sequences are learned, great variations can be expected in the repertoires between individuals, groups, and cultures, with varying reasoning skills being the result.

Word association sequences that function in social reasoning can be learned from various sources. There are interpersonal word association sequences (rules) embedded in religious formulations, as the following examples indicate: "Be kindly affectioned one to another with brotherly love: in honour preferring one another: not slothful in business: . . . Provide things honest in the sight of all men. . . . If it be possible, as much as lieth in you, live peaceably with all men. . . . Therefore if thine enemy hunger, feed him. . . ." (Romans, 12). Social reasoning sequences are learned that are idiosyncratic to one's family, friends, socioeconomic group, culture, the literature read, and so on. Cognitive theorists have stated that human behavior is rule governed, not determined by the stimuli of the situation. The present approach is not antagonistic with this cognitive conception. It is necessary, however, to stipulate what internal rules are— for example, language response sequences—how they are acquired, and how they have their effects. The latter is the subject of the next section.

THE VERBAL-MOTOR REPERTOIRE AND PROBLEM SOLVING. In many cases the significance of reasoning behaviors lies in the instrumental behaviors they finally elicit. One of the important repertoires in language discussed above is the verbal-motor repertoire. It was stated that the child learns a vast repertoire of behaviors in which a verbal stimulus, provided by oneself or by someone else, elicits a motor response. It is this repertoire that comes into play in many problem-solving situations. The stimulus events in the problem are first labeled, let us say. The labeling in combination with other stimuli then elicits sequences of reasoning responses. The reasoning responses finally culminate in the elicitation of some overt motor response. Ordinarily, if the labeling and reasoning have been isomorphic with the events involved, the overt behavior finally elicited will be appropriate to those events and the problem will be solved—or at any rate be responded to adjustively.

There are several studies that demonstrate these principles clearly. Palkes, Stewart, and Kahana (1968), for example, trained children in talking to themselves prior to taking a Porteus maze test. In a variety of tasks the children were trained to use simple verbal commands such as "Look and think before I answer." Hyperactive boys were used as subjects. The boys who had received the training attained better intelligence measures on this test.

Meichenbaum and Goodman (1971) gave the children more extensive training in labeling and verbal response sequences. The following are examples, cited in the study, of the verbal responses the subject was trained to make:

> I have to go slow and be careful. Okay, draw the line down, down, good; then to the right, that's it; now down some more and to the left. Good, I'm doing fine so far. Remember go slow. Now back up again. No, I was supposed to go down. That's okay. Just erase the line carefully. . . . Good. Even if I make an error I can go on slowly and carefully. (Meichenbaum and Goodman, 1971, p. 117)

Children who had been trained to make these verbal responses in solving problems were then given the Porteus maze test and the performance part of the WISC intelligence test. In one study the children were second-graders with behavior problems, in the other the subjects were impulsive kindergarten and first-grade children. The children who were trained in the verbal reasoning and problem-solving behaviors performed better on the intelligence tests, and this advantage was maintained in a follow-up measure one month later.

It should be noted that the efficacy of the verbal reasoning requires that the motor behaviors to be elicited by the reasoning sequences have been learned by the individual. For example, the doctor could label the patient's defect veridically, he could reason correctly that this means that such and such an operation could be performed, but he might not have the instrumental skills himself. (Actually, each of the components of a reasoning act—the labeling, the verbal reasoning sequences, and the instrumental act—may be performed by different people.) It may also be the case that two individuals with the same reasoning processes may respond differently because they have learned different motor responses to the same language sequences. Two children faced with a beautiful cake may both label it veridically by saying "Um, that cake looks good." This and the cake may then elicit the verbal response sequence "But mother said not to touch the cake in the refrigerator." However, if one child has been trained to respond to the verbal stimuli "not to do such and such" by "desisting" behavior he will leave the cake alone. If the other child has not learned this aspect of the verbal-motor repertoire, or if he has learned to perform behaviors in the face of such verbal stimuli, he may respond quite differently.

It can be added that problem situations are usually those for which the individual has no ready instrumental responses that resolve the situation. Thus, for the beginner learning to drive, performing the chain of responses necessary to get the car in motion appropriately constitutes a problem. As has been indicated (Meichenbaum and Goodman, 1971, p. 117; Staats, 1963, pp. 89–90) verbal labels and reasoning sequences will be employed in this task to mediate the performance. Later, however, the motor responses are directly learned to the stimuli in a sequence and the labeling, and reasoning responses drop out as unnecessary and inhibiting. Many of our skills are acquired in this manner, which is another general reason why language repertoires are such an integral part of intelligence.

*Emotion words and actions.* The emotional properties of words play an important role in the way language functions in reasoning, thinking, and planning acts (Staats, 1968c). This occurs because of the several A-R-D functions that emotional words have.

To elaborate, a frequent result of a reasoning sequence is that it culminates in a word, or set of words, that elicits an emotional response. If the emotional response is positive, the reasoning sequence will tend to elicit positive (striving for) actions. If the emotional response is negative, the reasoning sequence will tend to elicit negative (striving away from or striving against) actions (Staats, 1968c).

An example will illustrate the process. Two students may receive the same invitation to a party. The problem may reside in the fact that each faces a midterm examination very soon. This situation, let us say, elicits the same reasoning sequences in each of the students. They each describe the party in positive emotional terms. They also describe the impending test, however. And additional stimuli in the situation, the amount of free time they can obtain for study before the exam, and so on, also elicit verbal sequences. These various sequences of labeling responses, conditioned sensory responses (images), word association sequences, and so on, then elicit the final sequence "If I go to the party, it will cut down the time I have for study and increase the chances that I will get a lower grade." The extent to which the last phrase in the reasoning sequences, "increase the chances that I will get a lower grade," elicits a negative emotional response will be a determinant of the overt instrumental behavior of the students—the going to the party or the staying home and studying. Other things equal, the student in whom the reasoning sequence elicits the more negative emotional response will be more likely to avoid going to the party.

It can be added that if words elicit more generally negative emotional responses in one person than in other people, then this person will more generally tend to avoid actions that risk aversive consequences. He will be spoken of as having a "cautious personality." The effects of the emotional meaning words elicited by the individual's reasoning sequences will be apparent in all of the reasoning he engages in, in all the various concerns of his life, from technical, professional activities to those of a social nature (Staats, 1963, 1968c). When the person's language is idiosyncratic in these respects, as in the others, his overt behavior will have this characteristic. Other behavioral approaches to personality and clinical psychology are recognizing the importance of emotion-eliciting words in the control of behavior (Bandura, 1969, pp. 584–587; Mischel, 1971).

## Self-language and imagination

It has been suggested that the individual learns productive language and can generate sequences of word responses. It can be briefly noted that in so doing the individual can generate sensory and emotional experience for himself. That is, word responses the individual makes himself will elicit the emotional responses or sensory responses in him usually elicited by such word stimuli. Thus, aspects of daydreaming and dreaming can be considered to involve emotional and sensory experience elicited by the individual's own language production.

The act of imagining may parallel previous experience. The individual in this case may simply rehearse descriptions of things he has experienced and in this way relive emotional and sensory responses he has had before. Or, the language-imagination act can piece together parts of the individual's experience into new combinations, and novel images can result. Moreover, the previous experience that gives rise to later images may not be primary. The experience that resulted in the language production could occur through reading or some other form of language.

It should be emphasized that once the individual has learned to produce language sequences, he can to some extent provide for himself the various stimulus functions that others can provide for him. His language stimuli can serve as an emotion eliciter, an image eliciter, a reinforcer, and a directive stimulus. The individual can condition himself through his own language responses (Staats, 1968c). Language, it is suggested, is at the heart of self-reinforcement, self-direction, self-satisfaction (positive emotion elicitation), self-punishment (guilt, and so on), and self-sensation (imagination).

## The self-concept: A cognitive-emotional personality trait

As has been indicated, a single act of reasoning can include attentional responses, perceptual responses as in labeling, complex sequences of verbal responses, emotional responses, and perhaps overt instrumental responses, and more—with interaction between elements. Various learning principles may be involved. To consider such a complex act in some unitary way would preclude understanding of the basic principles and behaviors of which the act was composed.

It is suggested that it is necessary to analyze complex human behaviors, separating components. However, it should be realized that in actual life occurrences these components may not be easily distinguishable. Human behavior gains its complexity through the building up of more and more complex constellations of behaviors. It is suggested that when the constellation has reached a certain level of complexity, composed of repertoires that have been permanently learned and thus provide continuity, such constellations of behaviors tend to be labeled as aspects of personality.

The construct of the *self* or of the *self-concept* has been referred to in various personality theories as a determining personal process. It is said that each person has a psychological organization consisting of interests, attitudes, values, beliefs, and knowledge which constitutes one's perceived self. This system heavily determines one's behavior, since this self-concept is a salient feature of one's field (environment). Moreover, one's primary motivations are to maintain and to enhance the self (Snygg and Combs,

1949). The important point in the present context, however, is to stress that such personality constructs can be considered to refer to complex constellations of repertoires the individual has learned, including in prominent perspective the basic language repertoires.

This requires some elaboration. To begin, it has already been indicated how the child learns to label various objects and events, including the behaviors of others and social events in general. Social and behavioral events are also stimuli, and the person will acquire an extensive repertoire for labeling such stimuli. It has also been said that the child will learn to label his own behavior. Some of this will be through continued "training." As one illustration, the author knows of a case where the mother repeatedly over the years indicated that one son had "mechanical interests" just like his deceased father. Since the father was revered, statements drawing a similarity between father and son were also quite reinforcing. This experience had the effect of establishing the labels concerning the child's mechanical interests and abilities in his repertoire.

In addition to such direct training, once the child has acquired a repertoire of labeling responses, he will label his own behavioral stimuli on his own. He will observe himself in many situations, he will have observations of himself in comparison to others, he will experience his internal feelings in those situations, and he will also observe the way people respond to him absolutely and in relationship to the way they respond to others. He will label these stimuli, many times with labels which themselves have emotional value. These experiences are complex, taking place over a long period of time. They produce a complex of interrelated behaviors, of a heavily verbal nature and with strong emotional characteristics, called the self-concept (Staats, 1963, 1968c, 1971a).

This constellation of behavioral repertoires can play a central role in determining how the individual will react in many different situations. For example, the self system may be expected to play the same role in directing the individual's reasoning in self-related situations as has been described for other reasoning repertoires. As the individual is confronted by situations that are problematical and call for some action and in which the individual's sensorimotor skills, interests, knowledge, beliefs, social competencies, and so on, are relevant, what the individual does will depend upon the course of his reasoning. For example, the boy who has acquired self-concept statements that he is mechanically inclined like his father will reason that he should major in engineering when he begins college.

In the process of helping determine the individual's actions, the emotional characteristics of his self-statements will be very important. As in any type of reasoning, the emotional response to the sequences will determine whether "striving for" or "striving away from" actions are finally elicited. When the individual labels himself in relationship to some decisions in terms that elicit positive emotional responses, his reasoning will incline him toward a positive decision. And the converse is true. Let us say that a job opening arises that calls for certain skills. The individual who describes himself in negative emotional terms as not exceptionally capable or bright, who considers himself to lack creativity and initiative, who describes himself as fearing positions of leadership, who considers many others to have a more positive effect when dealing with people, and so on, will be likely to reach a negative decision concerning whether or not to accept a position that demands skill in such areas. The emotional responses elicited will elicit "striving away from" behaviors. Actually, the individual involved could have the same skills as another individual who described himself more positively (had a more positive self-concept) and as a consequence of the positive emotional responses would apply for the position. It may be added that an individual who has a long history of such positive reasoning will ordinarily engage in and receive experience in many activities, jobs, and challenges that finally provide him with greater skills than the individual who, because of his negative self-attitudes and self-reasoning, has hung back in many situations. This is another example of behavioral interaction principles in personality development.

It should be noted also that the individual's self-concept, as this constellation of emotional and language repertoires is reflected overtly, will help determine the way that other people will respond to him. The individual who describes himself in negative terms, other things equal, will be responded to differently from the one who describes himself somewhat (but not unrealistically) positively or who does not speak of himself at all. In each of the first two cases the individual can provide others with terms for their reasoning and decision making with respect to him.

To indicate again something of the complexity of behavioral interaction principles, it can be added that the individual's emotion-eliciting statements about himself, in affecting others, will also have a return effect upon him. Part of the individual's self-concept will result from the way that others behave towards him. If the individual generally says overly modest things about himself or behaves that way by remaining silent, then people will not respond to him as one with as many positive attributes as he actually has. And their response to him will be further experience in maintaining his overly modest self-concept. Other behavioral accounts have utilized elements of this analysis of the self-concept (Bandura, 1969; Mischel, 1971, 1973).

Finally, there is further compatibility between the behavioral analysis and the humanistic theory that focuses on the self. Snygg and Combs (1949) and others have indicated that the individual strives to enhance and maintain the self. This can be interpreted in terms of the principles presented herein. Others may say things about us that elicit a positive emotional response, and we may say these things ourselves. As the general principle states, we strive for positive emotional stimuli and are reinforced by them. We strive away from negative emotional stimuli, verbal or otherwise, whether spoken by others or ourselves. It is suggested that the principles of the A-R-D system are involved in striving to enhance and maintain the self. For example, the individual may be considered to be striving to maintain and enhance his self-concept when he does not accept criticism of his behavior. This can be seen more generally to be avoidance of negative A-R-D verbal stimuli. This is not to say that the individual will never label himself with negative emotional stimuli, however, or accept such labeling by others. For he will usually also have been trained, to varying degrees, to label events veridically. Thus, when he is himself involved, he will ordinarily label himself as positively as possible within the constraints of the events involved. (See also the later discussion of defense mechanisms in Chapter 8.) Experimental hypotheses based on the learning principles involved are derivable from such considerations.

## WORLD VIEW

One final illustration will be given of the social behaviorism rapprochement that integrates behaviorism and cognitive theory. The illustration will also indicate that complex aspects of human behavior can be treated. There has been in anthropology an explanatory concept called "world view." It is suggested that there is a determining relationship between man's conceptions of the world and the way he behaves. In commonsense terms, the world view involves the individual's conception of the universe, the ways of the world, and his own and other men's position in the world. The world view also includes many philosophical and religious beliefs, myths, and prescriptive statements of modes of social interaction. It has been suggested that the world view helps determine complex social actions; for example, the world view of the peasants of developing nations will in part determine the way they will respond to opportunities for economic growth and social progress (Redfield, 1952).

Such observations can also be seen to be entirely congruent with the present formulation. What has been called the world view may be seen as constellations of behavior—especially language behaviors—of great complexity. It is thus suggested that the individual acquires in his vast learning history very complex systems of behaviors tied together by language that apply to many of the situations he faces and to the actions he will ultimately make with respect to those situations. He learns conceptions about the physical world, the biological world, the social world. He learns very complex forms of thought (language), many of which are imbued with emotional value and are linked with classes of actions and instrumental skills. This will include subsystems that may be related with varying degrees of closeness. As one example, the individual will have a conception of causation. He may have learned sequences of language that state that the events of the physical world are determined by supernatural forces, as occurs in most theological systems. The individual may also have similar sequences with respect to individual and social events—wherein overt behaviors are attributed to internal and external divine forces. Such a language-emotional subsystem, a view of man's nature, will affect the way the individual behaves toward others in many situations. It will also help determine the individual's decisions with respect to events ranging from voting for politicians, to favoring a war or social welfare policy, or raising children.

Another example can be taken from a matter of present concern. It has been suggested that the traditional concept of intelligence is part of a very general world view (Staats, 1971a). The learning conception of intelligence suggests a change from the traditional views, stating that the person's intellectual abilities are learned rather than biologically given or gained from divine inspiration or personal and familial righteousness. It can also be suggested that whether or not the learning conception of intelligence will be accepted will depend upon other aspects of the individual's world view. If the individual already has a world view that human behavior is divinely determined according to biological inheritance, then he will not easily accept the learning conception of intelligence. Thus, in many respects the various aspects of the world view are interrelated, and the subsystems may be seen to be linked in the larger system.

A subject of such complexity and significance as the present one cannot be dealt with adequately in such a brief passage. The relationships of the concept of world view and the way it is employed in anthropology and the economics of developing nations should be analyzed in detail, employing the learning aspects of personality, and the resulting analyses should be utilized in accounting for and dealing with the types of behavior (such as economic) that are of concern. The present discussion aims only to indicate the relevance of the present principles for concepts considering the most general characteristics of man.

It is thus suggested that human cognitive characteristics derive from repertoires of behaviors the individual has learned. Previously learned complex repertoires are the reason the human can learn so readily, learn vicariously, imagine things he has never seen, hypothesize concerning worldly circumstances, and reason, think, and plan—and perform all the other cognitive acts that are so common. These repertoires are largely of a language nature but also involve other personality repertoires. More complete accounts are available (Staats, 1968c, 1971b), as is the developmental description of the acquisition of the repertoire (Staats, 1971a).

In concluding this chapter, it will only be added that the present discussion has dealt with language as an intrapersonal activity rather than a type of interpersonal interaction. However, most social interactions involve language. This, actually, is a primary function of language—being a central mechanism of interaction. Additional references to language will be found in later chapters concerned with social interactions, as well as in sections concerned with other aspects of intrapersonal activity.

# 6

# *The instrumental personality system, imitation, and observational learning*

IN DESCRIBING the emotional-motivational system, emotional responses were said to be significant in good part because of the way they direct overt instrumental behaviors. The same was true in considering the language-cognitive personality repertoires; much of their significance was found to inhere in the way in which they provide stimuli that direct overt instrumental behavior.

The descriptions of the emotional-motivational and language-cognitive systems considered instrumental behaviors rather generally. It should be noted, however, that the individual's instrumental behavior repertoires also constitute essential aspects of his personality characteristics. What we actually do in terms of overt sensorimotor behavior is frequently the most significant aspect of our activity. Thus, for example, while our emotional response to an individual may determine whether or not we are aggressive to him, what he will experience is our aggressive instrumental responses. These responses can differ quite widely, ranging from subtle verbal responses of a snide nature to violent instrumental outbursts of murderous quality.

Individuals and groups differ significantly in the instrumental repertoires they possess. An adequate account of human behavior must be capable of considering these repertoires—how they are acquired, how they function in the individual's adjustment, how they affect his further learning, and so on. To begin, the various areas of psychology that have been concerned with the study of sensorimotor behavior will be mentioned briefly.

## Motor skills

The study of motor, or perceptual-motor, skills has taken its impetus from several sources. One of the interests of experimental psychology has been in motor learning, and there has been an applied interest in discovering the most efficient way of performing industrial and military tasks. As an outgrowth of World War II, large groups of experimental psychologists studied basic and applied questions concerning motor skills at various installations. Concepts and methods from psychology, for example, the psychological testing of intellectual characteristics have been utilized in this study. If there were motor skill abilities, perhaps it would be possible to ascertain levels of ability by tests, which would then be predictive of success in relevant industrial or military activities. Perhaps there are motor abilities that underlie general physical proficiency, such as a general athletic ability. In studies conducted to ascertain whether there are such primary motor abilities, however, the results suggest there is no general motor ability, or general athletic ability, although there do appear to be clusters of tasks that demand more specific skills such as manual dexterity (Hempel and Fleishman, 1955).

Concepts and methods from the field of experimental psychology have also been employed in the study of motor skills. This has had a long history, beginning with the investigation of individual differences in reaction time. The force of reaction that can be exerted in different kinds of movement has also been studied, as has the precision and rapidity of repeated responses (Gagné and Fleishman, 1959).

Motor skill sequences have also been the subject of study. The strategy of experimental psychology has been to simplify so that laboratory investigation is possible. Thus, the tasks employed in most motor skill research are simplified. Rather than complex sensorimotor performances, various standardized tasks have been developed. The pursuit rotor apparatus, for example, consists of a rotating turntable with a small disk embedded near its periphery. The subject's sensorimotor task is to keep a stylus in contact with the rotating disk. There are also mechanisms that present varied stimuli which demand adjustments of several responses.

These types of apparatus have been employed, along with some of the basic principles of experimental psychology, to study aspects of motor skill learning. The importance of learning trials in acquiring such skills has been shown (Wulfeck, 1942), for example. Schedule, or time distribution, of practice has also been studied. Learning trials may be massed in a continuous period of time, or they may be spaced, with intervals of time intervening. Generally, distributed practice produces better learning than does massed practice. Adams (1954), for example, found that as the between-trial interval increased from zero through 3, 10, 20, and 30 seconds, skill on the pursuit rotor was learned more rapidly at each step of lengthened interval, and a higher level of final skill was reached with each additional step interval.

Another area of traditional concern in the study of motor learning (and paired associate verbal learning as well) is that of transfer of training. It has been recognized that the learning of a skill may facilitate the learning of the skill in a different situation, or the learning of another skill. This is referred to as positive transfer. The case where the first skill interferes with later learning is called negative transfer. In general, similarity across stimulus situations produces positive transfer, as does similarity across the responses involved. The more similar the second situation is to the first and the second response is to the first, the greater the expectation of transfer. Negative transfer also involves similarity from first task to second task, but in this case interference occurs. For example, having previously learned a response to a stimulus situation may make it less easy to learn a different response to the stimulus. Many experiments have tested the principles involved in positive transfer (for example, Gagné and Foster, 1949) and negative transfer (for example, Lewis and Shephard, 1950).

Thus, although the study of motor skills has had impetus from a concern with applied problems (the selection of individuals for military and industrial positions or training), it has had also a major influence from experimental psychology. It may be noted, however, that the latter influence has been toward an interest in the discovery of general principles that apply to all sensorimotor skills. There has not been an orientation that provides an impetus to the use of learning principles in the analysis of particular sensorimotor repertoires or in the study of the acquisition of such repertoires. Thus, although there has been a great deal of published research (especially in the 1940s and 1950s, when many studies with the pursuit rotor and other experimental tasks were conducted), there has not been a great deal of movement in the direction of theoretical-experimental analyses of complex, functional human sensorimotor repertoires. As an example, there have been many studies in which the *transfer of training* has been investigated using a mirror drawing apparatus in which the individual must relearn his sensorimotor writing skills. But what writing learning consists of, or what learning conditions are involved in the child's learning to write, has not been systematically studied in this area.

## Social-personality theory and instrumental personality repertoires

It is suggested that areas concerned with human personality should be centrally concerned with the study of complex, functional instrumental repertoires of human behavior. It is true that there has been a good deal of interest in such areas as aggression and imitation, although the instrumental behaviors involved are not usually the focus of attention. Interest in these areas of behavior arose in the context of traditional personality concepts. Thus, it has been said that individual variations in aggressiveness are due to biological characteristics. Imitation also is commonly considered to vary as a function of the individual's inheritance—to be an unlearned tendency. However, it is possible to consider these areas in terms of the learned sensorimotor skills involved. Thus, reference to aggression and imitation can be made to indicate the need for studying aspects of the instrumental personality repertoire. Aggression will be considered first; imitation (or modeling or observational learning) is the subject of the next section.

## Aggression

To some extent elementary principles have been applied to both aggression and imitation by the social learning theorists. The frustration-aggression hypothesis (Dollard, Doob, Miller, Mowrer, Sears, Ford, Hovland, and Sollenberger, 1939) suggested that aggressive behavior resulted from frustration. A number of studies were conducted to test the hypothesis. For example, when Sears, Hovland, and Miller (1940) kept a group of college students awake all night, preventing them from eating, talking, or amusing themselves, it was found that the subjects evidenced aggressive behavior in the form of nasty remarks and other behaviors. Zander (1944) showed that children who were frustrated with insolvable problems also displayed instrumental aggressive behavior.

The original frustration-aggression hypothesis dealt with aggressive behavior as an unlearned response to frustration (see Bandura and Walters, 1959, for later accounts). However, this was broadened later to indicate that aggressive behavior can be learned. N. E. Miller (1948), for example, showed that two rats could be trained to fight each other when they were reinforced for doing so (by turning off the electrified grid on which they stood only when they fought). Walters and Brown (1963) also showed that boys who had been reinforced for punching an automated doll later demonstrated more instrumental aggressive behavior than control subjects did. A number of studies have been conducted to show how learning conditions can influence the display of aggressive behavior. For example, Miller (1948) showed that learned aggressive behavior would generalize from the original object to new objects.

However, there has not been a comprehensive behavioral analysis yet of what constitutes aggression, how such repertoires are learned, and how such repertoires function in the individual's learning, adjustment, and further personality development. It is suggested that aggression is not a single behavior but a complex repertoire. The stimuli involved in aggression include those that could be called frustrating, as well as other stimuli with other functions.

## A-R-D PRINCIPLES AND AGGRESSION

A general framework for the consideration of aggression can be provided by the personality concepts presented thus far. This can be illustrated very briefly. First, rather than the relatively narrow term *frustration*, it is suggested that the more general term *negative emotional response* can be considered basic in many cases in the determination of an aggressive response. Frustration theory suggests it is the thwarting of some action or goal that is the basic instigator. Although this circumstance is involved in many experiments, it is suggested that thwarting is but one way of eliciting a negative emotional response. For example, painful stimuli in general—those that can be considered to elicit a negative emotional response—have been shown to be effective elicitors of aggression (Ulrich and Azrin, 1962).

It should be indicated that when the individual experiences a noxious situation which elicits a negative emotional response which in turn elicits some aggressive behavior, the noxious situation serves in the role of a directive stimulus. This is not an unlearned behavior but a learned one. The nasty comments that the college students made in the Sears, Hovland, and Miller study, for example, were learned behaviors. It is suggested that such aggressive behaviors are learned through the principles of reinforcement. Thus, negative emotional stimuli, or at least their removal, can serve as the reinforcement for the acquisition of sensorimotor aggressive behavior, as the Miller (1948) study showed.

Positive emotional conditions and positive reinforcement can also be involved in the learning of the aggression personality repertoire, however. For example, the child learns aggressive behavior many times through the reinforcement provided by the "hurt" response of someone else. That is, the child can learn behaviors that will annoy, disturb, frighten, tease, or hurt another, because these behavioral stimuli are paired with positive emotional stimuli. The child may get the attention he wants, for example, when he annoys his mother. Through such experiences, the signs of someone else's annoyance will become positive A-R-D stimuli. The behaviors involved in such learning are also likely to be those termed aggressive.

There are individual differences in the extent to which such stimuli of another's "hurt" will be positive or negative A-R-D stimuli for the person, an important aspect of personality. Studies have shown that the hurt of another person can be reinforcing (Feshback, Stiles, and Bitter, 1967), especially when that person has annoyed the individual, and it has been shown that observing another in pain evokes an emotional response in the observer (Berger, 1962). Individual differences of the A-R-D system in these aspects should be systematically studied, as well as the principles involved and the conditions of learning. In the latter category, for example, it is posited that the stimuli of "hurt" of another should elicit a *positive* emotional response in the person when the person has a negative emotional response toward that individual. The reverse should also be true—the hurt of another should be a negative emotional stimulus when the person has a positive emotional response for the individual. Naturalistic observations support these principles, but systematic study is necessary to indicate their validity. Some experimental studies that illustrate the relevance of application of the A-R-D principles to the study of aggressive behavior will be summarized in the next chapter.

This account has been presented here to indicate that analysis of an area of behavior employing the A-R-D principles usually means that a repertoire of instrumental behaviors—those behaviors elicited by the directive (D) stimulus—are centrally involved. Although the negative emotional value of a stimulus may influence whether aggression occurs, the aggressive responses themselves are instrumental (including verbal) responses.

## INSTRUMENTAL AGGRESSIVE BEHAVIORS

Aggressive behavior can be defined as nonaccidental behavior that is aversive to someone else. It is possible to hurt another person in various ways. Some overtly aggressive behaviors result in physical pain. These behaviors, which can involve instrumental and cognitive skills of very complex sorts, are learned by different individuals to different degrees and extents. There are individuals who simply have not learned physically assaultive behaviors and who will not display such behaviors even under conditions that would ordinarily elicit them. There are other individuals who have learned a wide-ranging repertoire of aggressive sensorimotor behaviors in response to many situations. Aggressive behaviors may be elicited in such individuals when the situation would not elicit them in most people.

It is also possible to hurt another by verbal means. Criticism, insults, verbal teasing, and so on, are examples of such aggressive behaviors. Some individuals have well-developed skills of this type and display them frequently in various situations. Others will have acquired relatively few such behaviors. These are important areas of personality difference that require behavior analysis and systematic description. The learning conditions that produce such repertoires need to be systematically studied, as well as the extent to which the repertoires are adjustive or maladjustive.

Study is also needed of other instrumental behaviors of aggression. Withholding something that another person wants or needs may cause them hurt and thus be an aggressive behavior. The individual may also take something away from another person as a means of hurting the person. Another aggressive behavior is that of doing the opposite of what someone has indicated he would like. Very important personality characteristics reside in these types of behavioral repertoires. They have sometimes been considered under different labels—such as *negativism* in children, as in the example above.

It can be noted that in some of the examples it has been suggested that a behavior is aggressive because it is performed in order to hurt someone else. The aggressive behavior is defined by its "goal." A person could do the opposite of what someone else wants, not to hurt them, but because of other behavioral determinants. This would not be aggressive behavior.

Including "intention" in the definition of some types of aggression requires some further explanation. That is, it is necessary to indicate that intention does not refer to an undetermined mental state. Intention as used herein has a behavioral definition. As an example, let us take the child who has learned to find the discomfort of his parent has positive reinforcing value. Once the stimulus of parental discomfort has become positively reinforcing, the stage is set for the child to learn a repertoire of behaviors that achieve this consequence. The learning history involved provides the aggressive skills of the child, such as negativism. In this sense, he may do the opposite of what the parent requests, because of his past experience.

In addition, intention may involve language behaviors such as reasoning and planning, as will be indicated. In these cases the goal of the behavior may clearly be seen to be to hurt the other person and thus be aggressive.

## LANGUAGE-COGNITIVE REPERTOIRES IN AGGRESSION

The previous chapter described the manner in which the language (and image) repertoires of the individual could constitute reasoning and planning acts. These reasoning sequences, and the emotional responses they elicit, could then determine what the individual would actually do. Various types of overt behaviors can be directed by covert reasoning sequences and the emotional responses they elicit, including aggressive behavior. The "goal" of the individual in "intending" to hurt another person may thus reside in his reasoning.

As an example in Chapter 8 will indicate, the reasoning that directs the aggressive behavior need not be realistic. As long as the reasoning sequences concerning the person have sufficiently intense negative emotional value, and as long as the individual has the aggressive instrumental behaviors in his repertoire, the reasoning sequences can direct the overt behavior, even though the reasoning sequences do not coincide with actuality.

It can also be noted that language may play a role in aggression in serving as the mechanism by which negative emotional responses toward other people are learned. Moreover, the aggressive instrumental behaviors can be learned on the basis of language. For example, the guerrilla soldier or terrorist may learn to make a bomb by following the instructions of a manual.

In conclusion, this brief description of aggressive behaviors has demonstrated that there are repertoires involved in the personality trait of aggression which the individual has learned. It is essential to realize that an important aspect of understanding human behavior resides in the description of the instrumental repertoires involved in such personality traits. The following section, on imitation will provide additional detail in the analysis of another personality repertoire which is important in human adjustment and which also includes instrumental repertoires.

## Imitation (modeling or observational learning)

Imitation is generally considered to occur when one organism performs an action and an observing organism then performs a similar action. Imitation in man and other animals has been known for a long time and has always been interpreted within the concepts prevalent at the time. Thus, imitation has been considered to be instinctual in some species, like apes and man. Many individuals today consider imitation to be an inborn propensity in man and individual differences in the quantity and quality of imitation to be the result of biologically rooted personality characteristics.

In the 1920s imitation came to be considered within the principles of learning as they were known at that time. Allport (1924), for example, considered language development to involve the learned imitation of the parent's speech. The principle of learning involved was simply that of contiguity. It was thought that the child learns to repeat syllables in babbling. Later, when the parent makes a similar sound, the child learns to repeat it.

Miller and Dollard (1941) advanced a social learning theory of imitation considered as instrumental conditioning. Subjects reinforced for making a response that was like that of a model would come to imitate the model. For example, one of two children introduced into a room had been told which of two boxes contained a piece of candy. He was allowed to open the box and get the candy while the other child watched. Then the second child was given a turn to play, and if he went to the same box he also found a piece of candy. At first most of the second children did not imitate the leader. By the third trial they had usually learned the imitation, even when the box with the candy was mixed with three other boxes rather than just one.

Other learning theories have included imitation as an integral constituent (Mowrer, 1952; Staats, 1963). The manner in which imitation is involved in language learning has been systematically considered, as have many of the behavioral mechanisms involved in imitation (Staats, 1963, 1968c, 1971a).

In more recent years Albert Bandura (1969; Bandura and Walters, 1963) has become the leading social learning theorist. He has studied the manner in which learning takes place through imitation and has made imitation the focus of a more general theory. Bandura calls the area modeling (or observational or vicarious learning) rather than imitation and, as will be indicated, considers modeling principles to be basic—on the same level of theory as the elementary principles of conditioning. Bandura's (1962) conception of modeling was originally based upon Mowrer's (1960) theory of learning and has been called the secondary reward theory of imitation (Mussen and Parker, 1965). This conception can be characterized by a study in which children who had been reinforced in the presence of one adult and not another later imitated the former adult more than the latter (Bandura and Huston, 1961; Bandura, Ross, and Ross, 1963a). The conception was that the reinforcement procedure made the one adult a reinforcing stimulus, thus making what he did reinforcing. Since doing the same things as the adult was reinforcing, the children later imitated this adult.

Bandura's approach to modeling has gone through successive stages of development. The most recent statement (Bandura, 1971) coincides in various ways with some of the principles of the conception of the basic behavioral repertoires that have been outlined (Staats, 1963, 1968c, 1971a). Bandura's approach has been considered to be at least quasi cognitive in nature and has been well accepted by cognitive theorists. It will thus be of interest in describing the present conception of modeling to illustrate the similarity that Bandura's social learning approach has come to have to the analysis founded in basic conditioning principles. The major basis for a divergence in the approaches will then be outlined.

## ATTENTION: STIMULUS CONTROL OF RECEPTOR-ORIENTING BEHAVIOR

In introducing the concept of the basic behavioral repertoire, one of the importance repertoires described was that of attention (Staats, 1968c, pp. 408–427). It was suggested that the sensory organs are located in "anatomical structures whose 'movement,' in many cases, results in variation in what stimuli will affect the organism" (p. 408). For example, depending upon the movements of the body, neck, and eyes, the individual will see (attend to) different visual stimuli. Sometimes attentional behaviors may be complex, such as obtaining and focusing a pair of binoculars.

The fact is that the individual is usually faced with a multitude of stimuli. These stimuli may be considered to be in a competition for the individual's "receptor-orienting" responses. In the same situation, one stimulus may gain the attention of one person while another stimulus may gain the attention of another.

It is suggested that such differences in attention characteristics are a function of learning. "An individual who has, for example, been reinforced in the past for looking at a particular color, in contrast to other colors, will 'attend" to this color more than to others" (Staats, 1968c, p. 409). This training could take place in the process of the child's learning to name colors. There may also be competition between the senses in terms of what type of stimulus controls attention. One individual may have had experiences where he has learned to respond more strongly to auditory than to visual stimuli, although considerable overlap would be expected.

"The important thing to suggest here is that what we call attention, a very basic form of behavior, is subject to influence by the individual's conditioning history" (Staats, 1968c, p. 412). At the basic level, attention skill acquisition has been assumed to depend on maturation. However, it is suggested that even such basic forms of attention as focusing the eyes to obtain a clear image, cocking the head to hear a sound distinctly, running one's hand on a surface to feel texture, or managing a liquid in the mouth to taste it sensitively are learned attentional behaviors. Although focusing the lenses of the eye in obtaining a clear image takes place automatically and seems to be an inherent skill, the possibility that such skills are learned should be considered. There are learned skills (such as those of balance) that are automatic, yet learned.

To continue, however, some of the most important considerations in analyzing attention concern the stimuli that control attention responses and the learning that is involved. The individual not only learns attention behavior under the control of the stimulus itself, but stimuli in other sense modalities can come to control the attentional response. "Thus, as an illustration, the *sound* of an approaching car for the hiker along the road will control turning of the head so that the eyes are brought into position where they can sense the approaching car" (Staats, 1968c, p. 409). The infant will at an early age be seen to learn to bring objects into contact with his mouth—an attentional response—under the control of having seen the objects. Each of these examples involve cases where an attentional response in one modality has been learned under the control of stimuli in another modality.

A central aspect of attention, moreover, concerns the manner in which verbal stimuli control attention behavior. This is one of the aspects of the verbal-motor repertoire described above. The child must learn to respond appropriately to such verbal stimuli as "Look to your right," "Look up,"

"Look at the third one from the left," "See the red one," "Notice that the letter *m* has two humps while the letter *n* has only one." The same learning must also occur in the other senses, for example, "Feel the rough spot," "Listen to the squeak."

Because of the child's learning, the verbal stimuli of different people will come to have differing attentional controls for him. "For example, one's employer, wife, mother, children, the grocery-store man, an expert, and so on, will have differential amounts of control over the individual's behavior, including his attentional behaviors" (Staats, 1968c, p. 415). Later in life complex and subtle social stimuli will differentially control attentional behaviors to verbal stimuli. It may be added that the repertoire of attentional behaviors under appropriate stimulus control is basic to early cognitive and other types of learning. For example, the teacher relies upon verbal instructions to control the attentional behaviors of her pupils. "Where the repertoire has not yet been developed by the pupil it must be trained or the child will not learn" (Staats, 1968c, p. 418).

It can be added that there are finer skills involved in attention than those that have been discussed, skills that are usually referred to as discrimination. Two similar stimuli may be involved where the situation demands that one be responded to, or the situation may involve making a different response to one stimulus than to another.

> [T]he child if properly trained acquires general skills such that when he is faced with a problem of discrimination he makes scanning eye movements, and comparing eye movements looking from one to the other of the objects involved, and so on. Furthermore, many stimuli are of such a nature that they must be scrutinized in fine detail before the important stimulus is seen that controls the relevant behavior. It may be suggested that the detailed scrutiny of stimuli, the comparing of stimuli, and so on, are attentional behaviors that must be learned. (Staats, 1968c, p. 421)

In the present context it is important to realize that attentional and discrimination skills, and the stimuli that control them, are constituents of a basic instrumental behavior repertoire that is central to imitation.

> An important aspect of the imitation repertoire is that the child will have learned in certain situations to look at the actions of other people *and the stimuli that are controlling those actions*. With proper training, a child when in a situation where his own behavior is not controlled by the stimuli—that is, problematical situations where the child has no learned responses that solve the problem—will look at the ways that others behave. These attentional responses are very important to his learning, for if there is someone else who has learned a response the child will see (or hear, and so on) this response.
>
> Thus, basic to an imitational repertoire are the attentional behaviors that allow the child to sense the actions of another person, and the relevant stimuli that control the person's response. Two children could

be in the same problem situation, see someone else solve the problem, and yet profit differently because one closely observed the action and the controlling stimuli for the action and the other did not.

We learn to observe other people's behaviors when we have had the appropriate conditioning history. A child may be instructed, for example, to "Watch how I do it." . . . A child who has had a rich experience of this kind—where he observes and imitates and is then reinforced when he behaves in kind—will acquire a rich attentional repertoire for observing other people's behavior, as well as a rich imitational repertoire. (Staats, 1968c, pp. 426–427)

Bandura has utilized this analysis of these processes of attention as behaviors basic to imitation.

One of the main component functions in observational learning involves attentional processes. Simply exposing persons to modeled responses does not in itself guarantee that they will attend closely to them, select from the total stimulus complex the most relevant events, and perceive accurately the cues to which their attention has been directed. An observer will fail to acquire matching behavior at the sensory registration level if he does not attend to, recognize, and differentiate the distinctive features of the model's responses. Discriminative observation is therefore one of the requisite conditions for observational learning. (Bandura, 1971, 16–17)

## Sensorimotor repertoires and imitation

The actual behaviors of imitation are motor responses, instrumental verbal and nonverbal responses. The person watches the individual do or say something and then, or later, does or says something similar. The stimuli produced by the imitator's behavior match the stimuli produced by the model's behavior—which is a partial definition of imitation. For the imitative act to occur it must be a response that the observer is able to make; he has to have the sensorimotor skill involved.

The dependence of the imitative act upon the sensorimotor skill being in the individual's repertoire may be seen easily in everyday life. For example, it is interesting to ask a child under three years of age to imitate oneself in closing one eye, but one eye only. The instructions and the stimulus of the model will ordinarily already control the child's behavior of attempting the response. But a young child is not likely to have learned the response involved. He cannot close one eye without closing the other —or he will only be able to imitate the response very grossly with superfluous facial movements (usually in a very amusing way).

This is an example where the response itself is not in the child's repertoire. Imitation, however, involves not only making the response, but also making it under the stimulus control of the model. Sometimes a deficit in imitative skill lies not in being unable to make the response but in being unable to do so under the control of the model stimulus. As an illustration, the parent may ask the young child to "Look to the side like I do," while keeping the head pointed straight ahead in demonstrating the response to be imitated. The three-year-old child will ordinarily have learned to attempt to imitate by instruction but will be unable to perform the imitation. The deficit is that the response involved has not yet come under the control of the imitation stimuli. The child can perform the response, but only under the control of an object that moves across his visual field.

Concern with imitation has not involved the systematic study of the sensorimotor repertoires that enable the individual to imitate others in a wide variety of instances. The recent research on modeling has simply taken for granted that children can imitate. The fact is that the sensori-motor skills involved are many times learned only in long (if many times informal) training programs, and the understanding of imitation should involve theoretical and experimental analysis of this type of learning. In fact, the sensorimotor repertoires are quintessential in imitation.

A good illustration of sensorimotor imitation learning is provided in research the author has conducted on children learning to copy and write. The study was begun with one child (the author's daughter when three years old) and then extended to several additional three- and four-year-old children. Later, the procedures and principles were verified systematically with 11 additional four-year-old children who were culturally deprived (Staats, 1968c).

The child learning apparatus depicted in Figure 2.5 was employed to reinforce the child with a marble for his writing responses. The marbles could be exchanged for backup reinforcers in the manner already described.

> The child was given a black crayon and asked to write any letters he knew. Then he was asked to write his name. None of the children displayed any skill in writing. In later sessions the child was given instruction in holding the crayon and drawing lines. He was given a sheet of paper with a line on it and shown how to trace on the line. He was then given trials in tracing. After a number of these tracing trials, during which he learned to hold his crayon and write with it, he was introduced to the letter *a* in the large (1¼ inch) size. He was given trials on tracing this letter. Then he was given a sheet with a large letter *a* on it, asked to copy the letter, and instructed what to do. Following training on this, a medium-sized (½ inch) letter was introduced, and the child was trained in copying this. Later, the primary-sized *a* was introduced as the model stimulus and copying trials were given.
>
> Each time the child was given the letter to copy he was also instructed to copy the *a*. This was done for all the letters. After the trials in copying the letter *a*, the child was given a blank sheet of paper and asked to write the letter *a*. These various types of training trials were continued

until the child had attained some proficiency in copying and writing the first letter of the alphabet. Then the letter *b* was introduced and the child was given a few tracing trials (on the large-sized letter). Following this, he was given additional trials in copying both the letters in sequence, first in large-size, then medium-size, and finally primary-size print. This procedure was followed whenever a new letter was introduced. Learning trials were continued on the new letter as well as all the preceding letters, until the experimenter decided that the child had learned to write the new letter to a sufficient degree of skill to introduce the next letter. The standard of skill was increasingly high as the child's general level of writing skill increased in several dimensions. The child continued to receive training on all the letters that he had already learned to write. It should be noted that after the child had learned to write the letter *d*, the crayon pencil was exchanged for a regular lead pencil. (Staats, Brewer, and Gross, 1970, pp. 63–64)

The children each made between 500 and 1,000 responses, so it is not possible to show the learning process in detail. The general course of the learning can be seen, however, by systematically sampling the child's progress. Although the IQs of the children varied by 41 points, the course of learning under the conditions of reinforcement were very similar, and the learning process can be characterized with the record of one child who had an IQ of 104 (see Staats, 1971a, pp. 103–119, for the records of two additional children). The numbered samples in Figure 6.1 were selected from designated points in the child's training. The response numbered 1 was given to the first request to write any letters the child knew. The second response is the child's first tracing of the entire letter *a*, after 166 prior tracing learning trials. A couple of trials later, the first copying of the letter *a* is made, as shown in response 3. As can be seen, the letter is backward and slanted, and the lines do not meet. The first copying of the medium-size model stimulus, on the 208th training trial, is shown in response 4. The training trials have resulted in improved imitation. However, when the model stimulus is reduced to primary-size type (see response 5) on the 301st training trial, the letter is imitated less well. By the seventh sample, on the 330th training trial, there is improved ability to imitate the letters *a* and *b*. Further improvement is shown in samples 9, 11, 13, and 15, with gradual decrease in the size of the copied letters and in the precision of imitation. In producing the letter imitation skills shown in sample 15, 611 training trials were involved.

It should be noted that in the samples, 7, 9, 11, 13, and 15, the final letter is being imitated for the first time. The record thus shows that the child increases in his ability to imitate (copy) new stimuli the first time they are introduced. A general learning acceleration was also involved in the acquisition of this imitation repertoire. The author tabulated the results of the 11 culturally deprived four-year-old children and found that an average of 288 learning trials were necessary to learn to imitate and

**1**                 **2**

**3**                 **4**

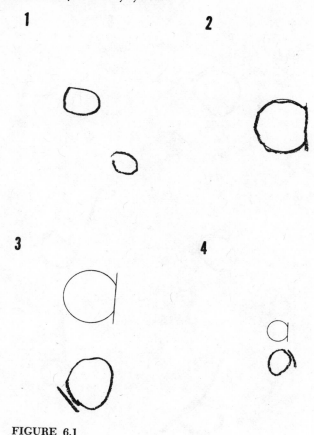

**FIGURE 6.1**

write the first four letters. The last four letters learned, in contrast, required a mean of 76 trials.

The children were learning four times as rapidly after the training as they had at the beginning. This occurred within the brief training process described, which lasted an average of fifteen hours and forty-two minutes. . . . Actually, the children's performance indicated they could copy newly introduced letters much better as the training progressed. . . . It would be expected that if this type of training had continued, the children's imitation skills would have approached the point where a new letter, or some other stimulus of that type, could have been imitated precisely on the very first trial. If one compared the child who had had that training with one of the same age who had not, one would get the strong impression that there must be something truly different about the children's internal "imitational abilities." (Staats, 1971a, p. 121)

**5**                     **6**

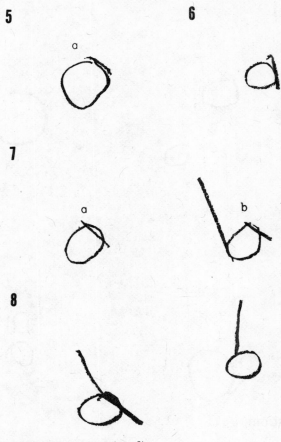

**FIGURE 6.1** (*continued*)

This process of learning demonstrates very clearly that imitation ability does not come about through innate propensity, dependent upon the maturation of the child. Four-year-old children with IQs ranging down to 88 learned the imitation skills of copying prior to the time the skills would ordinarily develop. The skills can be produced in the child at earlier ages than this (see Staats, 1968c, for additional results with two three-year-old children). Moreover, the course of the learning shows that the imitation repertoire is only gradually acquired. Hundreds of responses are necessary for the child to approach the fine sensorimotor skills involved. Ordinarily, the child has many of these learning trials in informal circumstances, like drawing with a stick in the dirt, or coloring with crayons, and so on. The process of learning has not been systematically observed in an experimental

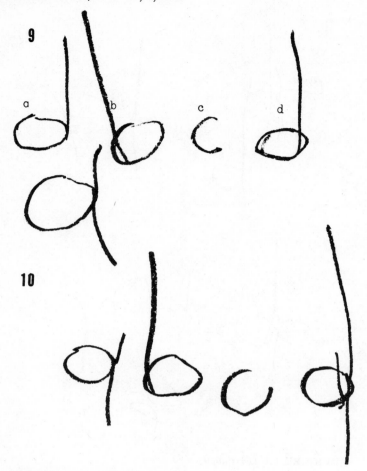

**FIGURE 6.1** (*continued*)

and longitudinal manner. When the process is observed, however, the nature of the learning involved can be specified.

The results described involved particular stimuli (the letters), and particular responses (the copying). It is suggested that such sensorimotor units are representative of the various sensorimotor imitation skills that the child acquires. The end result of such training is a repertoire composed of various sensorimotor units. The child, following the example, will come to be able to copy easily any number of different line configurations, including all the letters of the alphabet as well as others. In general, it is such sensorimotor units that compose the central element in modeling. It should be emphasized that the child does not have to learn these units in

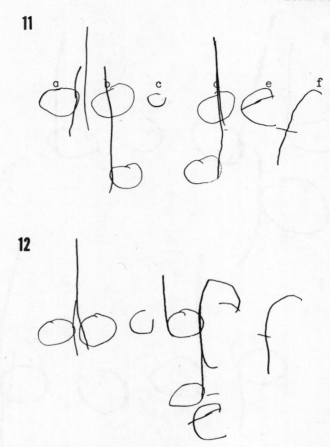

**FIGURE 6.1** (*continued*)

the act of imitation. He may have learned to bend down by himself, for example, in the act of reaching for an object on the floor. But once the sensorimotor skill is there, provided he has also learned to imitate when directed to do so, the sensorimotor unit may be elicited by imitational directions. We learn many acts through nonimitative experiences that are available to us in imitation when the appropriate situation occurs.

It can be added here that although the social learning theory of modeling has not provided analyses of or experimentation upon the sensorimotor units that form one of the bases for imitation, the concept of these units has been partly incorporated. Bandura (1969, 1971) refers to the sensorimotor units as motoric reproduction processes. "The rate and level of observational learning will be partly governed, at the motoric level, by the availability of essential component processes" (1971, p. 22).

a     b   c   d   e   f   g   h

13

14

FIGURE 6.1 (*continued*)

## LANGUAGE AS A MEDIATOR OF MODELING

If imitation occurred only in the presence of the model stimulus, this class of sensorimotor behavior would be much less significant than it is. However, as has been partially suggested, there are mechanisms by which the imitative act can be displaced from the model stimulus. The language mechanisms described above play an important role in this type of imitation.

Language repertoires that are essentially involved are those in which verbal responses are elicited, as in labeling, reading, word associations, and the like. In Chapter 5 the manner in which the child learns a large

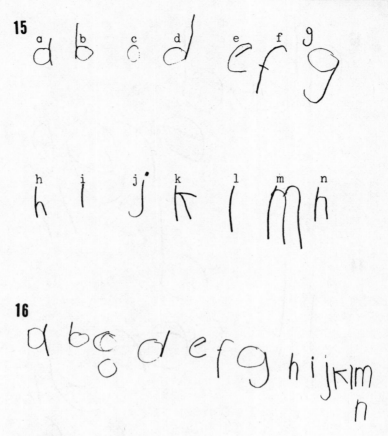

**FIGURE 6.1** (*continued*)

repertoire of words that label objects, people, actions, events and so on was described. Once learned, this repertoire can be the basis for new verbal learning in the individual as he has new experiences.

Thus, the visual stimulus of a boy pulling a girl's hair will come as a directive stimulus to control the appropriate verbal response sequence of the observing child, "He pulled her hair."

The example may be considered to be a form of communication. Its importance may be seen more clearly by indicating that it is possible to establish new sequences of verbal responses in a communication interaction by presenting a series of environmental objects to the individual. A silent film would be an example of a situation in which a sequence of nonverbal environmental stimuli each of which controlled verbal responses would result in a new complex verbal response sequence—which in this case would more or less accurately parallel the story of the movie.

**17**

FIGURE 6.1 (*concluded*)

> This may also be considered as communication: the process depends upon previously acquired S-R mechanisms and produces new verbal S-R mechanisms. (Staats, 1968c, pp. 128–129)

The other language repertoire that is important in the present context is the verbal-motor repertoire. It was suggested in Chapter 5 that the individual learns a large repertoire of verbal-motor units. The various verbs, as one class of words, become capable of eliciting appropriate instrumental responses, and instrumental responses are also under the control of other types of words. Sequences of words may be combined that elicit sequences of behaviors, sometimes in very complex and skilled ways. The author has also suggested that communication—the presentation of verbal stimuli —can result in learning verbal response sequences at one time. These verbal response sequences can then elicit sequences of overt instrumental behaviors at a later time (Staats, 1963, pp. 194–198). "[A]lthough a communication stimulus may have been effective in establishing a verbal response sequence, this sequence may not mediate motor responses until some later time when other stimuli in the situation also tend to control the same motor responses" (Staats, 1963, p. 197).

It was also suggested that imitation can be considered as "a sensory-motor skill in which certain social stimuli directly (or mediated by language or symbolic behaviors) come to control imitational behavior" (Staats, 1968c, p. 442).

> [As an example], the observing child may make verbal responses to the social events he experiences, and the verbal response sequence that is thus established may later mediate imitational behavior. The observing youngster may say to himself "When he starts the car he turns the ignition key first," and so on, in describing his father's complete act. Later, the youngster may again make these verbal responses, and the verbal stimuli produced could control his own imitation behaviors of starting the car. (Staats, 1968c, p. 445)

In this example, the verbal labeling repertoire is involved in the imitation act, as well as the verbal-motor repertoire. Bandura (1969) has interpreted imitation in terms of these principles and repertoires, calling the process symbolic mediation and referring to retention processes that bridge the lag between the modeling stimulus and the modeling act.

> The . . . representational system, which probably accounts for the notable speed of observational learning and long-term retention of modeled contents by humans, involves verbal coding of observed events. . . . After modeled sequences of responses have been transformed into readily utilizable verbal symbols, later performances of matching behavior can be effectively controlled by covert verbal self-directions. (Bandura, 1969, pp. 133–134)

## IMAGES AND THE MEDIATION OF IMITATION

As was indicated in Chapters 2 and 3, the conditioned sensory response or image has important functions in human behavior (Ellson, 1941; Leuba, 1940; Mowrer, 1960; Paivio, 1971; Sheffield, 1961; Staats, 1959). Images function in reasoning and problem solving where the images mediate the overt instrumental behaviors in problem solution. Images are frequently aroused by language stimuli, but this is not necessarily the case.

> As an illustration of the function of such responses in problem solving, let us suppose a person has performed a number of activities in various places in sequential order, only to find at some later time that he is missing some object he had been carrying at the beginning of the sequence. . . . [T]he sequence of activities would have resulted in a sequence of conditioned sensory responses (constituting what would be called his "memory" of the activities). The process of finding the missing object would then consist of the elicitation of the sequence of conditioned sensory responses. This alone may solve the problem since one of the conditioned sensory responses may be the visual response to the object sitting in one of the places the person had visited. Then his sensory response would mediate the motor response of going to that place and looking for the object. Or the sequence of the conditioned sensory responses might mediate the retracing of each of the previous activities and searching for the object at each place. Again, problem solving would depend upon the sequence of reasoning responses, that is, the sequence of conditioned sensory responses. (Staats, 1963, pp. 216–217)

It has also been proposed that such conditioned mechanisms function in imitation acts. The following example is one of negative imitation, which, as will be further indicated, involves the same principles as positive imitation.

> At any rate, in the example a child has just observed another child take a bicycle away from a third child and push him down. Let us also say that an adult then punishes the offending child. Such a sequence

of events will produce a sequence of conditioned sensory responses in the observing child. Thus, later when the sequence of conditioned sensory responses is elicited the child could describe what had happened —that is, the sensory responses (images) would control the appropriate verbal responses. Or, in certain other circumstances the conditioned sensory responses could be elicted and mediate avoidance of performing actions like that of the previously punished child. This would then be a case of negative imitation. (Staats, 1968c, p. 446)

Bandura, in his more extensive treatment of the published literature on imitation, also has adopted these principles and concepts in his social learning theory of imitation.

> Imagery formation is assumed to occur through a process of sensory conditioning. That is, during the period of exposure, modeling stimuli elicit in observers perceptual responses that become sequentially associated and centrally integrated on the basis of temporal contiguity of stimulation. If perceptual sequences are repeatedly elicited, a constituent stimulus acquires the capacity to evoke images (i.e., centrally aroused perceptions) of the associated stimulus events even though they are no longer physically present (Conant, 1964; Ellson, 1941; Leuba, 1940). . . . The findings of studies cited above indicate that, in the course of observation, transitory perceptual phenomena produce relatively enduring, retrievable images of modeled sequences of behavior. Later reinstatement of imaginal mediators serves as a guide for reproduction of matching responses. (Bandura, 1969, p. 133)

Aronfreed (1968) has also proposed a social learning theory of modeling that has centrally involved a concept that may be seen as equivalent to images. The concept is that the observer forms a cognitive template of the model's behavior. Attached to this representation of the model's behavior is an affective value that controls imitative behavior. Bandura has critized this view because "it does not specify in sufficient detail the characteristics of templates, the process through which cognitive templates are acquired, the manner in which affective valences become conditioned to templates, or how the emotion-arousing properties of templates are transferred to intentions and to proprioceptive cues intrinsic to overt responses (Bandura, 1969, p. 131). However, in his own approach, Bandura has adopted these specifications from other learning analyses, and the same analyses can be appended to Aronfreed's social learning theory as well. Moreover, Aronfreed has made very productive applications of the theory to the development of conscience and moral behavior in the child.

## SELF-REINFORCEMENT MECHANISMS AND SOCIAL LEARNING THEORY

Verbal and image mechanisms are relevant to modeling and observational learning in other ways. Although the following analysis pertains to the A-R-D personality system as well as to instrumental behavior, it has been centrally incorporated in social learning theory and is thus relevant here.

Mowrer (1952) described bird's vocal imitations as "self-reinforced behavior." More generally, it was suggested the individual's behavior produced a stimulus, many times within himself, which had reinforcing properties (Mowrer, 1960a; Staats, 1963, pp. 95–97). Extension of this analysis involved the manner in which matching stimuli generally become reinforcing and the way matching instrumental behaviors can be learned.

> . . . Conditions are [ordinarily] arranged so that the child learns to "match his behavior to others—or, more precisely, to match the stimuli his responses produce to the stimuli produced by someone else. Basically, the principles involved in matching behavior are stimulus discrimination and the conditioning of reinforcers.
>
> . . . Ferster (1960) has shown that pigeons can be conditioned to peck a key which matches another key in color. It would be expected on the basis of this type of training [that] matching stimuli would become conditioned reinforcers as well as [directive stimuli]. It should be possible to establish a variable stimulus and a standard stimulus such that a matching of the variable stimulus to the standard would constitute a [positive directive stimulus] and a missmatch [a negative directive stimulus]. . . .
>
> In addition, if a rat was reinforced when two white circular forms of equal size were presented and not when one of the forms deviated from the other in size, the animal would soon discriminate two matching circles from two nonmatching circles. In fact, if presented with a device constructed so that the variable circle could be adjusted by the rat himself, he could be trained to adjust the variable circle until it equaled the other one by presenting reinforcement only when the rat had adjusted the circles to match one another. Thus, the rat should be able to learn to match a stimulus produced by his own responses to a stimulus produced by someone else.
>
> In both of these suggested types of training, the visual stimulus of the equal circles would be expected to become a conditioned reinforcer because the organism would receive reinforcement in the presence of this stimulus. Thus, after this training it would be expected that it would be reinforcing for the organism to match the circles, even if no other reinforcement was made contingent upon the act. (Staats, 1963, pp. 123–124)

This suggestion that matching stimuli can become reinforcing and affect the learning of imitation has been agreed with (Baer, Peterson, and Sherman, 1967; Baer and Sherman, 1964) and criticized (Bandura, 1968b; Gewirtz and Stingle, 1968). Several experiments bear upon the question; the first follows the above suggestions very closely. Parton and Fouts (1969) conducted a study to test the possibility that similarity (matching) as a stimulus consequence is a reinforcing event with five-year-old children. One group of children could make a response that would produce a light stimulus like that of the standard, or another response that produced a disparate light. The control group could make either of two responses, neither of which produced a matching light stimulus. The first group of children learned the response that produced matching. The second group did not learn either of the responses which failed to produce matching stimuli. In addition, Baer and Sherman (1964) showed that children who were reinforced for imitating the behavior of a puppet on three responses later pressed a bar in nonreinforced imitation of the puppet more than they had done so before they were reinforced for imitation. Other studies have shown also that children who have been reinforced for imitation will imitate new responses for which they have not been reinforced for imitating (Baer, Peterson, and Sherman, 1967; Lovaas, Berberich, Perloff, and Schaeffer, 1966; Lovaas, Freitag, Nelson, and Whalen, 1967; Metz, 1965).

The concept of matching stimuli as a type of self-reinforcement has been further elaborated, in addition to its relevance for modeling. (Staats, 1968c). The learning analysis included additional self-reinforcement mechanisms involving language and cognitive processes of other types—for example, images (conditioned sensory responses).

"Children differ dramatically in the standards of performance which they set for themselves. Some children decide that performance just a bit above average for their class is adequate to meet their standards; others demand of themselves the top position in the class. The child's commitment to work and persevere will be a function of the standard that satisfies him" (Kagan, 1965, pp. 558–59). The observation that children find different standards to be reinforcing certainly appears to be the case. However, the present analysis would suggest that the individual differences that we see in children's standards, and the consequent work and study habits created by this source of reinforcement, are produced through their individual experiences according to the principles of conditioning. The child does not set standards for himself spontaneously: although he may carry them "within," they are learned and it is important for a human learning theory to indicate the mechanisms involved. Thus, as one example, we may observe a boy alone practicing a certain "move" in football until he is "satisfied," that is, until he has attained some internal standard. The standard stimulus within, as one example, may be a conditioned sensory response (image) which the boy has acquired from observing a more skilled player. The novice actor may practice a walk or gesture in the mirror until he attains the standard of an internal image. The painter or musician acquires similar sensory standards through a long history of conditioning and continues working on a piece until the standards are matched and reinforcement is produced. In other cases the internal standards are of a verbal nature. The individual continues responding until his own behavior elicits in him verbal responses that match some standard the individual holds. Thus, in a simple example, the child might continue to practice reciting a poem, after missing one word, because he says to himself "Last time I recited and missed a word I got a 'B' instead of an 'A.' " When at last he recites perfectly he says to himself "Now I will get an 'A' " and is reinforced. The subject of verbal reinforcers, which is relevant to the present discussion of internal standards, [has been] treated [in Chapter 5]. It should be indicated here, however, that *internal standards that have reinforcing value will to that extent also have discriminative* [directive]*stimulus value and will control behaviors that strive to attain the standard*. (Staats, 1968c, pp. 454–55)

Bandura has adopted these concepts of the self-reinforcement system and has employed the analysis in recent works. His term "self-evaluative consequence" is analogous to the concept of verbal self-reinforcers and the term "self-generated anticipatory consequence" refers to the directive function of internal A-R-D stimuli in eliciting imitative behavior.

Not all human behavior is controlled by immediate external reinforcement. People regulate their own actions to some extent by self-generated anticipatory and self-evaluative consequences. At this higher level of psychological functioning, people set themselves certain performance standards, and they respond to their own behavior in self-rewarding or self-punishing ways, depending on whether their performances fall short of, match, or exceed their self-imposed demands. (Bandura, 1971, p. 47)

Although specific patterns of self-reinforcing responses can be acquired observationally without the mediation of direct external reinforcement, undoubtedly the valuation of performances that fall short of, match, or exceed a reference norm results partly from past differential reinforcements. Thus, for example, parents who expect their children to exceed the average performance of their group in whatever tasks they undertake will selectively reward superior achievements and punish or nonreward average and lower level attainments. Differential achievement levels thus assume positive and negative valence [emotional value] and the performance standard common to the various activities is eventually abstracted and applied to new endeavors. That is, a person for whom average performances have been repeatedly devalued will come to regard modal achievements on new tasks as inadequate and attainments that surpass modal levels as commendable. Once the evaluative properties of differential accomplishments are well established, adequate or inadequate matches are likely to elicit similar self-reinforcing responses irrespective of the specific performances being compared. At this stage the whole process becomes relatively independent of external reinforcement and the specific contingencies of the original training situations, but it remains dependent upon cognitive evaluations based on the match between self-prescribed standards, performance, and the attainments of reference models. (Bandura, 1969, p. 34)

Cautela has employed verbally elicited images, of the type described in Chapter 5 and above, as emotion-eliciting and reinforcing stimuli in conducting therapy (1967, 1970). For example, he has had the subject imagine an emotionally negative stimulus (an example might be vomiting) while he also imagined the stimulus that was too positive for the individual and needed change—for example, a compulsive behavior. Cautela (1970) reported successful treatment of cases of test anxiety, obsessive-compulsive behavior, maladaptive approach behavior, self-control, and so on, employing images as reinforcers. Bandura (see 1969) and Kanfer and Marston (see Kanfer and Phillips, 1970) have conducted a number of experiments that support the general concept of self-reinforcement. Bandura and Kuypers (see Bandura, 1969), in an early study showed that children who observed models tended to match the self-reward patterns of the model. Bandura and Whalen (1966) found that children who had failed on a preceding task rewarded themselves less often in a later task than did children who had succeeded on the preceding task, even when the children were equally competent. Kanfer and Duerfeldt (1967) found that inappropriately giving oneself self-rewards was related to age, the presence of an adult model who took undeserved rewards, and so on, suggesting the importance of learning in self-reinforcement. Experimentation has not yet been conducted on the elaborated concepts of internal standards of evaluation and sensory and verbal reinforcers.

## ORIGINAL BEHAVIOR THROUGH IMITATION

"Analyses which treat an individual's behavior as a function of his experience, as does the present one, are frequently criticized on the grounds that while they can perhaps account for behavior for which the individual has been specifically trained, they do not indicate how the emergence of original or novel forms of behavior may occur" (Staats, 1963, p. 236). "Any conception of human behavior that cannot deal with original behaviors, it must be stated, is inadequate" (Staats, 1968c, p. 168). Bandura also has recently considered traditional learning accounts of imitation to be seriously limited because of "the failure of these formulations to explain how novel responses are learned to begin with" (Bandura, 1971, p. 8). Moreover, he sees the need for introducing concepts into modeling theory that can produce novelty in human behavior. "[T]he purpose of a theory of observational learning is not to account for social facilitation of established responses, but to explain how observers can acquire a novel response that they have never made before as a result of observing a model" (Bandura, 1971, p. 11).

The concept developed to indicate how learned processes could produce originality was simple in principle (Staats, 1963). The concept was that an individual could independently learn to respond to each of two (or more) stimuli. Having done so, if the two stimuli occurred in some compound configuration, the stimuli could call out an integrated response composed of the two response components. (This S-R mechanism is illustrated in Figure 2.9(b), Chapter 2.) The response combination could be quite novel, never having occurred before to the individual and never having been seen by him.

The simple example given in Staats (1963) was in the area of language learning. It was suggested that the child could learn to label various men with the response "Man," and he could independently learn to label running organisms with the term "running," but never have done so in relation to a man. The two stimuli, however, could occur together as a compound stimulus, a running man. The child in this circumstance might make the novel response of saying "Man running," or "Running man," although he had never said this before.

Whitehurst (1972) has conducted an experiment that directly tests the possibility that such novel behavioral compounds can emerge from independent component learnings. The subjects were two two-year-old children. They learned new word responses to a color and also to a figure. Then a colored figure was presented to the children. The children proceeded to employ the two-word response compound to the compound stimulus. The verbal compound was novel, as was indicated by Whitehurst.

Bandura began using this mechanism in accounting for the production of novel behaviors in the context of imitation in a 1965 experimental paper. He observed that an imitation task in which there are complex stimuli presented by the model will elicit various compounds of responses.

> The component responses that enter into the development of more complex novel patterns of behavior are usually present in children's behavioral repertoires as products either of maturation or of prior social learning. Thus, while most of the elements in the modeled act had undoubtedly been previously learned, the particular pattern of components in each response, and their evocation by specific stimulus objects, were relatively unique. (Bandura, 1965, p. 594)

Complex and important cases of originality may arise on the basis of the compound response mechanism (Staats, 1963, 1968c). Examples have been given of new behavioral skills arising when individuals consult an instructional manual, as in learning golf or the skills of a mechanic. The following illustrates the process in an area of artistic behavior.

> It seems that after a suitable repertoire of responses to verbal stimuli has been established, novel skills involving chains of motor behavior can be produced entirely on a verbal basis. An athlete or dancer, for example, acquires many new skills and sequences on the basis of such verbal-motor associations. A choreographer needs only to present a verbal sequence to a trained dancer, each stimulus unit in the sequence controlling a response unit. As a result, the response sequence will be acquired. The dancer, of course, has acquired a much more extensive repertoire of such

verbal stimuli and movement associations than the nontrained individual and thus can much more quickly learn new sequences of dance steps. (Staats, 1963, p. 181)

In this context it is interesting to note that there is evidence to support the theoretical analysis. An "alphabet of motion" for dancers has been invented and employed in recent years; Cullen (1966) reports that Rudolph Benesh uses a kind of shorthand to indicate the exact position of each body part for each note of a musical score. The alphabet can be seen as a set of language stimuli that when presented to a trained dancer can elicit the movements designed by the choreographer.

> In notating dance movement he uses a five-line stave which is printed immediately below the musical score. On this "human stave" he can indicate by means of symbols where the dancer is on stage, the direction he is facing, the position of his limbs, details of hands, head and feet, plus the direction of his movement. (Cullen, 1966, p. B–1)

Bandura has employed the verbal-motor mechanism as a general type of modeling which may involve acquiring new behaviors, including some of those described above. The social learning theory account may be summarized as follows:

> As linguistic competence is acquired, verbal modeling is gradually substituted for behavioral modeling as the preferred model of response guidance. People are aided in assembling and operating complicated mechanical equipment, in acquiring social, vocational, and recreational skills, and in learning appropriate behavior for almost any situation by consulting the written descriptions in instructional manuals. Verbal forms of modeling are used extensively because one can transmit through words an almost infinite variety of behavioral patterns that would be exceedingly difficult and time-consuming to portray behaviorally. Moreover, since verbal descriptions is an effective means of focusing attention on relevant aspects of ongoing activities, verbal modeling often accompanies behavioral demonstrations. (Bandura, 1971, p. 41)

One of the very important types of behaviors produced through novel behavior compounding occurs in the development of new words in the child's language acquisition. The manner in which this occurs was described in the preceding chapter, which suggested that after the child can imitate the basic sounds in the language (phonemes), he can put the sounds together in new combinations. When the child repeats the new combination he has vocalized a new word, which can then be learned. As was indicated, this mechanism for language learning—which markedly accelerates the process—has been shown to work effectively in the author's child learning laboratory where preschool children learned to say *Drosophilus Melanogaster* through successive acquisition of additional syllables.

The social learning theory of modeling has also utilized the concept, since this type of language development is also explanatory in this area of study. It is interesting that the basic sounds are considered in this account to be due in development to innate factors.

> [P]ersons can produce a variety of elementary sounds as part of their natural endowment. By combining existing sounds one can create a novel and exceedingly complex verbal response such as *supercalifragilisticexpialidocious* (Bandura, 1969, p. 146). In instances where observers lack some of the necessary components, the constituent elements may be modeled first; then in a stepwise fashion, increasingly intricate compounds can be developed imitatively. (Bandura, 1971, p. 22)

## SOCIAL LEARNING AND SOCIAL BEHAVIORISTIC THEORIES OF MODELING

The social learning and the social behavioristic approaches have been described in discussing some of the aspects of imitation and observational learning. Although the two approaches began differently, they now overlap in certain specific principles, concepts, observations, and analyses. The social learning approach has incorporated many of the social behavioristic characteristics.

Notwithstanding this present overlap, contemporary social learning theory (Bandura, 1969, 1971) began in a different tradition which is reflected in basic differences that remain. In this context it may be noted that while Bandura criticized Aronfreed (1968) for introducing cognitive concepts for whose acquisition the principles are not stipulated, his successive developments of his social learning theory have only partly corrected this same drawback. While he has adopted learning analyses in part, he still introduces observational learning, for example, as a basic principle—one that is not to be explained in terms of more basic principles. In sum, social learning theory is presently a hybrid. It includes learning principles and learning analyses, especially from the clinical area (behavior modification) and in parts of its treatment of modeling. But its contact or concern with the basic field of learning is not close or deep. Moreover, this orientation is mixed with a non-behavioristic characteristic in which mental concepts are introduced that are not defined by behaviors or the determinants and effects of those behaviors. For example, Bandura uses the concept of linguistic competence in the above quotation, without stipulating what this competence consists of, how it is acquired, and how it has its effects—all of which require complex analyses and empirical support. This is at variance with the hierarchical theory orientation of social behaviorism, as will be indicated in proposing elaborations of the study of modeling and observational learning.

MODELING AND OBSERVATIONAL LEARNING: BASIC LEVEL OR PERSONALITY

LEVEL OF THEORY? The social behavioristic paradigm has indicated that there are different levels of theory development that must be recognized in the study of human behavior. It has been suggested that it is important to indicate what the elementary principles are, as well as the derived human learning principles and S-R mechanisms, the personality level of study, and so on. Moreover, the paradigm must indicate how study in these several levels of the paradigm are to be related. When the levels of theory are recognized and the ways of interaction are indicated, certain issues are clarified and guidelines for further development emerge.

It can be added that when viewed in the context of the hierarchical theory levels described, there are differences between the social learning theory approach and the social behavioristic theory approach. Imitation, or modeling, has been the central focus of Bandura's empirical and theoretical work. The development of the theoretical aspects of this work has involved the successive incorporation of different behavioral concepts and principles into the social learning account. The result has been a much more general theory. However, the basic characteristic of the approach, in terms of the present methodology, stems from the original mission of the approach—which was to account for aspects of human behavior employing the imitation or modeling principles.

Thus, in the social learning theory construction task, modeling principles have been considered to be basic. Modeling is proposed as a third type of learning, of the same generality as classical conditioning and instrumental conditioning.

> Considering the prevailing influence of example in the development and regulation of human behavior, it is surprising that traditional accounts of learning contain little or no mention of modeling processes. If the peripatetic Martian were to scrutinize earth man's authoritative texts on learning he would be left with the belief that there are two basic modes of learning: People are either conditioned through reward or punishment to adopt the desired patterns, or emotional responsiveness is established by close association of neutral and evocative stimuli. (Bandura, 1971, p. 2)

In proposing modeling as a third type of learning, a major thrust of modeling research has been to show that instrumental learning can take place through modeling or vicarious learning (Bandura, 1969), as can classical conditioning (Bandura, 1969). This latter type of conditioning, called vicarious classical conditioning, will be elaborated a bit to exemplify the point. The major question involved in this discussion is whether or not modeling principles are basic, like the elementary conditioning principles.

VICARIOUS CLASSICAL CONDITIONING. In his most complete statement of the social learning approach Bandura (1969) has sections on higher-order classical conditioning as well as vicarious classical conditioning. They are considered to be different basic principles, with higher-order conditioning playing a relatively minor role in human behavior. On the other hand, heavy significance has been implied for Bandura's suggestion that emotional response conditioning takes place widely in man through vicarious processes. This suggests that man operates according to special modeling principles that, while basic, are of a higher nature than the principles of conditioning. Examination of the empirical support for the process of vicarious classical conditioning as a separate basic principle suggests otherwise, however.

The experiments employed to define the principle of vicarious classical conditioning involve subjects who watch another subject being subjected to an emotion-eliciting experience (Barnett and Benedetti, 1960; Berger, 1962). Berger, for example, instructed one group of subjects that a model whom they were observing would receive a painful shock whenever a light dimmed. The dimming of the light was preceded by a buzzer. The model simulated pain by jerking his arm when the light dimmed. Other groups of subjects were (1) told the model would make a voluntary arm movement, but he was receiving no painful stimulation, (2) told the model was being shocked, but he did not simulate pain, or (3) were not told the model was being shocked, and he did not withdraw his arm. The measure of vicarious conditioning in the observers was that of the peripheral emotional response of sweat gland activity in the hands (GSR), with the buzzer as the conditioned stimulus. The subjects who were informed the model was receiving painful shock and who saw the model simulate pain were most strongly conditioned to the peripheral emotional response. Bandura also cites cases of the extinction of emotional responses through vicarious conditioning (1969).

These phenomena can be considered in the following terms. The occurrence of vicarious classical conditioning depends on the behavior of the model in the context of the situation (including the experimental instructions), acting as an emotion-eliciting stimulus for the subjects. That is, if the sight of the model's "painful jerk" elicits a negative emotional response in the observer, he will be conditioned to respond negatively to the contiguously presented buzzer. This would be a case of classical conditioning.

One must ask, in addition, why this social stimulus elicits an emotional response in the observer. It is suggested that this elicitation depends on the previous conditioning history of the observer. The pain of another person is a stimulus—perhaps visual, as when the person grimaces, struggles, jerks, or goes limp; perhaps auditory, as when the person screams, begs, moans, or shouts for succor. For the most part, it is suggested, the pain stimuli of another person elicit a negative emotional response in the observer because these stimuli have previously been conditioned stimuli themselves. Ordinarily the observer has experienced other people (or himself) perform such "painful behaviors" at the same time the observer is experiencing negative emotional responses. This results in the stimulus of someone being in pain eliciting a negative emotional response. This type of conditioning may involve directly elicited negative emotional responses, but much of the conditioning takes place through language. For example, let us say that an adult admonishes a child for hurting another child by saying the following: "Look, you have made him cry. That was a bad thing to do. You should be ashamed." The words would elicit a negative emotional response in the one child in the presence of seeing another child cry. With this type of experience the sight and sound of someone crying will be a conditioned stimulus that elicits the negative emotional response.

Thus, it is suggested that vicarious classical conditioning of emotional responses depends upon the stimulus of someone in pain or discomfort (in the negative case) eliciting a negative emotional response in the observer. This negative emotional response, however, is itself learned on the basis of past conditioning. Understanding the process demands understanding of the learning.

Centrally, however, vicarious classical conditioning is thus only a special case of higher-order classical conditioning. It does not involve a new principle. It is not on the same level of theory as the basic principles of conditioning; it is derived from them. Vicarious classical conditioning depends upon the previous conditioning of the observer. This can be realized readily by reference to cases where the sight and sound of others in pain does not serve as a stimulus that elicits a negative emotional response, but just the opposite. There are sadists, for example, for whom such stimuli elicit positive emotional responses. This occurs because of the atypical conditioning history of the sadist, it is suggested.

Analysis of the empirical results as cases of higher-order conditioning has implications that differ from the introduction of terms suggesting new principles such as vicarious conditioning or observational learning. Furthermore, it is suggested, when analyzed in terms of higher-order conditioning, the related theoretical developments become relevant in a manner that has additional implications. One such implication is that the A-R-D principles should apply to the area of vicarious classical conditioning. Thus, the sight of someone in pain should serve not only as an emotion-eliciting stimulus but also (1) as a negative reinforcing stimulus and (2) as a directive stimulus that elicits avoidance behavior, including that of stopping the condition causing the pain to the model.

In addition, studies not thought to be examples of vicarious (higher-order) conditioning can be seen to be of the same type. For example, Chapter 7 will describe a special case of higher-order conditioning in which the emotion-eliciting stimuli are words. A word eliciting a negative emotional response will, when paired with another stimulus, result in the stimulus coming to elicit a negative emotional response also. Vicarious classical conditioning is actually just like this, since a word is also a social stimulus produced by a "model." It may also be suggested that conditioning through words accounts for an important type of communication which includes positive as well as negative emotional conditioning.

Finally, it appears that this type of communication can take place on a nonverbal level, and nonverbal communication may be analogous to the studies called vicarious conditioning. For example, Rosenthal, Archer, Kiovumaki, DiMatteo, and Rogers (1974) have conducted a study in which they attempted to isolate the features of nonverbal communication. Using film they presented different parts of the stimuli that compose the person, for example, only his face, only his body from neck to knees, a compound of face and body, only his voice, a compound of his face and voice, and so on. Half of the scenarios used involved communicating positive affect and half involved negative affect. The authors in their first report were interested in the characteristics of the observers that were associated with differential sensitivity in responding appropriately to the nonverbal communication, that is, in indicating correctly the direction of the affect being depicted in the scenario. They found that positive and negative affect could be communicated by the partial social stimuli. Moreover, they found that females were better than males at responding to the stimuli that elicited emotional responses, especially when negative emotional responses were involved.

With respect to the latter, Rosenthal et al. (1974) suggest that "this superiority on the part of females in reading negative nonverbal cues may have had survival value" (p. 2). However, this result could also be interpreted in terms of different learning histories of males and females. The "ability" to respond, it is suggested, rests upon having had a conditioning history such that the behavioral cues of the model elicit corresponding emotional responses in the observer, even when only some of the behavioral cues are presented. Supportive of this learning interpretation is the finding that students enrolled in courses in small-group dynamics, a seminar in nonverbal communication, and couples with small children all responded better to the nonverbal communication cues. In any event this nonverbal communication process may be seen as an example of learned emotional elicitation. In general, concepts like vicarious conditioning interpose a schism between basic learning principles and human behavior study, thus hindering hierarchical theory development.

MODELING PRINCIPLES ARE DERIVED, NOT BASIC PRINCIPLES. The social behavioristic paradigm considers modeling differently than the social learning view does. Modeling concepts, principles, and observations must be considered in terms of their appropriate level of theory construction, it is suggested. This consideration provides implications for the development of modeling theory and for the general task of understanding human behavior.

It is suggested that the failure of the traditional learning theories to deal with modeling is justified because they were concerned with elementary or basic principles of learning—and modeling does not involve such principles. The consideration of modeling (imitation, observation learning, or vicarious learning) has its place at the personality level of the theory. Modeling theory is not to be placed in competition with the basic principles of conditioning. Rather, modeling in human behavior derives from, or is based upon, the elementary principles of conditioning, as well as upon the principles of the S-R mechanisms.

Furthermore, there are also analyses of basic behavioral repertoires (personality repertoires) that are more elementary than modeling principles; that is, modeling competence is based upon the prior acquisition of those basic behavioral repertoires. As has been indicated, the basic behavioral repertoire of attention and discrimination skills underlies the individual's modeling abilities. The same is true for various aspects of the individual's basic language repertoires, such as labeling, the verbal-motor repertoire, the verbal-image repertoire, the A-R-D system, the verbal-emotional repertoire, and so on. Moreover, modeling depends upon the individual's basic instrumental behavioral repertoire, or such sensori-motor units as those of imitation.

Thus, it is suggested that modeling is generally a comparatively advanced personality process that is founded upon more basic processes. Recognition of this suggestion has certain implications for further study.

THE NEED FOR BEHAVIOR ANALYSES. When the focus is upon indicating the importance of modeling as a basic principle and the use of modeling to describe human behavior, the theory gives impetus to the study of how behaviors are learned *through* modeling. The theory does not give impetus to analysis of the basic behavioral repertoires that *underly* the modeling process. However, when the levels of theory are separated and the personality level is recognized, the task becomes one of analyzing in theory and experiment the basic behavioral repertoires of which personality is composed, including the repertoires that modeling depends upon.

Thus, the present approach not only calls for recognition that attentional and discrimination repertoires are basic to modeling; it also requires that these repertoires be described in detail and dealt with in experimentation. The same is true of the other repertoires. It is not enough to recognize that linguistic competence is involved in modeling, that verbal sequences provided by someone else, or elicited through reading or primary experience, can mediate imitation. The basic language repertoires involved must be analyzed in detail, and the principles involved in their acquisition and function must be made explicit and demonstrated empirically. The A-R-D system is also basic to various aspects of modeling and requires analysis. The social behavioristic approach to modeling thus suggests additional theoretical and experimental extension.

LEARNING THROUGH MODELING. The focus of experimentation in the field of modeling has been upon how modeling processes operate, or upon how additional learning in the child can occur through modeling. Thus, for example, the Bandura, Ross, and Ross (1963) study showed that the extent to which the model is associated with rewards will influence the extent to which the model is imitated. As another example, Gerst (1971) had four groups of children watch a film sequence. One group verbally described what they had seen. Another group imagined what they had seen. Another provided summary labels of what they had seen, and a control group performed unrelated arithmetic problems. It was found that the three groups who performed the labeling rehearsals or imaginal rehearsals (which probably included labeling also) showed greater ability to reproduce what they had seen than did the control group. This study indicates the importance of verbal and imaginal processes in modeling.

Other studies have demonstrated how children can learn through imitation. Bandura and Harris (1966) showed that children who were exposed to a model who employed certain grammatical forms of language later used more instances of that grammatical form than did a control group. As has been indicated, and as this study further demonstrates, important aspects of language learning occur through observational learning. In general, it is important to continue to study how the child can learn through modeling. Bandura's experiments and those stimulated by these experiments have provided very important data in this significant area of behavior (Bandura, 1969, 1971).

Further analysis should be given to what is being learned in such instances, however, to isolate the general principles involved. Leaving the grammatical form unanalyzed in terms of elementary S-R mechanisms, for example, does not indicate what the processes are that account for grammatical speech. In general, thus, there is a need for analysis—one must ask what it is that is learned and whether there any general types of learning that are involved.

It is suggested that when some of the mechanisms involved in modeling are considered, it can be seen that different types of learning may be involved. For example, it has been suggested that novel behavior can be produced by modeling. The model stimulus may be a compound of stimuli, each of which elicits a response in the individual that was previously learned. The modeled behavior, thus, may be a compound that is original, never before made by the individual. However, what has been learned in this type of modeling? It is suggested that there are two types. First, when a compound (or sequence) of behaviors is elicited from an individual, they will be associated with one another. In a sequence the stimulus produced by the first behavior will come to elicit the next behavior. The dancer who makes a sequence of responses under the choreographer's instructions learns the sequence. Under repetition the sequence will come to occur as a unit. This is one type of learning that can occur—the separate responses come to be associated together and form a higher unit of response.

In addition, however, there is a change in the controlling conditions for the compound response. In the beginning it may be the model stimulus of someone demonstrating the behavioral compound, or a pictorial stimulus, or some verbal form of communication, such as written instructions. An important part of the learning that should generally be understood, however, is that new stimuli come to control the new behavioral compound. The goal is not only that the dancer come to learn the new sequence but also that he will be able to perform the sequence even in the absence of the choreographer's instructions.

This process was involved in the writing training described above. The child was first given training in imitating (copying) the letter. When he was copying the letter the name of the letter would be presented. For example, the experimenter-trainer would say, "This is an *a*; let me see you copy the *a*." The child was also trained to say the letter as he copied it. In this manner the letter-writing response, learned as an imitation, came under the control of the auditory presentation of the letter itself. Improvement in writing under the control of the child saying the letters *himself* is shown in Figure 6.1 above in samples 6, 8, 10, 12, 14, 16, and 17. The total training trials involved in each of these samples was 313, 338, 396, 451, 494, 614, and 778. By the end of this training, which for this child required 15 hours and 15 minutes, the child was writing the letters *under the control of his own saying of the letters* with considerable skill. Later, of course, children must also learn to write under the control of various stimuli—such as environmental stimuli, or images, in the act of describing an experience.

To summarize, the child's imitation skills first result in the response being evoked by the model. However, the response, in being evoked, is conditioned to other stimuli that are present. *Later, such stimuli can serve to elicit the response in a manner that is no longer imitation.* In many cases, as will be indicated, the later eliciting stimuli are those the individual himself provides. Thus, the behavior that began as an imitation comes to be under the control of the individual. It is in part the acquisition of new stimulus control over acts that accounts for the significance of imitation.

These illustrations indicate the suggestion that the nature of the learning that is involved in the modeling act should be isolated. Only in this way will additional understanding of the processes emerge and ways of dealing with the processes be established.

IMITATIVE BEHAVIOR, INNATE OR LEARNED. In the same way that social learning theory considers modeling to involve basic principles, it has not stipulated whether the processes involved are innate or learned. Bandura has only suggested that both innate and learned skills may be involved (1971).

It is suggested that this lack of specification springs from the fact that the social learning theory—by considering modeling as basic—does not provide impetus for the study of the acquisition phases of the basic behavioral repertoires that are actually involved. The hierarchical theory conception of social behaviorism indicates that the basic behavioral repertoires must be studied to understand modeling. It has been assumed by many theorists that such skills as imitation come about through biological maturation in humans (and in other primates). When an imitation skill is dealt with in its original form, however, the manner in which it is learned can be seen.

It was in part for this reason that the development of copying letters was demonstrated. With no observations of this type of behavior development it has frequently been assumed that it emerges full blown, through maturation. The study of this behavior development shows, however, that a long learning process is required—one that involves thousands of learning trials. This is true whether the learning commences when the child is two, three, four, or five years old or whether the child has an IQ of 89 or 130 (Staats, 1968c; Staats, Brewer, and Gross, 1970).

It is suggested that each of the basic behavioral repertoires involved in modeling and other advanced skills must be studied in an experimental-longitudinal way, so that the manner in which the behavioral repertoire is learned is the topic of investigation. This type of research will provide a conception of child development which can be added to the research showing that modeling is effective in producing advanced learning. It will be only through experimental and theoretical specification of the basic behavioral repertoires involved that explanation will be provided, that individual differences in modeling will be understood, and that principles and procedures will be provided for doing something about the problems of development that arise in this area of behavior. The need for such research will be further discussed in Chapter 10.

## Instrumental personality repertoires

Social learning theorists, while originally eschewing the concept of personality, are beginning to recognize the behavioral interaction conception (Krasner and Ullman, 1973; Mischel, 1973), as has been indicated. This has involved recognition of the principles of behavioral interaction. However, the principles, important as they are, represent only the skeleton of a full conception of personality. The meat of the conception resides in theoretical and experimental behavior analyses of the major personality repertoires. The social learning theories (Bandura, 1969, 1971; Krasner and Ullmann, 1973; Mischel, 1971, 1973) have not made such analyses. It has been the purpose of the past three Chapters to introduce the possibilities of dealing with several of the major personality repertoires. The present chapter indicates that one important realm of personality lies in the individual's sensorimotor instrumental skills.

We know from naturalistic observations that there are large differences between people in their instrumental personalities. There are vast areas of such skills in athletics, music, art, science, warfare, work, and so on. These sensorimotor repertoires help characterize personality differences among individuals, groups, and cultures. For example, it would seem that one of the advantages that Western European groups have in competition with groups in undeveloped countries lies in the realm of sensorimotor repertoires. Children in our culture, to illustrate, learn to play with mechanical toys at an early age. They learn to ride bicycles and later drive cars, and they have experience with various mechanically operated objects. These and other activities provide them with different skills than a Bedouin child, for example, who may have had practically no experience with mechanical or motor-driven machines. As a consequence, considerable difference would be expected in the rapidity and skill with which a boy from our culture would learn to operate military aircraft in comparison to a Bedouin boy.

Instrumental behaviors are also involved in individual social behaviors of various kinds. For example, the man's or woman's appeal as a sexual partner will in part depend upon sensorimotor skills in this area of interaction. There are sensorimotor skills that characterize the individual's recreational characteristics, and hence other aspects of his life—his aggressive characteristics, his dependency characteristics, and so on.

Naturalistic evidence clearly indicates the importance of sensorimotor skills in the study of human behavior. However, very little is known of the elements involved in such repertoires. There are no analyses of such skills in terms of specific stimuli and responses, as there are for the skills involved in playing a musical instrument, for example. There is little specification of how sensorimotor skills are acquired. The children's performance in learning the sensorimotor skills of writing letters was described to indicate the form of early sensorimotor learning and to illustrate that such learning can be studied. It is suggested that behavior analyses of other sensorimotor skills are needed. Moreover, on the bases of the analyses, research is needed to study the manner in which the skills are acquired.

## Instrumental, cognitive, and emotional personality repertoires: A tripartite but interactional approach

There have been various schemes for classifying personality. The manner in which Freud divided personality into the id, ego, and superego has been described, as one example. The past three chapters have suggested that there are three important general realms of personality, described as the emotional-motivational system, the language-cognitive system, and the instrumental or sensorimotor skill system. It should be noted that this account is intended to be descriptive. It should not be taken to mean that there are structures within the individual that correspond to this division. The three categories also are not proposed as a classificatory system that is based upon principles, such that there are clear and definite separations between them, or that all behaviors of man may be sorted to one or the other category.

Rather, it would appear that the emotional-motivational, language-cognitive, and instrumental behavior repertoires are intertwined in complex interactions. Each of the major personality systems described actually involves aspects of the other personality repertoires. This was seen clearly in describing imitation. Instrumental skills are involved, as are language-cognitive repertoires. As will be indicated further in the next chapter, moreover, the emotional-motivational system is involved in modeling.

This type of interaction is generally the case. As has been suggested, behavior analysis is necessary to provide the theoretical basis for dealing with human behavior in research, for treatment purposes, and for personality measurement—as will be indicated. In working toward more adequate analysis of personality, however, it is helpful to consider separately the three realms of behavior repertoires, at least as a first approximation.

# Social behavior and social
# interaction principles

THE EMPHASIS to this point has been upon the manner in which the basic conditioning principles and various personality repertoires are relevant to an understanding of the individual's behavior. It should be understood, however, that although basic principles may be studied by focusing on the response of the individual organism to the experimental situation, it is necessary also to indicate how the principles operate when humans interact. Much human behavior occurs in interaction with others. Thus, much of what is learned is through social interaction, and the learning sees expression in interaction.

A general theory of human behavior must elaborate the basic principles to provide a model of how people affect one another. The elaboration of learning-behavior principles should also lead to empirical hypotheses that can be subjected to test. Then the theoretical structure within which social behavior is treated requires extension to the various areas that have been of concern to investigators in social psychology and the social sciences.

## Attitudes and their learning functions

The way in which basic learning principles can be utilized in making analyses of personality repertoires has been indicated. It will be suggested that the same principles, and the concepts developed in the personality analyses, also have relevance for the consideration of social behavior. A feature of the present approach is that the same principles are used to account for areas of psychology usually considered with different concepts.

One of the central concepts in traditional accounts of social interaction has been that of attitudes. Kiesler, Collins, and Miller (1969) suggest that the concept of attitudes can be traced to Carl Lange's 1888 use of the term *aufgab* to describe the subject's task attitude in an experimental task. These authors, however, give credit to sociologists (Thomas and Znaniecki, 1927; Blumer, 1939) for the characteristics which launched the concept as an important one in social psychology.

Murphy, Murphy, and Newcomb (1937) also treated attitudes as a central concept in the field of social psychology. They defined an attitude as "verbalized or verbalizable tendencies, dispositions, or adjustments towards certain acts" (p. 889). The concept of attitudes has remained in its central role in the area of social psychology, as will be indicated. At this point, however, it is relevant to consider the development of the concept within the learning theory context.

## LEARNING THEORY AND ATTITUDES

Leonard Doob (1947) first began to extend formal learning theory to the consideration of attitudes. Utilizing Hull's (1943) learning theory, he suggested that attitudes are anticipatory responses which can mediate overt behaviors. Hull's theory was that on receipt of a reward (positive reinforcement) by an animal at the end of a T maze, it would make a goal response—like the consumption of food. Parts of this goal response would be associated with stimuli in the goal box and with stimuli preceding the goal box. These anticipatory goal responses also would have stimulus characteristics that could be conditioned to the instrumental response of running to the goal box.

In Doob's account attitudes were conceived of as anticipatory goal responses that could in turn elicit instrumental responses. This classic analysis of attitudes, based as it was on Hullian principles, was complex and did not clearly specify the principles of classical conditioning and instrumental conditioning and the interactions of the principles. Moreover, it was presented as an abstract analysis, without illustration by actual examples of human interactions or conditions that defined attitudes and their effects in naturalistic or experimental situations. Thus, additional analysis was needed before the learning model could generally generate research in social behavior. Improvements in the basic learning theory were also needed. But Doob had made an important first step.

Doob's conception of attitudes was used in an early learning study. (Eisman, 1955). The complex theoretical structure was followed in the study and the procedures were correspondingly complex and difficult to relate. However, evidence was provided that supported the Hullian analysis.

The present author's account received an impetus from the previous Hullian analyses (Doob, 1947; Osgood, 1953). However, a two-process learning theory was employed rather than the one-process learning theory of Hull (1943). Moreover, the process of conditioning was seen to be basically one of classical conditioning rather than instrumental conditioning (Staats and Staats, 1958). Hull's concept was that of the goal response, which suggests readily only the positive case. However, attitudes can be negative as well as positive. Moreover, basing the concept of attitudes in the concept of the goal response is limiting in terms of considering the many types of circumstances in which attitudes are conditioned in everyday life.

The classical conditioning concept of attitudes, on the other hand, relates the concept of attitudes to that of emotions. This makes the classical conditioning model more general in several ways. First, it takes the concept of attitudes out of the narrow confines of animal research, on which Hullian theory is based. The concept of attitudes is then related also to Pavlovian conditioning, to the physiological studies of emotion, and to the various animal and human learning studies involving various types of emotional responses. Moreover, the concept of emotional response has a negative as well as positive case.

Based upon this concept of attitudes, the present author planned a series of studies to investigate the principles involved. The most basic study tested the possibility that when an unconditioned stimulus was paired with another (to be conditioned) stimulus, the latter would come to elicit the physiological emotional response involved (Staats, Staats, and Crawford, 1962). Language stimuli were employed as the conditioned stimuli, since the author felt that these stimuli are generally involved in human attitudinal (emotional) learning. Thus, one word among the many that were repeatedly presented to the subjects was paired systematically with either electric shock or loud noise, either of which elicited an emotional sweat gland response. The word alone came to elicit the emotional response through the conditioning. Moreover, the intensity of the conditioned emotional response was correlated with the intensity of the rating of the word on a seven-point attitude scale indicating degree of pleasantness and unpleasantness. This type of attitude conditioning has been shown also by Zanna, Kiesler, and Pilkonis (1970) and Maltzman, Raskin, Gould, and Johnson (1965). It should be noted that such basic studies, which deal with physiological emotional responses, provide connecting links for relating the concept of attitudes to the concept of emotions.

The importance of words as the means of producing attitude conditioning was also studied (Staats and Staats, 1958), employing the author's method of changing attitudes through language conditioning, a form of higher-order classical conditioning. In the procedure subjects were presented with national names like *Dutch* and *Swedish,* one at a time on a screen, with instructions to simply look at the names. Each time a name appeared on the screen a word would be pronounced by the experimenter, which the subjects were supposed to listen to and repeat. Some of the national names were paired with nonemotional words like *with, key, chair, paper, up,* and so on. With one group of subjects, however, the national name *Dutch* was paired only with positive emotional words such as *beauty, win, gift, sweet, rich,* and so on. With this group of subjects the national name *Swedish* was paired with negative emotional words such as *thief, bitter, ugly, sad, cruel,* and so on. The positive and negative conditioning procedures were reversed with another group. After the conditioning the subjects rated how they felt toward the various national names on the seven-point *pleasant-unpleasant* attitude scale that had been shown to be an index of physiological emotional responding. It was found that the subjects had been conditioned to either positive or negative attitudes toward the two national names, depending on the direction of the language conditioning.

Later studies showed that the strength of conditioning is a function of the number of conditioning trials (Staats and Staats, 1959), that both negative as well as positive attitudinal responses could be conditioned (Staats and Staats, 1959), that the principle of extinction applies to attitude conditioning (Carlson, 1970), and that deprivation affects the conditioning in the expected manner (Staats, Minke, Martin, and Higa, 1972), and so on.

The classical conditioning principles have been verified also with various socially conditioned stimuli. The author's first study employed nonsense syllables (see C. K. Staats and A. W. Staats, 1957), to control for the previous learning of the subjects. Personal and national names have been used, as indicated, and attitudes have been classically conditioned to the names of actual people (Early, 1968), as well as to racial color names (Harbin and Williams, 1966). The principles and procedures have also been employed to condition subjects' attitudes toward informational items (Bugelski and Hersen, 1966; Carriero, 1967) and concepts (D. E. Coleman, 1967). It was suggested that it would "be possible to pair an individual's picture with word reinforcers and thus produce the attitude conditioning" (Staats, 1964b, p. 335). Byrne and Clore (1970) have employed this hypothesis and the language conditioning procedure to condition attitudes to pictures of people.

The procedures that have been described for producing attitude conditioning have involved primarily words and to some extent primary aversive stimuli. Other emotion-eliciting stimuli have been used for attitude conditioning. For example, pictures and films and tape recordings of individuals and their behaviors can have emotional value (Staats,

1964b, pp. 334–35). One study has shown that negative emotional responses elicited by pictures of wounds can be conditioned (Geer, 1968). Stalling (1970) showed that statements with which subjects agree will elicit positive emotional responses and statements with which subjects disagree will elicit negative emotional responses. He produced attitude conditioning using such statements, as have Sachs and Byrne (1970) and Byrne and Clore (1970). It has also been found that items from interest tests can serve the emotion-eliciting function and produce attitude conditioning (Staats, Gross, Guay, and Carlson, 1973). It would be expected that the actual stimuli referred to by the items also could serve the same function, for example, reading a book, seeing an art show, going to a movie, playing cards, participating in athletic activities, and so on.

As has been indicated in describing the basic learning principles, classical conditioning can take place in the instrumental conditioning situation. That is, when a behavior is followed by a reinforcing stimulus, that behavior will occur more frequently later. However, the reinforcing stimulus is also an emotion-eliciting stimulus. Thus, the emotional response it elicits should be conditioned to the other stimuli in a situation. This principle has been shown to produce positive attitudes in children (Lott and Lott, 1960) and adults (Scott, 1957).

## ATTITUDE FUNCTIONS

The usual focus in the study of attitudes is on the nature of the attitudes and the manner in which attitudes are formed. It is of equal importance, however, to indicate how attitudes function. In describing the learning theory of the emotional-motivational system it was indicated that stimuli that elicit an emotional response will thereby also have two other functions. An emotional stimulus will serve as a reinforcing stimulus when it is applied following an instrumental behavior. Moreover, an emotional stimulus will serve as a directive stimulus, tending to elicit a class of approach behaviors in the positive case and a class of avoidant behaviors in the negative case. When the concept of attitudes is equated with the concept of the emotional response, the principles of the A-R-D theory become relevant. Attitude stimuli should thus serve as reinforcers (Staats, 1963, 1964b), and as directive stimuli (Staats, 1968d).

THE REINFORCING FUNCTION OF ATTITUDES. Staats (1964a) reported a study that employed the attitude words of the language conditioning procedures as reinforcers in a simple two-response discrimination task. Examples of positive attitude words employed were *family, laughter, honest, home, hero, angel, smile, brave, famous,* and so on. Examples of negative attitude words were *thief, ashamed, sad, hurt, foolish, fat, guilty,* and so on. It was found that when one of the two alternative responses was followed each time by presentation of a positive attitudinal word, the subjects were conditioned to make this response more frequently than a group of subjects receiving nonattitudinal words. When the response was followed by presentation of a negative attitude word, the subjects were instrumentally conditioned to make the response less frequently than the neutral word subjects (Finley and Staats, 1967).

Golightly and Byrne (1964) also conducted an early study of the reinforcing value of attitude words. They did not isolate the identifying nature of the attitude words—the emotional or affective property—but rather considered attitude statements that agreed with the subject's to be positively reinforcing and those that disagreed with the subject's attitude statements to be negative reinforcers. These investigators employed a discriminative task also, reinforcing responses to one stimulus feature and not to other stimulus features. The agreed-with statements were found to strengthen response to the stimulus. Classical conditioning and the formation of attitudes were not considered by Byrne. Stalling (1970) related Byrne's work with attitude agreement-disagreement to the classical conditioning attitude theory, however.

In general, it is suggested that social stimuli that have attitudinal properties for an individual will have reinforcing properties for him also. If the attitude stimulus occurs after the individual has made an instrumental behavior, the instrumental response will be reinforced in a behavioral sense.

THE DIRECTIVE FUNCTION OF ATTITUDES. It is paradoxical that in the definition of attitudes the emphasis is upon the attitude as an internal state or process. However, rarely is this internal process ever observed or measured. The internal process is inferred. The event that is actually observed is the instrumental behavior that is made toward the attitude stimulus. This involves the directive function of attitude stimuli—their ability to bring on approach or avoidance behaviors.

Stipulation of the attitude (A) and the directive (D) functions of attitude stimuli, and of the principles involved, leads to explicit understanding of the principles. Thus, for example, attitude rating scales, which are usually considered to be measures of the attitude (emotional) response to the stimulus, can actually be seen to measure the directive value of the stimulus. That is, making a rating response to an item involves making an instrumental motor response of writing, or of making an instrumental verbal response, such as in answering an item orally. The instrumental response may serve as an index of the inner attitudinal response, but the two are not the same.

Thus, for example, a study was described in which a word was paired with electric shock to condition a sweat gland response to the word. Following this treatment the subjects also rated the pleasantness or unpleasantness of the word on a seven-point attitude rating scale. It should be noted that the two measures dealt with the different functions of the attitudinal word stimulus. Physiological equipment was employed to measure the actual emotional response, the A function, and the rating scale measured the value of the word stimulus for directing approach-avoidance instrumental behavior. The placing of a mark on the scale to indicate the word stimulus was unpleasant may be considered to be in the class of avoidant responses, which would include other types of responses such as moving away from an attitudinal stimulus, saying negative things about the stimulus, and so on.

The study thus actually concerned classically conditioning an emotional response to a stimulus and measuring the extent to which this changed the directive value of the stimulus. The results showed that the negative emotional conditioning resulted in the stimulus coming to be a directive stimulus eliciting avoidance behaviors (a type of transfer of control).

The fact that the extent of the emotional conditioning and the extent of the instrumental rating response were correlated indicates that the rating behavior can serve as an index of the underlying emotional (attitudinal) response. It is suggested that it is because the attitudinal and the directive functions of stimuli generally covary that the latter is used as a measure of the former.

However, the difference between them should be noted, since the traditional usage of attitude scales as indicants of the underlying attitude has tended to suggest that the two are one and the same. This leads to various misunderstandings. For example, one of the traditional types of study of attitudes involves selecting a group of subjects on the basis of attitude scales and then relating the "attitudes" to some type of social behavior. The attitude measuring instrument is taken to be more basic, to tap the internal process more directly, than the social behavior measure. The present analysis suggests that both are examples of instrumental responses under the control of the emotional value of the attitudinal stimulus. Neither of the two measures is more basic than the other.

As another example, D. T. Campbell (1947), in investigating his attitude theory that attitudes involve an affective component, a cognitive component, and an action component, had subjects respond to statements about different racial and national groups. He measured the affective component by like-dislike items, the cognitive component by the subjects' statements about the morality and intelligence of the groups, and the action component by a social distance scale which involved statements concerning the closeness of social relationship one would tolerate in the members of another group. In terms of the present analysis, these measures do not represent different types—such as affective, cognitive, and behavioral. They are all items that measure approach-avoidance behaviors, in verbal form, in part under the directive control of attitudes toward the racial and national groups. They are simply different instances of the same classes of instrumental behaviors. As members of the classes of approach or avoidance behaviors, these several behaviors would be expected to indicate the emotional, reinforcing, and directive values of the attitude stimuli involved.

THEORY INTEGRATION THROUGH USE OF THE A-R-D PRINCIPLES. A central function of a classical theory involves the manner in which it can integrate diverse concepts and observations. The A-R-D theory should be considered in terms of this characteristic. A useful example occurs in the area of attitudes. In the field of social psychology the study of attitudes is one in which many studies have been conducted and many concepts suggested, many of them disparate from one another and quite unrelated, and most of them unrelated to the basic principles of learning. This has been true even though most of the theorists in the field of attitudes include the conception that attitudes are relevant to emotional or affective responses. The manner in which the A-R-D analysis may be employed to integrate the diverse concepts of attitudes as well as the observations that underly these concepts would require a great deal of space. The following excerpt, however, will illustrate the integrational possibilities. The excerpt is of a passage that employs the A-R-D theory to interrelate empirical concepts and observations of human behavior ordinarily treated as disparate, antagonistic, or unrelated. Greenwald (1968) begins by quoting some of the variegated definitions of human attitudes:

> An attitude is a mental and neural state of readiness, organized through experience, exerting a directive or dynamic influence upon the individual's response to all objects and situations with which it is related (Allport, 1935).
>
> . . . an attitude is a predisposition to experience, to be motivated by, and to act toward, a class of objects in a predictable manner (Smith, Bruner, & White, 1956).
>
> Attitudes are predispositions to respond, but are distinguished from other such states of readiness in that they predispose toward an evaluative response (Osgood, Suci, and Tannenbaum, 1957).
>
> An attitude is a disposition to react favorably or unfavorably to a class of objects (Sarnoff, 1960).
>
> . . . attitudes are enduring systems of positive or negative evaluations, emotional feelings, and pro or con action tendencies with respect to social objects (Krech, Crutchfield, and Ballachey, 1962).
>
> Attitude is the effect for or against a psychological object (Thurstone, 1931).
>
> Attitude is . . . an implicit, drive-producing response considered socially significant in the individual's society (Doob, 1947).

Greenwald then goes on to integrate these definitions of the concept of attitude, within the principles of the A-R-D theory, as follows:

Staats' formulation of the three functions of attitudinal stimuli will provide the basic language to be used here for comparing attitude definitions (see Staats, 1968d). The three stimulus functions of attitude objects identified by Staats are: (a) conditioned stimulus function—objects elicit emotional reactions; (b) reinforcing stimulus function—exposure to an attitude object may function as a reward or punishment; and (c) [directive] stimulus function—the attitude object serves as a signal for the performance of a variety of instrumental responses.

The first two definitions listed above (Allport, 1935; Smith et al., 1956) identify attitude broadly as a readiness to respond. These and other similar definitions may be said to stress the [directive] stimulus function of the attitude object with minor emphasis on the conditioned stimulus function (since it may be assumed that a portion of the response tendencies in relation to an object may be emotional).

Definitions that refer to readiness to respond but specify an evaluative dimension (Krech et al., 1962; Osgood et al., 1957; Sarnoff, 1960) appear to lay approximately equal stress on conditioned and [directive] stimulus functions. That is, the tendency to respond favorably, say, may include the tendency to respond with positive emotional reactions (e.g., affection) and instrumental responses (e.g., "striving for" to use Staats' term).

Attitude definitions that refer to affective reactions to an object (Thurstone, 1931) or to "implicit drive-producing" reactions (Doob, 1947), have apparently focused the conception of attitude on the conditioned stimulus function.

None of the attitude definitions sampled here makes explicit reference to a reinforcing stimulus function. (pp. 362–63)

It should be noted that in each case the attitude concept was based upon systematic observations of human behavior, some obtained in controlled laboratory study, but the empirical concepts referred sometimes to different aspects of the observations and were not unified by a common set of theoretical principles or related to a general theory. Thus, the conceptions that have remained disparate and competitive can be considered within one set of principles.

## Social stimuli

The A-R-D model integrates the concepts of personality and learning principles in the consideration of social interaction, providing a central place for the concept of attitudes as emotional responses. Each person in an interaction can be considered to have, as an important personality repertoire, a previously acquired emotional-motivational system. Part of this emotional-motivational system involves attitudes towards social stimuli, as well as the social behaviors the attitudes mediate. The social behaviors would involve large numbers of instrumental and verbal responses of various kinds and various degrees of complexity. As mediators of the social behaviors, the emotional-motivational characteristics of one individual in response to another will strongly influence the character of the interaction.

It should be added that social interaction represents more than the functioning of previously acquired personality repertoires. The individuals' behavior will be modified by the nature of the interaction according to the principles of classical and instrumental conditioning. The individuals in an interaction can learn new attitudes. For example, in a sexual interaction that is positive the person may learn new positive attitudinal responses. In addition, the person's instrumental behaviors of love making can receive reinforcement by the actions of the partner. Moreover, new instrumental behaviors can be elicited by the partner and learned, also in a manner that may include the action of reinforcement. In these general ways many new attitudinal, cognitive, and instrumental behaviors may be acquired in social interaction.

These are abstract principles, however. It is necessary to indicate what it is about a person that can function as a stimulus for another. For one thing, the individual is a physical stimulus object that has multiple characteristics (Staats, 1963), as will be indicated. For another, the individual, as he behaves, produces stimuli. It may be added that there may be additional stimuli present in a social interaction that are not part of the parties to the interaction but that also affect the interaction and its outcomes.

## THE PERSON'S PHYSICAL ATTRIBUTES AS SOCIAL STIMULI

A person as a physical stimulus object can serve as an unconditioned stimulus in several ways. Like other physical stimuli, the individual is a complex visual stimulus. He can thus serve as a stimulus which elicits visual sensory responses in another person that are conditioned as images. On this basis the person may experience visual memories of the individual in imaginal form. Such conditioned images may serve to affect the later behavior of the individual in any of the types of social interaction to be described—for example, conformity, imitation, attraction, leadership, and so on. The individual may also be an auditory, tactual, or thermal stimulus, and so on. As an example, the male who has felt the body of a woman may have conditioned kinesthetic and tactual sensations that enable him to image what she feels like.

Social stimuli typically elicit emotional responses along with their role in eliciting sensory responses that can be conditioned. A pretty girl not only elicits a visual sensory response, she also elicits emotional responses in the male viewer. The tactual stimulus of a kiss or caresss is an unconditioned stimulus that elicits emotional responses. Genital contact with the genital organs of the opposite sex or contact of the sexual organs in other ways also constitutes a strong emotion-eliciting unconditioned stimulus.

Because of the importance to humans of learning, there are few unconditioned stimulus functions of people that have not been influenced by learning history. While a caress, for example, has unconditioned stimulus aspects as a physical stimulus, part of its functions will undoubtedly stem from the conditioning history of the recipient. This can be seen in differences between individuals, between groups, and between cultures.

As has been indicated, a stimulus that elicits an emotional response will also function as a reinforcing stimulus. This is true, also, of the physical characteristics of people. The reinforcing function of people for each other can be seen even in naturalistic situations. The "double take" response is a very good example. A frequent occurrence that has been the butt of many jokes is the way a man will glance at a buxom female passerby and then the reinforcement of the visual stimuli will immediately strengthen his various attentional behaviors so he takes another look—the "double take." In general it would be found that handsome people reinforce attentional behaviors and thus are attended to more strongly than plain people.

It has been indicated that emotional stimuli will also serve as directive stimuli for a large class of instrumental behaviors—either "striving for" or "striving against" behaviors, depending upon the value of the emotional stimulus. It follows that people who as physical stimuli elicit an emotional response will also have a directive function as well. Using the examples that have already been employed, the individual who is physically beautiful will elicit "striving for" (approach) behaviors in others to a greater extent than will the individual who is plain. Thus, a pretty girl will elicit a greater number of phone calls and requests for dates, more frequent "body contact" responses, invitations to social events, attention, and so on.

The physically handsome individual would be expected to control many other types of responses that were not directly related to sex, however. He should be listened to more, approached more in a physical sense, imitated more, followed more, voted for more, and so on, as was indicated in preceding chapters.

## THE PERSON'S BEHAVIORAL ATTRIBUTES AS SOCIAL STIMULI

At this point it becomes important in discussing social interaction to indicate the manner in which the *behavior* of parties in a social interaction, as contrasted to their physical characteristics, constitutes an important variable. The behavior of the individual can have emotion-eliciting properties for another person. The spanking the parent gives to the child elicits a negative emotional response, for example. There are various behaviors that elicit negative emotional responses. Selfish behavior is another example, although in principle this is similar to punishment in that it takes positive A-R-D stimuli away from the other person. Driving a car too fast may be aversive to one's passengers. Talking loudly, being devious or untrustworthy, being cold and remote, beating another at games, being dependent, and so on, may all be aversive to others. There are many types of behavior that elicit positive emotional responses as well.

Again, behavioral stimuli that elicit emotional responses in others will also have reinforcing value for those others. When the individual asks a person to turn down the radio, the behavior of doing so and removing the aversive stimulus has reward value. It is the reward value of the helping behavior of the person that reinforces and maintains thereby the verbal behavior requesting action in others. In general, we learn social behaviors of many kinds because these behaviors are reinforced by the types of behaviors they elicit from others. As an example, it has been shown that the behaviors involved in relief from stress (a sigh of relief) can serve as a reinforcing stimulus to strengthen another person's instrumental helping responses that provided the relief (Weiss, Boyer, Lombardo, and Stich, 1973). Lombardo, Weiss, and Buchanan (1972), in a study which also provides support for the present principle, showed that yielding after prior disagreement was a behavior that had reinforcing properties.

The directive function of emotional stimuli pertains to behavioral stimuli that one person provides another in social interaction. Let us say, as an example, that the child sees a teacher severely admonish another child. The behavior of the teacher will come to elicit a negative emotional response in the observing child. The teacher's behavior will thus tend to control the class of "striving away from or against" behaviors of the observing child, perhaps including the behaviors of being good and thereby avoiding admonishment.

## OTHER STIMULI IN SOCIAL INTERACTION

The focus has been upon the manner in which the stimuli of one person elicit emotional responses in another. This can result in classical conditioning and instrumental reinforcement, as well as in directing approach or avoidance behavior. It should be noted that the particular responses that the person makes in interaction ordinarily will be a function of conditions in addition to those of his A-R-D system. For example, past training in terms of responding respectfully to people, may determine the nature of the individual's response to a person, not the negative attitude that the person elicits. To illustrate, Rokeach and Kliejunas (1972) found students' attitudes toward cutting class was a stronger determinant of their incidence of class-cutting behavior than their attitudes toward their professor.

Situational stimuli, other than those associated with the individuals involved in the interaction, will also play a role in the process of the interaction. For example, an experiment by Griffitt (1970) may serve to exemplify this process. Subjects in an uncomfortably warm and crowded room, which would be expected to elicit a negative emotional response, gave attitudinal ratings of another person that were significantly more negative than subjects who experienced a more comfortable temperature in a less crowded room.

In Chapter 3 it was suggested that the individual's personality repertoires influence how he will behave, but that situational stimuli also have an influence (see Figure 3.1), and the effects can be in opposition. The example used was LaPiere's (1934) study in which expressed attitudes about Chinese people did not correspond with actual behavior exhibited toward a Chinese person. This discrepancy has been noted in other studies; Wicker (1969) reviewed a number of studies in which verbally assessed attitudes did not agree with overt behavior.

In the context of discussing the effects of attitudes on overt behavior, it was originally suggested that the individual may learn verbal behaviors that do not agree with the nonverbal behaviors that have been learned. "Another and more common reason for an apparent discrepancy between the way an individual talks (or how he responds on an attitude test) and how he acts, may be the introduction into the situation of additional stimuli that are also related to the overt behavior" (Staats, 1963, p. 351). The example given was of a person with a negative attitude toward a political billboard who does not deface the billboard because of the presence of a policeman. Rokeach (1966) has suggested similarly that interacting attitudes should be considered; that the attitude toward the situation has an effect, as well as the attitude toward the social stimulus. Bandura (1969) agrees with these interpretations: "A theory that predicts attitudinal responses on the basis of both subject and situational variables would undoubtedly have greater predictive power than one relying solely on subjects' evaluations of the attitude object in undefined contexts" (p. 599). Baron, Byrne, and Griffitt (1974) also agree: "Failure to account for situational differences between the verbal and behavioral assessments of attitudes may account for many findings revealing attitude-behavior inconsistencies . . ." (p. 214).

It can also be noted here that research has shown other results that may be interpreted to be inconsistent with the principles that have been described. For example, Aronson and Linder (1965) had subjects interact with another person who was actually a confederate of the experimenter. Later, a subject overheard the confederate evaluating the subject, in (1) a consistently favorable way, (2) a consistently unfavorable way, or (3) in a manner that commenced with negative remarks and progressed later to favorable comments. The subjects were then assessed for the extent to which they had positive attitudes toward the confederate. As expected from the conditioning principles, the subjects in the first group had more positive attitudes than the subjects in the second group. However, the most positive attitudes were displayed by subjects who received the third treatment.

As Aronson and Linder noted, these results are inconsistent with a simple reinforcement theory; because more favorable comments were given in the first condition than in the third condition, however, the third showed the greater attitude conditioning. Aronson and Linder explained this result by suggesting that the subjects in the third condition were first made anxious by the unfavorable remarks. When the confederate's statements then became more positive they not only received reward, but their negative emotional responses were reduced. It should be noted that this interpretation is not in opposition to an expanded learning account. Deprivation of an A-R-D stimulus increases its value. And the condition of unfavorable comments could have effects analogous to deprivation. There are other learning principles that could also be relevant in this context (see, for example, the principles of behavioral contrast, Herrick, Myers, and Korotkin, 1959; Reynolds, 1961a, 1961b; Staats, Finley, Minke, and Wolf, 1964).

Aronson and Linder also thought it possible that their results could be explained by what the subjects said to themselves. The subjects in the first group said that the confederate was an indiscriminate judge, and the subjects in the third group said that the confederate was a discriminating judge. This too is a possibility, and one that is not inconsistent with the present interpretations—as a later section on communication will indicate. Thus, while the present chapter is primarily concerned with the operation of A-R-D principles in social interaction, it should be remembered that other principles may be involved. For example, the individuals in an interaction will have acquired language-cognitive systems that are likely to be elicited, in addition to attitudinal elicitation. Where the study does not control the subjects' language-cognitive reactions, A-R-D manipulations may not have clear effects. The studies to be cited in the following sections isolate the operation of the A-R-D principles. Many other studies, however, require more complex analyses.

In this context, the concept of "reactance" may also be mentioned. Brehm (1966) has proposed that the dependence of an individual upon another person arouses hostility in the person (see also Berkowitz, 1973). In the present terms such dependence can be considered a negative

A-R-D stimulus, since it involves aversive effort on the part of the person. Considered in this light, some research literature that has been considered to be inconsistent with a learning interpretation can be seen to be compatible. For example, Jackson (1972) has shown that giving a rewarding stimulus (a soft drink) to a subject resulted in the subject responding less to the experimenter's influence than occurred when no reward was involved. It is suggested that there are situations in life where a gift, a seemingly positive A-R-D stimulus, actually elicits a negative emotional response. This can occur in various ways: the gift may be a "put-down," as in the case of charity; the gift may be a bribe, with immoral and illegal involvements; the gift may have directive controlling power for effortful behaviors that outweigh the positive aspects of the gift (reactance), and so on. Research in cognitive dissonance (Festinger, 1957) can also be considered in these terms. Again, it must be emphasized that there are cases where experimental results may be used to suggest the introduction of new concepts when actually more detailed analysis in terms of the behavioral principles is possible. While it is not feasible to treat all the research literature herein, it is suggested that much that seems to be inconsistent with the present principles is not.

## Social behavior areas of study

The organization and development of social psychology has to a large extent been influenced by the areas of social behavior studied. Experimental tasks and studies using the tasks have tended to yield enclaves of research—for example, areas such as conformity, attraction, prejudice, social perception, imitation, leadership, helping behavior, imitation or modeling, impression formation, persuasion, communication, and so on. There has also been a tendency for separatistic theories to develop in the context of the experimental tasks utilized in the areas of study. Consideration of the different areas in terms of the A-R-D principles, however, provides a unifying systematization.

The author first attempted to show such possibilities in an analysis in 1964 entitled "Social Interaction, Attitude Function, Group Cohesiveness, and Social Power." An excerpt from this account will be presented to provide an example of the manner in which theoretical analyses can be made at this level from which experimental work can be projected.

> It is suggested that a change in the reinforcing [A-R-D] value of an individual will increase verbal and motor approach or companionable responses, respectful responses, affectionate behavior, following behavior, smiling, pleasant conversation, sympathetic responses, and the like.
>
> It is important to indicate that there are two ways that the reinforcing [A-R-D] value of a stimulus can be increased. Reinforcing value may

be increased by various learning processes. . . . In addition, however, the reinforcing value of a stimulus for an individual can be increased by depriving the individual of that stimulus. Thus, both of these operations should also have the effect of increasing the $^DS$ value of the stimulus. . . . As an example, it would be expected that . . . imitation behavior . . . would [occur] more strongly if the child [was] first . . . deprived of people. . . . It also follows that the $^DS$ value of a reinforcing stimulus could be reduced by reducing its [A-R-D] value. This could be done by satiation operations . . . .

It can be suggested that [there are] several different ways that the reinforcement value of a social stimulus could be changed through conditioning. For example, the same process could be accomplished via moving or still pictures. It would thus be expected that if an adult was shown to "possess" many reinforcers in a film, his own [A-R-D] value as a stimulus would be increased, and consequently his $^DS$ value for controlling following behavior would also be increased. . . .

In addition, it would also be expected that the change in the reinforcement value and $^DS$ value of a person could be changed purely through the function of language. . . . It would not even be necessary that the individual himself be paired with the word reinforcers. The same end could be achieved by pairing his name with the word reinforcers [A-R-D words]. . . . It would also be possible to pair an individual's picture with the reinforcing words and thus produce the conditioning and its effects. For these reasons, the change of reinforcer and $^DS$ stimulus value and the ways these influence group cohesiveness and other social interactions could be made via mass communication media —film, radio, television, newspapers, magazines, and books—as well as through face-to-face communication.

It should be possible to demonstrate the validity of these implications and the basic analysis in systematic experimentation. For example, it should be possible to enhance the leadership (social power) of an individual . . . by increasing that individual's [A-R-D] value through direct or language conditioning of the various types mentioned. This could also be accomplished by "labelling" an individual with a name that has positive or negative reinforcement value, such as "doctor." It should also be possible in these various additional ways to change imitational control . . . as well as . . . group cohesiveness.

The learning analysis also appears to have other implications. For example, concerning group prejudice, if a group's distinctive physical features, name, dress, language, or other behavior, elicits little positive [A-R-D] value, the members of the group will have little social controlling power. That is, members of the group will not be listened to, imitated, followed, chosen as a companion, and so on, or even hired, promoted, and so forth to the extent that members of groups will who have more reinforcement value. If the group, or its name, elicits negative [emotional] responses its members will be more likely to be insulted, avoided, opposed, arrested, rejected, fired, and so on, and in extreme cases, killed, lynched, and fought against. Verbal passages, films, or direct experience that presents members of the group, or its name, in contiguity with negative reinforcers would be expected to enhance the undesirable behaviors controlled by (made to) members of the group.

Matters of concern in the field of propaganda and advertising can also be incorporated into this learning analysis. One of the common techniques of the political editorials, for example, is to present the name of a political figure in contiguity with positive [attitude] words when the writer favors the politician or with negative [attitude] words when the writer is opposed. Each one of the pairings of name and [attitude] word may be considered to be a conditioning trial that changes the [A-R-D] value of the political figure's name and person. When that name or its owner appears on later occasions, the individual who has been so conditioned, taking the negative case as an example, will speak against the political figure and his principles, vote against him, fight against him, and so on (Staats, 1964b, pp. 333–36).

Some of the research that tests the principles of the A-R-D model in areas of social behavior, or that can be seen to involve the principles, will be summarized in the following sections.

## ATTRACTION

The field of attraction research refers, basically, to the fact that people will approach other people in a variety of different ways and to differential degrees. The behaviors involved in attraction studies are customarily some form of "striving for" social behavior, such as voting for, dating or indication of liking for dating, the physical proximity (social distance) chosen, sociometric ratings, attitude ratings, questionnaire ratings, and so on. Some form of manipulation of attitudes in the subjects is basic in attraction studies. This will be seen to be general to the various areas of social behavior. Usually the effect of the attitude change is related to change in the directive function of the attitude stimulus. As will be seen, however, the effect of attitude value on reinforcing value has been treated also.

An early study of attraction within the context of learning principles was conducted by Lott and Lott (1960). They had groups of three children play a game in which success could be manipulated for each individual child. The rewarded children later chose more members of their group on a sociometric test than did children who were not rewarded. In terms of the A-R-D principles, it may be said that rewarding the child conditioned a positive attitude in him toward his two playmates. The positive attitude then elicited the later "striving for" behavior of selecting the playmates on the sociometric test (Staats, 1964b).

Another experimental study was recently conducted by Griffit and

Jackson (1970). In their study business school subjects had to make personnel selection recommendations on candidates for a position. The experimenter manipulated information concerning the intellectual performance (behavior) of the candidates. The applicants' grade point average was given, as well as his performance on a test of scholastic ability. Higher grade point averages and higher scores on the test of scholastic ability are behavioral characteristics that elicit positive attitudes in the subjects. It would be expected, according to the A-R-D model, that candidates described with these positive emotional behavioral characteristics would have positive directive value and would be recommended more highly than subjects with less positive descriptions. The results supported this expectation.

This study demonstrates the manner in which verbal labels, when applied to another individual, may change the person's attitude toward another individual and thus his instrumental responses to that individual. A study that directly tested this hypothesis employed 40 photographs of women. These photographs were shown to subjects with either positive emotional labels, such as "social workers"; neutral labels, like "cocktail waitresses"; or negative labels, like "alcoholics." The subjects gave an attitude rating on a pleasant-unpleasant scale. It was found the subjects' attitudes were determined by the emotional value of the label employed for the group of photographs (Staats and Higa, 1970).

Early (1968) conducted the first study to show that overt instrumental behavior toward a person could be changed by changing the subjects' attitude toward the person through conditioning. Using the language conditioning procedures, she conditioned positive attitudes in the members of a class to the names of two socially isolated children in that class. As a consequence of the isolate children's names (and hence the isolate children themselves) eliciting positive attitudes, it would be expected the children would be approached more. The incidence of approach behaviors was tabulated and the expectations were supported.

It has been suggested that the physical and behavioral characteristics of the individual will elicit attitude responses in another, and this will help determine the instrumental behaviors (attraction) made to that individual (see also Staats, 1963, 1964b, 1968d, p. 56, and p. 389). Sigall and Landy (1973) note that:

> Physical attractiveness is a variable which clearly has important interpersonal consequences. Over the past few years, these consequences have been investigated by several social psychologists. Two general findings which have emerged are that (a) good-looking people have greater social power—they can be more persuasive, their evaluations have more impact —than their unattractive counterparts (e.g., Mills and Aronson, 1965; Sigall and Aronson, 1971; Sigall, Page, and Brown, 1971), and (b) all other things being equal, physically attractive individuals are liked better

than unattractive individuals (e.g., Byrne, London, and Reeves, 1968; Walster, Aronson, Abrahams, and Rottman, 1966). (p. 218)

Walster, Aronson, Abrahams, and Rottman (1966) arranged a "computer dance" for which participants had been assessed for physical beauty. It was found that physical beauty was the only factor that seemed to affect liking and desired contact, among all the other variables on which the experimenters had data. Byrne, Ervin, and Lamberth (1970) also tested the effect of physical beauty on attraction. Male and female subjects were introduced and spent 30 minutes together. Later, they were independently assessed for positive instrumental responses to each other. The results showed that physical beauty influenced such instrumental behaviors as (1) how close the couples stood to each other in speaking to the experimenter, (2) memory of the other's name, (3) talking to each other socially after the experiment, (4) desire to date the other in the future, and (5) and ratings of the other as a desirable spouse. This experiment supports, incidentally, one of the main suggestions of the A-R-D model, that is, that the positive attitude response actually has directive value for a whole repertoire of "striving for," or approach, instrumental behaviors. Landy and Sigall (1974) have shown, as another example, that essays supposedly written by pretty girls (as shown by pictures) were evaluated more highly than were essays written by less physically appealing girls. Murstein (1972) has shown that similarity of physical attractiveness is evident within couples who evidenced their "approach" for each other by being engaged or going steady. Sigall and Landy (1973) have also found that a male with an attractive partner is evaluated positively, and a male with an unattractive partner is evaluated relatively negatively.

Byrne and his associates have tested other hypotheses suggested in the A-R-D model. For example, the suggestion that attitudes would control voting behavior has been verified in a study by Byrne, Bond, and Diamond (1969). It was also suggested in the original analysis that the social interaction effects could result from mass media presentations. Byrne and Clore (1966) supported this expectation, employing three different modes of stimulus presentation: a colored movie, a tape recording, and verbal presentation. Attraction responses were affected by the several types of attitude conditioning. Pandey and Griffitt (1974) have recently shown that subjects with positive attitudes toward a person will help (perform work for) that other person more than subjects who have neutral or negative attitudes.

The A-R-D model also has other implications for attraction. For example, it has been suggested that an attitude stimulus would function as a reinforcing stimulus. Liked (or attractive) people should thus have more reinforcing value than people toward whom the attitude response is not so positive. In this context one hypothesis tested was that the names of people toward whom subjects had already learned attitudinal responses could also serve as reinforcers. Names such as *Ernest Hemingway, Bill Cosby, John F. Kennedy, Robert Frost, Albert Einstein,* and *Carol Burnett* were selected because they elicited a positive attitude rating in a sample of subjects. Neutral names were used in a control group; examples of there were: *Pancho Villa, Paul Tillich, Alger Hiss, Alexi Kosygin, Richard Daley,* and *Chou En-lai.* For one group of subjects one of two responses was followed each time by a positive attitude name. For the other group the neutral names were used as the reinforcing stimuli. The conditioning was significantly greater in the first group (Staats, Higa, and Reid, 1970). Since names appear to function in the same way as the person does himself, the results suggest that people as stimuli may function as reinforcers in affecting what instrumental behaviors are learned.

This principle receives further support from a study conducted by Lott and Lott (1969) which employed the previously described suggestion that photographs would have the same A-R-D functions as the actual person. The subjects were school children and the photographs were of peers, either liked, neutral, or disliked. Subjects whose instrumental response was followed each time by a picture of a liked peer learned the response in greater strength than did subjects for whom neutral pictures were the reinforcers. The neutral pictures served as positive reinforcers more effectively than did the pictures of disliked peers.

Another line of research that shows the effect of attitude value of the individual on his ability to reinforce another person involves the similarity between the two people. Sopolsky (1960), in a study that introduced this research area, first had subjects complete a questionnaire on behaviors they desired in social situations. Then the subjects worked with another person who was supposedly either like them on the questionnaire or unlike them. The other person was actually the experimenter, who reinforced the subject with a verbal reinforcer for certain types of responses in a later experimental task. First, the experimental manipulation produced positive ratings for the similarity condition and negative ratings for the dissimilarity condition. Thus, the directive value of the experimenter was changed. Second, the efficacy of the experimenter's social reinforcement was positively related to his described similarity to the subject. Spires (1960) conducted a similar study employing the word *good* as the verbal reinforcer, with substantiating results. The experimenter was described in positive attitude-eliciting terms or in negative terms, rather than through the similarity-dissimilarity method, and this affected his reinforcement value.

Byrne has focused his work in the field of attraction primarily in terms of whether the participants are similar to or different from one another, especially in terms of agreeing or disagreeing attitudes. He began his work in a cognitive theory framework (Byrne, 1961, 1962). Golightly and Byrne (1964), as has been described, applied the principle of reinforcement—showing that agreed-with attitude statements could function as positive reinforcers and disagreed-with attitude statements could function as negative reinforcers in instrumental conditioning. Since that time Byrne and his associates have conducted a number of studies that show the reinforcing properties of attitude statements as dependent upon their similarity to the subject's attitudes (Byrne, 1969). Stalling (1969) indicated, however, that it was not similarity that was crucial, but the emotion eliciting properties of the attitude statements, as the A-R-D model suggested. Byrne's work on attitude similarity may thus be considered as a subclass of the more general treatment of attitudes.

Except for the focus on attitude similarity Byrne has in recent years elaborated his "reinforcement model" of attraction so that it has progressively come to be more like the A-R-D model. Although they call it the reinforcement model, Byrne and Clore (1970) have come to recognize the classical conditioning function of agreed-with and disagreed-with attitude statements, and they have also begun to acknowledge the directive function of attitudes in the model. These successive additions bring the varied and productive research findings of Byrne and his associates into strong support of the general A-R-D model, in the area of attraction and similarity.

PREJUDICE AND AGGRESSIVE BEHAVIOR. This study of attraction has concerned only one half of the A-R-D model—the manner in which *positive* emotional responses tend to elicit a large class of "striving for" behaviors, and so on. On the other side, the model suggests that negative emotional responses elicited by an individual will tend to elicit a large class of "striving against" instrumental responses. "These responses, both verbal and motor, will include cruel behavior, derisive or insulting comments, obstructionistic behavior, antagonistic responses, oppositional voting, and the like. . . . An increase in the negative attitude value of an individual through conditioning would be expected to insure that these overt responses would be more strongly controlled" (Staats, 1964b, p. 334).

Prejudice may be considered to refer to negative emotional responses elicited by social stimuli. The term is especially appropriate when there is a class of people involved toward whom the person has learned a negative attitude. Any individual who is identified as a member of that group will then elicit the negative attitude. Aggressive behavior may be considered to refer to the instrumental behaviors elicited by the negative attitudes. The behaviors are part of the general class of "striving against" behaviors that has been described.

A number of studies indicate that the A-R-D principles also operate in this area of human behavior. For example, Cooper (1959, 1969) has shown that disagreed-with statements concerning other groups elicited more intense emotional responses, as recorded by physiological apparatus (galvanic skin response), than did agreed-with statements. This work is relevant to the work of Sapolsky and Byrne and ties in with the findings of Stalling (1969), Sachs and Byrne (1970), and Byrne and Clore (1970) that such statements can be employed to condition attitudes to new stimuli.

Rankin and Campbell (1955) and Porier and Lott (1967) have shown that persons with stronger negative attitudes toward black people, as indicated on an attitude scale (which actually involves the directive function), responded with a more intense emotional response (GSR) when touched by a Negro. Westie and DeFleur (1959) showed the same effect using photographs of Negroes. These studies show clearly that minority group members, towards whom people in the majority of the culture have expressed negative attitudes, actually elicit negative physiological emotional responses. Again, such studies help provide basic evidence linking the concept of attitudes to the study of emotional responses.

Berkowitz has focused upon the "classical conditioning" model of aggression which uses some of the A-R-D principles. Thus, he used the present author's language conditioning method of attitude formation to study aggressive behavior.

> Extending the earlier research by Staats and Staats (1958), Berkowitz and Knurek (1969) conditioned subjects to dislike a certain name by pairing this name with unpleasant words. When this attitudinal training was completed, each subject participated in a brief discussion with two fellow students, actually the experimenter's confederates, one of whom bore the critical name. Each of these people later rated how friendly or hostile the subject had been towards him. As these ratings indicated, the subjects had exhibited greater hostility toward the person having the disliked name than toward the confederate bearing the neutral name. The unpleasant affect evoked by the name had generalized to the man carrying this label. (Berkowitz, 1970, p. 107)

This study supports several aspects of the A-R-D model. Besides showing language conditioning of attitudes and the effect of the attitude toward the individual, the study illustrates how a negative attitude tends to elicit a whole class of hostile behaviors, since the ratings of hostility were based upon different behaviors of the subjects. Furthermore, the importance of stimuli associated with the individual—especially verbal labels—is clearly demonstrated.

A study by Berkowitz and LePage (1967) has shown how other attitude-eliciting stimuli in a situation can also exert directive control over aggressive behavior. In the procedure the subjects were first conditioned to have a negative emotional response to the experimenter's confederate by having the confederate administer an electric shock to them. Then the subjects were given an opportunity to shock the confederate. For half the subjects weapons were placed on a nearby table; for the other half there were pieces of sporting equipment. The subjects in the presence of the weapons were more hostile to (that is, gave more electric shocks to) the confederate.

In another study (Geen and Berkowitz, 1967) it was additionally shown, among other findings, that negative attitudes were conditioned toward an individual who insulted the subject. Following this negative attitude conditioning based upon the behavior of the confederate, it was found that subjects administered more electric shocks than subjects who had had a neutral interaction with the confederate.

Berkowitz and his associates have conducted a number of other significant studies of aggression that support the A-R-D model of social interaction. A systematic analysis of these studies in terms of the principles and concepts of the general three-function learning theory would help elaborate the theoretical body. This analysis could suggest additional research as well as contribute to a comprehensive account of aggression.

One more study will be mentioned to indicate the manner in which negative attitudes of a person toward another can affect the course of social interaction by influencing the extent to which the second can reinforce the first. Insko and Butzine (1967) found that an experimenter who had a negative interaction with subjects prior to attempting to condition them through reinforcement was less effective in delivering social reinforcement with such subjects than he was with subjects who had not experienced the negative interaction.

Various experiments can be organized within the three-function learning theory to account for social interaction (either attraction or aggression), with various implications of the model for research. As one example there have been studies of social distance, that is, the closeness of relationship people will allow themselves with respect to one another (Bogardus, 1925). A recently completed study (Kapenberg, 1973) conditioned subjects' attitudes toward a label, using the language conditioning method. Then the subjects indicated how close in physical distance they would place themselves in conversing with someone with that label name. Subjects conditioned to positive attitudes had a closer social distance score than subjects conditioned to negative attitudes.

## COMMUNICATION AND PERSUASION

As has been indicated, communication has been considered in simple terms as a form of classical conditioning (Mowrer, 1954; Staats and Staats, 1958). More complex forms of persuasive communication have been considered within the context of learning theory. Thus, for example,

Hovland, Janis, and Kelley (1953) have conducted a series of studies of variables influencing the extent to which persuasive communications affect attitudinal change. Some of these are coincident with expectations derived from an A-R-D analysis. To illustrate, the effectiveness of the message as a function of attitude toward the communicator has been recognized in a number of studies, as shown by the fact that persons with high prestige value will be agreed with more frequently than those with low prestige value (Lewis, 1941; Sherif, 1935). Hovland, Janis, and Kelley (1953) extended such findings in showing that there was greater acceptance of the communication when the communicator had high credibility, due to either trustworthiness or expertness. The manner in which the trustworthiness and expertness characteristics of a person come to be directive stimuli for "agreeing with" (or following) behavior has been suggested in terms of the reinforcement that this type of behavior gains (Staats, 1963, pp. 328–330). Hovland, Janis, and Kelley additionally found that whether or not the communicator was seen as trying to further his own ulterior ends also affected acceptance of the communication. It is suggested that such results as these indicate that there are language-cognitive behaviors of the *communicatee* that are also operative in communication in the naturalistic situation.

> [O]ne qualification should be made to account for the seeming inconsistencies in response to language stimuli that seem to occur. For one thing, conditioning does not occur the same to individuals who have learned different meanings to words. Furthermore, language conditioning occurs in the context of other conditioning experiences. . . . It is also true that because of the person's past training he may respond to the efforts of the propagandist by saying to himself "What this man is saying is not true, such and such is really a good program (or person)". In such a case the individual's past training will result in verbal behavior that actually counters the conditioning provided by the communication source. . . . [T]*he same conditioning results from language stimuli whether the stimuli are produced by someone else, or by oneself.* Thus, when the individual says to himself, as in the above example, that such and such "is really a good program," the positive attitudinal response to "good program" is conditioned to the name of the program. It is in part by showing that our own language (behavior) affects our later behavior that a rapprochement between learning theory and cognitive theory (and other naturalistic conceptions of man) may be achieved. By such means we can begin to understand how it *appears* that we determine our own behavior by our thoughts and decisions (behaviors), and yet still consider human behavior within a scientific determinism—as lawfully caused rather than spontaneous. (Staats, 1968c, pp. 39–40)

The suggestion that there are such language-cognitive behaviors on the part of the communicatee that affect the acceptance of the communication receives support from other results of Hovland, Janis, and Kelley

(1953). They found a delay of three to four weeks after presentation of the communication would dissipate the effects caused by credibility of the source of the communication in that the increased persuasion of a high credible source and the decreased persuasion of a low credible source vanished. This result would be expected if the delay decreased what the communicatee said to himself about the communicator. The effect of the communication would then depend upon the communication itself.

Kiesler, Collins, and Miller (1969) have indicated some of the criticisms of the approach, including the lack of close relationship between the analysis of Hovland, Janis, and Kelley and learning theory:

> Few of the experiments reported in *Communication and Persuasion* even manipulate the classical variables that appeared in the major contemporaneous presentations of S-R theories of learning. . . . To push things still further, it can be argued that even the theoretical terms invoked to explain their findings as well as the language on which they are based are not consistently drawn from the S-R vocabulary. (p. 116)

Recently, a beginning has been launched in the study of the types of persuasion findings of Hovland, Janis, and Kelley within the confines of the A-R-D learning principles. Granoff (1973a) has classically conditioned attitudes toward the topic of a persuasive communication and measured the extent to which the conditioned subjects agree with the communication: "When the communication was consistent with the attitudinal value of the topic, agreement with the communication occurred; and when the communication was inconsistent with the attitudinal value of the topic, disagreement with the communication occurred" (p. 14). In another study subjects were also conditioned to positive, neutral, or negative attitudes toward the names of different groups of foreign people. These names were then presented in pro and con communications which the subjects could choose to write about.

> The positive attitude topic controlled acceptance (writing an essay in favor) of the pro communications and rejection of (writing an essay against) the con communications. The negative attitude topic controlled acceptance of the con communications and rejection of the pro communications. In the neutral control group, there were approximately equal frequencies of acceptance and rejection for both pro and con communications. (Granoff, 1973b, p. 10)

These studies show that manipulation of attitudes toward aspects of a message helps determine the instrumental acceptance or rejection of the communication; this involves the directive function of the communication. Additional research findings may be interpreted within the principles of the three-function learning theory. As one example, Hoffman and Maier (1964) showed that the emotional value of verbally presented solutions to a problem affected the extent to which a group of subjects would accept the solutions. Butler and Miller (1965) demonstrated that the extent to which a person received communication messages from the members of a group was influenced by the person's power to reinforce (the extent to which he elicited positive attitudes in the group members). Kelman and Eagly (1965), on the other hand, found that when the communicator elicits negative attitudes in the communicatees, they will respond to him by misperceiving his message and displacing his message away from their own position—an excellent example of a "striving away from" behavior.

R. F. Weiss (see 1962, 1968) has conducted a series of studies derived from the learning theory of Hull (1943) and Spence (1956). His theory has been that a subject reading a communication first experiences the sight of the communication. This constitutes the stimulus. When he reads the opinion part of the communication, this constitutes his response. The argument—the information which supports the opinion—is considered to be the reinforcement. The result is that the subject learns the opinion to the communication. Following this model, Weiss has proceeded to conduct studies that test various implications of the learning theory, for example, the effect on learning of the delay between the response and the reinforcement. This work has made general methodological, research, and theoretical (Weiss, 1971) contributions to the extension of learning principles to social psychology.

It should be added that research in the area of persuasion and communication, within the context of learning theory, has only begun. The theoretical basis for this area of study, besides the A-R-D principles, should include a learning analysis of language. In general, it should be emphasized that communication is not a unitary process. For example, communication can be based upon the verbal-motor repertoire where the communicator presents words that directly control the behavior of the communicatee. It can involve imitational speech where the communicatee repeats something the communicator has said and thereby learns of the sequence of verbal responses involved. Communication can also involve the learning of attitudinal responses to new words and hence to the objects labeled by those words. When the communicator employs an emotional word as a reinforcer to strengthen or weaken the behavior of another, it can be said that a communication act has occurred. All of the basic language repertoires may be involved in communication, as well as all of the learning principles (Staats, 1963, 1968c). (Examples of some of these types of communication will be given in the section below on imitation.) This multi-process analysis of communication suggests additional research in this area.

To focus solely on the attitudinal value of the message or of the communicator is to oversimplify, and it has been done here only for purposes of illustration and because of space limitations. The multiple facets of communication can be seen in the fact that some studies have not found attitude manipulations, such as source credibility or prestige, to predict the effects of communication (Fine, 1957). It should also be noted that communication and persuasion can involve negative as well as positive attitudes. The process of training soldiers to participate in physical aggression will ordinarily include the conditioning of negative attitudes toward the "enemy." It is suggested that this is an example of negative persuasion. Research should be conducted in communication and persuasion phenomena of a negative sort.

## CONFORMITY

In a group situation, when individuals learn of the behavior of other group members their behavior is subsequently modified in the direction shown by the other group members. Studies of such conformity behavior are very reliable and have constituted an important part of social psychology from early times to the present. It would be possible to select many studies that show the effects of A-R-D variables on conformity.

As one example, a study of Kidd and Campbell (1955) can be seen to manipulate conformity behavior by conditioning attitudes toward the members of the group. Although this study was not conducted within a learning theory formulation, it can be seen as a counterpart to the study of Lott and Lott (1960) in which the individual's positive attitude is conditioned toward group members by rewarding him (presenting him with a positive A-R-D stimulus) in the presence of the group members. This was done by Kidd and Campbell (1955) by indicating to some groups that they had been "successful" in solving a group task:

> Persons in groups given three successes and no failures showed significantly more movement of the second estimate in the direction of the purported group average [conformity] than did either those in the three-failure groups, or those in a control situation involving no actual group experience. Persons in groups given two successes and one failure were intermediate." (pp. 392–93)

Persons in the three-failure group showed less conformity than the control group—an important type of "striving away from" behavior—although the results were not quite significant statistically.

The directive function of attitude stimuli also should have an effect in controlling nonconformity behavior. A study by Vaughan and Mangan (1963) can be interpreted in these terms. The attitude value of the *material* being estimated was varied. When the subjects' values for (attitudes toward) the material were high they conformed less to group behaviors than when their attitudes were less intense. A study by Snyder,

Mischel, and Lott (1960) may be similarly interpreted, although it also was conducted in the context of a cognitive theory of conformity.

## IMITATION AS AN INTERPERSONAL INTERACTION

Imitation or modeling was treated in the last chapter in the context of personality, describing some of the repertoires of which imitation is composed. The manner in which the emotional-motivational system affects imitation and the manner in which imitation may be considered as an interpersonal behavior were not dealt with in any detail. These topics are relevant to an understanding of both modeling and social behavior.

To begin, it is suggested that imitation is one type of social interaction involving a model and at least one modeler. The event that is of interest in the imitative interaction is the way in which the modeler behaves in a manner that is on a dimension of similarity to the characteristics of the model. The modeler may act, speak, or dress like the model. Some of the other behaviors that Bandura (1969, 1971) categorizes as modeling are more usually considered under the headings of aspects of language and communication or in other social interaction areas such as leadership, persuasion, and so on. It should be realized that the same principles and mechanisms are relevant to more than one area of study.

In the modeling interaction, as in the other types, the personality behaviors of the modeler with respect to the model are important. It is suggested that the modeler's emotional response to the model—and hence the reinforcing and directive functions of the model—influences the course of the modeling interaction and the learning that ensues for both the model and the modeler (Aronfreed, 1968; Staats, 1968c). Extension of the A-R-D principles to the modeling interaction suggests various new implications.

EMOTIONAL VALUE OF THE MODEL IN CONDITIONING NEW ATTITUDES. The model may have emotional value for the observing person. For example, as an elicitor of an emotional response, he may produce new classical conditioning in the observing person. Any new stimulus paired with the model will come to elicit the same emotional response he elicits. It is for this reason that advertisers select athletes and other notable and attractive people, who elicit positive emotional responses in many viewers, to pair with their products. The pairing results in positive emotional conditioning to the product in the viewer. (This event could be labeled as communication rather than modeling, or vicarious conditioning, of course.)

EMOTIONAL VALUE OF THE MODEL AND HIS REINFORCING VALUE. Models can serve as reinforcing stimuli in conditioning instrumental behavior in the observing person to the extent that the model elicits an emotional response in that person. This reinforcement by the model may be done verbally. For example, the known athlete who visits an aspiring athlete and says "If you keep working out like that you will be a better player than I am" will have provided a reinforcement contingent upon the novice's behavior. The model with strong emotion-eliciting, and hence reinforcing, value can reinforce the other person in many ways. (Again, this is also an example of a persuasive communication.)

It should be added that the model can also elicit a negative emotional response. In this case the stimuli he applies contingent upon the observing person's behavior will result in a decrease in the strength of the observing person's instrumental behavior. In general, positive words from a person eliciting a negative emotional response, or negative words from a person eliciting positive attitudes, would be expected to act as negative reinforcers. These principles also must be systematically investigated.

EMOTIONAL VALUE OF THE MODEL AND HIS DIRECTIVE VALUE. The principle outlined in describing the A-R-D theory was that a stimulus that elicits a positive emotional response in the person will thereby elicit a class of "striving for" behaviors. These behaviors include those of doing what the model does. The experiment of Bandura, Ross, and Ross (1963a) may be interpreted in this way. That is, one model in the experiment was paired with positive emotion-eliciting stimuli and thus came to elicit a positive emotional response. Later, when the children who had been so conditioned were exposed to this model and another model who did not elicit a positive emotional response, the positive emotional response to the one model controlled the children's imitation behavior.

Following this principle, it would be expected that increasing the positive emotional value of the model would increase his ability to elicit imitative behavior. As another example, Lefkowitz, Blake, and Mouton (1955) dressed the model either in high-status clothes (suit and tie, and so on) or low-status clothes (soiled, worn, and cheap). It was found that the high-status model, the model eliciting more positive emotional responses, was imitated to a greater extent than the low-status model in violation of traffic signals. Models with high ethnic status (Epstein, 1966) and with high competence (Gelfand 1962, Mausner, 1954a, 1954b), have also been shown to control stronger imitation than models with less positive characteristics.

Guay (1971) employed specific derivations from the A-R-D theory in testing the possibility that conditioning emotional responses to the model would result in increased model effectiveness. The children in the experiment were reinforced in a task for selecting one of two colors. This procedure was intended to condition a positive emotional response to that color. Later, the children were exposed to a puppet who performed several distinct acts. The color of the puppet was varied for different children. When the color was the same as the one for which the children had learned a positive emotional response, the puppet's acts were imitated in various ways to a greater extent than in the other condition.

NEGATIVE IMITATION. Thus far the discussion of the A-R-D model of imitation, in following the characteristic of the field, has dealt only with the positive case. It has been suggested, however, that there is a negative case as well, based upon negative attitude responses and the functions that negative attitude stimuli can have (Staats, 1968a, p. 446; 1971a, pp. 123–24).

> It would also appear that the child learns, conversely, not to imitate some people and some types of actions, and so on. It may be suggested that one of the important mechanisms involved here is analogous to that just described, except rather than learning to imitate people with positive reward value, the mechanism involved is learning not to imitate when the person has negative attitude value. Nonimitating behavior may also involve negative imitation—that is, doing the opposite of what someone else has done. Thus, it is suggested that the child ordinarily receives training in which in the presence of nonrewarding models he is reinforced for non-imitating or negative imitative behavior.
>
> For example, a young boy who as is customary has been admonished or teased (both punishing) for imitating girls will have less reward value for girls. In this process the boy will also have learned not to imitate girls, and more generally not to imitate people who elicit in him negative attitudes. (Staats, 1971a, p. 123)

Burdick and Burnes (1958) have conducted a study that can be interpreted in these terms. Subjects filled out opinion scales on two topics and were then given a pleasant interaction and lecture on one of the issues. They again filled out the opinion scales and also rated the speaker. The lecturer came back in again, berated them extensively, and did not speak about the second issue. Again the subjects filled in the opinion scales and rated the speaker. Thus, one issue was never dealt with by the speaker. On this the subjects maintained their opinions over the three tests. After being lectured to pleasantly the first time, and having positive attitudes conditioned to the speaker (as shown also by the way they rated him), they imitated the opinions he had expressed in his talk. After the negative attitude conditioning (as shown also by the way the speaker was rated), the subjects changed their opinions on the one topic in the direction opposite the speaker—a case of negative imitation. It should be remembered that the speaker had not lectured at all upon the topic in the second interaction but had only been unpleasant.

In addition, a study derived from the A-R-D theory and involving negative attitude conditioning has been conducted to investigate the effect upon imitation in children (Martin and Staats, 1973). Children were conditioned to negative attitudes toward a color. They were then faced with a choice of imitating a puppet of that color or a puppet of a color toward which they had no negative attitude (a neutral stimulus). It was found that the children imitated the puppet toward which they had a negative attitude significantly less frequently than they imitated the neutral puppet. These preliminary results require more systematic test.

The A-R-D theory, as indicated, provides a structure for various additional theoretical extensions and experimental expectations, a few more of which may be briefly mentioned. For one thing, it would be expected that the emotion-eliciting value of the model would interact with the individual's previous matching learning. That is, because of our previous learning, when we have a positive emotional response to the model, a discrepancy (nonmatching) between the behavioral stimuli of the model and our own behavioral stimuli will generally have negative A-R-D properties. When the model elicits a positive emotional response in us, matching his behavioral stimuli is reinforcing. This is the condition that the above experiments support.

Other implications arise, however, when the model is a person who elicits a negative emotional response in the observer. When this occurs, matching behavioral stimuli should have negative value, and *nonmatching* the model's behavior should be positive. It has been assumed by some investigators that similarity has positive value. This has been central in Byrne's approach to the study of attraction, for example, where agreed-with statements are employed as positive reinforcers (Golightly and Byrne, 1964). It is suggested, however, that whether or not similarity is positive (in the three A-R-D senses) will depend upon the type of emotional response that is elicited by the model.

Some evidence to support these suggestions can be obtained from a recent study by Kian, Rosen, and Tesser (1973). They found that attitudes and similarity interacted in affecting reinforcement value of attitude statements paired with the source of the statements. The results showed that "Similar attitudes from liked sources are more facilitative than either dissimilar-liked or similar-disliked attitude-source combinations" (p. 366). Moreover, it was found that dissimilar attitude statements from disliked sources act as positive reinforcers. These authors considered this result to be congruent with a cognitive consistency theory and incongruent with the approach of Byrne. The result is in agreement with expectations derived from A-R-D theory, which includes analysis of negative imitation learning.

CONFORMITY AND IMITATION. Finally, in the interest of unification, imitation and conformity can be briefly compared. The difference between the traditional experiments in these two separate areas appears to depend upon whether the group's behavior controls the subject's behavior, in contrast to an individual model controlling the subject's behavior. The first is called conformity; the second is called imitation. It is suggested that the same principles apply to both, and to the next topic as well.

## LEADERSHIP

Leadership refers to a social interaction in which there is influence by the behavior of one party and compliant or following behavior on the part of another. Indices of leadership involve the extent to which compliant behavior is elicited in terms of both the number of followers and the incidence of following behavior, as well as the generality of following behaviors that can be elicited.

It would be expected that the A-R-D qualities of a person would affect the extent to which he would have control over the following behavior of others. There have been many investigations concerned with the personal and behavioral traits of leaders, and many of these can be interpreted within the A-R-D theory. For example, Fiedler (1967) has found that leaders who are more permissive and more considerate are more effective leaders, at least under moderately stressful conditions. Raven and French (1958) have indicated evidence that an elected supervisor controls work output more effectively than an assigned supervisor. In both cases, it seems likely that the leaders were, at least in part, more effective because they elicited a more positive attitude in their followers. It may be added that where obedience in following is the only "striving away from" (avoidance) behavior preventing the receipt of punishment, increase in negative attitudes toward the leader may increase his effectiveness.

As physical stimuli that elicit emotional responses, people can also serve as directive stimuli and directly elicit instrumental behaviors. It can be suggested that the physical features of the individual may contribute to leadership-follower characteristics. An example in support of this was recently reported in the newspapers, which reported that the beginning salaries of graduates of business schools were correlated with their height. Taller graduates secured higher salaries, independent of their abilities. It would be expected that leadership in business—as well as in various other pursuits—would be affected by such physical features. Physical features eliciting positive emotional responses in the members of the society would be an asset; those eliciting negative emotional responses could be a handicap.

It has been suggested (Staats, 1963) that behavioral characteristics also contribute to leadership. Followers (and imitators) of a physically skilled boy will have their behavior reinforced more highly than followers of one who is not so skilled. The same would hold true in the classroom or in any realm of intellectual endeavor. This analysis of the development of leadership has been corroborated in a study by Marak (1964). He had groups of five individuals engaged in a task of attempting to predict the responses of another person to a personality test. The situation was rigged so that one of the five subjects would be correct. The situation involved a number of trials, and group discussion. It was found that when the group was rewarded with money for being correct, group members began to be influenced in their own choices by the person who was correct.

More importantly, they began more to accept (reinforce) his statements in the discussions. And the incidence of leadership attempts of the individual who was made the most able in the task also increased. The results thus substantiate several aspects of the A-R-D theory analysis. That is, ordinarily the individual who has the most able behavior will himself be rewarded. This means the individual will be associated with positive A-R-D stimuli, and positive emotional responses will be conditioned to him in people who observe the association. As the individual acquires positive attitude value for others, he will acquire additional reinforcement value and directive stimulus value. He will be able to influence the others more, and they will follow him more. It may be added that if their behavior is then followed by reinforcing stimuli—if their behavior, whatever the area, is successful—then additional A-R-D value will be conditioned to the "leader." Moreover, as they reinforce his leadership behavior, they increase its frequency, thus training him to lead more.

Long-term interactions of this sort can help produce leadership behavior. It may be added that negative experience with a leader may produce negative attitude conditioning. Stang (1973) found that when a leader's speech was aversive in length, the leader was liked less than when his speech was moderate in length. The negative attitudes would be expected to lead to "negative following," in a manner analogous to that of negative imitation.

The various aspects of the learning-behavior theory that apply to leadership cannot be considered here (see also Staats, 1963, pp. 328–330). In fact, only part of the importance of A-R-D principles has been mentioned. The negative side of leadership—the control of obedient behavior through negative emotions (fear)—has not been treated, theoretically or experimentally. Actually, the various principles when extended to more general social phenomena may also be seen to link leadership with social change, as will be indicated to some extent in the "Social Evolution" section of Chapter 14.

It should be noted, finally, that what has been described thus far has involved leadership-follower social interactions in which the actors are in face-to-face contact. The principles, however, apply to situations in which the follower group is much larger and the leader no longer has face-to-face contact with his followers. In such cases, however, it is necessary to include analyses of the means by which the leader's stimuli are conveyed to the followers. Language and the mass media ordinarily play a very large part in this type of social interaction, and this requires explicit analysis.

## SOCIAL PERCEPTION

In social psychological research, it has long been thought that there is selective perception based upon the perceiver's values. A large number of studies have been conducted to ascertain whether stimuli will be perceived more easily to the extent that they correspond with the value system of the individual. Postman, Bruner, and McGinnies (1948), for example, selected subjects on the basis of their measured values, using an inventory (Allport, Vernon, and Lindzey, 1951). The value areas were theoretical, aesthetic, economic, social, political, or religious. Words that represented each of the value areas were then presented to the subjects. The word presentations were first very brief, with a gradually increasing tachistoscopic presentation beginning at .01 seconds and increasing in hundredths until the subject indicated he recognized the word. The subjects recognized words for which they had high values most easily.

As will be indicated later, the author has suggested that such tests as the Allport, Vernon, and Lindzey study of values actually measure the individual's A-R-D value for certain classes of stimuli. It might be expected on this basis that the individual's attentional responses would be better controlled by stimuli that elicit positive emotional responses in him (Staats, 1968c, 1971a). This interpretation is supported by experiments that show that a word may be presented so briefly that the individual will not be able to identify the word—yet it will elicit emotional responses in the individual nonetheless (see Lazarus and McCleary, 1951). It might thus be expected that words that elicit negative emotional responses in the individual would be attended to less, or even avoided, and thus would not be recognized so quickly.

McGinnies (1949) conducted a study in which subjects took a relatively long time to recognize taboo words, and he interpreted his results in terms of conditioned autonomic responses which led to avoidance of recognizing the words. A number of studies were conducted to test other possibilities, for example, that word familiarity was involved, the individual being more familiar with words he positively valued. Howes and Solomon (1950) also interpreted the selective perception results as due to the reluctance of the subject to report words when they are emotionally negative, as has been suggested by some studies.

Research continues within this technology, with a good deal of support for the expectations that the A-R-D value of a stimulus will affect the quickness with which it is perceived and reported. It may be suggested, however, that a more definitive research of this type may lie in studying how manipulations of the A-R-D value of stimuli affect the attentional behaviors of the observer. It would be expected that if we took a neutral stimulus and paired it with a stimulus that elicited a positive emotional response, using any of the methods that have been described herein, that neutral stmulus would control attentional responses better than would a neutral stimulus not so treated. Then, in a situation in which the individual is confronted by a number of stimuli, all "competing" for his attentional responses, the stimulus that has higher A-R-D value will be attended to. This heightened directional control of attentional responses could be measured in various ways, including recogniton measures.

An experiment by Nunnally, Duchnowski, and Parker (1965) can be interpreted in terms of these principles. Children played a spin-wheel game in which the wheel had nonsense syllables instead of numbers. When the pointer stopped at one syllable the children received two pennies, when it stopped at another syllable the children received nothing, when it stopped at a third syllable the children lost one penny. This procedure is analogous to pairing a positive emotional stimulus, a neutral stimulus, and a negative emotional stimulus with different verbal stimuli —as was done in the attitude conditioning studies mentioned above. Later, attention measures were taken on the children by measuring the number of times they would press a button to light up a nonsense syllable in a viewing device. The child attended to the positive emotional stimulus the longest, the neutral stimulus the next, and the negative emotional stimulus the least.

The study thus showed that the extent to which a stimulus elicits a conditioned emotional response will help determine the strength of the stimulus in directing attentive behavior (the D function). This study can be considered to form the empirical basis for a firm interpretation of selective perception as it has been important to social psychology. The use of the classical conditioning principles makes possible study of how selective perception can be *produced*. Alteration of the emotional value of stimuli affects the extent to which the stimuli control attention. Additional experiments should be conducted to show the effects of the A-R-D values of stimuli in controlling attention in a multiple stimulus situation where there is "competiton" for attentional control.

A recent experiment has shown how the social stimulus of a person can affect the way that other individuals will attend in the presence of that person (Dweck and Reppucci, 1973). Although A-R-D theory was not employed in the study, the findings are predictable from this approach. In the study children first had experiences in which they were successful in solving problems presented by one adult and failed in solving those presented by another adult. Later, "a number of children failed to complete problems administered by the failure experimenter when her problems became soluble, even though they had shortly before solved almost identical problems from the success experimenter and continued to perform well on the success experimenter's problems" (p. 109). It is suggested that the negative attitude conditioned to the failure experimenter would also serve as a directive stimulus that would lead the children not to attend to other types of materials presented by this experimenter. This hypothesis could be tested experimentally.

It can also be noted that it would be important to test further the effects of the classical conditioning of emotional value on attentional responses when the stimuli are more social. That is, it is important to realize that attention to people is a function of their A-R-D value. And the extent to which we attend to another person is an important determinant of how much that person can influence us in various ways—imitation, conformity, and leadership interactions, as examples.

## IMPRESSION FORMATION

In a related area of research, social psychology has long been interested in person perception. "Person perception focuses on the process by which impressions, opinions, or feelings about other persons are formed" (Secord and Backman, 1964, p. 91).

One of the methods for the study of impression formation stemmed from the work of Asch (1946). He read a list of six adjectives that supposedly described a person as, for example, industrious, intelligent, and so on. The subjects then wrote a personality sketch of the person and selected one of each of 18 pairs of adjectives that best described the person. The adjectives originally paired with the person were shown to affect the way the subjects later felt about that person. This may be interpreted straightforwardly in terms of the emotional responses elicited by the adjectives being conditioned to the person, in a type of language conditioning of attitudes.

Asch, however, showed several other effects. For one thing, a set of adjectives could have a different effect with the change of only one. For example, the adjectives *intelligent, skillful, industrious, warm, determined, practical,* and *cautious* would have different effects when including the word *cold* rather than the word *warm.* Asch also showed that the order of the presentation of the adjectives had an effect. For example, the words *intelligent, industrious, forceful, critical, stubborn, envious,* would have a more positive effect than would the same words in reverse order. These effects have been corroborated in other studies.

It is interesting to note that Chalmers (1969) has been able to account for the order effect in impression formation within a classical conditioning of attitudes model involving a more elaborated set of assumptions set into a quantitative expression.

> These findings serve to lend experimental support to the basic assumptions to be employed in the model herein developed with respect to judgmental responses. First, it was determined that the extremity of evaluative ratings of a nonsense syllable (construed as a conditioned stimulus, CS) varied with the number of times it was paired with words (construed as unconditioned stimuli, UCS) of uniformly high (or uniformly low) evaluative intensity (Staats and Staats, 1959). Secondly, words of initially low evaluative intensity, when paired with words oppositely polarized in intensity, decreased in rated extremity (Staats, Staats, and Biggs, 1958).
>
> If translation is made to the impression formation situation, the strength of the evaluative meaning response [attitude] that is conditioned to a (CS) person by the presentation of a particular (UCS) adjective

will depend upon the strength and direction of the prior meaning response (the impression) cumulated to the (CS) person. (Chalmers, 1969, p. 452)

Chalmers went on to elaborate his weighted average model of generalized order effects from the basic conditioning principles. He derived concepts of solidity of the impression, facilitation of the impression by similar affective adjectives, and interference of impression formation by dissimilar adjectives. Chalmers then demonstrated how his model would account for the empirical findings concerning order in impression formation. Byrne and Nelson (1965), Anderson (1962, 1965, 1971), and Kaplan and Anderson (1973), have also developed quantitative statements of the principles of impression formation.

The action of A-R-D principles in an impression formation situation of relevance to everyday life may be seen in a study by Mitchell and Byrne (1973). The college student subjects were given the facts of an actual university disciplinary case. They were to serve as jurors and decide the disposition of the case, which involved theft. The jurors were also provided with "information" on the supposed offender. This information was manipulated either to show agreement with the jurors' attitudes, in one group, or disagreement, in another group. The defendant was judged guilty more frequently and was sentenced to a more severe punishment when the jurors' attitude differed from those of the defendant, in comparison to the case when there was juror-defendant similarity.

## Elemental behaviorism and attitudes

As has been indicated, B. F. Skinner has emphasized instrumental conditioning and employed classical conditioning very little in considering human behavior. This coincides with his emphasis on observable behavior, since most classically conditioned responses are internal to the organism. This approach has been followed in considering various areas of human behavior, including the study of attitudes. Bem, for example, has suggested that "An attitude is an individual's self-description of his affinities for and aversions to some identifiable aspect of his environment" (1968, p. 197). By affinities for or aversions to something, however, Bem refers to the overt instrumental behavior displayed. His analogy is that the individual says he likes brown bread because he has observed his behavior of eating it, although Bem recognizes that there may be a weak influence of internal affective states on the self-description.

At any rate, it is the self-description—an instrumental behavior—that Bem considers to be the attitude. Bandura suggests that the concept of attitude be abandoned altogether (1969). He prefers treating attitudes as a form of instrumental behavior, while recognizing there is a need to

indicate how internal affective states are related to cognitive and instrumental behaviors. Ferster and McGinnies (1971) summarize this elemental behavioristic view very cleraly:

> Some psychologists, such as Staats (1967), have viewed an attitude
> essentially as a conditioned emotional response. We prefer, however, to
> define an attitude as a class of behaviors under the control of some
> social referent, such as a group, a religion, or a political party. Staats has
> suggested that the measure of the strength of an attitude is the physio-
> logical, "emotional" response that the subject makes. Actually, if we
> take a behavioral approach toward attitudes, recognizing them as com-
> prising an operant [instrumental] repertoire, then the physiological signs
> to which Staats points are concomitants or symptoms rather than the
> attitude itself. Thus, the strength of an attitude is reflected in the fre-
> quency of the component performance. It only incidentally happens to
> be true that high frequency behaviors that are either positively or nega-
> tively reinforced also tend to be accompanied by physiological changes
> (p. 367).

The elemental behavioristic approach has an immediate appeal in terms
of its simplicity. Parsimony is certainly a worthy aim. However, the empirical evidence that has been summarized in the preceding sections suggests that restriction to instrumental conditioning principles in this area
of human behavior, as in others, makes it incomplete. Many of the
studies cited, for example, employed classical conditioning procedures
with the attitude stimulus, not instrumental conditioning procedures involving the reinforcement of some instrumental behavior. Many cases of
human behavior involve such classical conditioning circumstances, and
yet they result in changes to the subject's instrumental cognitive and
motor behaviors. These findings, and much that is yet to be described in
various areas of study, require a theoretical and experimental analysis that
indicates the principles of attitudinal change and the relationship of attitudes to the instrumental behaviors the attitudes control. Thus, the issue
here is substantive, not one to be decided by the aesthetic appeal of
parsimony.

## Cognitive mechanisms and social interaction

In this brief presentation it has only been possible to deal with the
A-R-D mechanism and its principles within a few areas of study in social
psychology. It should be emphasized that there are other areas of social
behavior with extensive research literatures that could be considered, such
as helping and altruistic behavior (see Aronfreed, 1968; Isen and Levine,
1972; Weiss, Buchanan, Alstatt, and Lombardo, 1971). Moreover, the
*various* personality repertoires the individual acquires will operate in his

social interactions with others. As one example only touched upon so far, language-cognitive repertoires—all of them—constitute a major basis for social interaction. Thus, two individuals who have large verbal-motor repertoires, each of whom can speak, may have many interactions on this basis. The words of one can constitute directive stimuli for the other, and vice versa. Much social interaction involves these repertoires.

The A-R-D properties of language are also heavily involved in social interaction, as has been suggested in part. Words can act as emotion elicitors, reinforcing stimuli, and directive stimuli in a myriad of ways in social interactions.

The word associations one person has to what another person says may affect the behavior of the first person. Moreover, two individuals with different word association structures—as in grammatical style, slang, in-group jokes, and so on—will relate differently than would be the case where their language is more similar.

As another example, it has been said that individuals differently learn to label internal stimuli (feelings). Such differences will affect various social interaction processes. The individual who can label how he feels, and thus provide cues upon which another person can act, will have different social interactions than the one without such skills or the one who labels his feelings incorrectly.

It has been indicated that a learning conception of communication has been elaborated to include the various repertoires of language (Staats, 1963, 1968b). The learning analysis of communication requires detailed extension to the social psychology of communication as well as to the various ways that social interaction can be affected through communication. Before leaving this topic, the manner in which language and cognitive mechanisms can bridge a delay between social interaction stimulation and the response of the individual will be mentioned.

## DELAYED LANGUAGE AND IMAGE STIMULI IN SOCIAL INTERACTION

The author (Staats, 1963) has indicated how a communication received at one time and learned as a verbal response sequence might not immediately have any overt behavioral effect. As an illustration, a soldier who reads a leaflet instructing him how to surrender may do nothing at the time, but later, in more adverse conditions, may behave in accord with the communication. Humans do not have to respond immediately to a stimulus. The language responses elicited by the situation can be readily acquired as a sequence. These learned behaviors can be retained and at a later time can provide the stimuli upon which the individual acts.

Nonlanguage cognitive mechanisms can also affect delayed social interaction. The same effect can occur on the basis of the individual's conditioned sensory responses. As a result of experience which elicits sensory responses in the individual, he may learn a set of images. Later, these image responses may provide the stimuli that direct his overt behavior. For example, the boy who meets a pretty girl at a social event he attends with his fiancée may only have acquired a conditioned sensory response. At a later time, however, when he no longer is engaged, he may recall the image of the pretty girl. This may then control "striving for" social interaction behaviors which had formerly been inhibited. Such learned images can have long-term effects upon conformity, imitation, leadership, and other social behaviors.

These examples are meant to indicate that the products of a social interaction may be cognitive and may show no effect upon instrumental behavior. Such cognitive products, however, can provide the basis for later behavior. The later behavior then is a function of the earlier interaction, with the cognitive repertoires in between. As these and other examples in the present book indicate, inclusion of the language-cognitive principles removes the necessity of a schism between the learning approach and cognitive theory. This can be done in a way not dealt with by elemental behaviorism or by social learning theory. The former rejects consideration of personality concepts. The latter sometimes refers to such concepts, especially cognitive concepts, but does not explicate what is involved in terms of learning principles or basic behavior repertoires.

## Role stimuli and symbolic social interaction

Social philosophers have noted that there are general characteristics in social interaction. Freud, for example, indicated that the nature of the relationship of the child with the parents would help determine the child's responding to other people. Freud noted that his patients would respond to him and other authority figures as they had to their fathers.

Such naturalistic observations can be considered within the explanatory principles of a learning-behavior theory. This involves the principle of stimulus generalization. Responses conditioned to a stimulus will be elicited by similar stimuli. When a child learns certain A-R-D stimulus functions to his father, similar individuals will serve those same stimulus functions. Men in general are similar along various stimulus dimensions. Particular men will be more similar to the father than will others and should more strongly elicit the same types of responses.

This analysis can be elaborated to consider the concept of social roles, which has been important in social psychology and the other social sciences. To deal with this concept it is necessary to introduce another learning principle. When a behavior is reinforced in the presence of different stimuli that have a stimulus component in common, that common stimulus component will become abstracted from the whole and will by itself serve to elicit the response. Hull (1920) called this concept formation, and Staats (1968c; Staats, Brewer, and Gross, 1970) has suggested that this learning mechanism is operative in various significant human behaviors.

In the present context it is suggested that there are component stimuli that are common to groups of people. We learn in our social interaction experience to respond in certain ways to these stimulus components. Thus, to employ the above example, a man is a complex stimulus. As a stimulus, however, he has certain features that are common to other men. In comparison to women, boys, or girls, he tends to be bigger, heavier, and stronger, to have a deeper voice, to be anatomically different, to wear different clothes, to behave differently. Moreover, these stimulus components will be similar within the class of men.

In addition, however, in a culture the child will be reinforced for responding in certain ways to men and in different ways to women, boys, and girls. The component stimuli that are similar to men would thus be expected to come to elicit those responses that have occurred in the presence of men but not the other classes of people. Such social component stimuli—stimuli which are common to a group of people and which come to elicit certain responses in others—may be considered to be role stimuli. As the example indicated, role stimuli may be those physical stimuli of a class of people, such as men. Or, the role stimuli may reside in the behaviors of the class, for example, musicians. Thirdly, the role stimuli may be those of other stimuli associated with the class of people. The essential point is that if certain responses of the child are learned in the presence of certain stimuli that are components of some groups of people, those component stimuli will come to elicit those responses.

This means that a totally new individual who as a complex stimulus includes those components will elicit those previously learned responses. The totally new individual will be responded to in ways that are appropriate for the class of which he is a member. It is not necessary for the responding person to have had previous experience with the individual. We respond in certain systematic ways to policemen, long-haired youths, teachers, physicians, professors, babies, authority figures, and so on, on the basis of previously learned behaviors that are elicited by component stimuli that are part of the individual.

It should be noted that the characteristics of role stimuli may vary over individuals and groups. Moreover, the behavior learned to role stimuli may also vary. For example, ghetto blacks have different experiences with policemen than do middle-class whites. The component stimuli that identify policemen thus come to elicit different responses in the members of the two subgroups of our society. There may also be differ-

ences in experience as a result of geographical location, familial idio-
syncrasies, personal history, and so on.

## SYMBOLIC ROLE STIMULI

As has been indicated, the basis for the role stimulus may be common
physical characteristics, such as the physical differences in men, women,
boys, girls, fat people, sick people, and so on. Sometimes the role stimuli
reside in behavior. For example, in England different ways of responding
to individuals are learned, depending upon the speech of the individual.
An individual who speaks in a cockney dialect will be responded to
differently than one who speaks the cultured English of Oxford or Cambridge. Table manners, behaviors of expertness, authoritative mannerisms,
and so on, are additional role stimuli.

Other role stimuli may reside in the symbols attached to the individual.
One example is that of titles. The individual introduced as "doctor" will be
responded to in part on the basis of that symbolic role stimulus. Many
occupational labels can serve this function. Uniforms also serve to elicit
standard behaviors, especially in the military. Clothing, jewelry, cars,
houses, and various other stimuli associated with an individual can perform as role stimuli and commonly affect the way that other people will
respond to him. Kretch, Krutchfield, and Ballachey (1962) summarize
roles into five categories: "(1) age-sex groupings, (2) biological or family
(kinship) groupings, (3) occupational groupings, (4) friendship and
interest groupings, (5) status groupings" (p. 310). Different subcategories
are recognized within each major one, for example, within the first there
could be the roles of infant, boy, girl, adult male, adult female, old male,
old female, and so on.

## PERSONAL ROLE STIMULI AND THE SELF

The concept of the self has already been referred to and analyzed, at
least in summary form, within the present theoretical orientation. This has
been done in part because the concept is such an important one in social
theory as well as in theories of personality. The self was a central aspect
of George Mead's classical approach to sociology, for example.

> In asserting that the human being has a self, Mead simply meant
> that the human being is an object to himself. The human being may
> perceive himself, have conceptions of himself, communicate with him-
> self, and act toward himself. As these types of behavior imply, the hu-
> man being may become the object of his own action. This gives him the
> means of interacting with himself—addressing himself, responding to
> the address, and addressing himself anew. (Blumer, 1966, p. 538)

In the context of the present concerns, for example, it is important to indicate that the individual himself responds to his own role stimuli. In this way the individual's role, his role stimuli, can be considered to be a determinant of his behavior. To elaborate, when an individual achieves a particular "role" in society, the various role stimuli are applied to him. He is given the title of that role, for example, that of Doctor. He will wear the uniform, in this case the physician's white jacket. He will emit the behaviors that he has learned, for example, the specialized language of the physician. These are all role stimuli that are active for the individual himself as well as others. In a sense he will behave toward himself as the role stimuli dictate. His planning, reasoning, and decision-making behaviors will include the effects of the role stimuli. He will also behave in general under the control of the role stimuli. In various situations he will respond as a doctor would: he may become more dominant in behavior; he will take precedence in social interaction in circumstances where formerly he would have given way to someone else; he will act in a way that would be described as more mature, more self-confident, perhaps more pompous, and so on.

In addition, as has been indicated, the individual's role stimuli will affect the behavior of others with whom he interacts. Because of his role stimuli the physician is treated quite respectfully in our society. The respectful behaviors of others, however, are also effective stimuli for the physician. For example, the physician will be listened to more, and his opinion statements will be rewarded more than those of an individual without his role stimuli. Such reinforcement will also have the effect of increasing the strength of his dominant, self-confident behaviors. Thus, the role stimuli will have an effect upon the behavior of the person himself. This conception helps bridge the gap between humanistic approaches to human behavior and elemental behavioristic approaches.

Other relationships between cognitive theory and learning theory in the realm of social interaction should be considered in a more complete treatment. At this point, however, social behavioristic principles and analyses will be elaborated to serve as an approach to the consideration of abnormal behavior.

# *Personality and abnormal behavior*

THE TRADITIONAL APPROACH to abnormal behavior was in terms of method, like that in the field of personality generally. The dominant conception was that human behavior was a function of internal personality processes or structures. These processes or structures were considered to be largely, or in part, dependent upon the individual's organic makeup. When behavior was seen to be abnormal, it was inferred that the individual's internal personality structures or processes must be abnormal. It was concluded that the abnormal behavior was a reflection of abnormal personality processes, of mental disease.

It is characteristic that if the main aspects of a conception are accepted, its various methods and strategies will also be accepted. In the present case, considering abnormal behavior to be the reflection of underlying personality disturbances or disease made the methods of medicine relevant. In medicine it had been useful to describe and classify the various symptoms of disease. Once a constellation of symptoms had been described and differentiated from other constellations, the physician could diagnose individual cases and thereby better deal with the illnesses.

The same methods thus were adopted for dealing with abnormal behavior. Since it was thought that there must be mental diseases associated with different types of abnormal behavior, the medical classificatory approach was applied. Different types of abnormal behaviors were described by Kraepelin (1923), for example, and a classificatory system was developed.

244

> Kraepelin was able to differentiate a set of symptoms which he thought were characteristics of dementia praecox, schizophrenia, in contrast to another set of symptoms which constitute the syndrome of manic-depressive psychosis. . . . Kraepelin's aim was to define disease entities based on definite organic pathology, and he always felt that dementia praecox was ultimately based on faulty metabolism. . . ." (Arieti, 1959, p. 456)

Kraepelin's (and others') assumption of organic disease underlying abnormal behavior is not accepted by many people in the field today. The term *functional psychosis* is employed, for example, to indicate that the abnormal behavior is a functional rather than an organic disturbance. Nevertheless, strong biological orientations are still present in the field—especially the concept that abnormal behavior may be the result of genetically determined physiological abnormalities. At any rate, Kraepelin's classification of mental disorders constitutes the basis for official taxonomic systems.

Thus, whether or not the fields of psychology and psychiatry in general accept the disease conception of abnormal behavior, the approach to the study of abnormal behavior is still heavily influenced by that conception. The description of abnormal behavior is still conducted according to the classification system dictated by that conception.

A general conception of human behavior, however, should extend also to abnormal psychology and to development of a system for the description of abnormal behavior. Thus, in making an initial theory of various areas of human behavior, one of the present author's goals was to advance a preliminary taxonomy of abnormal behavior.

> Actually, all the principles that have previously been described would seem to be potential contributors to abnormal behavior as well as to normal behavior. Nevertheless, in an abbreviated description of abnormal behavior it seems useful to organize the discussion in terms of behavioral deficits, . . . inappropriate behaviors, improper stimulus control, and inadequate reinforcing systems. This is not meant to suggest that any deviant behavior pattern actually fits into merely *one* of these categories. (Staats, 1963, p. 466)

It was recognized that this nosology, although useful, was abbreviated and expedient, even when several subtypes of the major categories were included. This categorization of abnormal behavior in terms of these types of problems has since been followed in other behavioral attempts in the field. For example, Bandura (1968a) has utilized this classification system, calling it a social learning approach to abnormal behavior. Kanfer and Saslow (see 1965) developed some similar categories in a schema for psychiatric diagnosis to be employed in the place of psychological tests. Suinn (1970) also employed the classification system in a recent textbook on abnormal psychology. Since the classificatory system that was originally suggested was only an initial summary, it was not intended to be complete or closed. Further development of the behavioral conception of human behavior to include a personality level allows further elaboration of the system for describing abnormal behavior. Abnormal psychology has not been based upon the general conception, however, and many of the necessary observations by which to complete a behavioral system of abnormal psychology description and classification are not available. The task thus must remain an open-ended one. However, some of the steps in the direction of a more complete schema can be indicated as a means of indicating the possibilities for further development.

As will be described, a psychology of abnormal behavior must include more than just a description of abnormal behavior. For one thing, it is important to know the conditions that maintain a certain type of behavior, especially in the case of a behavior that appears to be nonfunctional or maladjustive. It is also important to know the learning conditions that produced the abnormal behavior in the first place; this is sometimes included in the etiology of a "mental disease." A behavioral interaction theory of abnormal behavior must also include consideration of the interactions that occur, and these may be of several kinds. There is behavior-environment interaction in which the behavior affects the environment which in turn affects the behavior, and so on. There are also problems of human behavior that arise in institutional-individual interaction. Finally, there are interactions between different aspects of the individual's personality (basic behavioral repertoires). That is, because of the characteristics of one type of basic behavioral repertoire, the individual may be disposed to develop certain characteristics in other basic behavioral repertoires. The human learning and behavior interaction principles previosuly described must be elaborated in the context of abnormal behavior. The following sections will attempt to exemplify the manner in which these principles and concepts can be employed in developing a social behavioral abnormal psychology.

## Basic personality repertoires and abnormal behavior categories

To begin, it is suggested that only two categories are necessary to consider abnormal behavior generally. Either behaviors that are necessary to the individual's adjustment are absent, or behaviors that interfere with his adjustment are present. Thus, a major classification begins with the two categories—deficits in behavior and the presence of inappropriate behaviors. In the original classification there were additional categories—improper stimulus control and inadequate reinforcing systems—but these categories are actually subdivisions of the major dichotomy. Moreover, there must be more than just those two subdivisions.

**TABLE 8.1**

**Categories of abnormal behavior**

| BASIC BEHAVIOR (PERSONALITY) REPERTOIRES | SYMPTOMS | |
| --- | --- | --- |
| | *Deficit behavior* | *Inappropriate behavior* |
| *A-R-D system* (emotional-motivational) | Flat emotionality of schizophrenia<br><br>Lack of motivation of neurasthenia<br><br>Lack of achievement motivation of cultural deprivation<br><br>Childhood autism deficits in affection for parents | Psychosomatic disorders<br><br>Sadistic pleasures<br><br>Phobias<br><br>Sexual disturbances involving aberrant sex objects; homosexuality, fetishes<br><br>Autistic children's preoccupation with self-stimulation |
| *Language-cognitive system* | Mental retardation deficits<br><br>Autism deficits<br><br>Lack of verbal-motor control in psychopaths<br><br>Cultural-deprivation deficits | Paranoid delusions<br><br>Defense mechanisms<br><br>Antisocial conceptual systems like racism<br><br>Stuttering, gesticulation substitutes for language in nonspeaking children |
| *Instrumental repertoires* | Lack of social skills<br><br>Lack of approach behavior to the opposite sex<br><br>Lack of work skills<br><br>Enuretic lack of toilet skills<br><br>Lack of recreational skills<br><br>Lack of imitational and observational skills | Violent social behavior (rape, beatings)<br><br>Aversive social behavior (arrogance, cruelty, overly demanding, selfishness)<br><br>Bizarre actions of psychotics<br><br>Self-stimulation, repetitive behavior, self-destructive behavior of autism |

The behavioral interaction conception of personality that has been presented, even in its early development, provides a basis for a more complete set of categories than the author originally presented. It has been suggested herein that there are basic behavioral repertoires that are the constituents of personality. Three major areas—the emotional-motivational, the language-cognitive, and the instrumental systems—were outlined in some detail. These require inclusion in the classification.

The general descriptive system of a behavioral abnormal psychology may be exemplified as shown in Table 8.1. As the table indicates, the major categories are deficits in behavior and inappropriate behaviors. The individual may not have acquired behaviors that he requires for adjustment to his social or physical environment. Or he may have acquired behaviors that are inappropriate in that they handicap his adjustment to his social or physical environment. This does not mean that one deficit in behavior, or even one inappropriate behavior, would mark an individual as abnormal.

These two types of maladjustments can occur in any of the three personality repertoire areas. Thus, the table makes categories of the emotional-motivational, the language-cognitive, and the instrumental personality repertoires. Examples are provided within each cell to illustrate some of the categories and concerns of abnormal psychology. These examples require elaboration, as the following sections will illustrate.

### DEFICIT EMOTIONAL-MOTIVATIONAL SYSTEM

The first cell deals with deficits in the emotional-motivational (A-R-D) system of the individual. One example that is prominent in abnormal psychological symptomatology concerns the deficits apparent with people called simple schizophrenics.

#### FLAT EMOTIONALITY IN SCHIZOPHRENIA

In simple schizophrenic reactions, the patient evidences a gradual narrowing and waning of interests, loss of ambition, emotional indifference. . . . He no longer cares whether he passes or fails in school and is no longer concerned about his friends or family. He may show periods of moodiness and irritability, but becomes increasingly emotionally indifferent and seclusive. . . . [T]here is little or no interest in the opposite sex. (J. C. Coleman, 1950, p. 245)

These descriptions concern the individual's A-R-D system. A later chapter, will present evidence that what we refer to as interests actually consist of stimuli that elicit positive emotional responses that will as a consequence serve as rewards which maintain the individual's behavior, and that will also serve as directive stimuli that serve as goals for which the individual strives. The interested, motivated person—among his other assets—has learned strong emotional responses to important stimuli in his life. These stimuli, represented in some form (usually through language), serve as goals for him and elicit extended, arduous, and continuing behaviors from him. Thus, the positive emotional quality of the "idea" of receiving a high grade in a course may serve as a goal to the student in the sense that it elicits vigorous studying behavior. Furthermore, these emotional directive stimuli (or their representations, as in language and imagery) will serve to reward the individual's present behavior and thus maintain it. The student may say to himself after a rigorous evening of study, "I sure put in a good day's work, which should help in getting a good grade on the midterm exam."

When circumstances are such that the individual does not learn to respond emotionally to stimuli that are necessary if his behavior is to be adequately maintained (motivated), or if his life circumstances undo aspects of the learned A-R-D system, serious behavior deficits can occur. When the individual loses interests, ambitions, and goals and is emotionally indifferent, there are no strong sources of reward and directive control to maintain adjustive behaviors. Such deficits can be primary constituents of the behavioral constellations labeled *simple schizophrenia*.

INADEQUATE EMOTIONALITY IN CHILDHOOD AUTISM. As the description of simple schizophrenia suggests, the individual may have shown at one time less severe deficits in behavior. His emotional-motivational deficits, having been at one time of a lesser degree, may become progressively more serious. Childhood autism and childhood schizophrenia, on the other hand, involve early deficits, including those in the emotional-motivational system. The autistic child is said to have disinterest in his surroundings, to have no emotional response to people, to be unconcerned with the parents' social approval or attention, and so on.

To some extent, this statement is exaggerated. It is not entirely the case that the autistic child is unconcerned with social approval and attention. A number of studies have shown the severely disturbed child's abnormal symptoms can be lessened by not socially reinforcing them (Wolf, Risely, Mees, 1964). A better description of such abnormally behaving children is that their A-R-D systems are poorly developed, along with other deficits in behavior that would ordinarily secure attention and approval.

A-R-D DEFICITS IN NEURASTHENIA. Another example of nonadaptive behavior patterns that seem to involve deficits in the emotional-motivational system is that of neurasthenia—classified as a neurosis. This problem usually occurs in young adults and housewives. The major description is of physical and mental fatigue in the face of everyday duties.

> In fact, one of the most significant things about the neurasthenic's fatigue is its selective nature. Many of these patients show ample energy and good endurance in playing tennis, golf, or bridge or in doing anything else which really interests them. In the face of occupational and

other routine activities, however, they are usually monuments of listless-ness, lack of enthusiasm, and general "tiredness." (J. C. Coleman, 1950, p. 169)

The description of neurasthenia, which frequently is attributed to other causes, appears to heavily involve deficits in the individual's emotional-motivational system. The fact that there is nothing wrong with the indi-vidual's "energy" is shown by the fact that he may show high physical activity and endurance, but only when consequences of strong A-R-D value are present. When no reinforcement is present—in work situations —the individual's behavior is described by others as listless, unmotivated, lackluster, lazy, and disinterested, and by himself as fatigued. It is sug-gested that the stimuli that ordinarily have reward value, especially in the sphere of work, do not have this value for the neurasthenic.

## INAPPROPRIATE EMOTIONAL-MOTIVATIONAL SYSTEM

In beginning this section, an example can be employed to indicate the overlap that occurs between the deficit and inappropriate categories. The example is that of homosexuality, which may be considered to involve both a deficit in emotional response to appropriate stimuli and emotional response to inappropriate stimuli. Although not discussed in the present terms, Coleman describes sexually deviant behavior as "sexual patterns which are considered abnormal in regard to the choice of sexual object, such as homosexuality and sadism" (1950, p. 404).

THE A-R-D SYSTEM AND SEXUAL DEVIATION. The principles of the A-R-D system appear to constitute strong explanatory concepts for understanding abnormal sexual behaviors. It should be remembered that when an object elicits a positive emotional response in the individual, the object will exert control over the individual's approach behaviors—and obtaining the object will reinforce these approach behaviors. Many sexual abnormalities occur because objects that should have sexual A-R-D value for the indi-vidual do not. Homosexuality is a case in point. One of the important aspects of this problem is that members of the opposite sex do not elicit in the person sexual emotional responses. The male homosexual is not aroused by the stimuli of females. This has been shown neatly by Freund (1963), who employed photographs of nude people, children, adolescents, and adults. The subjects who viewed the pictures were males, homosexual and heterosexual. The measure of their sexual emotional response was the blood volume of the subject's penis—a measure of tumescence. In gen-eral, individuals who were homosexual demonstrated emotional sexual response to pictures of males, and not to those of females. The opposite was true of heterosexual males. It was also possible to detect sexual emo-tional arousal to adolescents and children. McConaghy (1967) has substantiated these findings employing moving pictures of male and female nudes.

The behavior of the person follows the dictates of his emotional-motivational system. It could be hypothesized, for example, that the reinforcement and directive value of the sexual pictures would follow their emotion-arousing value. This could be shown in the following experiments. The reinforcement value of such pictures could be tested by the length of time the individual would continue visual inspection. One could also test the extent to which the subject would "approach" such pictures. This could be done with a visual apparatus in which the picture would first be shown in very small size. The subject would have to perform some response, like turning a wheel to enlarge the picture. Sexually emotionally arousing pictures would be expected to elicit approach behaviors more than those that are not arousing.

As has been indicated, it is the effects that the A-R-D stimuli have upon the instrumental behaviors of a person—the reinforcing and directive functions of the stimuli—that constitute a central part of the problem. When members of the same sex elicit sexual emotional responses, the person is likely to learn behaviors that lead to sexual contact with members of that sex. The pedophiliac is a person in whom children elicit sexual emotional responses and who is thus led to attempt sexual contact with children. The person for whom pain and violence are positive sexual re-inforcers will be likely to develop instrumental behaviors that culminate in such events. The individual in this society is considered sexually aberrant if his sexual A-R-D system and consequent behaviors include persons of the wrong sex, wrong age, wrong social relationship (brother-sister, father-daughter, and so on), the sight of others in pain, animals, and so on.

## FETISHISM AND INAPPROPRIATE A-R-D STIMULI

*Fetishism* is a deviation in which the object of choice is a part of the human body, or an object, commonly a shoe [or other object of clothing]. In extreme cases, the sight or touch of the shoe is sufficient to arouse an orgasm; the individual thus does not require a human mate for a sexual experience. Fetishism appears to be more common with men than with women. The object may be lace, silk, perfumes, or soap, the hair, the eyes, the hands, or ears. (Suinn, 1970, p. 314)

Fetishism may also be considered in terms of inappropriate A-R-D system development. Once the individual has learned an inappropriate sexual emotional response to an object, the behavior of approaching and gaining access to the object will tend to be learned. Raymond (1956) cites the case of a man for whom baby carriages elicited sexual responses. This man derived sexual reinforcement (including orgasm) by contacting (roughly) or soiling the carriages and sometimes their occupants or the women pushing them. He was repeatedly hospitalized because of the behavior elicited by the fetish objects. It is interesting that although married, his wife had less sexual A-R-D value for him than the fetish objects. Examples by which one can learn emotional responses to unusual objects will be described in a later section.

PHOBIAS AS INAPPROPRIATE A-R-D STIMULI.   Another category of abnormal behavior that can be mentioned is that of phobias, which can be considered to involve negative emotional responses. Phobias are inappropriate in terms of the particular stimuli that elicit the negative emotional responses or in terms of the intensity of emotional elicitation. As an example of the former, one has a phobia when the stimuli of being outdoors elicits a negative emotional response (anxiety). The negative emotional response may be perfectly normal when elicited by certain stimuli of danger. It is inappropriate because the stimuli of being outdoors does not normally elicit such a response. This is thus an example of *defective stimulus control.* It is not the response that is abnormal; the abnormality lies in the stimuli that elicit the reponse.

Moreover, because the emotional response one makes to a stimulus also dictates how one responds otherwise, the fear of being outdoors is abnormal in the instrumental behavior that results. That is, if the emotional response is intense enough, the individual will be driven to stay at home—by escape from the negative emotional response—and this is ordinarily maladaptive.

The abnormality of the phobia may also reside in the intensity of negative emotional response elicited, not only in the type of stimulus that elicits the response. Thus, for example, many people find that snakes and insects and such elicit a negative emotional response in them, but the emotional response is not of great intensity. Most people find that flying on airplanes elicits, at least at times, a negative emotional response. When the response is intense enough, however, it will prevent the person from flying and may be a problem.

PSYCHOSOMATIC ILLNESSES AS INAPPROPRIATE EMOTIONAL STIMULI. Closely related to phobias in principle is the area of psychopathology referred to as psychosomatic illnesses, which may be described as follows:

> [I]t has been repeatedly demonstrated that psychological factors, such as anxiety, fear, anger, and other emotional conditions, seriously affect the general resistance of the organism to disease, and in some instances may even bring about actual tissue pathology . . . for example, peptic ulcer, high blood pressure, asthma, acne, migraine, overweight, and even the common cold. (J. C. Coleman, 1950, p. 205)

It is suggested that the principles of both classical conditioning and instrumental conditioning can be involved in such problems.

> There is . . . a recent tendency in psychosomatic medicine to regard many somatic disorders such as allergies, peptic ulcers, mucous colitis,

hypertension, etc. as conditioned response patterns of hidden origin. . . . There are, for example, many indications that bronchial asthmatic attacks can be provoked by means other than direct contact with the specific noxious agent, i.e., various conditioned stimuli may be effectively substituted for the original agent. . . . (Franks, 1961, p. 477)

SELF-STIMULATION IN AUTISM AND FANTASY IN SCHIZOPHRENIA. It is not uncommon for normal individuals to stroke some part of their body for a moment or two—to pull an ear lobe repeatedly, for example, or to twist a lock of hair. It is also not uncommon for normal individuals to spend some time in daydreaming—to image rewarding events. However, these types of behavior can occur to the exclusion of behaviors ordinarily maintained by other rewards. One of the prominent symptoms of autism is self-stimulation, where the child frequently engages in an activity that provides himself with some bodily stimulation. A prominent symptom of some schizophrenics, which has been considered typical, is that of fantasy, again where the individual provides his own stimulation.

These types of behavior, in the monotonous frequency displayed in these cases, appear to be aberrant in and of themselves. However, there is good reason to consider the behaviors in the context of deficits in the individual's A-R-D system. To elaborate, a normal individual can provide reward for himself by self-stimulation, either through physical contact as in stroking, masturbating, rocking, and so on, or by the use of imagery and covert language repertoires. The strength of A-R-D value of such behaviors, however, must be considered in terms of relative strength. Ordinarily, there are many other sources of reward available to the individual. If his learning history has been usual he will have learned strong sources of social rewards, work rewards, recreational rewards, heterosexual rewards, and so on. Such A-R-D stimuli are usually stronger than self-stimulation rewards. Behaviors that attain such rewards will thus become strong for the individual.

When the individual has not learned such A-R-D stimuli, or the behaviors which obtain those rewards, then weaker A-R-D stimuli, and the behaviors that attain these will be in a relatively stronger position. It is thus suggested that the autistic child engages in self-stimulation because his A-R-D system has very limited development and he does not have stronger A-R-D stimuli that replace the self-stimulation reinforcers. The schizophrenic engages in such a rich fantasy life because competing rewards are not available through other behaviors.

These possibilities should be studied further. One type of evidence that supports the hypothesis can be seen in findings obtained by Lindsley (1956), in the context of other interests. Lindsley was interested in the manner in which psychotic patients responded to different types of partial reinforcement schedules. He used cigarets and so on to reinforce the behavior of pulling a plunger, recording the frequence of the behavior.

He found that when the behavior was strong, because of the manner in which it was reinforced, the hallucinatory behaviors occurred less frequently. When the lever-pulling behavior was weak, the hallucinatory behaviors occurred more frequently. It may be concluded that strong lever-pulling behavior apparently squeezed out the hallucinatory behavior. In the present context, it is suggested this provides support for the expectation that strong sources of reward would make the A-R-D effects of self-stimulation much less influential. The self-stimulation behaviors can be seen to reflect behavioral deficits rather than as an intrinsically abnormal type of activity. At any rate, the various principles of A-R-D theory and the various aspects of the A-R-D personality system must be considered in the context of abnormal behavior.

## Deficit language-cognitive repertoires

A brief summary of some of the language-cognitive repertoires was presented in Chapter 5, which also gave a few examples of how these repertoires are assessed on intelligence tests and how they function in the individual's reasoning, problem solving, planning, decision making, and so on. It may be added that deficits in the language-cognitive repertoires will make the child's adjustment defective. Furthermore, these language repertoires are basic to the child's further learning. When he has deficits in the basic behavioral repertoires of language, he will not learn well in school. He will not be able to learn to read, which is a repertoire that is basic to much additional cognitive learning, and he will not be able to learn other intellectual skills. The child will be retarded. If he has severe deficits he also will not be able to participate in social interactions and learn social skills.

MENTAL RETARDATION, CHILDHOOD AUTISM, AND LANGUAGE-COGNITIVE DEFICITS. Mental retardation is customarily defined as "a defect of intelligence existing since birth" (American Psychiatric Association, 1952, p. 23). It is generally considered that this psychological defect rises from "genetic cause or is induced by disease or injury" (Jervis, 1959, p. 1289). The diagnosis rests primarily upon poor performance on intelligence tests, the severity of the mental retardation depending upon the child's score. As will be indicated, however, a learning etiology must be suspected in any case where no detectable biological cause can be shown (see Staats, 1971a).

Cases of childhood autism and schizophrenia appear also to involve primary deficits in basic behavioral language-cognitive repertoires, such as those that have been described herein (and more completely in Staats, 1971a). Jervis (1959) notes that "It may be estimated that roughly one third of the children variously labeled as cases of infantile autism or childhood schizophrenia show, with standard intelligence tests, intelligence

quotients below 70 and may be therefore classified as mentally defective"
(p. 1312).

PSYCHOTIC LANGUAGE AND LANGUAGE-COGNITIVE DEFICITS. One of the
author's observations in working with clinicians of a traditional personality
view was that they were interested in the disturbances in what adult
psychotics said as a means of gaining understanding of the patient's
underlying mental illness. However, as has been indicated, deficits in
language repertoires may be at the seat of the individual's problems, since
various types of adjustment depend upon those repertoires.

The patient who has not acquired, or who has lost, integrated labeling
skills and reasoning repertoires cannot be expected to reason appropriately.
The following is an excerpt from a psychotic patient's language, an answer to a question concerning what she was thinking about.

> She never did like chairs say . . . I must get out of this office say.
> . . . I am going to strike you say . . . I don't know say . . . I would
> be terribly angry with you say . . . Keep your tongue between your teeth
> and be terribly angry with you say . . . Do the dirty and I would beat
> you down say . . . Shout yes for that say . . . My children are like that
> . . . Yes immediately . . . Sit down Gordon . . . Nan Gordon my
> favourite sister say . . . He's wasting his time . . . He's wasting his
> moments. Come off the seat you know he will kill you dead for that
> . . . Stone dead to the ground to the floor everything . . . Get engaged
> to me . . . I am sorry to bring you here. (Freeman, Cameron, and Mc-
> Ghie, 1966, p. 90)

This patient further demonstrated various deficits in her language
repertoires. She usually did not respond to simple instructions (verbal-
motor deficits), was incapable of conducting a simple conversation (deficits in various repertoires), showed "lack of coherence due to faulty
grammatical construction," used words incorrectly, and was unable to
attend to questions asked of her or statements made to her.

The author has observed a number of patients in hospitals who have a
major problem because of deficits in their language-cognitive repertoires.
Deficits in language repertoires are also involved in what is called the
emotionally disturbed child, the autistic child, cultural deprivation (a
group form of retardation), the impulsivity (lack of planning and verbal
control) of the psychopath, and types of schizophrenia. Cohen, Nachmani,
and Rosenberg (1974) have recently demonstrated a disturbance of communication in schizophrenics that may be considered in terms of various
of the basic language repertoires, especially that of complex labeling.

## INAPPROPRIATE LANGUAGE-COGNITIVE REPERTOIRES

As was indicated in Chapter 5, the basic behavioral repertoires of language are involved in the individual's reasoning and planning and in the

actions that are mediated by these cognitive activities. When an individual's language repertoires are awry (or inappropriate), his reasoning will be awry, and the behaviors that are mediated will be awry. Various of the repertoires of language may be involved: the individual may label events incorrectly, his word association sequences may not parallel reality, the emotional responses that his words elicit may be unusual. Maher (1966) provides an excerpt of schizophrenic language, taken from spontaneously written documents, that illustrates these various deficits:

> If things turn by rotation of agriculture or levels in regards and "timed" to everything; I am re-fering to a previous document when I made some remarks that were facts also tested and there is another that concerns my daughter she has lobed bottom right ear, her name being Mary Lou. . . . Much of abstraction has been left unsaid and undone in this product/milk syrup, and others, due to economics, differentials, subsidies, bankruptcy, tools, buildings, bonds, national stocks, foundation craps, weather, trades, government in levels of breakages and fuses in electronics too all formerly "stated" not necessarily factuated. (p. 395)

It should be added that various studies have been made of schizophrenic language in the context of different conceptions of language; an excellent review has been provided by Maher (1966). For example, Ellsworth (1951) considered schizophrenic language to be like that of children. He found that both children and schizophrenics used more nouns and pronouns and fewer verbs and adjectives than subjects of the same age as the schizophrenics. Other formal aspects of language have been employed to compare schizophrenics and normals, for example, the ratio of the number of different types of words used to the total number of words used (Baker, 1951), the ratio of the number of verbs to the number of qualitative descriptions (adjectives, nouns, and participles of verbs) (Mann, 1944), and so on. Maher, McKean, and McLaughlin (1966) used a computer program with which to analyze samples of spontaneously written schizophrenic language. "The group judged to exhibit language pathology showed a tendency to use more objects per subject, fewer qualifiers per verb, fewer different words, and more varieties of negative words" (Maher, 1966, p. 402).

Hunt and his students (Hunt and Arnhoff, 1956; Hunt and Jones, 1958) studied the extent to which deviance in language can be reliably judged. It was found that clinicians showed high agreement in judgment when the language consisted of responses to the vocabulary test on the Wechsler-Bellevue Intelligence Scale (Wechsler, 1944). Furthermore, Hunt and Jones (1958) found that judgments of schizophrenicity of language was largely made on the basis of the communicability of the person's verbal response.

A general suggestion that arises in the context of this type of study is that the methods might be productively employed in conjunction with a learning theory of language. There are meaningful categories in the various language behavior repertoires that have been described. It would be expected that differences between normal language and schizophrenic language would occur within these various repertoires. It should be noted that the learning theory also suggests the functions of the repertoires in the individual's adjustment, and deficits in the categories would be meaningful for understanding abnormal behavior. Moreover, since the language repertoires are based upon a learning analysis, an etiology is also implied, as are treatment procedures.

PARANOID DELUSIONS AND INSTRUMENTAL BEHAVIOR AND INAPPROPRIATE LANGUAGE-COGNITIVE REPERTOIRES. A case of abnormal reasoning and the manner in which maladjustive behavior may be mediated is appropriate here. The case is that of a young man who had placed a bet on a horse and won (according to his account) but was told by his bookies that he had bet on another horse.

> The acute paranoid attack came the day after the patient had picked a violent quarrel with the racing bookies over a bet and had threatened to assault them . . . As he [later] sat pondering alone in the hotel lobby, nursing his wrath . . . it flashed upon his mind, he said, that bookies were notorious for having national gangster "protection,". . . This thought frightened him badly. . . . The more he thought, the more he felt in danger of his life. . . .
> The very next day things began to happen, he noticed a number of rough-looking strangers hanging around the hotel lobby. They seemed to be watching him closely and waving signals to one another. The patient himself became watchful and apprehensive. . . . His vigilance grew. . . . He saw strangers and loiterers everywhere. . . . In a near panic, he barricaded himself in his hotel room against surprise attack. (Cameron, 1959, p. 521)

As this case illustrates, this man verbally describes (labels) a situation in a manner that elicits intense negative emotional responses (anxiety) in him. The anxiety further elicits reasoning sequences which mediate his actions in that he begins to attend to the presence of others in ways that he had not before. He scrutinizes people that he formerly did not notice. His anxiety emotional responses, and his reasoning sequences, lead him to label the people he notices as "gangsters." This behavior then additionally elicits anxiety responses and additional "irrational" reasoning. Since the man has no social interactions with others who will point out the unlikely nature of his reasoning at the beginning, the abnormal thought and emotional responding continue to grow as the man attempts to "escape" by fleeing across the country—always seeing evidence of surveillance that adds to his delusion. The unreal reasoning sequences thus elicit unreal, and maladjustive, instrumental behaviors.

DEFENSE MECHANISMS AND LANGUAGE-COGNITIVE REPERTOIRES. An example will be given to demonstrate a type of inappropriate language in abnormal behavior, as well as to illustrate the possibilities of unifying some of the observations and concepts of psychoanalysis within the social behaviorism theory. The concepts at issue are called the "ego defense mechanism." Freud suggested that the part of the individual's personality called the ego has defense mechanisms by which it defends against anxiety. The defense mechanisms were seen to achieve this purpose, but without resulting in any change or improvement in the situation that caused the anxiety. One of the defense mechanisms is projection.

> Projection is a defensive reaction by means of which we (1) transfer the blame for our own shortcomings, mistakes, and misdeeds to others, and (2) attribute to others our own unacceptable impulses, thoughts, and desires.
>
> Projection is perhaps most commonly evidenced in our tendency to blame others for our own mistakes. The student who fails an examination may feel sure the examination was unfair, the erring husband may blame his moral lapse on the girl "who led me on." "It wasn't my fault, he hit me first" or "If I hadn't taken advantage of him he would have taken advantage of me," and so it goes. (J. C. Coleman, 1950, p. 86).

Such responses can be seen clearly to consist of language behaviors. Moreover, although projection behaviors may not change the negative situation—past events such as failed examinations, moral lapses, and so on, cannot be expunged from the past—the language behaviors can change the social situation the individual faces because of his mistake. The erring husband may placate the offended wife by means of his "projection" language behavior, the failing student may decrease the punishment of his disappointed parents, and so on. However, when the individual's statements noticeably differ from reality, other persons' positive attitude to the individual will lessen and he will be responded to in different and more disadvantageous ways than would otherwise be the case.

The abnormalities of such defensive behaviors also occur because as the individual's language deviates from reality it becomes less adjustive to him in mediating his own behaviors. It must be remembered that the individual's own labeling and verbal response sequences also constitute his own reasoning and planning and enter into the mediation of his own problem-solving behavior. As a example, take the student who fails on the examination but says that it was unfair rather than giving it a more veridical label acknowledging his own inadequate preparation or lack of background. In doing so he supplies himself with a poor basis for planning improvements. When the individual mislabels reality to himself, preventing appropriate personal anxiety or social disproval, he may provide himself with reasoning sequences that are maladjustive in other ways.

There are other types of defense mechanisms. Examples are rationalization, repression, and denial. Each of these involves language, and each serves to remove the anxiety that more veridical verbal descriptions would elicit, as well as to allay social disapproval and the anxiety this elicits. However, when the individual's language sequences largely become of this type, his language sequences can no longer serve their role in adjustive reasoning and planning. As has been indicated, in order to reason about the realities of the world, both physical and social, the individual's language system must be isomorphic with the events of the world. Any large divergence from isomorphism leaves the individual without the central human mechanism for adjustive reasoning, thinking, planning, and decision making. The theory of language that has been described (Staats, 1963, 1968c, 1971a, 1971b), it is suggested, is basic to a learning-behavior theory of abnormal behavior including mental retardation, autism, neurotic, psychopathic, and psychotic disorders.

## DEFICITS IN INSTRUMENTAL REPERTOIRES

The many instances of instrumental skills involved in abnormal behavior can also be considered as a lack in the instrumental repertoire or the presence of inappropriate instrumental behaviors. Several examples of such behavior listed in Table 8.1 above and will be "discussed here, with deficit behavior being considered first.

DEFICITS IN TOILET SKILLS. Toilet skills constitute a good example of deficits in behavior that are of concern in abnormal psychology. An analysis of the behaviors and training involved in toilet skills will be made in Chapter 10. It is relevant here to indicate that it is important to the child to acquire these skills at an appropriate age. Most children are trained in appropriately retaining and expelling urine and fecal matter by the age of three, but some will display toilet incontinence at much later ages. The development of toilet skills is often thought to be a function of biological maturation; learning analysis shows, however, the skills involved and the training conditions necessary to produce the skills. Inappropriate toilet behaviors are frequently associated with other childhood disturbances such as mental retardation, childhood autism and schizophrenia, emotional disturbance, and so on.

DEFICITS IN WORK AND SOCIAL SKILLS. As will be indicated in a later section, there is frequently a reciprocal relationship between behavioral repertoires—the absence of one behavioral skill may be related to the acquisition of another. This is pertinent in the realm of deficits in instrumental behaviors. Many times the deficit in behavioral skill is overlooked because the person demonstrates inappropriate behaviors that are more serious from the immediate standpoint of society, even though it is the skill deficit that is more fundamental in the disturbance. A person who has deficits in social skills and work skills and cannot get and maintain a job will be faced with environmental pressures that may lead to behaviors considered inappropriate by the society. A man who does not have the social skills and the sexual skills to achieve a relationship with members of the opposite sex faces sexual deprivation that can lead to deviant sexual acts.

Psychotic patients in hospitals frequently demonstrate severe lack of instrumental behaviors which partly constitutes their abnormality. The following description of a hospitalized schizophrenic patient provides an example which involves deficits in social behavior repertoires, although this is not the sole cause of the abnormality.

> The patient had a brother and sister. . . . so much older than she, that they had never been her playmates. By children of her own age, she was rather unnoticed than disliked. She preferred above all to stay at home and help her mother with domestic duties—"a regular little old woman," her mother used to call her. . . . The patient had always been particularly shy with boys, whose company she persistently avoided. When she reached puberty, this shyness increased to such a painful degree that her avoidance changed to aversion. In consequence, she was more than ever passed over in the social activities of her peer culture. (Cameron and Magaret, 1951, p. 504)

Sometimes a patient will have such a total lack of the various work and social skills that he will end up in permanent hospitalization. Individuals like this may live at home with parents and create no problem in a sheltered environment but be completely helpless when the parents pass on.

## Inappropriate instrumental repertoires

The various types of psychopathology that involve inappropriate instrumental behaviors include temper tantrums, which often are considered to be significant symptoms of psychopathology, especially if they are unusually severe or continue past childhood. The child in a tantrum may kick and scream and strike out at others in an uncontrolled rage.

> A sixty-two-year-old unmarried lawyer controlled his household, in which he lived with his less competent siblings, by the same kind of temper tantrums which he had used to control his parents. . . . So infantile were his attacks of rage, when he was frustrated in a hospital to which he had come for treatment of an unrelated illness, that the nursing staff at first reported them as convulsions. (Cameron and Magaret, 1951, p. 146)

Autistic children and schizophrenic patients, child and adult, frequently display unusual instrumental behaviors. These can range from bizarre posturing and grimaces to violent episodes. Aggressive behavior that is antisocial in criminals and psychopaths can also be considered to include inappropriate instrumental behaviors. Also, as was indicated in the context of phobias, sometimes a behavior is inappropriate not in and of itself but because of the stimuli that elicit the behavior.

## CONCLUSIONS

The present approach can only be illustrated with the examples given above. The general suggestion is that psychopathology—abnormal behavior—is composed of deficit and inappropriate personality repertoires. The various classic categories of "mental disease," it is suggested, are composed of the individual's personality repertoires. It is thus the task of a behavioral abnormal pychology to analyze categories and cases of psychopathology in terms of these personality repertoires. This must be done in a specific and detailed manner. It is not possible to consider systematically the etiology of a "mental illness" without this specification. Etiology, or understanding of causes of abnormal behavior, is central, however. Learning conditions play a causative role, it is suggested. The next section will indicate some of the principles involved.

## *Learning and maintaining environments and abnormal behavior*

In criticism of psychoanalytic theory, it has been said that "All treatment of neurotic disorders is concerned with habits existing at *present*; their historical development is largely irrelevant" (Eysenck, 1960, p. 11). Skinner's functional analysis of behavior approach has led to the same type of conclusion. Lovaas (1966) states the following, for example:

> The experimental laboratory design and the objective of isolating functional relations place restrictions on certain kinds of questions that the investigator may want to raise concerning abnormal behavior—for example, about the parent's role in the etiology of childhood schizophrenia. Answers to such questions, though often intriguing, entail so much confounding that they are meaningless in a functional analysis of abnormal behavior. (pp. 111–12)

This rationale has led to a deemphasis in the behavior modification treatments of abnormal psychology (Ullmann and Krasner, 1969), and clinical psychology (Bandura, 1969; Kanfer and Phillips, 1970) on specification of the repertoires that constitute psychopathology and the learning histories involved. The cumulative-hierarchical learning concept, however, and the behavioral interaction principles include as a major tenet that the learning history of the individual is central in understanding his personality repertoires, as well as specific instances of behavior. Moreover, it is only through understanding the manner in which learning histories can produce normal and abnormal personality repertoires that it will be possible to provide appropriate experiential conditions for the child, and appropriate therapy to those with personality problems. Our aim must be knowledge that will allow us to prevent the occurrence of abnormal behavior, it should be noted. This is of even greater importance than knowledge of how to treat abnormal behavior after it has arisen.

## DEFICIT LEARNING ENVIRONMENTS

The previous section suggested that abnormal behavior can be categorized in terms of deficits in personality repertoires and inappropriate personality repertoires. Learning environments that produce personality disorders can also be characterized as having either a deficit or inappropriate character. Several examples of psychopathology that involve deficit learning environments will be given first.

DEFICIT ENVIRONMENTS AND LANGUAGE-COGNITIVE DEFICITS AND INAPPROPRIACIES. A four-year-old "emotionally disturbed" child the author treated had a host of inappropriate behaviors, including gaping deficits in the boy's language-cognitive repertoires. He had no functional language. The deficit learning environment was constituted in the following way. The child had parents who were very occupied. Both parents worked— different shifts, long hours. When they were home they had chores to do and personal activities to look after, and they slept most of the time. They took good biological care of the child, and both had the conception that this was largely what was necessary, expecting child development to be largely biological growth. Moreover, both parents tended to be laconic; the father particularly was occupied with his own interests.

These conditions prevailed during the first two years of the child's life, and he did not develop speech. A housekeeper was then hired to be with the child, but she spoke Spanish and she had two of her own children with her. The boy did learn some Spanish words during a year and a half under this circumstance, but when the parents moved his development of the Spanish language ceased, and he was again without language. His mother recognized that he was retarded in language development and attempted to provide training for him. Her conception of language, however, was that it was only necessary to get the child's language development "primed" or started. Thus, she recited nursery rhymes to the child and prayers and gave the child training in imitating her. He thus acquired some imitational skills, but since no other language training was involved, he did not learn the other aspects of language.

Verbal imitational skills by themselves are not a functional form of language (see Staats, 1971a), and in fact this condition is one of the recognized symptoms of autism, called *echolalia*. A child with echolalia, which was the case with this boy, will respond to someone else's speech by imitating it, even when it is inappropriate to do so. When directed to do something, having no verbal-motor repertoire, the boy would repeat what was said instead of doing as directed. Deficits in the various language repertoires make a child appear "mentally defective," or crazy.

DEFICIT ENVIRONMENTS AND PSYCHOPATHIC BEHAVIOR. As another example, let us take the case of what is called variously social deviation, psychopathic personality, social immaturity, acting out, or by other terms. In the case to be cited the young man was diagnosed as a case of "constitutional psychopathic inferiority."

Casual conversation, as well as formal tests, suggested that he was highly intelligent, well-educated, and particularly adept at mechanical tasks. . . . Toward the antisocial behavior which brought him to the county jail the patient showed a consistently moralistic attitude. He readily admitted the forgery, stated that he knew his behavior was wrong, and said that he was sorry.

Lengthy interviews with the patient's wife and his mother, and a painstaking search of hospital and jail records, revealed at last that this man's behavior was by no means unusual for him. . . . It was only the almost fierce protection of his reputation by his mother and his wife that had kept his behavior a secret from his acquaintances, and even from other members of his own family.

The background of this selective retardation in social development was one of early maternal overindulgence and general family protection. . . . His mother and sisters idolized him. In the fights, arguments and misunderstandings with peers and teachers—which are an inevitable part of the socialization of any child—the patient was consistently made to feel that his behavior was justifiable, and that others were always in the wrong.

When the patient was in the sixth grade, his teachers began sending home value reports hinting that he might be untrustworthy. These reports were greeted by his mother and sisters with indignation and dismissed as false. . . .

After three years in the merchant marine the patient . . . married a woman older than he, a college graduate and former school teacher. . . . [The patient regaled his wife with untrue stories of his accomplishments and vocation during the first months of his marriage.] At this point he was discharged from the merchant marine as undesirable, and returned with his wife to his parental home. [Here his wife discovered the untruth of his stories, but excused them by calling them "slightly exaggerated."]

The patient's wife was at first received by his mother and sisters with coolness and some hostility. . . . In this atmosphere of tension [that ensued] . . . the patient had his first so-called "spell"; he left home without explanation for a three-month period. A private detective agency employed by his wife located the patient in jail in a nearby state, where he had forged checks to the amount of several thousand dollars. In this emergency, the patient's wife and mother joined forces and toured the country for four months, visiting every town the patient had visited, and making good every worthless check he had written. The patient was then released from jail and returned to his home town. . . . (Cameron and Magaret, 1951, pp. 193–95)

This example includes several different principles that are worth noting, including an account of how deficit learning conditions can result in deficit personality repertoires and abnormal behavior. To elaborate, it is evident in this case history that the patient has not had experience that would produce some necessary developments in his A-R-D system and in his language-cognitive repertoires. For one thing, as has been indicated (Staats, 1963, pp. 384–586), the child has to receive experience in which he learns not to do things that are aversive to others—for example, not to take things that belong to others, not to do things that hurt someone else. Only through appropriate experience does the child learn "socially controlling stimuli."

Moreover, it has been suggested that the child must learn a repertoire of verbal stimuli that elicit negative emotional responses and hence control the class of instrumental avoidance behaviors.

> The words *don't, dangerous, bad, stop, forbidden, mean, unkind, unlawful,* and so on, must come to elicit "escape" responses and weaken behavior, or the individual may encounter harmful or socially aversive consequences of far greater severity than that involved in the aversive verbal stimuli. . . .
>
> The establishment of aversive verbal stimuli seems to have even greater significance for the adjustment of the child. When the child learns reasoning sequences in addition to verbal aversive stimuli, he then has a means of avoiding behaviors that would be followed by punishment without requiring the administration of controlling stimuli by others. . . . It would seem that the efficacy of one's reasoning in certain situations would depend in part upon the existence of verbal stimuli in the individual's repertoire which were of an aversive nature. . . . The cautious man who "anticipates" the aversive consequences of certain actions, the socially sensitive man who "anticipates" socially aversive consequences of certain actions, and so on, would seem to do so, at least in part, because of training that had established for them effective verbal aversive stimuli, as well as the necessary reasoning verbal response sequences. (Staats, 1963, pp. 397–98)

Part of the psychopath's problem resides in the deficit environmental conditions that leave him with deficits in these language-emotional repertoires. If there is no understanding of the repertoires involved, the behavior of the psychopath may be a mystery. Thus, in the above case, authorities were puzzled by the young man's behavior, as were those who knew him. Indeed, as would be expected in the present approach, the patient's behavior "was inexplicable even to himself" (Cameron and Magaret, 1951, p. 196).

## INAPPROPRIATE LEARNING ENVIRONMENTS

INAPPROPRIATE REINFORCEMENT PROCEDURES AND CHILDHOOD PSYCHOTIC BEHAVIOR. The person can learn all types of psychopathology from inappropriate environmental circumstances. He can learn inappropriate aspects of his language-cognitive system, inappropriate aspects of his A-R-D system, and inappropriate instrumental behaviors. This can occur through the various learning principles and learning conditions that have been described. The case of the four-year-old boy who had no functional language may be used to indicate how certain inappropriate, "bizarre" behaviors can be learned in childhood.

On the opposite side of the ledger, the child had already acquired repertoires of "skilled" abnormal behaviors. He had learned "uncontrolled" motor behaviors of getting things for himself, such as in ransacking the refrigerator and food shelves to secure things that he wanted but was not supposed to have. He had also learned a number of abnormal behaviors that would get him out of situations aversive to him. Thus, when he was taken places that were boring to him—such as the grocery store, shopping tours, sightseeing excursions, and so on—he would behave in ways that the parents thought indicated he had "gone out of his mind." He might scream, look and act wildly, run around in a frenzy, go lifeless, and so on. When he behaved in a sufficiently "crazy" manner, convincing them he was indeed in a crazy spell, his parents would take him home, thus rewarding him by taking him out of the aversive (boring) situation. He would also go into these "crazy" behaviors when put into other situations he did not find rewarding. Thus, he would not go to sleep at night, he would not allow himself to be trained in any manner, and so on. As a matter of fact, as soon as a situation took on any of the flavor of training he would escape the situation. (This indicated that training situations had been aversive for him in the past.) The parents would let him have his way once they were convinced he was "out of his head," that is, was crazy. They thus were teaching him, inadvertently, to behave more and more bizarrely, since they only gave him what he wanted (reinforced him) when he behaved that way. He also employed these bizarre behaviors when he was being punished, because he could avoid the punishment when it was thought he was "out of his head" and not responsible.

This is a brief analysis of how a child comes to acquire behaviors that will lead him to be diagnosed as psychotic, autistic or emotionally disturbed. (Staats, 1971a, pp. 309–310).

INAPPROPRIATE ENVIRONMENTS AND A "NERVOUS BREAKDOWN." Cases that are cited in abnormal psychology literature are unlikely to involve analyses of specific behavioral repertoires or the learning conditions that influenced these repertoires. Adult behavior, as indicated, involves complexes of various repertoires and a complex learning history. The following case, thus, refers to the general learning conditions in the young man's cognitive system.

> One college student, for example, developed behavioral disorganization, with tension, sleeplessness, religious preoccupations and suicidal thoughts in a setting of overwork, fatigue and academic failure. But it was obvious that his difficulties had begun long before this. For he had been trained throughout childhood and adolescence, by his mother, in self-examination regarding the fundamental tenets of his religion, and by his father in ambitions which far exceeded his abilities. The patient described his mother as a fanatically religious person who repeatedly discussed with him controversial and insoluble problems of sin and punishment, evil and damnation. He described his father as a brilliant but unyielding man who would tolerate nothing less than perfection in his son. In college, the boy began to question for himself the tenets of his mother's faith, and at the same time he found himself unable to meet the academic standards his father set for him. His training had provided him no defense against religious uncertainty and no tolerance for imperfection in himself. "I am affected with a troublesome conscience," he said, "and I have the unfortunate attitude that winning the game is more important than playing it." With his supporting attitudes thus threatened, the patient grew tense, depressed, suicidal and finally seriously disorganized in his behavior. (Cameron and Magaret, 1951, p. 58)

THE REINFORCEMENT PARADOX. Frequently abnormal behavior is so bizarre or purposeless, or even self-defeating or self-punishing, that it is incomprehensible. In such a case it is very difficult to see how the behavior could have been learned. Thus, it is customarily (sometimes without basis) claimed that the behavior is constitutionally caused. This is one of the reasons, it is suggested, that childhood autism has been thought of as a constitutionally based psychopathology; that is, the autistic child is seen to perform self-harmful behaviors that result in painful stimulation. This appears to contradict the principle of reinforcement itself and suggests that the autistic child functions according to different principles than normal people do. The autistic child may batter his head against solid objects, producing lacerations and contusions and the possibility of brain damage. Some autistic children have to be restrained to prevent them from biting themselves to the bone, and so on.

How could such behavior be learned through the principle of reinforcement? There are several principles that can be suggested in considering how such behaviors can be acquired. First, the principle indicated in the previous example of the four-year-old with no language provides one explanation. This little boy learned bizarre behaviors through the action of negative reinforcement—the relief from aversive conditions. When he behaved in a crazy manner he escaped punishment and other aversive situations. A child can learn one pain-producing behavior—like head banging—as an escape from another aversive circumstance.

Secondly, behaviors can be learned gradually, through reinforcement, that would not be learned if at the beginning only the final version of the behavior were reinforced. Thus, for example, if one would reward a child only for fully developed head banging it would be ineffective in producing the behavior. However, if the training program commenced with relatively mild tapping of the head against an object, and this was followed by solicitous attention of a parent, and the requirement for the force of the tapping was gradually increased, then some very unusual behavior could be produced. The child will adapt to the painful stimulus so that it is not as aversive as it would be to the child who had no progressive experience.

Such a training program would probably arise inadvertently, but not so uncommonly, at least in principle. The usual parent is concerned about the care of the child. When he sees the child doing something which is potentially harmful or which scares the parent because it appears unusual, the parent will attempt to change the child's behavior. Many parents will do this by positive means, especially if they think the child has something wrong with him. This means they will give the child what he wants, or they will give him attention, be solicitous, and then try to get him to do something less upsetting. Children learn all kinds of inappropriate behaviors in this way, because in their efforts to prevent the behavior the parents reward the behavior. The whining child receives a cookie, the child by a furious temper tantrum gets what he wants, the child who stops breathing gets attention, and so on.

Sometimes the parent vaguely realizes that the behavior should be discouraged and attempts not to reward it when it occurs. The parent may try not to give the child what he wants when he has a tantrum but relent when the child is in danger of hurting himself, or when the parent thinks there must be something wrong with the child to carry on so violently, or some such consideration. This state of affairs may mean that a gradually increasing standard is applied to the child. He is not rewarded until he behaves in a sufficiently violent manner. Under such an inadvertent training program, extremes of behavior can be produced that are unusual and difficult to understand.

An efficient, although again inadvertent, method for training self-punishing behavior occurs when there is a combination of punishment and sympathy. Thus, the parent of a retarded child may use some form of punishment in trying to train the child, albeit with ambivalence and guilt. When the child then performs a bizarre or self-punishing response, the parent feels responsible for the "spell" and relents and gives the child attention. Such means can produce abnormal behavior skills.

LEARNING A FETISH, AND A CASE OF PEDOPHILIA. Fetishes are frequently so unreasonable that they are not easily considered in terms of learning. The author had an early opportunity to observe the development of the following case—except for the first conditioning trials.

A two-year-old boy had had experience in which he was reinforced by affection while sitting in his mother's lap (and that of a nanny) while the child fondled the parent's ear. It was a mutual exchange of reinforcing attention. When the father noticed the habit, which came to extend to himself and to some extent to others, he suggested that the behavior be discouraged. This was done to some extent, but the behavior was not considered important, and the treatment was not systematic. It involved pulling the childs' hand away some of the time, especially when the father was around. Within a couple of years of this inadvertent training the child had an incipient fetishistic behavior. He bothered other children by touching their ears, sometimes as a caress and sometimes in order to bother them and get attention. The behavior by this time was well learned and the boy had developed skills of a surreptitious nature with respect to the father.

The undesirable nature of the behavior was accepted at this time by the mother, and systematic retraining of the child was conducted by her as well as by the father. At this time punishment was needed. This case did not result in a full-blown fetish, but it is easy to see how such an innocent behavior could result in a sexual fetish. That is, as described, the child learned to rub others' ears under the control of affectional rewards. He was then punished for contacting ears in this manner at some times and at others times was rewarded. Thus, the behavior and the ears became even more salient, and he also developed surreptitious ear-rubbing skills that appeared to be incidental brushings. The only lacking ingredient was that of attaching sexual A-R-D value to the ears and ear rubbing. This could have occurred if at the onset of puberty the boy had masturbated while fantasing (imaging) ears and ear rubbing. Sexual A-R-D value would in the process be conditioned to both stimuli, resulting in the learning of behaviors that would lead to ear rubbing.

Annon (1971) has described the learning history of a case of pedophilia, wherein a young man was compulsively attracted to girls between the ages of 6 and 12, which provides support for such expectations. As an 11-year-old, the patient's first sexual experience occurred in the context of playing a "strip tease show" game with other preadolescent girls and boys in a small town. The boys would then chase the nude girls and finally indulge in mutual manipulation of genitals when the quarry had been run down.

This experience was discontinued when the patient's family moved to the city. After puberty he learned to masturbate and employed an image of a nine-year-old girl from his previous experience. At age 16 he had further experience in mutual masturbation with a ten year old girl. For the next five years she was his main fantasy, as well as other girls of this age he would see during the day. He had additional experience of petting with a 12 year old when he was 20.

On the other hand, at the age of 19 he had a repulsive and unfulfilled attempt at sex with a 45-year-old prostitute. From then on he repelled approaches to sexual intimacy by mature females. In his twenties he became a medical technician. While working with children between the ages of two and four, he would fondle them sexually and later masturbate to the fantasy of the children. Later, he had the same experiences with 12 year olds. He sought out families as friends in which he could have sexual contact with preadolescent girls involving mutual genital stimulation. He was so attracted by young girls that he was in danger of exceeding normally accepted practices. He read books about pedophilic sex, went to libraries at times when he could observe young girls, and so on. Moreover, he had continued to acquire very negative attitudes toward adult females and adult sex. He was worried because his sexual fantasies were becoming sadistic. "[H]e would use imagery suggested by his reading historical accounts of various tortures being inflicted on women. He pictured himself ripping and tearing at female anatomy such as their arms, breasts, or hair. All this imagery he saw as the continued spread of his sickness" (Annon, 1971, p. 392).

This patient was sure that he was abnormal, sick, and loathsome. It is not too difficult to see how a person with this life history of conditioning, if he had also acquired the instrumental behaviors of violence, and under the press of his fear and loathing, could on the one hand come to perform acts of violence that would be considered bizarre and bestial. On the other hand, he could have ultimately committed unacceptable sexual acts with children.

It is important to understand how such A-R-D systems and their consequences can arise in terms of learning history. This individual, left in the small town, would have had sex experience with maturing girls and learned a normal sexual A-R-D system. When his life was changed by moving from the town, his sexual A-R-D system development was "fixated" and then elaborated in an unfortunate direction, based upon imagery paired with masturbation. This case suggests that the employment of "normal" imagery is important in masturbatory activity because of the A-R-D conditioning that results. This is a point in favor of "normal" pornography as opposed to "abnormal" pornography, since these can be sources of sexual imagery.

## MAINTAINING ENVIRONMENTS AND THE NEUROTIC PARADOX

It has been said in various ways that psychological treatment of a patient is ineffective if the patient is then returned to the circumstances in which the problem arose. Both the disturbed child who is returned to parents and again displays his abnormal behavior and the criminal who returns to society and reverts to his old ways are examples that have led to such statements.

In general terms, it is suggested that there is in many cases of abnormal behavior an environment that maintains the behavior. The abnormal behavior is not a disease entity that unfolds under its own impetus; it is supported by the environment that continues to help determine the individual's behavior. Although it may be important to understand how the individual has learned his abnormal behavior, a closely related topic, it is equally important to have knowledge of the present maintaining environment.

Frequently, there is mystery concerning abnormal behavior because the nature of the maintaining environment is not understood. Mowrer (1950), for example, posed a quandary that he called the "neurotic paradox." His question concerned how behavior that was self-defeating, nonfunctional, and even aversive is maintained. Why is the behavior not unlearned if it does not lead to positive consequences?

It is suggested that the maintaining environment must be inspected in such cases. For example, a number of years ago the author observed, while training in a Veteran's Administration mental hospital, that the professionals, in interacting with their patients, were actually maintaining the abnormal symptoms in which they were interested. This will be referred to in greater detail in a later chapter. It is relevant to say here that like the parents who inadvertently create inappropriate behaviors because they do not understand the principles of reinforcement, professionals can also create problem behaviors in their patients. The author analyzed a specific case in these terms where the therapist's interest in a confused speech symptom of a schizophrenic patient, and the therapist's consequent attention to it, was maintaining the speech symptom (Staats, 1957a).

Many behavior modification studies that followed this analysis have added support to this conception. Ayllon and Michael (1959) showed, for example, that the attention of the nurses in a psychiatric ward maintained various abnormal behavior symptoms. Allen, Hart, Buell, Harris, and Wolf (1964) showed that the attention of teachers was maintaining the isolate behavior of a nursery school child; the teachers were concerned about the child's loneliness. Harris, Johnston, Kelly, and Wolf (1964) demonstrated that a regression—crawling instead of walking in a four-year-old child— was being maintained by teachers. Unusual crying in a nursery school child was shown to result from the same types of maintaining conditions (Hart, Allen, Buell, Harris, and Wolf (1964). Numerous studies can be interpreted to demonstrate that "Sympathetic feelings, no matter how well intentioned, combined with a poor conception of human behavior, can actually create [and maintain] rather than remove problems of overt behavior and internal emotional states" (Staats, 1973, pp. 219–20).

FUNCTIONAL, BUT ABNORMAL, BEHAVIOR: SOLVING THE NEUROTIC PARA-
DOX. It is thus suggested that there may be environments that maintain
abnormal behaviors. This pertains to deficits in the environment that
maintain deficits in essential behaviors. It also pertains to inappropriate
environments that maintain inappropriate behaviors. What may appear
to be a paradox, in that the behavior seems to be nonfunctional to the
individual and thus, according to behavior principles, should not be maintained, may actually occur according to principle.

The individual's behavior may be maintained, for example, by conditions that exist in a small group which contrast with the conditions in the
larger group. Thus, the intransigent child who disrupts the classroom may
be reinforced by several other children who are also maladjusted in the
classroom. The behavior, however, may be considered to be abnormal by
the general group. At another example, the parents may overprotect and
overindulge the child so that he is actually faced with deficit learning
conditions in his home. He may not learn self-care behaviors, social behaviors, group play behaviors, and so on. In the larger group outside of
the home, however, the deficits are considered abnormal. In both examples
the maintaining practices of the smaller group produce behaviors antithetical to the standards of the larger group. This is the way that seemingly nonfunctional behavior is actually functional and is maintained.

As another type of example, the individual may reinforce someone
else's behavior, even though he dislikes the behavior and considers it inappropriate. For example, an individual may be overly selfish and demanding in social interaction. He may dominate the conversation and center
it upon himself. He may insist on getting his way, doing the things he
wants, and so on. Persons with whom he interacts may give way to his
pressure, thus providing immediate reinforcement for the behavior. However, they will also be conditioned to dislike the individual, and their
later actions toward him will be determined by their negative attitudes.
They will avoid him, be uncooperative with him, vote against him, not
recommend him highly, and so on. These are all behaviors that will be
aversive to the individual, but in time too far removed to affect the maladjustive behaviors that produce the conditions.

There is also a strong cultural characteristic of giving special privileges
to individuals who are ill or who suffer some handicap. Someone who is
labeled mentally ill also fits into this category. The special treatment given
such individuals, however, may simply be another circumstance in which
the individual is reinforced for undesirable behaviors. When this is done
to the extreme, the individual's behavior will be extreme.

Sometimes the immediate reinforcement of the undesirable behavior,
with the aversive consequences of the behavior only occurring later, is
inadvertent. Thus, for example, many criminal behaviors may be immediately reinforced and the more strongly aversive consequences may

come only later. The man who commits rape attains immediate sexual reinforcement. The aversiveness he suffers in formal and informal punishments may far outway the positive reinforcement—but the positive reinforcement may help maintain the behavior.

The general point here is that the task of removing the paradox and mystery of abnormal behavior lies in the analysis of what the behavior consists of. On the basis of such analyses it is possible to make a detailed analysis of the learning environments that produced the behavior and the maintaining environments that ensure the existence of the behavior in the face of superficially discouraging conditions.

## Behavioral interaction (personality) principles and abnormal behavior

The analysis of abnormal behavior cannot be approached as a simple task, with a conceptually simple set of tools. In addition to the basic learning principles, it is necessary to include a knowledge of personality repertoires, as has been indicated. Furthermore, it is necessary to include the principles of interaction that have been described. While it is important to indicate how the environment can produce behavior, it is also important to indicate how behavior affects the environment. It has been said that the individual's behavior is an independent variable or cause, as well as a dependent variable. This conception is as relevant for abnormal behavior as it is for normal behavior. The individual's display of deficits in normal, adjustive behaviors, or his display of inappropriate, maladjustive behaviors, may affect the physical or social environments in ways that in turn affect the individual's further behavioral development. Examples of these types of behavioral-environmental interactions, as they are relevant for understanding psychopathology, will be described here.

### TYPES OF BEHAVIORAL-ENVIRONMENTAL INTERACTIONS

Table 8.2 provides a categorization of interactions between the individual's behavior and the environment. There are (1) deficits in behavior that lead to environmental (experiential) deficits for the individual, (2) deficits in behavior that lead to inappropriate environmental conditions, (3) inappropriate behaviors that lead to deficit environmental conditions, and (4) inappropriate behaviors that lead to inappropriate environmental conditions.

DEFICIT BEHAVIOR AND DEFICIT ENVIRONMENTS. Several examples of the effects deficit personality repertoires may have in producing deficit environments are given below.

*Toilet skill deficits and educational deficits.* Toilet skills are not com-

**TABLE 8.2**

Behavior-environment interaction and abnormal behavior

|  | Deficit environment | Inappropriate environment |
|---|---|---|
| Deficit behavior | Example: Deficit in language-cognitive repertoirees leads to deficit environment of institution for retarded child | Example: Deficit toilet skills leads to aversive social treatment |
| Inappropriate behavior | Example: Inappropriate behavior leads to social ostracism | Example: Criminal behavior leads to brutalizing environment of prison |

plex skills, nor is the training to effect them difficult or complex (Staats, 1963, 1971a). Notwithstanding, the parent may fail in effecting the training, and when this occurs the deficit in the child's skill may have environmental effects that are very significant. In most public school systems, for example, the child must be toilet trained to be accepted into regular kindergarten. Because of this stricture, an incontinent child will not receive the normal learning opportunities. The deficit in his behaviors will lead to severe deficits in the environment that can yield severe intellective deficits in the child.

*Language deficits and social environmental deficits.* As another example, the four-year-old boy who had been diagnosed as emotionally disturbed may be mentioned again. One of the central problems of this child was that he had glaring deficits in his basic language behavioral repertoires, as has been noted. Words did not control his behavior; his verbal-motor repertoire was absent. He had no speaking repertoire other than an inappropriate imitation. Words did not elicit emotional responses or other "meaning" responses in the child; they did not serve as rewards. The child did not score at all on an individually administered intelligence test.

It is easy to see how this constellation of behavioral deficits would result in environmental deficits that hindered this child's continued behavioral development. The fact that the child had not learned a repertoire of behaviors under the control of words (as in following directions) strictly limited and distorted his further social, intellectual, and emotional learning. It was quite evident that the child could receive little tuition from a teacher in nursery school whose major training contact would have been on a verbal level. For example, stories the teacher told were meaningless sounds to this child.

Moreover, his deficits prevented him from gaining normal social learning with classmates. A child might say to him, "Let us play cars. You go get the red car." Such a verbal stimulus would elicit no appropriate response from the boy. Or a group of children might ask him to play house and suggest that he play some role. When this elicited no appropriate behavior, the children would get someone else to play. Much of even four-year-olds' play takes place through the use of language. This boy was unable to participate, to gain the social rewards of play, and in the process to learn important new intellectual and emotional responses. The deficits in language repertoires produced such serious deficits in the social and intellective environments that it would have to be expected that he would continue to accumulate deficits in his repertoires and continue to merit an "abnormal" classification on this basis alone.

*Intellectual deficits and institutional deficits.* It should be added that there are institutionalized ways by which the individual's behavioral deficits can result in severe deficits in his environment. An example is that of the retarded child. Because he has not acquired basic behavioral repertoires that make him "intelligent," he is frequently placed in an institution which is very deficit as a learning environment. Many institutions offer only custodial care. Even in better ones the child will not gain the rich environmental training the normal child receives. The institutionalized child will have primary associations with children who, like himself, have severe deficits in behavior. Since a great deal of learning occurs from child to child, this alone will represent a severe environmental deficit that will enhance the difficulty.

### Deficit behavior and inappropriate environments

*Toilet skill deficits and inappropriate parental reactions.* There are also cases in which a deficit in the child's behavioral skills will result in an environment that is inappropriate in producing inappropriate learning in the child. The case of the deficit in toilet skills, for example, was said to lead to a deficit environment. In addition, however, such deficits may lead to other unfortunate consequences. First, as the child grows older his ordure becomes more unpleasant for the parent to deal with in cleaning the child. This constitutes negative attitude conditioning trials for the parent with respect to the child. Whatever behaviors the parent has learned to negative attitudes—impatience, punishment, criticism, and so on—will tend to be elicited as a consequence.

If the child is not toilet trained past the usual age, the parent will suffer additional negative experience. The parent is likely to experience social disproval himself, to be embarrased. These experiences may be intensified if the child has difficulty in preschool or in school or in the homes of friends; both the parent and child will become the target of social censure. Usually the parent, with his conception of biological development, will conclude there is some deficiency in the child. The parent is likely also to attempt to solve the problem by some training method. Without explicit information—which the parent is not likely to obtain—he will ordinarily continue to fail. This failure to successfully train a child in what is a simple behavioral skill will be additionally aversive to the parent. In such cases it is typical in our culture to resort to some form of punishment. The punishment may be physical, or it may be admonition to try harder, not to behave like a baby, and so on. It may also involve restrictions, loss of rewarding activities, and the like.

This, then, constitutes a case where a deficit in the child's behavior—through a deficit in the training he has received—produces an inappropriate social environment. Punishment should not be involved in toilet training; it elicits negative emotional responses incompatible with eliminatory responses. Such training will not be successful; it will, in fact, produce other behaviors that are undesirable. For one thing, the more the parent administers punishment of whatever kind, the less the parent will be a positive emotional stimulus for the child. The child thus will learn negative social attitudes, one aspect of which is to make adults ineffective in the future training of the child, which will lead to further deficits in the child's learning. In the specific case of defective toilet habits, however, the child will also learn to avoid the toilet training situation if he is punished. Once he has learned a fear response to the situation, he will have difficulty eliminating; he will avoid going into the bathroom, especially if the parent is there; he will stay out of the parent's way at times appropriate for elimination, and so on. This, of course, will insure that the child has more toilet accidents, receives more punishment, and will learn to avoid the toilet situation even more strongly.

There are many types of cases where deficit behaviors lead to inappropriate environments that train the individual to additional deficits or inappropriate behaviors. As another example, the underachieving child in school is subject to social conditions that are aversive and create negative attitudes. The young man who has a deficit in masculine social and athletic behaviors may meet social derision from girls that produces negative attitudes in him and prevents him from obtaining the experience with girls that would give him heterosexual emotional-motivational desires. The general point is that the individual who fails to behave in the manner that his group will reward will meet with aversive circumstances that may constitute an inappropriate environment in various ways.

INAPPROPRIATE BEHAVIOR AND DEFICIT ENVIRONMENTS. As Table 8.2 above indicates, there are also inappropriate behaviors that result in the individual experiencing deficits in his environment. The person who begins to display behaviors that are considered to be psychotic may be placed in a mental institution. There he will have an environment that may be deficit in many ways; he may have no work demands, for example, and his work skill development may cease. He may have little or no social interaction, and his social behaviors may cease to develop, or they may deteriorate. In being institutionalized, the person with criminally inappropriate behavior may be placed in an environment that is deficient in the very experiences he requires.

Even without institutionalization, inappropriate behavior will ordinarily lead to various forms of social ostracism. The individual who behaves inappropriately with others will have fewer opportunities for social interaction. This may constitute a severe deficit in learning that is necessary for the individual.

INAPPROPRIATE BEHAVIOR AND INAPPROPRIATE ENVIRONMENTS. The abnormal behavior of an individual that leads to commitment to a mental institution may also result in an environment that is inappropriate as well as deficient. In a mental institution the person associates largely with others who display inappropriate behaviors of various kinds. The patient in such an institution is thus likely to have many inappropriate experiences. The same is true of the individual apprehended for criminal behavior. He is placed with other individuals whose inappropriate behaviors constitute an inappropriate environment for him. A common criticism is that the first offender has experience in prison with hardened criminals who subject him to various types of inappropriate experience, such as forced homosexual activity. He will also usually learn negative, antisocial attitudes by virtue of the brutality of a prison, as well as additional skills and values of criminality.

## SOCIAL INTERACTION AND ABNORMAL BEHAVIOR

Several of the examples given thus far are actually cases of social interaction. It should be understood that the principles of social interaction are central in understanding abnormal behavior.

PARENT-CHILD INTERACTION. An explicit example of a social interaction that includes the elements common to other social interactions is the parent-child relationship. In the realm of abnormal behavior, for example, a mother rewards dependent, childish, demanding behavior in a child. In this illustration, this then sets the conditions for the father to have aversive interactions with the child, since the father has negative emotional responses to such behaviors. The father in his interactions with the child is punishing in various ways to the extent that the child also learns negative attitude responses to the father. The mutually negative attitudes, and the behaviors elicited that are punishing for the other, can be considered to be inappropriate. The interactions of father and child can add little happiness in either direction and yield much unpleasantness.

RECIPROCAL NEGATIVE ATTITUDE INTERACTIONS. The general principles of social interaction are relevant in the above example, as well as in other settings. When an individual is aversive to another, the second person will learn a negative attitude toward the first individual. This attitude will then elicit negative behaviors toward that individual. This, in turn, will condition the first individual to negative emotional responses to the second person. If they are in repeated interaction, each negative action toward the other may result in a returning negative action. Repeated interactions of this kind can yield two individuals who have intensely negative attitudes toward each other and who consequently display very negative "striving against" behaviors that are inappropriate in the aversiveness that each experiences.

NEGATIVE ATTITUDE INTERACTIONS IN MARRIAGE PROBLEMS. *Reciprocal negative attitude interactions in marriage problems.* In a marriage, the two people will customarily have positive emotional responses to each other at the time of marriage and presumably, at least in part, during their marriage. Let us say, however, that one has a behavior that is aversive to the other, like flirting. By flirting, the wife may elicit negative emotional responses in the husband. He, in turn, may then do something negative to the wife, under the action of his own negative emotionality. This action, in its aversiveness, may then contribute to more frequent actions on the part of the wife that are aversive for the husband. The negative attitudes may prevent each from behaviors that are rewarding to the partner. The wife or husband with negative attitudes may not be as considerate, or may not participate positively in sexual activities. A marriage can degenerate into a social interaction in which the class of striving against responses is the most dominant behavior.

NEGATIVE ATTITUDE INTERACTIONS IN SEX RELATIONSHIP. The root of many marital sex problems may be considered to reside in the sex relationship as a social interaction. As a brief example, let us say that the husband, on getting married and under a state of sexual deprivation and anxiety, ejaculates too soon and is generally not a satisfying lover. The wife who is also anxious and dissatisfied, shows her disappointment. She then acts to elicit increased anxiety responses in the husband during the next sexual encounter, and the social interaction is again mutually aversive. Continued interactions of this type may lead to a deficit on the husband's part; his anxiety may make him impotent. The aversiveness on the part of each for the other may be entirely inadvertent, but in an interaction process may result in negative attitudes of each towards sex.

POSITIVE INTERACTIONS AND THE DEFENSE MECHANISMS. Social interactions may be other than aversive and lead to abnormal behavior. A husband, for example, may over a period of years describe in a highly inaccurate way his social relationships at work. His wife may reinforce this behavior with attention and solace. The verbal behavior may also serve the purpose of a rationalization in that it removes some of the aversive implications of the husband's lack of advancement. In so doing, however, the experience may strengthen the reasoning system that leads the husband to behave aversively to his fellow workers. If the rationale grows to the extent that it is delusional, and his wife shares the delusion, the social interaction would have produced a *folie à deux*.

Deficits of behavior and inappropriate behaviors can thus emerge from a social interaction. The principles of social interaction are important in considering certain cases of abnormal behavior, and there has not been systematic consideration of such principles in the study of such behavior.

### INDIVIDUAL-INSTITUTIONAL INTERACTION

This is also a large area for consideration within the context of behavior analysis of the interaction between the individual and the social institution in the production of abnormal behavior. It has been suggested that the individual has a personality, which consists of the complex behavioral repertoires he has acquired. Chapter 14 will indicate in outline form that social institutions also can be characterized in terms of behavior principles. One example mentioned is that a social institution may also be considered to have an A-R-D system. Also, it will be suggested that the conjunction or disjunction of the A-R-D systems of the social institution and the individual must be studied for the social problems produced. Within the brief outline in Chapter 14 there lies a great deal of significance for understanding both normal and abnormal behavior. This may be elaborated a bit in the present context by the use of an example.

The school system may be considered to have an A-R-D system. It does not, for example, utilize money in its A-R-D system, as do certain social institutions, such as an industry. Moreover, its rules of delivery of its A-R-D stimuli are different than those of other social institutions. It delivers its reinforcers for obedience and academic learning excellence, in contrast to a military service institution that delivers reinforcers for obedience and physical courage, or to a university institution that rewards its faculty for independent work and creativity.

The A-R-D system of a social institution, however, is complex, and knowledge of its effects would be expected to require a detailed analysis. The A-R-D system operating in our schools, for example, includes practices intended as formal rewards, such as grades and honors, as well as informal practices. Thus, for example, the athletic equipment and facilities of a school are not systematically administered as rewards. However, they do constitute strong A-R-D stimuli for many students and affect the behavior of the students. The social approval of the teacher is not systematically applied, nor are story telling, recesses and free time, access to books for pleasure reading for some children, and so on. Furthermore, the A-R-D system has not been systematically described in behavioral principles in terms of how it functions. Thus, for example, in any class in which there is relative grading—where the success of the child is measured, at least in part, in terms of how he compares to his classmates—the system insures that while some children will be strongly positively reinforced, others will be punished by the institutional signs of failure.

In any event, the child has an A-R-D system, and the school has an A-R-D system. If the two do not match, the child may develop undesirable behavior as a consequence. Let us say, for example, that the adult (such as the teacher) does not have positive A-R-D value for the child. In the institution, however, the positive A-R-D value of the teacher may be a prominent part of its A-R-D system. As a result of the disjunction, the child will not respond emotionally positively to the teacher, his attentional and work behaviors will not be under the directive control of the teacher, he will not be rewarded for academic endeavor and success by the teacher's approval. Thus, the child will not learn. As a consequence of this the child will come to receive negative A-R-D stimuli in terms of the derision of his classmates, the disdain of the teacher, social punishments, and so on.

If the experience is intense enough the child can learn very negative attitudes toward the school, teachers, successful students, books, and so on. Such attitudes may be abnormal in terms of their intensity, in terms of not enhancing the child's adjustment, and also because they will induce behaviors that will be highly undesirable. In one case that has been described (Staats and Butterfield, 1965), there was evidence of this in that the child baited teachers and students, fought in school, disrupted class, and the like. Later, he participated in vandalizing a school. Such a child will also "escape" the school in various ways, such as daydreaming and absenteeism, which further aggravates the deficits he already has.

## INSTITUTIONAL-INSTITUTIONAL INTERACTIONS

It can be briefly noted that the underlying conflict may be better described as involving two institutions, rather than an individual and an institution. That is, the individual's problems may be a function of being affected by contact with two conflicting social institutions (or "group memberships," as the term has been employed in social psychology). Thus, in the above example, the child has been reared in a social institution, the family. In this experience, primarily, he acquires his A-R-D system. It may be because the A-R-D system of the family is incongruent with the A-R-D system of the school that the child has behavior problems.

As another example, let us say that a boy has been exposed to a religious institution through his family. As a result of his experience with both institutions he acquires values (part of his A-R-D system) that prevent him from participating in wars or in military institutions. He may face severe conflict when he is exposed to the conflict between the religious institution and other institutions of the society.

These examples are intended to indicate that the approach recognizes the importance of social conditions in human behavior. Moreover, it calls for social analysis of individual problems of human behavior. Most importantly, however, it is possible to make these analyses within the same set of principles that has been utilized in the learning-behavior theory.

### BEHAVIOR-BEHAVIOR INTERACTIONS: DIRECT AND SETTING

It has been indicated at various places in the present book that there are behavior-behavior interactions. The individual may reason, let us say, which involves sequences of language behaviors. But these language behaviors then elicit some overt action on the part of the individual. This is a direct behavior-behavior interaction. An example has already been given in which the individual's delusional reasoning that his bookies were going to kill him led to a long sequence of behaviors attempting to escape them. As another example, the child who has learned negative attitudes to school baits teachers and students, vandalizes the school, and so on. The negative attitudes directly elicit the overt abnormal behaviors.

It should be noted, however, that there is also another type of behavior-behavior interaction in which the interaction does not involve one behavior directly eliciting the other. Rather, the first behavioral state sets the conditions by which the other will occur. Examples of this have also been given; one example demonstrated that the child who does not have the basic language repertoires sets the conditions for his failure in school. He is not able to follow directions, to respond appropriately to stories, to interact with other children, and so on. The child's own personality is thus a determining condition for what he experiences and what happens to him in various ways. One abnormal behavior can be a determinant of another.

It is important to understand these types of behavioral interactions in understanding and dealing with abnormal behavior. Thus, the abnormal behaviors of the delinquent child who is apprehended for baiting students and teachers and vandalizing a school may be considered the problem. One might then attempt to treat these behaviors directly. However, if the child's basic problem is that he does not have the basic language-cognitive skills to succeed in school, then this is the primary problem to be dealt with. The analysis of abnormal behavior must consider the manner in which various aspects of the individual's personality interact in producing problems of adjustment for the individual.

In these behavior-behavior interactions there can be deficit behaviors that result in other behavioral deficits. Deficit behaviors can result in the individual developing inappropriate behaviors as well. On the other hand, inappropriate behaviors can lead to either deficits in behavior or the development of other inappropriate behaviors.

## Human learning principles and abnormal behavior

The need to discover human learning principles, in addition to those that have been demonstrated in the animal laboratory, has been stressed for the study of human behavior. Several of the human learning principles that emerge from or are important in the consideration of abnormal behavior will be mentioned briefly here.

### BEHAVIOR COMPETITION

The principle of behavior competition has been called the "competition between normal and abnormal learning" (Staats, 1971a, pp. 305–307). The general principle is that acquisition of one behavioral skill may militate against the acquisition of another behavioral repertoire.

Take, as an example, the case where the individual has learned the two-finger "hunt-and-peck" skills of typewriting. Once this sensorimotor skill has been acquired, the individual is less likely to acquire the more effective skills of touch typing methods because the two skills are in "competition," in a sense. Having the first skill makes the second comparatively less reinforcing. It means the hunt-and-peck typist must give up a relatively highly rewarded proficiency for a much more effortful and less rewarding behavior, for a long period of time. The individual with neither skill faces the same effort and lack of reward whether he types one way or the other and is thus more disposed to select the learning task that promises more eventual reinforcement. Several other examples will be given.

PATHOLOGICAL COMMUNICATION DEVELOPMENT IN COMPETITION WITH NORMAL SPEECH. The principle of pathological communication development vs. normal speech applies to many types of human behavior, especially problem behaviors. In stating this principle the author described a form of retarded speech development where the child gets what he wants by gesticulating, whining, grunting, crying, and so on. Once the child has attained skills of this type, if the parent attempts to delay giving the child what he wants in order to teach the child appropriate language, the aversiveness of the delay will lead the child to intensify his grunting-gesticulating-crying repertoire. Getting what he wants then will reinforce this more intense abnormal behavior and make language learning even more difficult (see Staats, 1971a, pp. 305–7). The skills of pathological communication displace the normal learning.

TREATING ABNORMAL BEHAVIOR BY MODIFYING NORMAL REPERTOIRES. The principle of behavior competition was further elaborated by illustrating how a child who misbehaved severely in school was treated for his behavior problems, not directly by dealing with his behavior problems, but by reinforcing him for working in a reading training program. Strengthening these work behaviors and the resulting increase in achievement resulted in a decrease in the child's misbehavior (Staats, 1971a, pp. 315–16). Ayllon and Roberts (1974) later designed a study to specifically investigate the principle of behavior competition. They showed that reinforcing academic behavior in a classroom resulted in a decrease in discipline problem behavior. Increasing the "normal" behavior, it is suggested, had the effect of decreasing the "abnormal" behavior, as the principle indicates.

The principle of behavioral competition has wide relevance for the consideration of abnormal behavior and its treatment (Staats, 1971a, pp. 305 ff.). One implication is that the first effort for the control (treatment) of abnormal behavior should be in insuring that the child learn normal behavioral repertoires. The normal behavioral repertoires will in many cases compete with (prevent) the development of abnormal repertoires. This rationale suggests that interest must be focused upon indicating what the normal repertoires are and in methods for producing them— rather than in the traditional concentration upon the abnormal behaviors themselves (Staats, 1970b). More will be said of this in the next chapter.

ADDITIONAL EXAMPLES. There are behavioral repertoires involved in being dependent, in being psychopathic (as the young man in the previously cited case history), in being helpless (as some retarded and psychotic individuals are), in being violent, in criminal behaviors, in being socially belligerent, and so on. Many of these repertoires result in the receipt of various kinds of reinforcers and have been well learned. The normal repertoires that would replace them would demand a long-term learning process, with relatively little reward. This can be the dynamic reason underlying retention of an abnormal behavior.

## CUMULATIVE-HIERARCHICAL LEARNING OF ABNORMAL BEHAVIOR SKILLS

The hierarchical conception of learning suggests that learning to be a human is a long-term affair. It involves a process in which a repertoire of behavioral skills is learned that is the basis for the learning of additional, more advanced behavioral skills. The acquisition of the new skills, in combination with others already acquired, then enables the acquisition of even more advanced behavioral skills. The acquisition of complex repertoires of great skill can only be understood in terms of the basic skills upon which the advanced learning is founded. This statement is couched in the positive terms that refer to the acquisition of desirable skills. But the same principles of cumulative-hierarchical learning, involving the various principles of interaction and competition in learning, also apply to the learning of abnormal personality. Some of the previous case examples have been presented to illustrate this, as in the following elaboration of the description of the four-year-old-boy who had no functional language.

> In addition, because words as stimuli elicited no imaginal-meaning responses in the child, nor any emotional-meaning responses, there were other ways that this child could not profit from the school program. Most children learn a great deal from stories the teacher tells them in preschool. To this child, however, the verbal stimuli of stories were so much nonsense stimulation. While the other children sat and listened to the teacher, their attentional behaviors reinforced by the imaginal and emotional responses elicited by the words, this child would be in a situation where he was receiving no reinforcement of any kind. The other children were not attending to him or playing with him. The teacher was not attending to him or rewarding him in any way.

> The manner in which deficits in behavior can produce abnormal behaviors can be well exemplified in this one situation; for if this particular child in the above situation sits quietly, he is in a state of deprivation for attention or any other rewarding event. While the other children are reinforced by the story, he has no source of reinforcement. Let us say, however, that he gets up and begins to wander about the room—which in itself is inappropriate in this situation—and finally bangs on the piano, or begins to run around in a frenzy, or does something else to attract attention. When this has occurred and he has disrupted the story telling time, the child is rewarded by removing the aversive situation and by gaining attention for his disruptive behavior. Thus, again, the one sure source of reinforcement for this child is through the display of some type of disruptive behavior. He will as a consequence learn new "abnormal" behaviors to this school situation. These behaviors will keep him from learning adjustive behavioral skills, however. Moreover, in the usual situation it is this type of behavior that will result in the child being placed in a class for emotionally disturbed children, where his opportunities for normal learning will be reduced and his opportunities for learning abnormal behaviors from other children through imitation will be increased.

> In conclusion, it may be suggested that the four-year-old child who has been described would ordinarily as a consequence of his grievous behavioral deficits encounter a continued social experience that would cumulatively add to his nonadjustive, abnormal behaviors. Moreover, he would cumulatively acquire behavioral deficit piled on top of behavioral deficits. After a long history of this kind, the character of the individual's behavior may be so different from normal that one could not easily see how it could have been acquired according to normal principles of development. Long-term analysis of the complex learning circumstances, in the context of the concepts of hierarchical learning, can indicate how such abnormal cases occur. Moreover, such analyses—because they are

based upon empirical principles—can be employed to avoid such circumstances, or to treat such problems once they have arisen. (Staats, 1971a, pp. 310–311)

## THE DOWNWARD SPIRAL OF CUMULATIVE-HIERARCHIAL LEARNING AND ABNORMAL BEHAVIOR

Following the principles of cumulative-hierarchical learning, an analogy may be posed for a behavioral kind of race. The race involves the rapidity and excellence displayed in the acquisition of successively advanced behavioral skills, and it pertains to a large extent all the way through childhood and into and throughout adulthood.

In this race the child who more quickly acquires certain behavioral repertoires also creates a situation that tends to accelerate his further acquisition of new repertoires. This occurs in two ways. First, the early acquisition of the basic behavioral repertoires will provide the foundation for further skill acquisition. In addition, however, the acquisition of a repertoire of desirable skills at an early time has social consequences. Society in general, and that segment of it which is personal to the child, will provide social rewards to the "winners" of the race at every level of skill acquisition. The informal "competition" involves the various types of skill the child must acquire—ranging from talking first, walking first, and so on, as gauged by the developmental charts, through later acquisition of academic and athletic skills as evidenced by grades, scholarships, honors, intelligence tests, and success in organized reports. The child who is advanced in the acquisition of skilled repertoires will receive social rewards more heavily than the not-so-advanced child. Such rewards will contribute strongly to maintaining the work and learning behaviors of the child for the many learning tasks that remain in the hierarchical task of human learning.

Thus, the child who is more advanced in acquiring skills and thus has the basic skills for the next task will also have created for himself a situation conducive to continued work and learning. When as a result he learns the next repertoire of skills, and more rapidly than other children, he enhances these salubrious conditions for further learning. Moreover, there are other conditions that will tend to support the "winners" in the race as they continue. The advanced child will have parents who are able instructors. Ordinarily they will continue throughout to provide the child with superior learning advantages.

In contrast, let us take the case of the child who does not gain his initial repertoires as rapidly as most other children because of poor learning conditions for one reason or another. In this case, in the race we can expect to see a downward spiral of relative performance. The child who acquires his basic behavioral repertoire more slowly than others is not ready as soon to succeed in the task of learning the next more advanced skill. Moreover, at any level, as a consequence of being a laggard in the learning race he will find himself in a less propitious social circumstance of reward for learning. Thus, his attentional and working behaviors will be poor, and his learning will be at a less rapid rate than would be the case in better motivating conditions. Moreover, it can be expected that these conditions grow progressively worse. The less advanced the performance, the less the reward. The less the reward, the less the maintenance of learning behaviors. The less the learning, the greater the decrease in the reward, and so on. The mental retardate can expect to find himself in such a downward spiral of relative progress in learning.

As has already been suggested, the downward spiral may be exacerbated by the conditions that result from retarded learning. The social consequences of being a loser in the learning race can create conditions by which the child learns undesirable behaviors that are considered to be abnormal. These behaviors will frequently be such that they interfere with the further learning of the repertoires of skills demanded by society. When this occurs, the downward spiral of relative learning is accelerated. In certain cases the learning of the child can switch almost entirely to the learning of undesirable behaviors that can be conceived of as being basic behavioral repertoires of abnormal behaviors. That is, the acquisition of some abnormal behaviors will enable the acquisition of others or will create the conditions in which other abnormal behaviors will be learned. This may include the production of a social situation that is conducive to learning abnormal behaviors. Sometimes the conducive situation for learning abnormal behaviors is an institution, as has been indicated for retardates and as is usually the case also for emotionally disturbed, psychotic, autistic, or delinquent children. "[I]t may be suggested that the severity and type of abnormal behavior that the person develops will be influenced by the point in the hierarchical development at which the child falls behind" (Staats, 1971a, p. 314).

One additional example will be employed to illustrate the cumulative and interaction principles in a case of long-term abnormal behavior development.

One morning, let us say, an individual walking along the street suddenly whips out a revolver and shoots several businessmen who are standing on a street corner talking in a group. Later, under questioning, the individual states that people are plotting against him, that he has been under surveillance, and that the group of men talking were part of the movement to kill him. Although seemingly "senseless" when only the final behavior is considered, such a behavior might be seen as the culmination of experiences all involving the operation of established principles of learning. Before suggesting a history which could lead to such bizarre behavior it will be useful to restate the extreme behavior in terms of the three links in social reasoning. First, suppose because of this individual's highly unusual personal history he labeled the businessmen as "Part of the group who is after him." This complex label then elicited the sequence "They are going to kill me, but I'll get them first." This sequence finally mediated the instrumental responses of drawing and firing the gun.

It may be difficult to see how such unusual labels, verbal response sequences, and instrumental behaviors could arise. That is, it is difficult to see how normal training processes could produce such deviant behavior. Let us suppose, however, that this individual as a child acquired behaviors in the home which were aversive to other children and adults—unreasonable, demanding, aggressive behavior. To the extent that the child's behavior was aversive, other children would respond with aversive behavior in return. Let us say also that when the child reported how other people had mistreated him, the parents said that other children were jealous of him, were wrong to treat him in that manner, and so on. The child thus had many experiences in which such events were labeled in terms of the other person's jealousy, perversity, and so on. The child was never trained to label correctly his own behavior as aversive and other people's unpleasant behavior as a response to his own behavior.

Let us also say that the child acquired, in addition to labels, verbal response sequences such as "People are evil," "You can't trust anyone," "If you have something they want they will try to take it from you," "Watch out for people and beat them to the punch if they try to get the best of you." Suppose, furthermore, as this individual grew older, violent behavior to others who did not "treat him right" had been reinforced. In this process, not only might fighting with other children be reinforced, but he might also acquire the verbal sequence "Fight to protect yourself; if someone mistreats you the best way to handle him is to fight," and so on.

Let us also say that because of this individual's aversive behaviors, verbal and nonverbal, he has many aversive interactions with people as he grows older. For example, when someone else is promoted over him (actually as a consequence of his behavior) he responds with aversive verbal behavior about and to the other person. Further, adding to the cycle, since he is aversive to many people, people may indeed "organize" against him, talk about him, get him fired from jobs. Thus, his social experiences may continue to be generally aversive. He cannot hold a job, thus many other primary and social reinforcers are withdrawn, and so on.

The individual labels each one of his "social conflicts" as he has been trained to do: "They are jealous of me, they are all against me." People in general could in this manner become very aversive for this individual and he might respond to everyone—his landlady, the grocer, his casual associates—in verbal and other ways that are aversive. Again, these people would respond aversively to him, talk with each other about him, plan retaliations against him, laugh at him, and mistreat him. In reality because they are a group they could be more aversive to him than he could be to them.

With this history, it might not be unreasonable that the individual would label his accumulated social experience by saying "Everyone is against me and is out to get me, "They hate me so they want to harm me," "They are plotting to hurt me." In such a history, the gradual road to extremely abnormal aggressive behavior might be seen. (Staats, 1963, pp. 387–88)

## Conclusions

It is suggested that the nosological system cannot be taken from an earlier time. As has been indicated, our conception helps determine what we observe, how we interpret it, and what we conclude. The classificatory system of abnormal behavior helps perpetuate the "old" conception, and it helps determine research findings and clinical practices, as will be indicated.

As a social behaviorism theory of human behavior is formulated and accepted, it calls for a reconceptualization in the area of abnormal behavior. The learning-behavior theory must be developed as a full theory of personality, to include description of the various important basic behavioral repertoires. The conditions of learning and the principles of learning involved, especially human learning principles, require specification, which must be carried into the realm of abnormal personality. We need a description of the various basic behavioral repertoires that lead to maladjustment. This description and classification should not be expected to follow the traditional categories. It is the behavioral repertoires that are central, not the traditional labels.

The examples given in the present chapter have illustrated the fact that different abnormal categories overlap in the behaviors and the principles of learning involved. Thus, as an example, deficits in basic language repertoires are common to mental retardation, autism, emotional disturbance, and schizophrenia in children. Distinction in classification comes in other behavioral repertoires the children exhibit. As another example, deficits in the emotional-motivational (A-R-D) system are exhibited in schizophrenia, psychopathic personality, neurasthenia, phobias, dependent personality, and so on. It is suggested that the A-R-D system is a more fundamental level of description than mental disease entities.

The important point here is that present systems of classification carry implicit suggestions that there is some common underlying process that determines the individual's behavior. They also suggest that there is some unity to the individual's problem. In doing so the classification systems take attention away from the behavioral repertoires that compose the psychopathology.

It is suggested, rather, that the full learning conception of abnormal behavior should be elaborated. This conception suggests that it is the personality repertoires and maintaining conditions that are important. The personality repertoires must be described, including the way they are learned and maintained, as well as the behavior-behavior interactions, social interactions, and individual-institutional interactions that are involved. It will be found, it is suggested, that such a description of the individual, his behavior, and his environment will provide understanding of the origin of the individual's problems, the present maintaining events, and the steps that must be taken in remediation of the problems. The purpose of the present chapter has been to outline a framework for this endeavor.

# Part three

## Functional behaviorism levels

# 9

## Social behavioral clinical psychology

THE "BEHAVIOR MODIFICATION" approach in clinical psychology is well known and accepted today. This acceptance, however, is a relatively recent development, having occurred in the past decade.

The history of the application of learning principles to clinical problems may be considered to have begun with Watson and Rayner's (1920) demonstration that emotions (fears) could be learned. The systematic development of learning theory as an approach to clinical psychology, however, began in 1950. The first, as well as the most sophisticated, systematic, and important, work was Dollard and Miller's *Personality and Psychotherapy* (1950). The authors employed learning principles of a Hullian variety in conjunction with a psychoanalytic theory of human behavior. They included many productive analyses, but the learning approach was not itself used in its complete role as a theory of human behavior—normal as well as abnormal. Psychoanalytic theory took precedence as the personality theory, and the methods of treatment suggested were traditional psychoanalytic therapy sessions. Thus, this effort did not stimulate the straightforward application of learning principles to the removal of abnormal behaviors and the production of normal behaviors.

Dollard and Miller's work did provide a context for an interest in using learning principles in clinical psychology. To some extent a book by Mowrer (1950) which did not link the learning theory and psychoanalytic theory closely and a paper by Shoben (1949) also contributed to this interest, as did the work of Rotter (1954). In the latter case, the extension

291

of learning principles led to laboratory studies rather than to clinical applications. However, the approach and its findings contributed to the demonstration that the principles were relevant to the human level. A number of clinical psychologists trained in the Rotter approach were positively disposed toward the behavior modification developments and were able to utilize and extend behavior modification findings.

Another relevant development grew out of the finding of Greenspoon (1950) that social reinforcement of a subject's verbal responses would increase the frequency of that type of verbal response. This finding could be seen to have relevance for consideration of verbal psychotherapy, since the psychotherapist's response to his client's statements could be considered as social reinforcement which would affect what the individual would say. This suggestion had implications for questioning the validity of patients' statements in psychotherapy when constructing theories of personality and psychotherapy; it might be the influence of the psychotherapist that was reflected in the patient's statements. The clinical psychologists who were most active in this research (see Kanfer, 1958; Krasner, 1958; Salzinger, 1959; as examples), however, were concerned largely with theoretical-experimental issues. The principle of reinforcement was useful in conducting the experiments and in making theoretical criticisms of psychoanalytic and client-centered techniques, but it was not used by the early experimental clinicians in treating problems of behavior. The model of the professional-researcher which was to continue to the 1960s was of a clinical psychologist who employed traditional psychodynamic methods in treatment but who also did research of a more basic nature.

Skinner's major influence upon the development of behavior modification in clinical psychology was of the type yielded by Watson's early interest or by Thorndike's interest in educational psychology. Skinner's statement that human behavior can be considered within the set of operant conditioning principles (1953) provided a general context but did not consider problems of human behavior. Skinner contributed to the criticism of psychoanalytic theory (1953), as did Eysenck (1952), in a manner that helped weaken exclusive reliance on this system of thought and its attendant psychotherapeutic methods. The term *behavior therapy* was employed by Skinner and Lindsley in a research project in which psychotic patients were studied in an operant conditioning apparatus (see Lindsley, 1956). However, the abnormal behaviors of psychotics were not given a behavior analysis, and there was no attempt to treat the abnormal behaviors. The main thrust of the work was to employ the psychotic's manner of responding to reinforcement schedules as a diagnostic tool.

Thus, at this time, and for some time following this, there was a foundation for an interest in learning theory in the context of clinical problems of human behavior. Additional constituents were needed, however, to begin work on problems of human behavior employing learning principles and procedures. A major need was the analysis in learning terms of the actual behaviors that constitute clinical problems (analyses that had been exemplified by Watson and Rayner, 1920, and also by Dollard and Miller, 1950) as well as the development of treatment procedures and research methods based on learning theory for obtaining explicitly stipulable results. There were two main lines of development that began meeting those needs—one which came to be known as behavior modification and which began in a general reinforcement theory framework but later was identified with operant conditioning, and the other which has been called behavior therapy and which developed in the context of Hullian learning theory. The first to be considered will be behavior modification, beginning with a description of the author's own experience in developing this area (see Staats, 1970b, 1970c, 1973).

## Naturalistic observations and experimental-naturalistic research

When the author undertook clinical training in the early 1950s as part of his preparation for the general study of human behavior, there were no efforts to apply conditioning principles directly to the treatment of clinical problems. There were general suggestions that learning principles be applied to human behavior, but there were no analyses of abnormal behaviors at a level that could serve as suggestions for specific treatment employing conditioning principles and procedures. In fact, there were no teaching laboratories where students could learn how conditioning principles could be employed to train even animals to functional behaviors. The author's first applications of learning, as a matter of fact, involved training a cat to useful behaviors (see Staats, 1968c, 1970a). This experience was very meaningful in demonstrating that conditioning principles are lawful and work even outside of the laboratory, a possibility that received very little attention at that time. Not until behavior could be reliably affected according to learning principles did their status as behavioral laws become convincing. At any rate, these demonstrations were impressive to the author and his associates in showing that learning principles could be applied to functional behavior. The value of this experience led the author later to develop an animal laboratory for teaching undergraduate students which became the basis for a behavior modification training program (see Michael, 1963).

This early application of learning principles to behavior in its naturalistic settings was not limited to work with animals, however. In a clinical training position in a Veterans Administration psychiatric hospital, the author had an opportunity to observe critically the psychodynamic (largely psychoanalytic) practices employed in dealing with problems of human behavior. One observation concerned the lack of relationship between psychodiagnosis, employing tests, and the actual behaviors of patients in which one was interested. Staff conferences were notable in including psychodynamic assessments of patients in which their psychosexual levels of development were inferred and the nature of their past experiences that could have caused psychic conflicts was suggested. But there was little discussion of the patient's actual behaviors or life problems, ways of interacting with others, and so on. Much time was spent with testing that gave no directives concerning what specifically to do with patients. It was evident that the focus of the psychodynamic approach was on the nature of the patient's alleged inner mental states and supposed Freudian motivations, on the past experiences that could have left unresolved conflicts or undeveloped inner personality structures. The patient's overt behavioral characteristics were considered unimportant except as indicators of those internal states.

Because of this orientation, essential types of observations were not being made. Thus, the everyday behaviors of adjustment were not of concern, although in the writer's view these were actually central. Clinical workers were sensitive to the aberrant behavior, the slip, that would reveal the nature of the individual's internal conflicts. The mission the psychodynamic theory set for the clinician was the discovery of internal conflicts through these lapses in the patient's psychological defenses. This mission leads away from the study of the normal behaviors the human must have to adjust to his life situation. It does not even lead to a focus upon the individual's abnormal behaviors which would show the manner in which the abnormal behaviors are learned but nevertheless prevent obtaining normal reinforcers (rewards) in life, prevent normal learning and the acquisition of normal behavioral repertoires, and generally inhibit adjustment.

The author's learning approach led him to make observations of the actual behavior of people, including their problems, and of the learning conditions that were affecting them. Analyses were made of the learning conditions that could be changed in order to also change the problem behaviors. One example that involved putting the analysis into actual practice can be summarized to indicate how behavior modification using conditioning principles can be accomplished outside of the laboratory or clinic. The case involved a graduate student who was scheduled to take a comprehensive oral examination. The author was concerned for her ability to emit confident and fluent speech in the presence of the grilling that can occur, having had difficulty himself at an earlier time. Furthermore, there was a good deal of fear on the part of the "subject."

The behavior analysis and treatment involved both instrumental and classical conditioning principles. Observations revealed that she had met vociferous argumentation and criticism when venturing an opinion in the orally competitive student interactions that customarily occurred. This was enhanced because the subject was several years behind her close associates. There was thus no opportunity for the subject to learn confident speech, since it was not reinforced. (For a more complete discussion of the importance of socially reinforcing fluent speech in cases of speech hesitancies, see Staats, 1971a). The environment was thus "abnormal" with respect to producing confident speech skills.

After making the observations and the analysis, the author enlisted his associates Jack and Betty Michael to aid in instituting the behavioral treatment. This was to consist of giving positive attention to the opinions of the subject until a point of view was completely expressed. It was also to include expression of approval for the view. Over a period of time, it became evident that there was an increase in the "confidence" and fluency of the subject in speaking in the group. It is interesting that the subject was completely unaware of the systematic nature of the reinforcement for her verbal behavior or its effects.

In addition to this, the author organized two "mock" oral examination sessions. In these the pretending faculty members were Jack and Betty Michael, Lloyd Brooks, and the author. Questions were asked in appropriate form. It was felt that this treatment would have the effect of both lessening fear to the orals situation and providing skill in answering series of questions. A recent study has reported success using a similar procedure for a similar problem (Suinn, 1968).

This was in 1956, at about the same time an article appeared in the *Journal of Abnormal and Social Psychology* describing an abnormal case that clearly lent itself to analysis in terms of learning principles. The case was that of a schizophrenic patient with confused speech which was characterized by the saying of the opposite of the "appropriate" response (Laffal, Lenkoski, and Ameen, 1956). For example, the patient would reverse language usages such as that of *yes* and *no*. As was the method at the time, this abnormal behavior was considered important for what it told about the patient's psychodynamic motivation. The present author, however, considered the "opposite speech" of the patient as important in and of itself. Speech is a central type of human behavior, and abnormal speech of any kind can be considered to be a severe problem. Rather than wondering what psychic problem the opposite speech indicated, his learning analysis was concerned with what conditions and principles could bring about and maintain such abnormal behavior.

The verbal conditioning studies had already shown that social attention, when delivered contingent upon the utterance of certain types of words, would increase the frequency of such utterances. Greenspoon (1950) had demonstrated, for example, that when the experimenter said "mmm-hmm" when a subject said plural nouns (in a task of simply saying any words aloud) the subject would be conditioned to say plural noun words more frequently. Other studies suggested that various kinds of social attention would act as a reinforcer in social interaction. Delivered contingent upon a type of speech, attention would result in the increase in frequency of that type of speech (Talbot, 1954; Verplanck, 1955). These studies were with normal subjects.

The author thus suggested in the behavior analysis of the schizophrenic's opposite speech (Staats, 1957a) that social reinforcement could be involved in the acquisition and maintenance of psychotic behavior. The evidence provided by the extensive protocols of interviews with this patient (Laffal, Lenkoski, and Ameen, 1956) corroborated this analysis in the specific case. It was clear that the normal speech of the patient was of little interest to the clinicians involved. As indicated, the psychodynamic conception deemphasized the importance of normal behavior in "understanding" the patient's internal psychic conflict.

It was when the patient uttered one of his *unusual* statements that the doctors became interested in what he was saying. Then they would perk up, ask him questions concerning why he said that instead of the other, did he really mean that, did he not mean the opposite, and so on. The present author's observations of patients in neuropsychiatric hospitals had indicated that there are many patients who receive very little social attention, interest, or approval. Many patients have sparse behavioral repertoires that do not merit attention and approval. Moreover, the other patients are not able to provide attention and approval because of their own behavioral deficits and inappropriacies of behavior. Furthermore, patients have low status; they are "abnormal" people and thus are for each other weak sources of social reinforcement. Contact with high-status, "normal" people and with the hospital professionals is limited. Thus, it must be expected that many patients are deprived (starved) of social attention and approval, and their behavior can be expected to be strengthened even more by social reinforcement than is usual.

The behavior analysis thus suggested that the social reinforcement provided by the professionals with whom the patient interacted would have an important influence in producing and maintaining the abnormal behavioral symptom of the schizophrenic patient. The clinicians, on the other hand, were interested in the abnormal "opposite speech" because they felt it revealed something about the inner nature of the patient's psychic problems, suggesting speech reversal was an expression of repressed hostility, and so on.

With this conception the therapists did more than give the patient social reinforcement in the way of attention for opposite speech. In their interest in exploring the symptom and the assumed underlying pathology, the clinicians gave the patient material reinforcement contingent on the abnormal behavior. For example, the patient was asked if he wanted a cigarette. This was done to provide a situation for eliciting the opposite speech. When the patient said "No" he was given the cigaret anyway, which he accepted and smoked. The author noted that "Giving the cigaret to the 'No' response is a reinforcement and would be expected to strengthen that response, that is, raise the probability that the patient would again say 'No' in the same situation." (Staats, 1957a, p. 268)

The author then went on to indicate how an analysis of an abnormal behavior in terms of learning principles provides a basis for suggesting treatment of the behavior.

> . . . Reinforcements such as these [social and material reinforcers for the abnormal behavior], without concomitant admonition, would be unlikely to occur in the normal person's social environment.
>
> Certain implications are derivable from this interpretation. If the opposite speech is maintained by positive reinforcement, then lack of such reinforcement should lead to extinction of such behavior. For example, withholding the cigaret should weaken the strength of opposite speech and giving the cigaret to correct speech should strengthen that type of response. (Staats, 1957a, pp. 268–69)

The author also suggested that there were other sources of reinforcement for opposite speech. Moreover, it was suggested that the principles of reinforcement applied generally to other types of abnormal language behavior of schizophrenics that serve to disrupt verbal interactions with others as well as the individual's own thinking.

> Another example of learning theory formulations applicable to schizophrenic speech was made in Dollard and Miller (1950). . . . Certain thoughts arouse anxiety [a conditioned negative emotional response, in the present terms]. Cessation of thinking those thoughts reduces anxiety. Thus, stopping thinking about that topic becomes a well-learned response. The same analysis can be applied to speech. Dollard and Miller give an example of a group of people who change the topic of a conversation because it arouses anxiety, and state, "people tend to learn to avoid unpleasant topics of conversation" (1950, p. 199).
>
> It could be said that the schizophrenic patient's verbal behavior, when it is not confused by reversal, elicits an anxiety response in him, perhaps because of its typical content. Confused ways of speaking, and perhaps even of thinking, would therefore be anxiety reducing. This rationale could be extended to obsessive thinking in addition to other types of confused schizophrenic speech and thought. In addition, for the schizophrenic, communication with others which is understandable probably introduces touchy subjects which arouse anxiety. Reversed verbal behavior and other confused speech may reduce anxiety when it produces breakdown of the communication and cessation of the anxiety producing subject matter or of the bothersome conversation itself.
>
> Why the schizophrenic's speech might elicit anxiety in him need not herein be elaborated, since the speech symptom is the relevant topic. It could be stated, however, that the unhappy life situation of an adult

schizophrenic probably elicits thought and speech which are not positive secondary reinforcers, but instead arouse anxiety. It is also probable that the lack of success of the schizophrenic's life behavior evokes verbal behavior from others which is anxiety producing for the schizophrenic. Verbal and nonverbal behavior of the schizophrenic which would avoid this anxiety would thus be well learned. (Staats, 1957a, p. 269)

This brief paper summarized the principles the author had developed which, as will be indicated, became foundations for the early behavior modification studies. The principles can be abstracted as follows:

1. The analysis criticized the psychodynamic approach specifically for its inappropriate use of reinforcement in producing and maintaining abnormal, psychotic behavior. It was suggested that the misconceptions of the clinicians not only did not help the patient, his psychotic symptoms could be learned in the hospital and maintained through the misdirected social reinforcement of professional personnel. This gave behavioral specification to the criticism of Eysenck (1952) that psychotherapy was not effective.

2. The paper on the schizophrenic's speech was an actual conditioning analysis of an abnormal behavior. Unlike the studies of verbal conditioning (see Kanfer, 1958; Krasner, 1958; Salzinger, 1959) or Lindsley and Skinner's study of reinforcement schedules with psychotics, the analysis of schizophrenic speech extended the principle of reinforcement directly to important functional behaviors. This was done to treat the abnormal behaviors themselves, not for academic purposes or purposes of diagnosis.

3. The analysis of the actual abnormal behavior that was of concern specified treatment expectations. That is, since the principles of learning are cause and effect principles, analysis of the abnormal behaviors in these principles suggested that if specific things were done, the behavior would be affected in a specific way. Thus, if the abnormal opposite speech was positively reinforced, it would be maintained. If it was not reinforced, it would extinguish. Normal speech, if reinforced, could be increased in frequency. Primary tenets of behavior modification have come to be rewarding desirable behavior and avoiding reinforcement of undesirable behaviors, as suggested in this analysis.

4. It was also shown that the naturalistic evidence of the clinic could be employed to validate a behavioral analysis. The principles were not restricted to the laboratory or to simple behaviors. Thus, the protocols of the clinicians' interaction with the patient appeared to corroborate the reinforcement principles. There was evidence that the undesirable opposite speech was being reinforced.

5. The further elaborations of the analysis of psychotic speech confusions and avoidance of verbal interactions suggested that learning principles were generally important for the analysis and treatment of abnormal behavior, as was indicated.

In concluding, it should be stated that this note is not intended as a complete analysis of the opposite speech of the schizophrenic patient. Perhaps it points out that hypotheses which apply to pathological language can be derived from a learning theory approach. At any rate, it is suggested that learning theory has reached a state where it has something to offer clinical theory *and practice*. (Staats, 1957a, p. 269)

This conceptual framework, along with the naturalistic behavior modification work described, provided a foundation for launching more formal studies. It should be emphasized, however, that a major step was involved in developing methods for such formal studies. This was done by Michael and his student Teodoro Ayllon, building on the foundation described.

## Growth of the behavior modification study

The first formal study by Ayllon and Michael tested the use of reinforcement and extinction principles in the modification of symptomatic behaviors of psychotic patients. The majority of the cases treated by Ayllon involved analysis of the undesirable abnormal behavior in terms of inadvertent social reinforcement given by the nurses who interacted with the patient. The treatment was generally of the type suggested with the opposite-speech patient—extinction for the undesirable behavior and positive reinforcement for desirable behavior, or just extinction alone. Thus, for example, one patient's compulsion to visit the nurses' office, in the face of kindly requests not to do so, was treated by withdrawing the nurses' attention (the maintaining reinforcement) when the patient visited.

The previous analysis of schizophrenic language and its treatment was given even more direct support through a case of a female patient who persisted in delusional talk about an illegitimate child and men who constantly pursued her sexually, none of whom existed. Like the clinicians who reinforced the opposite speech, "Some of the nurses reported that, previously [to the extinction procedures], when the patient started her psychotic talk they listened to her in an effort to get at the 'roots of her problem'. . . . These reports suggested that the psychotic talk was being maintained by the nurses' reaction to it" (Ayllon and Michael, 1959, p. 328). In treating this patient, sensible talk was socially reinforced and the psychotic talk was not reinforced. As the analysis of opposite speech had predicted, tally of the two types of behavior revealed the extinction of the abnormal speech and the increase of normal speech.

Another problem treated was that of several patients who demonstrated compulsive hoarding of magazines and other articles in their clothing—resulting, in one case, in skin rashes. Again, it was decided that the social reinforcement of being solicitously "detrashed" was maintaining the behavior. It is interesting to note that the opinion of a nurse dealing with the patients again was that the "behavior has its roots in the personality of the individual. The fact that he hoards too much indicates that Harry has a strong need for security. I don't see how we are going to change this need, and I also wonder if it is a good thing to do that" (Ayllon and Michael, 1959, p. 333). This evidence corroborated the general conception already outlined—that the traditional psychodynamic view led to the overt abnormal behavior being considered not as something important in itself but only as an indicator of internal personality problems.

This classic paper by Ayllon and Michael was very influential in greatly increasing interest in the application of learning principles to abnormal behaviors. A number of additional studies which were also conducted began to form the basis of a behavior modification literature. An example is the study that Isaacs, Thomas, and Goldiamond (1960) conducted on several mute schizophrenics. As indicated in the preceding analysis, schrizophrenics can learn not to speak because speaking is not positively reinforced, and in fact may be socially punished. The analysis suggests that mute schizophrenics could be encouraged to speak through positive reinforcement for speaking. The problem remains of getting the patient to speak so that he may be rewarded. Isaacs, Thomas, and Goldiamond used a procedure in which they reinforced, with pieces of gum, incipient oral responses of schizophrenic patients who had been mute for many years. Then they gradually reinforced only better and better cases of speech in the patients until the speech behavior was increased in frequency.

In another early but not widely recognized study, Williams (1959) treated a case of temper tantrums at bedtime in a 21-month-old child. This was done with the simple imposition of extinction. Prior to the behavioral treatment, whenever the parent would leave the room before the child fell asleep, the child would scream and rage. The parent would then return and provide attention and social reinforcement conditions which must have trained the child in his aberrant tantrums. It required one and one-half to two hours before the child would go to sleep under usual conditions. The treatment consisted of withdrawal of the sources of social reinforcement for crying—when the child was placed in bed, the parent did not return even though the child raged. The child cried and screamed for 45 minutes the first night, but within a week he was going to sleep in a reasonable fashion.

The principles and procedures developed to this point began to generate additional behavior modification studies. Ayllon and Haughton (1962) conducted a study in which they showed that psychotic patients who were feeding problems could be readily trained to self-care feeding habits, using food as a reinforcer. And Haughton and Ayllon (1965) showed that a psychotic behavior, the compulsive holding of a broom that was thought to be due to inner psychodynamic forces, could be manipulated using a cigarette as the reinforcer. Examples of other behaviors studied were stuttering (Goldiamond, 1962) and tics (Barrett, 1962). Zimmerman and Zimmerman (1962) employed social reinforcement contingent upon desirable behavior and no social reinforcement contingent upon undesirable behavior in a classroom. Baer (1962) showed that watching cartoons could be used as a positive reinforcer and that thumbsucking behavior could be modified in five-year-old children.

## Token-reinforcer systems and behavior modification

The present author, in using learning principles in his naturalistic settings and in conducting an active research program on language learning, quickly became convinced that the simple principles were relevant to functional, albeit simple, human behaviors. It was necessary to move on from that point, however. One area of central importance was in the development of reinforcement procedures that improved upon simple consumatory objects (like candy, food, and cigarettes) and social reinforcers (like attention and approval). The importance of the systematic manipulation of powerful, variable, and durable reinforcers was not understood yet. Skinner and his associates, in their teaching machine and programmed instruction work (which was Skinner's area of learning theory applications), depended upon "achievement" reinforcers. Learning new skills and moving ahead in tasks were considered sufficient reinforcers with children. "With humans, simply being correct is sufficient reinforcement" (Holland, 1960, p. 278). As a consequence of this orientation, those in programmed instruction did not systematically consider the question of appropriate reinforcers for children, nor did the early behavior modifiers who were of a Skinnerian orientation.

In his early clinical training the present author had worked in the late Grace Fernald's clinic at UCLA with "disturbed" children who had problems of learning that are usually associated with emotional or behavior problems.

> The thing that struck me was that these children did not learn because in a unit of time they made far fewer learning trials than did normal children. That is, their attentional and work behaviors were poorly maintained by the reinforcers that were usually effective. Under these conditions they learned new material at a very slow rate. (Staats, 1970b, p. 12)

It was evident that moving ahead, achievement, acquisition of skills, social approval, and so on, are all *learned* rewards. Whether or not they will be effective, and how effective, will depend upon the child's history of learning. The author had seen in the clinical situation that social approval and attention are not potent enough or durable enough as reinforcers to conduct behavior modification training of long duration in which the learning is arduous and repetitive (boring) for the person.

Moreover, it was just such behaviors that require study. While it was an important step to *begin* to consider simple behaviors that are symptomatic of psychopathology in terms of learning principles, it was also necessary to begin the study of complex, functional repertoires of human behavior. The major areas the author selected were those of the language-cognitive repertoire and the emotional-motivational system. A major commitment to the study of one of the realms of language—that is, reading—was made. Reading was selected because it involves arduous, repetitive learning trials, is fundamental to adjustment, and is related to personal and social problems that are severe and resistant to treatment.

The observation that a central problem in treating complex problems of behavior modification involves maintaining attention and work behaviors suggested the need for development of an adequate reinforcing system. The conception behind the token-reinforcer system the author developed was that of the money system. Money constitutes a ubiquitous, powerful, and enduring reinforcement system. Its characteristics are ensured by the fact that it is backed with a variety of reinforcers for all circumstances and states of the individual. The first token-reinforcer system utilized plastic discs of three different colors. The three different tokens were given different values, a half cent, a fifth of a cent, and a tenth of a cent. These tokens could be exchanged for an article the child had selected to work for. The item might be a toy, an article of clothing, a piece of sporting equipment—anything that the child had indicated he found was reinforcing and that was within reasonable cost. It was deemed undesirable for the child to select something of too great a cost, since he would then have to work too long before receiving the article.

The author tested this token-reinforcer system in the first part of 1959, with several children in a public junior high school.[1] The children had been selected by teachers as being problems in learning to read and behavior problems as well. The learning materials concerned reading and were designed to be simple to administer. Procedures were devised for recording the children's reading responses and their progress. More will be said later of behavior modification reading procedures. At this point it is only relevant to note that the child received the plastic tokens for his reading behavior. The system specified that the more desirable or effortful reading behaviors were reinforced with a token of high value;

---

[1] Judson R. Finley and Karl A. Minke, as undergraduate research assistants, administered the behavior modification procedures. Richard E. Schutz also participated in the conduct of the study.

lesser behaviors were reinforced with lesser value tokens. The child's daily accrual of tokens and his advancement toward his chosen backup reinforcer were charted for him so he could see his progress. The important result of the use of the token-reinforcer system was the immediate change in the behavior of the children from that which they ordinarily displayed in academic tasks. They became vigorous, attentive workers, and they learned well.

This was the first use of the token-reinforcer system in clinical and educational research and behavior modification. Work on the development of the token-reinforcer system and its use in behavior modification continued, as the results of the preliminary study had indicated the very extensive potentialities involved. The author communicated the efficacy of the reinforcement system to Jack Michael in the summer of 1959. Michael and Lee Myerson began to employ the concept and similar procedures in studying the maintenance of attention and work behavior of mentally retarded children at the University of Houston (see Michael, 1968). In addition, Patricia Corke and Samuel Toombs, students of Michael, attempted to use the reinforcer system with children in working with remedial reading. These studies, while not published, were communicated to others interested in applying reinforcement principles.

The advantages of the token-reinforcer system, and the possibilities for use of the system in working with *functional* behaviors of hospitalized patients (as opposed to the study of simple symptoms), were described to Ayllon in 1961. Ayllon and Azrin (1965) introduced the token reinforcement system in a whole psychiatric ward at Anna State Hospital, as will be described further on.

## CHILD BEHAVIOR MODIFICATION

While the dissemination of the token-reinforcer system and the behavior modification principles was beginning, the author went on to the formal demonstration of the need for token reinforcement in working with problems of child training and the treatment of behavior problems. The first step was to adapt the token system for work with young children, since many problems occur also at such ages and this is the appropriate time for prevention procedures.

The first published study employing token reinforcement involved application of the principles and procedures to four-year-old children (Staats, Staats, Schutz, and Wolf, 1962) over eight 40-minute training sessions. Three preschool children were given reading training under the no-reinforcement condition (although they always received social reinforcement). They worked a brief period and wished to quit. At this time reinforcement was instituted for correct reading responses, and they became enthusiastic, vigorous attenders to the task. Another three children were first given reinforcement, then no reinforcement, then reinforcement again after they wanted to quit. Again, the children would attend to the task and work when reinforced. When not reinforced, they fooled around and wanted to escape from the task. Reintroduction of reinforcement again improved their behavior. Their learning curves of words read followed the reinforcement manipulations.

> The experiment by Staats et al. [1962] was particularly significant because it demonstrated that with a token system and a variety of exchange items one is no longer dependent upon the power of a single backup reinforcer. That is, one is not limited to giving M&M candies whose power depends upon the momentary deprivation state of the child. Instead, the only limitation of backup reinforcer systems is the ingenuity of the experimenter. (O'Leary and Drabman, 1971, p. 380)

Dissemination of these results led to their use in other behavior modification projects.

> Initially, the teachers attempted to strengthen desirable classroom behavior and correct answers to academic materials by following such behaviors with remarks of approval and by ignoring inappropriate responses. Little, if any, improvement in sustained studying behavior was obtained under these procedures. Evidently verbal remarks in the form of approval and praise did not have reinforcing functions for these children. Consequently, a token reinforcement system similar to that used by Staats, Staats, Schutz, and Wolf (1962) was added. This procedure did indeed establish and maintain higher rates of effective study and greater cooperation. (Bijou, Birnbrauer, Kidder, and Tague, 1967, p. 316)

The first study (Staats, Staats, Schutz, and Wolf, 1962), however, while long by usual experimental practices, was actually brief in terms of the time required for the child to learn complex intellectual skills. Progress in terms of securing a more representative (longer-term) sample of reading learning was made in two additional studies with four-year-old children (Staats, Finley, Minke, and Wolf, 1964; Staats, Minke, Finley, Wolf, and Brooks, 1964). The children received automatically dispensed marbles as tokens. The token could be deposited for the immediate receipt of a plastic trinket or edible (peanut, raisin, and so on) reinforcer. Or the marbles could be placed in clear plastic tubes of different heights (and thus volume) so the child could work visibly for small toys he had previously selected (see Figure 2.5, Chapter 2). It was found that young children, supposedly with limited five-minute attention spans because of biological immaturity, would work for 40 minutes with good attention and vigor. The token-reinforcer system maintained the children's behavior with no noticeable decrement for eight weeks, during which time the children made hundreds of difficult and repetitive reading discriminations. The author concluded that the two studies indicated the approach and the reinforcement system had wide implications for behavior modification.

> [T]hese developments may be extended to the study of a number of types of significant behavior acquisitions, *e.g.*, speech learning, arithmetic learning, *etc.*, and to various special populations, such as deaf children, mutes, mental retardates, etc. Much operant research with humans has tended to involve only simple responses such as knob-pulling and button-pressing, and simple controlling stimuli. The present facility would seem to be useful in the study of the acquisition of complex responses of more immediate significance to human adjustment. This could also involve work which had remedial objectives, *e.g.*, remedial reading problems, the training of autistic children, general training problems in children resulting from deficient "motivation." (Staats, Finley, Minke, and Wolf, 1964, pp. 146–147)

This outline for extending the principles and reinforcement procedures tied in very well with the principles of behavior modification enunciated in the analysis of schizophrenic language, as supported by Ayllon and Michael's (1959) study and the others conducted up to this time. The suggestion that these various principles and reinforcement procedures could be used for remedial objectives was implemented very quickly. Montrose Wolf, who had been a research assistant on the studies with the four-year-old children, began extensive behavior modification work with young children at the University of Washington. For example, Wolf, Risely, and Mees (1964) undertook treatment of an autistic boy who had self-destructive temper tantrums and would not wear glasses necessary to retention of his vision. It was reasoned that inadvertent social reinforcement was maintaining the behavior, and extinction of the undesirable behavior and positive reinforcement for desirable behaviors was instituted, with success. Lovaas extended this work with autistic children (Lovaas, Schaeffer, and Simmons, 1965; Lovaas, Berberich, Perloff, and Schaeffer, 1966). Wolf's pioneering work and that of his colleagues supported the projection that behavior modification and token-reinforcer procedures had wide applicability to children's problems, and a new field of study was launched.

Wolf and his associates extended the behavior modification principles into the nursery school, again employing the finding that professionals can create and maintain problem behaviors through inadvertent reinforcement. It is suggested that a situation exists in education that is similar to that described in the context of psychiatric thought. There are concepts of child development that lead in many cases to solicitous attention being given to undesirable behaviors of children in the classroom. When a child performs a response in the classroom that the teacher is concerned about, the teacher is likely to give the child solicitous attention, especially with the young child. For example, a child who is isolated from other children is likely to be given such attention by the teacher, although this treatment would increase the frequency of the child seeking to be alone. Allen, Hart, Buell, Harris, and Wolf (1964) demonstrated this with a four-year-old nursery school child. The teachers began to reinforce the isolate child when he was in social activities with other children and ignored the child when alone. The results showed that improving the professionals' reinforcement practices removed the child's problem. Harris, Johnston, Kelley, and Wolf (1964) showed also that nursery school teachers were reinforcing a child for the regressive behavior of crawling instead of standing and walking. Change of the reinforcement and extinction conditions quickly solved the problem. Hart, Allen, Buell, Harris, and Wolf (1964) applied the same analysis to the problem of excessive and inappropriate crying in a nursery school child. Change to more appropriate rules of reinforcement successfully treated the problem.

It is suggested that our culture generally has the conception that a young child who is experiencing some difficulty should be treated solicitously. There are few distinctions in this conception. However, it applies in special force if the child is described as sick, in which case he many times will be treated solicitously (reinforced) no matter what he does. The conception also applies to children whose problem is their own behavior, even when it is the solicitousness of adults that maintains the behavior.

In the context of educational problems, as well as those of the mental health profession, it is suggested that it takes more than benign motives to help a child with a behavior problem. Sympathetic feelings, no matter how well intentioned, combined with a poor conception of human behavior can actually create rather than remove problems of overt behavior and internal emotional states.

The token-reinforcer system was added to the behavior modification principles in the treatment of other problems (O'Leary and Becker, 1967; Staats and Butterfield, 1965; Wolf, Giles, and Hall, 1968). Wolf, Giles, and Hall, for example, very successfully employed the system in a remedial classroom. The first use of a token-reinforcer program to control a large class of emotionally disturbed children was conducted by O'Leary and Becker.

Disturbed children who are institutionalized and culturally deprived children have been treated with the token-reinforcer procedures and behavior modification principles in long-term projects (Browning and Stover, 1971; Hamblin, Buckholdt, Ferritor, Kozloff, and Blackwell, 1971). Additional studies have employed various forms of the token-reinforcer system to work with delinquent and predelinquent children (Cohen, 1968; Meichenbaum, Bowers, and Ross, 1968; Staats and Butterfield, 1965; Tharp and Wetzel, 1969). These and other studies have generally demonstrated significant decreases in disruptive and inappropriate behavior as a result of the use of token-reinforcer programs (Kuypers, Becker, and O'Leary, 1968; Martin, Burkholder, Rosenthal, Tharp, and Thorne, 1968; Patterson, 1965a). It has been suggested, however, that the potential functions of the token-reinforcer system and the behavior modification principles with children have not yet been tapped (Staats, 1970b), and later discussions will deal with this topic.

## ADULT BEHAVIOR MODIFICATION

Ayllon went to Anna State Hospital in 1961 where he and Azrin began using the token-reinforcer system and the behavior modification principles in a project involving a ward of female patients. The experiments were conducted with reinforcement procedures that were developed so nurses and nurses' aides could implement the program. The behaviors in this project were of the functionally significant type that were adjustive—for example, self-care behaviors, making beds, working in a laboratory or a laundry, cleaning the kitchen, and so on, instead of the symptomatic behavior Ayllon had commenced working with. Tokens were delivered for the desired behaviors—for example, working six hours at the hospital laundry yielded 70 tokens. The backup reinforcers followed the multiple reinforcer rule, that is, a large number of backup reinforcers will prevent fluctuations in the efficacy of the tokens due to fluctuations in deprivation conditions of the individual.

Experiments were conducted to show the reinforcing value of the various backup reinforcers employed. This was the main research of the project. Several other studies were conducted exploring such things as the effect of satiation on a reinforcer and the effect of free access to popcorn on the reinforcing value of popcorn (see Ayllon and Azrin, 1968). This project did not provide understanding of psychotics, of their inability to adjust outside of the mental hospital, of the development of severe adjustment problems, or of the principles, methods, and procedures for treating such complex human behavior problems. However, the project has added a great deal to the original indications that abnormal behavior as well as normal behavior is subject to the principles of learning, and it has provided procedures that produce functional behaviors of the self-care type in psychotic patients. As a consequence, this project has been very influential in stimulating similar projects in other institutions and in improving methods for handling institutionalized psychotics.

As another example, Atthowe and Krasner (1968) extended the procedures of the token-reinforcer system to the mental hospital ward. This project applied the token-reinforcers contingent upon performance of a number of different desirable behaviors—for example, attending to assignments, self-care, going out on pass, and so on—and avoidance of undesirable behaviors, such as bed wetting. The 60 schizophrenic patients in a VA hospital were told they would receive one token for making a bed, two for attending a work therapy session, or whatever the contingency would be. The tokens could again be traded for a variety of desired things. In the project the incidence of behavior of the patients was first observed for 6 months before introducing the token-reinforcer system, and then the experimental program ran for 11 months. Significant increases were noted in such behaviors as leaving the hospital on passes, social interaction, following rules, attendance at group activities, and so on. Moreover, twice as many men were discharged during the program as had been discharged in the prior 11-month period. This was evidence of the therapeutic value of the program.

In addition to the token-reinforcer programs in entire psychiatric wards, a number of other behavior modification studies have been performed. Sherman (1965) has reported a controlled experiment over a number of sessions in producing verbal interaction behaviors in three long-term mute hospital patients. This study improved and systematized the procedures of Isaacs, Thomas, and Goldiamond (1960). Moreover, it was shown that reinforcement of nonverbal behaviors could result in the expected decrease in talking. Wilson and Walters (1966) used imitation procedures as well as reinforcement to reinstate speech in mute schizophrenics. Rickard, Dignam, and Horner (1960) have shown that withdrawal of attention (social reinforcement) when a patient emitted delusional speech, paired with social reinforcement for rational speech, could be employed to remediate this problem. They also demonstrated that interrupting delusional speech and social reinforcement of rational speech could be successfully employed. These therapeutic treatments could be generalized to the ward.

Various behaviors have been treated, as have various groups of individuals. Schaeffer and Martin (1966) successfully used the token-reinforcer system to treat apathy. Typical patient withdrawal was conceptualized as apathy. Token-reinforcers were employed to strengthen behaviors that replaced the "apathy" behaviors. The token system has been employed also with antisocial behaviors in a military setting (Coleman and Baker, 1969) and with prison inmates (Clements and McKee, 1968). Henderson (1969) has reported use of a token program in the transition between the residential institution and complete return to the community. The subjects were adults with seriously disturbed behaviors. Ingham and Gavin (1973) have employed a token system in treating stuttering.

Stuart (1969) has treated four couples with marital problems employing a variation of the token-reinforcer system. The couples provided tokens to each other in a program designed to strengthen behavior in each that was desirable to the other. The wife might reinforce the husband for spending time talking with her; the husband could give tokens for sexual intimacy. These are only examples of what is today a very active and productive field of research and clinical practice.

## NONPROFESSIONAL BEHAVIOR MODIFICATION APPLICATIONS

It has been suggested that central to consideration of abnormal behavior is analysis of the behaviors involved as well as the conditions that have produced and presently maintain the behaviors. When this has been done, the task of clinically treating the behavior may be quite straightforward, and in many cases it may be conducted by a nonprofessional person. This was apparent in the case of opposite speech in the psychotic patient. The treatment suggested was the withholding of social attention and other reinforcers for the psychotic speech and the giving of reinforcement to correct speech. This treatment could be conducted by anyone who dealt with the patient.

Suggestive evidence to support this suggestion was involved in the first behavior modification studies. In the Ayllon and Michael (1959) study nurses actually administered the reinforcement to the patients, and the authors likened the nurses to "behavioral engineers." In the present author's work with children with reading disabilities, undergraduate students served as the behavior modifiers when the first token-reinforcer system was applied. Recognition of the general implications involved a number of new extensions, as the following analysis of the behavior modification treatment of nonlearning in a juvenile delinquent suggests.

A final point should be made concerning the training procedures used in the present study. The procedures are very specific and relatively simple. Thus, it was not necessary to have a person highly trained in education to administer the training. . . . [T]he procedures could be widely applied or adapted by various professionals, for example, social workers, prison officials, remedial teachers, tutors, and so on. In an even more economical application, helpers of professionals could be used to actually administer the procedures; for example, selected delinquents (or prisoners) could administer the procedures to other delinquents. Thus, the procedures could be utilized in various situations, such as settlement houses, homes for juvenile delinquents, prison training programs, parts of adult education, and so on. All that is needed is a suitable system of reinforcers to back up the tokens. (Staats and Butterfield, 1965, p. 939)

One of the features of behavior modification analyses and procedures is that ready of application by nonprofessional people. This has led to a number of studies in various settings (Bernal, 1969; Evans and Oswalt, 1968; Patterson, Shaw, and Ebner, 1969; Ryback and Staats, 1970; Staats, Minke, Goodwin, and Landeen, 1968; Wahler, 1967; Wahler and Erick-

310                                                    *Social behaviorism*

son, 1969). This has been indicated in some of the studies already summarized, and further extensions will be suggested in a later chapter.

## Behavior therapy

The origin and early development of the modern behavior therapy
procedures for the treatment of anxiety were accomplished largely by
Joseph Wolpe. Wolpe employed Hull's learning theory as his basic
conceptual foundation, particularly the concept of conditioned inhibition.
According to this concept, each time a stimulus is presented and a response is elicited, an inhibitory process builds up (to some maximal
strength). The principle thus suggests that repetition of a response will
lead to an inhibitory process for the response, some of which is conditioned and becomes permanent. An individual required to repeat an
undesirable behavior—like stuttering, for example—would learn inhibition
for the behavior.

Wolpe also formulated a principle for the elimination of inappropriate
anxiety. Procedurally, if a response incompatible with anxiety was made
to occur in the presence of the stimulus that usually elicited the anxiety
response, the incompatible response would be conditioned to the stimulus.
The primary response that he has employed as the response antagonistic
to anxiety is that of relaxation. Practical considerations also led him to
have the individual imagine the anxiety-producing stimulus (although
the behavioral justification for this procedure was not indicated), since
it was usually difficult or impossible to have the actual stimulus in the
therapy situation. In addition, Wolpe had the subject list different occasions that involved the anxiety-producing stimulus in a hierarchy ranging
from the least anxiety-producing occasions to the most anxiety-producing
occasions. In therapy the patient would be led to go through the hierarchy
on an imaginal level while relaxing, progressing gradually from the least
to the most intense anxiety-producing situations (Wolpe, 1958).

Wolpe and others applied this "systematic desensitization" method to
a variety of cases. Phobias of various kinds were treated by means of the
procedures derived from a learning theory. Wolpe did the pioneering
clinical research necessary to corroborate the procedures as a therapy,
providing a basis for extensive research and clinical practice.

It should be noted that Wolpe's development of systematic desensitization took place in the 1950s and represented also the straightforward
extension of learning principles to the treatment of problems of human
behavior. This development had a strong impact upon the progress toward
a behavioral approach based upon the direct use of learning principles
in clinical treatment. Until after the first few years of 1960, however,
the development's impact was greater on English psychologists than upon
American clinical psychology.

Much of the significance of this work has been in the extensiveness with which problems of anxiety have been dealt with, as well as in the empirical evaluations that have been given to systematic desensitization. Wolpe worked with a specific set of procedures and specifically described them. Moreover, he assiduously recorded the results he obtained in employing the procedures in his clinical practice, and he was concerned with the empirical validation of his therapeutic efforts. He employed as the basic data patient self-reports of anxiety reactions to the anxiety-producing stimuli for which they had been treated. According to the rating procedure he employed, the patients had to have their anxiety problem and its related maladjustments reduced by 80 percent to be judged successful cases of systematic desensitization. According to Paul, of 85 of Wolpe's cases appropriate for consideration, 78 were successful (Paul, 1969, p. 73). This would yield a success rate of 92 percent.

Other case studies of a similar sort were conducted by Lazarus (see Wolpe and Lazarus, 1966). These results also supported the therapeutic effectiveness of systematic desensitization. However, the results were based upon the verbal reports of the patients. The treatment was conducted in the clinic, and many sessions and complex therapist patient interactions were often involved. Some treatments included more than 150 therapy sessions. Thus, the singular value of systematic desensitization in these case studies was not clearly indicated. However, systematic desensitization predicted specific types of results, which were obtained. Data on results were systematically collected, including follow-up data. The findings even at this point were a strong corroboration of the relevance of certain learning principles for clinical psychology.

Much additional extension and corroboration of systematic desensitization has occurred since the first clinical studies. Some has involved controlled laboratory studies and measures of the effectiveness of the desensitization procedures other than the self-reports of the patients. Lang and Lazovik (1963), for example, selected 24 subjects who had intense fear of nonpoisonous snakes. The assessment procedures included self-ratings of their fear, as well as an objective measure of how closely in linear measurement the subjects would approach a live snake, up to the point of touching it. Unlike the clinical studies, Lang and Lazovik's laboratory study included control groups. The experimental subjects received training in relaxation during the first five training sessions. This period also included training in hypnosis and construction of 20-item hierarchies of fears to snakes. Following this there were 11 sessions devoted to systematic desensitization.

Following the desensitization treatment, the experimental subjects' phobias were found to be significantly reduced. They increased significantly in the overt behaviors of approaching and touching the snake. Control subjects, including those who had experienced the relaxation and hypnosis training, did not show a reduction in their snake phobia. A six-month follow-up revealed that the experimental subjects had retained an advantage, although two subjects showed some deterioration of improvement. It should also be noted that five different therapists were utilized in the study, to free the results from the possibility that a particular therapist had an unusual charisma.

Paul (1966) conducted the first large-scale factorial study of systematic desensitization therapy, rotating the therapists through the different treatment methods employed. The anxiety or phobia was that concerning interpersonal performance in various social, interpersonal, or evaluative situations, especially anxiety concerning public speaking. Trained observers were employed to rate anxiety during a speech before an audience, and physiological measures of anxiety were also used. Both measures were taken before and after treatment and in long-term follow-up data collection. Subjects treated by desensitization showed less anxiety than subjects treated by insight-oriented methods or subjects treated in several types of control conditions. Also,

> Basing computations only upon [subjects] retained at the two-year follow-up, including the positively biased group of controls, the percentages of [subjects] showing significant improvement from pretreatment to two years after treatment termination were: systematic desensitization—85 percent; insight-oriented psychotherapy—50 percent; nonspecific attention-placebo—50 percent; untreated controls—22 percent (Paul, 1966, p. 117).

Other well-controlled studies of desensitization have shown the methods to be effective with various types of anxiety problems. For example, Moore (1965) has reported a study in which asthmatics were more successfully treated by desensitization than by relaxation alone or relaxation with direct suggestion. Emery and Krumboltz (1967) also conducted a study on text-anxious university students. Desensitized subjects showed significant improvement on self-ratings of anxiety before and during final exams and on pre-post test-anxiety scales. These and other systematically controlled studies give support to the efficacy of desensitization procedures, in addition to the support provided by case studies and less systematically controlled studies. There are many other studies in behavior therapy and behavior modification. Franks (1969) has provided an excellent compendium of the various findings and procedures that now exist.

## MODELING AND DESENSITIZATION

As has been indicated, Bandura's approach to modeling has been to show how changes in human behavior can be effected on the basis of modeling. Bandura, Blanchard, and Ritter (1969) have conducted a study to show that modeling procedures can be employed effectively to reduce the types of phobias that have been treated with Wolpe's desensitization method. Subjects were chosen who suffered from snake phobias that were inhibiting. The subjects were divided into four groups, one of which was a control group that received no treatment. Another group received Wolpean desensitization involving graded imaginal presentations of increasingly close contact with snakes, while the subjects maintained a relaxed state.

A third group had experience with films showing individuals in progressively stronger anxiety-arousing interactions with snakes. These subjects controlled the presentation rate of the films and could terminate the presentation at will or repeat scenes until the scenes no longer elicited anxiety whenever the film interfered with their maintenance of a relaxed state. These subjects were thus in control of their own observational learning.

A fourth group had first-hand modeling and guidance experiences. An experimenter (therapist) demonstrated the handling of snakes and in a graded manner gradually introduced the subjects into touching, stroking, and holding a snake. This was first done with gloved hands and then with bare hands. First the snake's body was involved, and then the snake's head and tail. This was done jointly, with the experimenter providing close guidance and support. Finally, the subjects were brought to the point where they would hold the snake in their laps, and so on. The various procedures were introduced to insure that anxiety was controlled.

Subjects were given a test of the strength of their instrumental avoidance of a snake before and after treatment. This included use of an unfamiliar snake after treatment. They rated their fear arousal in the acts involved. The results showed that the live modeling procedure produced better results than the film modeling procedure, which was slightly better than the standard desensitization procedure. All showed that treatment produced improvement more than the no-treatment control group condition.

This study suggests that fears can be treated by observing others experience the fear object or event without receiving negative experience. Setting an example may thus be effective in changing emotional responses. This can occur when the presentation is in the form of film, and to an extent this is commensurate with treatment using standard desensitization methods. The procedure that was called live modeling, however, must be considered to consist of more than just modeling. The subjects received instruction and physical guidance in approaching the snake. Where the subject would not stroke a snake, for example, he was led to place his hand on top of the experimenter's hand while the experimenter stroked the snake. Thus, he was receiving direct experience through this guidance. More than anything, the study shows that direct experience, more effectively than the indirect procedures of either kind, produced the greatest fear reduction. By this direct method 92 percent of the subjects were successfully treated.

## COUNTERCONDITIONING AND AVERSION THERAPY

Wolpe's desensitization procedures were described first in this section on behavior therapy because of their wide use. This process of aversion therapy may be described in the present terms as changing an undesirable positive emotional response by pairing the stimulus with a negative emotional unconditioned stimulus. For example, in a case that was important in stimulating interest in aversion therapy (Raymond, 1956), a fetish type of stimulus (baby carriages in the form of photographs) was paired with a stimulus that elicited a strong negative emotional response. The man was conditioned to a negative emotional response to the fetish stimulus.

As another example, Marks and Gelder (1967) conducted a study of aversion therapy employing five men who had fetishistic and/or transvestite emotional responses to women's clothing. Electric shock was paired with imagining the fetish stimuli and with actually interacting with the clothing objects. It was found that this treatment changed the attitude ratings of the subjects toward the fetishistic stimuli in the negative direction. The conditioning effect was also shown in the decrease of the penile erectile response to articles of women's clothing. Negative physiological emotional conditioning had to be conducted to each article of clothing (panties, skirt, bra, and so on) to obtain the results.

Aversion therapy concerns the negative side of counterconditioning. On the positive side, stimuli that elicit undesirable negative emotional responses can be changed to positive emotional responses through classical conditioning. The author, for example, changed his infant daughter's negative emotional response to being bathed (because soap had gotten in her eyes several times) to a positive emotional response. This was done by pairing the water stimulation first with candy and then with enjoyable play. Lazarus (1960) had earlier reported a case of a child who would not enter automotive vehicles who was also treated by pairing candy with vehicular stimuli.

This will serve as an introduction to the field of behavior therapy. Many other studies and clinical treatment procedures are currently being conducted in this active area of investigation. Moreover, theoretical developments have also been added (Davison, 1968a).

## *The psychodynamic and conditioning conceptions in clinical psychology*

It has been said that our general conception or paradigm determines our practices. This is true in clinical psychology as in other realms. There are general characteristics in the psychodynamic and conditioning conceptions that can serve as a useful introduction to discussion of the characteristics of the social behaviorism conception in clinical psychology.

In the interest of a brief summary, a few words may be said concerning the paradigm that has influenced clinical practices for many years and continues to do so. The theoretical basis for the psychoanalytic paradigm was influenced to a large extent by Freud's theory of personality development and function. According to psychoanalytical theory, the child biologically progresses through four stages of psychosexual development. Depending upon the child's experiences, largely with his parents, his personality structures are formed in each stage with certain characteristics. If the child-parent interactions are awry in one or more of the critical stages, the child's personality processes develop undesirably. The child then suffers psychological conflicts which manifest themselves in behavioral symptoms. Treatment of the behavioral symptoms depends upon resolution of the individual's psychological or psychodynamic conflicts.

The primary method for accomplishing treatment consists of the interview techniques that have come to be called psychoanalysis or, more generally, psychotherapy. The therapist and patient interact verbally in such a manner that the patient reveals the nature of his historical problems. Understanding the origin of the conflicts constitutes *insight*. The patient's conflicts thereby are resolved and his behavioral symptoms are no longer necessary.

The methods of psychotherapy are frequently concerned with the nature of the therapist-patient interaction. For example, Freud thought that the patient's reaction to the therapist was to a large extent like the patient's response to his parent of that sex (transference). Psychotherapy in general has also been concerned with giving the patient a chance to cathart—to express emotions that he cannot express elsewhere, thus relieving the patient of the pressures of the emotions. Moreover, psychotherapy methods have been concerned about the rapport the patient has with the therapist—which, as will be indicated, involves the attitudes the therapist elicits in the patient.

It should be noted that the theory of traditional clinical psychology was not stated in terms of explicit, elementary, cause and effect principles. Nor was an experimental methodology associated with the theory that provided the mechanism for continuing growth and empirical verification. This paradigm, thus, proved unsatisfactory to science-oriented clinical psychologists.

## THE CONDITIONING PARADIGM

The conditioning paradigm may be explicitly characterized as assuming that the theory for understanding human behavior consists of the basic principles and methods of conditioning described in the animal laboratory. The method is to go directly from the basic level and apply the principles and methods to problems of human behavior. The paradigm states that human problems are problems of behavior. The task is to identify the behavior—an undesirable instrumental behavior in the case of behavior modification, an undesirable emotional or instrumental behavior in the case of behavior therapy. When the identification has been made, then the conditioning principles and laboratory methods can be applied.

In contemporary times, the behavior modification and behavior therapy principles have been linked with an experimental philosophy influenced by B. F. Skinner's approach. This approach has tended to give precedence to experimental rather than theoretical analysis. The lack of theoretical analysis has led to some disillusionment with the behavioral approach.

> A third factor which militates against success with aversion therapy is our ignorance of the learning processes involved. For example, are we attempting to develop classical or instrumental conditioning? Last, and this is a serious puzzle, why should the patient refrain from carrying out the deviant behavior after he has left the clinic? He knows that if he cross-dresses (for example) in the safety of his home, no shock will be incurred. . . . [T]he connections among aversion therapy, psychological theory, and verified experimental data are tenuous. (Rachman and Teasdale, 1969, p. 279)

The characteristic of direct extension of the elementary conditioning principles to human problems has also been coincident with an emphasis upon the adaptation of basic conditioning apparatus for use with humans. There is still a tendency to value clinical methods that are directly analogous to animal conditioning procedures, as in counterconditioning employing unconditioned stimuli such as electric shock, or in using electric shock as a negative reinforcer (Lovaas, Freitag, Kinder, Rubenstein, Schaeffer, and Simmons, 1964). This aspect of the paradigm is also involved in the contemporary rejection of the verbal methods of traditional psychotherapy, as will be indicated.

A great deal has been written of the influence of the medical model on clinical psychology, where in considering behavior problems to be symptoms of mental disease various values concerning treatment were adopted that have been inhibitory (Szasz, 1961). This model is associated with the psychodynamic approach. It should be noted that there is also a model of treatment within the conditioning approach to clinical psychology. This model is derived from the animal learning basis of the paradigm; that is, in the basic laboratory the organism is the subject. His behavior is studied as a function of the manipulations the experimenter makes. The goal, in fact, is to find general principles that indicate that if such and such manipulations are performed, such and such behaviors will lawfully occur.

The conditioning paradigm for clinical psychology assumes the same model. The person with the problem is the subject. The therapist manipulates conditioning principles to produce certain changes in behavior in the person. The name *behavior modification* reflects this aspect of the paradigm. As will be suggested, in certain cases this manipulatory approach to treatment is appropriate. However, there is a need in many cases to recognize that the special characteristics of the human individual require a different orientation.

It may be added that because of its nature the conditioning paradigm has not related well to or utilized congruent concepts, findings, and procedures that are provided by other approaches. The manner in which the social behaviorism paradigm can provide a basis for a rapprochement that introduces a foundation for the needed generality will be outlined.

## The social behaviorism paradigm

The traditional psychodynamic approach, as has been suggested, considers the individual's internal personality characteristics to be important determinants of his overt behavior. In the realm of clinical problems, the seat of the problem is seen to be the individual's disturbed personality. It is the individual's personality conflicts that are the causes of the symptoms (neurotic or psychotic behaviors) the individual displays. Treatment, from this point of view, must involve dealing with the personality conflicts, resolution of which will result in cure of the symptoms.

The conditioning paradigm, on the other hand, deals only with symptoms, not with the underlying psychological problems considered to be primary. Such treatment of symptoms through behavioral means has been considered dangerous (Bookbinder, 1962). It has also been said that the conditioning therapies are doomed to be ineffective; unless the underlying psychological cause of the symptoms is treated, removal of the symptom will result only in its replacement by another symptom.

Theorists in the elemental behavioristic camp have rejected the traditional rationale. It has been suggested, for example, that the symptom *is* the problem, rather than being just an index of the underlying personality disturbance. "Cures are achieved by treating the symptom itself . . . . Symptomatic treatment leads to permanent recovery" (Eysenck and Rachman, 1965). There have been a number of behaviorally oriented studies that have attempted to show that removal of symptoms has not resulted in symptom substitution (Bergin, 1966; Davison, 1967; Eysenck and Rachman, 1965; Grossberg, 1964). It has also been definitively suggested that symptom substitution does not exist (Yates, 1958).

The issue of symptom substitution represents a difference that stems from the basic position of the traditional and conditioning orientations.

The traditional paradigm has been characterized; the conditioning therapies can be said to be based on the elementary learning principles. The conditioning approach views personality as the individual's collection of behaviors. From such a viewpoint behavior problems are seen as discrete and separate. The experimental emphasis of the approach and the value placed on objectivity have tended to orient the clinician toward isolation of specific behaviors that are accessible to observation in a manner that will reflect learning manipulations. Explicit behaviors tend to mean relatively simple behaviors. Elemental behaviorism does not lead toward analysis of past history, of how the particular behavior relates to the other aspects of the individual's personality, or of the social context for the individual that is unrelated to the specific behavior, and the like.

Such an orientation, and the treatments stemming from this orientation, do not satisfy the practitioner who is concerned with more general aspects of human activity. It is suggested that both approaches—the traditional and the conditioning—are partially correct, and both are necessary to a comprehensive clinical psychology.

On the one hand, as has been suggested, it was valuable to focus upon the actual behaviors involved in abnormal behavior. In fact, we need much more knowledge in this realm of study. The actual behaviors tended to be ignored in the psychodynamic orientation, in favor of concern with inferring the nature of the individual's internal psychological conflicts and motivations. As has been indicated, this led to unrealistic ways of dealing with patients.

It is also the case that the abnormal behavior itself may be the problem. This can be so even when the behavior is actually rather simple, when the behavior is isolated in the sense of not being an integral part of some personality system, and thus when specific treatment is recommended. Take, for example, the case where the child has not been toilet trained. As will be indicated in the next chapter, this behavior can be analyzed in straightforward terms to consist of sensorimotor skills, and deficits in such toilet behavior can be treated by straightforward learning procedures.

Toilet skills can be considered an isolated instrumental behavioral skill rather than indicative of anything about the nature of the individual's other personality systems. However, deficits in this behavior can result in serious consequences for the child's personality development and thus be a central problem of adjustment. This occurs because a deficit in such skills enters into a behavioral interaction process that will change the child's life, as was noted in Chapter 8. When the child does not display appropriate toilet behaviors, his relationship with parents and other family members will be affected adversely. Moreover, if the deficit continues into the school years the child will not be able to attend regular school, and his language-cognitive personality development may suffer. Thus, even though the child is developing normally until that point, when he does not acquire toilet skills he may be subjected to circumstances that disturb his personality development in various ways. The toilet behavior, as simple as it is, can be the problem that brings down a normal life structure that would have produced a normal individual. In such a case it is no mean accomplishment to be able to remove the problem.

There are many such problems. Whether or not the child behaves appropriately in school or is hyperactively inappropriate may involve simple behaviors. But the inappropriate behavior itself will seriously affect other aspects of the child's personality development. A phobia, to take another example, may involve a simple conditioned emotional response which is isolated and does not involve general emotional derangement. Nevertheless, the phobia may lead to undesirable behaviors—for example, when the person fears being outdoors and is thus restricted to staying home. The phobia may also be the source of secondary anxieties in the patient—for example, the fear that he is mentally abnormal.

These examples have been employed to indicate that it is often important to be concerned with the specific behavior itself, because the specific behavior *is* the central problem. Our behavior modification and behavior therapy principles and procedures have provided the needed emphasis on such problems. However, it is also true that there are cases of maladjustment that are more complex in form. They can involve *personality systems* that must be understood and dealt with if treatment is to be effective. Where such understanding is absent, treatment carried out as though an isolated behavior were involved may even lead to symptom substitution. The social behaviorism paradigm, in agreement with traditional clinical approaches, thus suggests that it is necessary to include a personality level in clinical theory as well as procedures for dealing with general personality changes. The next sections are devoted to indicating (1) how the basic three-function learning theory of social behaviorism provides a better conceptual system (than do the traditional learning theories) for considering the conditioning therapies, and (2) how the personality level is important in clinical theory and practice.

## THREE-FUNCTION LEARNING THEORY, A-R-D ANALYSIS, AND CLINICAL TREATMENT

As Rachman and Teasdale (1969) have indicated, there has not been a clear theoretical framework within the conditioning therapies for indicating the principles involved, for relating the various procedures employed, or for relating the procedures to psychological theory. For some time after the behavior therapy and behavior modification movement began there was no contact between the two types of clinical work. The conceptual relationship between the two areas has remained weak. The

three-function learning theory can serve as a basis for understanding behavior therapy and behavior modification, as well as for relating the two.
Behavior modification typically involves the use of reinforcement principles to increase or decrease the incidence of an instrumental behavior.
What is not realized is that the reinforcing stimulus is also an emotion
eliciting stimulus, so as the instrumental conditioning takes place, the
individual is also being conditioned to respond emotionally to the situation, with all this entails. This has not been understood, researched, or
employed in behavior modification work (Staats, 1970c).

Behavior therapy has been primarily concerned with classical conditioning procedures for changing the emotional value of stimuli (Davison,
1968a; Staats, 1968d), sometimes for this effect by itself, but usually for
other purposes. That is, many studies in behavior therapy actually demonstrate the three-function learning principles. The emotional response to a
stimulus is changed, but the concern is with the resulting change in the
directive value of the stimulus. *Change in the emotional value of the
stimulus also changes the instrumental behavior to the stimulus.* The traditional learning theories, such as Skinner's, do not provide a theory within
which to consider such "transfer of control," as discussed in Chapters 2
and 4. This is an excellent example of the importance of closely relating
the basic learning theory and human levels of study. The case of fetishism
(Raymond, 1956) previously mentioned may be used to illustrate the
present analysis.

> In this case study the aberrant behavior of an adult male involved
> contacting women's purses and baby carriages in a manner that was un-
> desirable. This instrumental behavior had resulted in repeated arrests.
> Treatment consisted of pairing the "fetish stimuli" with an aversive
> unconditioned reinforcing stimulus in a classical conditioning procedure.
> The result was that the instrumental behavior no longer occurred. It is
> necessary again to account for the change in the instrumental behavior
> through the same analysis. It is suggested that the fetish stimuli had
> come to elicit sexual emotional responses in the individual and thus to
> have sexual reinforcing value which as directive stimuli controlled the
> undesirable instrumental behavior. The emotional conditioning changed
> the reinforcing value of the fetish stimuli (in this case in a negative di-
> rection), and hence, the directive stimulus value controlling the unde-
> sirable approach instrumental behaviors (Staats, 1970c, pp. 144–45).

The A-R-D principles thus resolve the uncertainty that Rachman and
Teasdale allude to concerning whether aversion therapy (or counterconditioning or desensitization) involves classical or instrumental conditioning principles. The treatment involves classical conditioning of emotional responses. But the important adjustment effect is that the problem
instrumental behaviors of the patient are changed.

The field of behavior therapy in recent times has suffered from trying
to interpret its findings in terms of Skinner's theory, which does not relate

classical and instrumental conditioning. Since behavior therapy generally involves classical conditioning procedures, but its important effects are on instrumental behaviors, the basic learning theory employed must show the principles of interaction between the two types of conditioning, as the three-function learning theory does. A primary weakness of present day behavioral clinical psychology, in this and other areas, is that many of its practitioners employ a second generation basic learning theory that was formulated only on the basis of animal research, and which has not been changed in the face of the rich findings of the past several decades.

SYMPTOM SUBSTITUTION AND A-R-D PRINCIPLES. Behavior therapy has been criticized for dealing with symptoms rather than the underlying, basic problem of the individual. While it has been claimed by some that such treatment will lead to symptom substitution, this claim has been denied by others on both theoretical and empirical grounds. A sophisticated criticism has suggested that since behavior therapy states that neurosis consists of specific symptoms, behavior therapists cannot account for the *general* results of their treatments of *specific* behaviors (Breger and McGaugh, 1965). This is again a case where analysis in terms of personality repertoires indicates there is truth in both orientations. The Raymond (1956) case of fetishism can be used again as an illustration of the manner in which behavior therapy could lead to symptom substitution of an undesirable type. It also demonstrated how specific treatment can yield general effects, as Staats (1970c) notes:

> The problem, in the theoretical terms employed herein, was that women's purses and baby carriages had strong sexual reinforcing value for the patient, and thus strong directive stimulus value which controlled his instrumental behaviors of "handling" the objects. The treatment, which was successful, changed the attitudinal and reinforcing value of the stimulus objects and thus their directive stimulus value. The effect upon the patient's behavior had several facets. He no longer approached the stimulus objects. He did not masturbate while thinking of the objects, and his sexual behavior with his wife improved. When the treatment and its effects are analyzed more deeply, however, it may be seen how the success of the treatment depended upon certain factors which were not understood or controlled in the behavior therapy, and how, if these unassessed factors had been different, the treatment could well have resulted in symptom substitution.
>
> A more complete analysis of the treatment would include the following. The treatment changed the attitude and thus reinforcing value of the fetish objects. This, however, resulted in short-term "motivational" changes in the patient's A-R-D system, as well as in long-term changes. That is, let us say, that the patient had a hierarchy of sexual reinforcing stimulus objects in his A-R-D system, as would be expected. At the top of the hierarchy were the fetish objects. Somewhere lower in strength was the patient's wife. No doubt, as must be the case with people in general, other social stimuli (women, girls, men, boys) and other objects,

and perhaps animals, would also have sex reinforcing value for the patient. It would be expected that lowering the reinforcement value for the class of fetish objects in the individual's sexual reinforcing system would have several effects. It would raise the reinforcement value of the other reinforcers in the system relative to the stimuli whose value was lowered. Moreover, especially if the stimulus involved was the strongest reinforcer in the system, when a dominant sex reinforcer is no longer available (because of absence or because the individual no longer finds it sexually reinforcing) the individual to that extent suffers deprivation. Deprivation has the effect of raising the value of all the positive sex reinforcers in the individual's system.

On both of these bases it would be expected that the patient in this example, after treatment, would find his wife a strong sex reinforcer [and directive stimulus]—*provided she was a strong positive* [A-R-D *stimulus*] *in his system in the first place*. That being the case, it would be expected that the likelihood of sexual interaction between the patient and his wife would be increased. Each successful instance of such sexual interaction would be expected to increase further the [A-R-D] value of the wife through positive classical conditioning, a lasting result.

When this analysis is made, however, it can be seen that the treatment could easily have gone awry. What, for example, would have been the predicted outcome if small girls were the second strongest class of sexual [A-R-D] stimuli in the patient's system? Removing the fetish objects as sex reinforcers would have raised the sexual [A-R-D] value of little girls in the patient's [A-R-D] system. It would then be more likely that the possibility of aberrant sex behaviors involving children would have arisen as a symptom substitution. If members of the same sex were sexual reinforcers for the patient, then homosexuality could well have been precipitated or increased in incidence. The message is clear. An unsophisticated analysis of the factors involved in the individual patient's behavior, in this case that of the individual's A-R-D system, could lead to unsuccessful treatment and possibly to the production of even less desirable behaviors in a process reminiscent of classical symptom substitution. (Staats, 1970c, pp. 153–55)

This illustration has been employed to show how behavior therapy treatment could lead to symptom substitution, and also to answer the question of Breger and McGaugh (1965) concerning how general effects can arise from specific treatments. In this case the general effects involved the decrease in the approach behaviors to baby carriages as well as the generally improved marital relations of the patient, whereas the specific treatment involved only the pairing of baby carriages with an aversive stimulus. It is suggested, as a general principle, that whether the patient's problem involves only a specific response or changing the specific response will have other effects can only be ascertained by specifying the nature of the personality repertoires involved. Again a more complex conceptual system is necessary than that provided by an elemental behavorism and a simple basic learning theory.

## LANGUAGE-COGNITIVE PERSONALITY ANALYSIS AND CLINICAL TREATMENT

The place of analysis of personality repertoires in clinical treatment can only be illustrated. Let us take the case of deficits in the language-cognitive personality system, which have been said to be involved in mental retardation and autism. It has also been suggested that the token-reinforcer system and behavior modification methods can be applied to treating autistic children (Staats, Minke, Finley, and Wolf, 1964). The autistic child with temper tantrums who was treated so that he would wear glasses (Wolf, Risley, and Mees, 1964) was also treated by Risley and Wolf (1967) for deficits in his labeling language repertoire. Because the child already would imitate words, he was shown pictures and when he had performed an imitation of the action depicted he was reinforced. This was done until the child could label the pictures directly. Lovaas (1966) has also attempted to utilize reinforcement principles in training autistic children in language skills. This research contributed a great deal to the study of autistic children, but the therapeutic results were a mixed success.

> Each child who underwent treatment made measurable progress, even though the progress was slow and incremental and few children became "normal.". . . If the child possessed a verbal topography at the beginning of treatment, even if it was socially nonfunctional (such as echolalia), then we could help him substantially in developing a meaningful language. But we were less proficient in establishing new behavioral topographies in autistic children when none existed. For example, if the child was mute, his progress with behavior modification procedures was very limited (Lovaas and Koegel, 1973, p. 255).

The present author studied in detail the ways in which learning procedures could be used to produce language development in his own children, up through the advanced stage at which they had acquired a full reading repertoire (Staats, 1968c; see Chapter 10). Beginning in 1960, the various language repertoires were studied. It is suggested that this type of comprehensiveness must be involved in clinical treatment that aims to remediate severe deficits in language development. The language-cognitive repertoires were described in Chapter 5. A more complete treatment (Staats 1971a) has indicated the training procedures involved that the therapist or parent can use with the child who is not developing language normally. The procedures have been used in dealing with cases of language deficit, as in the example in Chapter 8 of the child who had not developed a functional language. In treatment, the child was trained in labeling objects and pictures, in labeling events (like "all gone," "bounce," "fall down," and so on), in labeling internal stimuli ("hurts," "tastes good," "running," and so on). He was also trained to respond to verbal-motor elements ("give," "stand up," "throw," "turn the page," "get," and so on); in using verbal instructions to obtain things he wanted; and in answering questions ("What's this?," "Where do you want this?," "Where do you want to go?" and so on).[2] In addition, in a series of sessions the mother was instructed by the author in the means for training the child in language skills at home. Her previous efforts had been limited to reciting nursery rhymes to the child and having him repeat them to her.

The results of the treatment were dramatic. This child, who had been socially isolated because of his nonfunctional language, now had a rudimentary language that enabled him to have friends both in a preschool class and at home. These experiences led rapidly to additional learning. The child had been considered to be hyperactive and autistic, by psychiatric diagnosis. In the treatment program he was no longer given tranquilizing drugs, and his IQ increased from zero to a score of 50, over a period of six months. The child attended 158 formal training sessions during which he made 2,476 verbal responses, for a mean of 15.7. The total amount of time spent in this training amounted to only 20 hours. In addition, informal training trials were involved in bringing the child from class to the child learning apparatus employed (see Figure 2.5).

There are presently several projects underway in which the analyses and training procedures relevant to some of the central language-cognitive repertoires are being employed in intervention programs. It has been suggested that language deficits are intelligence deficits (Staats, 1971a). These intervention programs are designed to remediate the language deficits that have already developed in mentally retarded children.

Bricker and Bricker (1974) have assembled an intervention program from various learning analyses that would be expected to have positive results. Moreover, they have developed a facility and program that involves various types of training and includes extension to home instruction and care. They have also developed a very interesting experimental situation studying the receptive language of young children.

The extent to which the cognitive approach has in certain characteristics become congruent with aspects of a learning approach to language can be seen in the language training program of Miller and Yoder (1974). Unlike the Chomskian-inspired focus upon syntax and multiple word utterances, the linguistic orientation is to the child's acquisition of semantic units. Miller and Yoder suggest that the basis of early language development is the semantic concept, which has been called the labeling response here. Within that concept, Miller and Yoder indicate

---

[2] Carl G. Carlson assisted in applying this treatment to the child in 1968. The author and his associates are presently elaborating this work for general application.

the relational functions and substantive functions to be learned by the child at the single-word, two-word, three-term, and four-term levels.

Miller and Yoder include sequencing as well as content in their analysis of a prototypical language intervention program. They suggest a functional criterion for sequencing, again in a manner congruent with the suggestions made within the learning theory. Frequency of occurrence of words in normal language is suggested as the criterion for functional importance. Another concern with function is indicated by Miller and Yoder's emphasis upon teaching the child language for communication purposes.

Unlike the cognitive accounts of language, the prototypical intervention program for work with retardates of Guess, Sailor, and Baer (1974) uses categories of language that have emerged from the learning theories. Thus, their program deals with labels (Skinner, 1957; Staats, 1961, 1968b, 1971a), verbal-motor unit training (Staats, 1971a), and so on. Within such categories of language repertoires, various subclasses of training are outlined. Guess, Sailor, and Baer have presented comments on methodology as well as principles and training procedures that will be valuable in the study and treatment of language problems with the mentally retarded.

When these several programs are operational and have accrued data it will be possible to evaluate the efficacy of their treatments. The preliminary indications suggest that the learning analyses of the various language-cognitive repertoires, and the training procedures, will provide support for the expected trainability of language development. Moreover, the evidence will cast light on the language-intelligence relationship that has been proposed.

## INSTRUMENTAL REPERTOIRE ANALYSIS AND CLINICAL TREATMENT

As has been indicated, Breger and McGaugh's (1965) statement of the need to indicate how behavior therapy and behavior modification could treat specific responses and obtain more general results was astute. It is also a paradox that both behavior modification and behavior therapy could involve different procedures and different responses and yet obtain the same treatment results. For example, a child with a school phobia could be treated by desensitization, in which various school stimuli are presented to the child while he relaxes and thus the child's anxiety response to such stimuli is removed (see Lazarus, 1960). Or the child could be treated by reinforcing him for the instrumental behaviors of going to school and remaining in school (Patterson, 1965b). In the first case the emotional (attitude) response is dealt with directly, and the indirect result is that the child's instrumental behaviors of approaching school are changed. In the second case the instrumental behaviors are dealt with directly, but the indirect (and unrecognized) result is that the child's attitudes are changed. Either treatment can be successful in this case. However, there are clinical problems where this is not the case.

> The instrumental behaviors that are ordinarily dealt with in such therapy as described above are actually already well learned. They are approach or avoidance behaviors that are very common. In cases where the instrumental behavior is not in the individual's repertoire, however, a simple change in the individual's reinforcer system would not be expected to result in immediate improvement. The patient would still require a training program (formal or informal) where the instrumental behavior has to be learned. Thus, as an example, making same sex stimulus objects (men or women) come to elicit negative emotional responses would not cure the homosexual's problems of behavior, if he did not have the appropriate heterosexual instrumental behaviors already in his repertoire. He would still have to acquire the very complex courting behaviors of various social, emotional, and cognitive kinds. As another example, changing the extent to which the general school situation elicits a fear response in the child will not successfully treat his school phobia if the fear response has arisen because of the receipt of negative reinforcing stimuli from his classmates because his cognitive repertoire is poor, because he is maladroit athletically, because he has an odd appearance, and so on. Although in some cases the school phobia is simply a problem of an inappropriate A-R-D system many other times this is complicated by inadequate social, cognitive, and sensory-motor instrumental repertoires of great complexity. . . .

> While this is not the place for a full discussion, it would seem that the complexity of the instrumental (and other) behaviors to be learned in these cases requires more than an ephemeral change in restricted aspects of the A-R-D system. In such cases, as will be further discussed, behavior therapy treatments could well be called symptomatic. (Staats, 1970c, pp. 145–46)

Annon (1971) has employed the theoretical framework within which to develop general clinical treatment for various sexual problems. One of them may be cited to illustrate the use of instrumental training procedures. The case was described in Chapter 8; he was particularly attracted to eight- and nine-year-old girls and had no interest in mature females. At one point his "abnormality" had brought him to the brink of suicide. The treatment included conditioning procedures to alter his emotional-motivational system in a manner to increase his sexual response to mature females and decrease his sexual response to young girls. In addition, the treatment program included provision for training the patient in heterosexual instrumental behaviors.

> He was started by being asked to smile at least once a day at any female that he might recognize from one of his classes. He was next asked to smile at unknown females. Next he was to start small comments with

fellow classmates, then asking them for coffee dates. Next he was asked to make "study" dates with girls for final preparation. When finals were completed he was asked to throw a party for some of the girls he had met and studied with. However, he not only asked four girls, but he also asked five of his male friends as well. The party had turned out well, and he had noticed three strong arousals to two girls in particular. . . .

The beginning of the new semester saw abrupt changes in a number of areas. Mr. Jones came in and enthusiastically reported running into a woman from London that he had met in Hawaii the year before. . . . He now saw her as a possible "test case" for him. . . .

That day he made a date . . . and he took her out that evening. They had a pleasant time at a motion picture theatre and when he took her home he gave her a kiss goodnight. He reported that he felt no arousal to the kiss, but it was "pleasant" and he was looking forward to taking her out again the next day. . . .

The next day . . . while they were kissing and petting . . . she went on to explain that it was her "fertile" time and that she did not have any contraceptives available . . . and they resumed petting. Through manual stimulation he brought her to a climax which was apparently highly satisfactory to her. . . .

He was happy to report that he had recently obtained the phone number of a new classmate, and he was planning to call her for a date. He said that he was now talking with many girls in his classes, and he was certainly not the "wallflower" of the past semester. (Annon, 1971, pp. 406–417)

This study exemplifies the need for consideration of instrumental personality repertoires in dealing with clinical problems. Many other studies have this same implication. Much additional consideration of deficits and inappropriacies in complex instrumental repertoires is necessary.

## Additional elements of the paradigm and further rapprochements with traditional psychotherapy

It has been suggested that a social behavioristic approach to clinical treatment must include as a conceptual basis more than the principles of classical and instrumental conditioning. Even at the basic theory level, the principles must be developed to show the interactions of the elementary conditioning principles. Moreover, such basic concepts as conditioned sensory responses must be added to the theory base. Productive work involving the use of images in clinical treatment is being conducted (Cautela, 1967, 1969, 1970; Davison, 1968b). As an example, Davison (1968) had the subject image sexual stimuli while masturbating. The masturbation is the self-produced unconditioned stimulus eliciting sexual emotional responses with intense reinforcement value. This treatment would be expected to condition the emotional value, and the reinforcement value, to the imagined stimulus. The individual should then respond similarly to the actual sexual stimulus as he does now to the imagined stimulus. Such treatments may be explicized and elaborated within the context of the learning analysis of conditioned sensory responses (Annon, 1971; Staats, 1968c, 1972).

In addition to the basic level of theory, the more advanced levels also have an integral place in clinical theory and treatment development. A very important part of this elaboration, as has been suggested, concerns the personality level and an abnormal psychology that is related to the personality theory. Clinical treatment, to be adequately founded, requires specification of the normal repertoires that people require for adjustment in their life circumstances. The emphasis of traditional abnormal psychology and clinical treatment has been upon the abnormal behaviors of the patient. This stemmed in part from the conception that it was psychic illness that was at the root of behavioral symptoms. Thus, the behavioral symptoms could give information about the illness, in a manner that normal behaviors could not.

This emphasis upon abnormal behavior in clinical psychology and psychiatry has involved a deemphasis of interest in normal behaviors that has been typical of clinical psychology. The social behavioral personality theory and the related abnormal psychology suggest, however, that knowledge of normal personality repertoires is necessary. This type of knowledge is important in prevention of the developmental deficits in personality repertoires that lead to abnormality in adjustment. It would also be expected to be requisite for the treatment of such deficits once they have arisen. In addition, it is necessary to have an abnormal psychology that specifies the principles and circumstances that can lead personality development astray and can result in inappropriate (abnormal) personality development. Social behaviorism suggests that the personality level of theory and an abnormal psychology are basic to behaviorally oriented clinical treatment.

Other aspects of the social behaviorism paradigm are also relevant. The cumulative-hierarchical learning principles provide a basis for considering the history of the individual in a central manner. These principles, along with the others, also provide a basis for the expectation that in many cases successful treatment will require long time periods. One of the criticisms of psychoanalysis has been the length of time involved, but the analysis of personality learning suggests that long periods and innumerable learning trials are usually also involved in this approach. Remediation of personality deficits or personality abnormality can be expected in many cases to require the same time as, or more time than, learning the normal personality repertoire in the first place.

It may be added that other principles—like social interaction principles or those involving behavioral interaction in personality development—are also needed. For example, without a statement of the social interaction principles it is not possible to deal with many problems of individuals that involve such interactions. Sexual and family problems may stem from long-term interactions in which there is repeated affect of one individual on another and in which habitual modes of interaction hinder solution to the problem. What may be demanded for treatment of such problems is a conceptual framework for understanding these interactions, not a set of procedures for effecting the conditioning of simple behaviors.

Evans (1972) has provided a good example of the use of behavioral interaction principles in the treatment of a clinical problem of the "fear of fear." To illustrate how one behavior may serve as the stimulus for another, he describes a patient who had once bitten her fingernails when anxious but had learned not to do so later on. However, under a stressful life situation she had again started to bite her fingernails. This behavior resulted in disfigured nails which in her line of employment produced a social stimulus that elicited a heightened anxiety response, which in turn increased the nail biting. In treating the patient, Evans suggested that she wear false fingernails as a means of breaking the behavioral interaction. The patient thought this a superficial treatment, but the removal of the social anxiety source lessened the anxiety response and hence the impetus to nail biting—thus resolving the problem.

Moreover, the conditioning paradigm has not had a means for relating its interests in human behavior to the knowledge of the social sciences. This has the effect of forcing consideration of problems of behavior as individual problems, although many of these problems are actually social problems. Some problems, as has been indicated, arise in terms of discrepant conditions between individuals and social institutions. In general, congruent concepts, findings, and procedures that are provided by *other approaches* may be important to the development of a social behavioristic clinical psychology. It is important that the approach include a philosophy that recognizes this, as the several sections that follow will attempt to suggest.

## LANGUAGE BEHAVIOR THERAPY

In the process of criticizing the major assumptions of psychodynamic theory (in a manner that gave impetus to the development of behavior therapy), Eysenck (1952, 1960a) rejected the methods of psychotherapy—those procedures in which the psychotherapist and patient interacted through language in a lengthy series of sessions. Behavior therapists and behavior modifiers have generally followed this lead.

Having seen language as central in human adjustment, however, the present author, in outlining behavioral clinical principles, included an outline of some of the mechanisms of "verbal learning psychotherapy" (Staats, 1963, pp. 509–11): "[I]t should be possible for psychotherapy, in any of the areas of behavioral maladjustment discussed herein, to take place on a verbal level. Deficit behavior, inappropriate behaviors, stimulus control, the reinforcer system, should all be accessible to change through verbal means . . . ." (p. 509).

References to these possibilities have been made by other behaviorally oriented theorists (Salzinger, 1969; Kanfer and Phillips, 1970). In these cases, however, elaboration of the conception was not made. In fact, doubt was expressed concerning the importance of language psychotherapy. "Improvement in conditioning techniques may eventually relegate interview therapy to the status of an *adjunct* for behavior modification techniques" (Kanfer and Phillips, 1970, p. 402).

It is suggested, however, that language is one of the most powerful methods by which human behavior is controlled and changed. Any general method for producing behavior change will have to include language methods as central. Moreover, individual and group language behavior therapy methods, as proposed (Staats, 1972), have the same theoretical and empirical legitimacy as straightforward conditioning procedures. Language behavior therapy also appears to offer avenues of treatment not available elsewhere.

To understand how behavior can be changed and treated on a language basis, it is necessary to understand the several repertoires of which language is composed. Each of the repertoires of language represents a mechanism for changing a wide realm of the individual's behavior or for changing general aspects of personality. The various language repertoires were outlined in Chapter 5. A more complete description of the ways language can be employed to produce behavior modification and behavior therapy has been presented in Staats (1972). A few examples can be mentioned here.

EMOTION MODIFICATION THROUGH LANGUAGE CONDITIONING METHODS. The manner in which emotional (attitudinal) responses to stimuli can be changed through language conditioning has already been described. When words that elicit a positive emotional response are paired with a stimulus, the stimulus also comes to elicit a positive emotional response. These principles combined with the other analyses of procedures for changing emotional responding, and analysis of the manner in which such changes will change the individual's instrumental approach and avoidance responses—provide the basis for behavior therapy through language. It should thus be possible to treat phobias and anxieties, the traditional focus of behavior therapy, through language conditioning (Staats, 1968c, 1970a).

Basic evidence that this is the case has been shown in the several studies reported in the discussions on social interaction (for example, Berkowitz and Knurek, 1969; Early, 1968). Very direct support has recently been provided in several studies (Hekmat, 1972, 1973; Hekmat and Vanian, 1971) in which Hekmat has pioneered the successful use of the language conditioning procedures to treat cases of snake phobia. He selected subjects with strong phobic responses and divided them into experimental and control groups. The 15 experimental subjects received 18 language conditioning trials in which the word *snake* (the $^C$S) was paired with positive emotional words (the U$^C$S). The control group had the same experience, but the word *peach* was substituted for *snake*. The experimental group showed positive attitude change significantly greater than the control group did toward the word *snake* on a rating scale. Moreover, the language conditioning treatment generalized to actual snakes, as measured by a standard test of proximity of approach to a live snake. As Staats (1972) notes:

> It may be suggested that the principles and procedures now require extension to the variety of clinical problems to which they are appropriate, and to extensive empirical test. For example, it has been shown that physiological anxiety responses to phobic stimuli can be reduced through systematic desensitization. Is it possible to change physiological anxiety responses through language conditioning procedures? The principles and findings would suggest so. . . . Moreover, since the language conditioning methods include a negative emotional (anxiety) conditioning potentiality, it should be possible to treat cases of inappropriately strong positive emotions—like fetishes, addictions, homosexuality, and so on—within the same procedures. (Staats, 1972, p. 176)

REINFORCEMENT MODIFICATIONS THROUGH LANGUAGE CONDITIONING METHODS. The three-function (A-R-D) learning theory suggests that the reinforcement value of words important to the individual's personality repertoires could be changed using the language conditioning methods. This been noted as follows.

*Producing verbal-reinforcer repertoires in language behavior therapy.* In an earlier section, it was indicated that a stimulus that elicited an emotional response would also serve as a reinforcing stimulus when presented contingent upon an instrumental behavior. Words as stimuli function similarly. Thus, in language behavior therapy as word stimuli are changed in their emotional value, the functional value of the word stimuli as reinforcers will also change. This has been shown very clearly in an experiment by Hall (1967). He presented emotion eliciting words to subjects repeatedly in a procedure expected to extinguish the emotional response. He then found that the words would not serve as reinforcing stimuli as they do prior to such extinction. Extrapolating, extinction of negative emotional responses to anxiety producing words in language behavior therapy would be expected to remove the negative reinforcing value of escape from those words (or from the actual events).

The message is clear. Changing the individual's emotional response to words, and thus to the actual people and events, will change his behavior toward them, for reinforcing value helps determine behavior (Staats, 1972, pp. 177–178).

Hekmat (1974) has recently corroborated this principle in a study published in *Behavior Therapy*. He employed the language conditioning procedures in which positive emotional words, with Group I, and negative emotional words, with Group II, were paired with the conditioned stimulus (which was the word "mmm-hmm," which is frequently employed in non-directive psychotherapy). He later used this word to reinforce the subjects' instrumental responding in a quasitherapeutic interview. The therapist's "mmm-hmm" served as a positive reinforcer for Group I subjects and increased the instrumental response occurrence, but had the opposite effect for Group II subjects. Hekmat concluded that the language conditioning procedures could be employed to change the reinforcing value of people, and other things, in psychotherapy, as had been suggested in the original language behavior therapy analysis.

It is interesting to note in another area the two pioneering texts of the language behavior therapy principles. In each case Hekmat has employed the principles and procedures in devising a clinically relevant technology. This is a direct answer to a recent criticism (following Skinner's philosophy) of the language behavior therapy analysis on the grounds that theory is an obstacle to the development of technology (Tryon, 1974).

LABELING, REASONING, AND LANGUAGE BEHAVIOR THERAPY. In considering abnormal behavior, it was suggested that from the individual who labels events incorrectly, or whose sequences of language responses (reasoning) do not parallel the relationships of the events involved, adjustive instrumental behaviors are not likely to be elicited. Moreover, it has been suggested that we learn our labeling and reasoning repertoires. As adults, these repertoires are changed to an important extent through the verbal interactions we call communication. Psychotherapy interactions are largely communication interactions. In his verbal interactions, the psychotherapist can provide the patient with new labels and new reasoning sequences. The psychotherapist can change the individual's labeling and reasoning concerning other people, as well as his labeling and reasoning concerning himself and his interactions with others. By changing these personality repertoires, the psychotherapist can change the individual's overt behaviors and thereby solve problems of human behavior.

The principles involved have not yet been subjected to systematic study. There is clinical evidence, however, of the treatment effect of changing the individual's labeling and reasoning repertoires (see Ellis, 1967). The study by Meichenbaum and Goodman (1971) described in Chapter 5 demonstrates the possibilities of explicitly improving adjustment through language procedures. These are only samples of the ways that personality and behavior can be changed through verbal means for therapeutic purposes, according to the learning-behavior principles. Several other implications of this approach will be mentioned in the following sections.

## CATHARSIS AND A-R-D MODIFICATION

As has been indicated, the schism between traditional and behavioral clinical psychology involves interview psychotherapy, as employed in psychoanalysis and client-centered therapy. Behavior therapy procedures appear antithetical to client-centered procedures (Rogers, 1951), for example. Client-centered therapy has made a great deal of the desirability of having the client talk in a warm, accepting atmosphere about his intimate concerns, including those topics he fears and is guilty about. Although client-centered therapy may not utilize learning principles in understanding this aspect of therapy, it is suggested that such procedures are in concert with the goal of changing the individual's inappropriate emotional response to his life situations. To see this it is only necessary to employ a learning theory model that includes the necessary analysis of the emotional properties of language. As part of this it must be realized that the negative emotional response to verbal stimuli will extinguish in the accepting client-centered atmosphere. As this occurs, the individual's response to the stimuli involved will also change in his actual life situation. Moreover, making the verbal stimuli less negative in this way will allow the client to use those words in his labeling and reasoning. It is important to realize that it is the same in operative principles to have a person relax when considering anxiety-producing topics in systematic desensitization as it is to have the individual speak of anxiety-producing topics in the "nonthreatening" interactions of client-centered therapy or in most of the other traditional psychotherapies, regardless of the differing theoretical frameworks.

This suggests that traditional psychotherapy procedures could be analyzed and founded in some cases upon explicit empirical principles. This would make it possible to investigate the therapy procedures using principles and procedures from behavioral psychology. For example, it would be interesting to investigate the possibility that the types of emotional behavior change achieved in behavior therapy could be induced in client-centered therapy.

## SOCIAL INTERACTION, TRANSFERENCE, AND RAPPORT

Traditional clinical psychology has included some recognition of social interaction factors, even though it has not stated them in terms of a detailed set of empirical principles. For example, psychoanalytic theory includes the concept of transference in its psychotherapeutic methods, and it has been usual in psychotherapy theory to be concerned with rapport. These concepts have not been precisely defined in terms of empirically testable principles—for example, the term *rapport* refers to the general "goodness" of the relationship between therapist and client. Nevertheless, there are clinical observations that underlie this concept—for example, that psychotherapy or psychological testing does not proceed well in cases where there is not "good rapport."

The conditioning paradigm, in rejecting traditional psychotherapy knowledge, has not treated such concepts of social interaction. The social interaction principles described in Chapter 7, as extended to consideration of abnormal behavior in Chapter 8, do provide a basis for considering social interaction conditions in the etiology of human problems. These principles are also relevant for clinical treatment. For example, it was said that the baby carriage fetishist was treated too simply when he was conditioned to a negative emotional response to the fetish stimuli. The nature of the man's sexual A-R-D system should have been ascertained. Further, it would be generally important in such cases to have knowledge of the man's social interaction conditions with his wife, including their sexual relations. Perhaps the reason she had less value as a sexual stimulus than the fetish stimuli resided in these interaction conditions.

In addition, the social interaction principles in the present approach suggest that interaction conditions between therapist and patient are important. The attitude the patient has toward the therapist—a central aspect of rapport—must be expected to influence the effect that the therapist can have. It is true that when one's contact with the patient is brief—such as in using some primary conditioning procedure—rapport variations are not crucial. This is also the case if the therapist manipulates some very strong variable, such as electric shock in a conditioning procedure.

However, many human problems are complex and require repeated interviews with the patient and cooperation in many ways. Even desensitization therapists have large numbers of interviews with patients. It can be expected that in such cases social interaction principles are important, and the clinician who wishes to understand the process of therapy must understand these principles. This includes understanding the attitudes (rapport) of the patient toward the therapist.

This also includes understanding of transference phenomena. Psychoanalytic theory has been productive in emphasizing the history of the individual and his relationships with his parents—even if one does not accept the analyses involved. In the same way, the concept that the patient transfers his emotional response to his parents to the therapist also points to an important area of study. It is suggested that this area of the psychotherapeutic social relationship may be productively approached employing the previous discussions of role stimuli. Specific research hypotheses could be derived from the theoretical body.

## INSIGHT, BEHAVIOR ANALYSIS, THE SELF-CONCEPT, AND LANGUAGE BEHAVIOR THERAPY

The goal of traditional psychotherapies has frequently been for the patient to gain insight into the nature of his psychological conflicts and their manner of historical origin. Behavioral clinical theorists have generally considered this to be a nebulous goal, without clear means of attainment or assessment. This rejection has coincided with the elemental behavioristic emphasis on overt behavior and rejection of mental (cognitive) processes.

The social behaviorism clinical paradigm provides a basis for rapprochement. In the first place, the method of behavior analysis is seen to underlie understanding of human behavior, as has been indicated. In order to treat someone's problems it is necessary to gain as full as possible an understanding of his relevant personality repertoires and the causative conditions involved in their formation and maintenance.

Moreover, the personality level of the approach gives the personality repertoires a causative role. If these can be changed, then the individual's general behavior will be changed. In addition, it has been suggested that many aspects of these personality repertoires can be changed through language, including language interactions for therapeutic reasons.

The fact is, man does not operate in large part on the basis of simple conditioning principles. To illustrate, if one wishes to modify the behavior of a dog, let us say in relation to turning on a switch, this may be done through an arduous process of instrumental conditioning involving some effort and skill on the part of the trainer. We can, however, simply verbally instruct the adult, or even the child who has acquired the usual language repertoires, and he will perform the act. The individual does not need some material reinforcement. He does not need to be shown a model doing the act. The word stimuli of the instructions suffice.

The same thing is true of more complex circumstances. Learning to label one's behavior realistically—one form of insight—can lead to beneficial changes in behavior. The selfish person in a marriage may have his behavior maintained by immediate reinforcement, for example, which maintains the selfish behavior. However, the behavior may contribute toward the inexorable dissolution of that person's marriage, a consequence the person considers more important than his selfish rewards. Insight into this problem, via labeling in language behavior therapy, may lead under these circumstances to a change in behavior that saves the marriage.

The lack of understanding of this level of personality theory in a conception of human behavior has limiting effects. Moreover, rejection of insight and related concepts has been responsible for a great deal of alienation among psychologists of various other approaches—humanistic, psychoanalytic and other psychodynamic approaches, client-centered, gestalt, cognitive, and so on (see A. T. Beck, 1970; Breger and McGaugh, 1965; Locke, 1971; and Wilkins, 1971). Other approaches that have recognized the significance of human personality have realized, albeit in lay language, that you can change the individual by changing his reasoning, his self-concept, his attitudes and emotions, his value system, and so on. Moreover, by talking to him you can lead him to acquire new skills, be influenced to no longer perform undesirable behaviors, and so on. It is wasteful, and sometimes insulting, to reinforce someone's behavior to get them to do something when you can tell them something which will immediately, and more surely, get them to do it. It is inefficient to treat a person as though he had not already learned an A-R-D system, an instrumental personality repertoire, and especially a full language repertoire—unless he has not.

The social behaviorism paradigm recognizes the importance of the individual's history, his complex personality repertoires, and his social interaction in the broadest sense. Moreover, the individual's language-imaginal-emotional representations of these events are causal factors themselves. These can be changed in communication interactions and thereby change the individual's adjustment. The paradigm provides a basis for expecting that changing the individual's labeling and reasoning sequences and self-concept system through verbal means to be veridical with reality, for example, can have therapeutic value. These are all characteristics that remove the justification for alienation of clinical psychologists who adhere to subjective approaches, as will be indicated further in Chapter 13.

In conclusion, it is suggested that a primary aspect of the social behavioristic approach is that there must be a rapprochement between the concepts and methods of the nonbehavioral approaches to psychology and those of behaviorism. This is as true in the realm of clinical psychology as in other areas. The empirical concepts and observations of various approaches to clinical treatment may be incorporated into the cause and effect principles of learning (which include human learning principles), with mutual benefit to both approaches. There is a fund of knowledge in traditional clinical theory and practice. The principles, procedures, and findings of our conditioning therapies are of great value. The possibility of breaching the separatism in these areas is thus important.

The contribution of social behaviorism to the practice of clinical psychology, in addition to its conditioning therapies, lies in the general conception of human behavior it provides, its analyses of human repertoires of behavior (normal and abnormal), and its methods for analysis, its empirical orientation and philosophy of science, and its experimental methodologies. But it is important to realize that contributions are to be made by other approaches, and a central aspect of the social behaviorism paradigm is to provide a basis for rapprochement.

# 10

## Social behavioral
## developmental psychology

ONE PROMINENT APPROACH to child development involves a philosophy that the development of the child depends upon his own nature. The following quotation from Maslow (1954) illustrates this approach.

> First of all and most important of all . . . man has an essential nature of his own, some skeleton of psychological structure that may be treated and discussed analogously with his physical structure, that he has needs, capacities and tendencies that are genetically based, some of which are characteristic of the whole human species, cutting across all cultural lines, and some of which are unique to the individual. These needs are on their face good or neutral rather than evil. Second, there is involved the conception that full healthy and normal and desirable development consists in actualizing this nature, in fulfilling these potentialities, and in developing into maturity along the lines that this hidden, covert, dimly seen essential nature dictates, growing from within rather than being shaped from without. Third, it is now seen clearly that psychopathology in general results from the denial or the frustration or the twisting of man's essential nature. By this conception what is good? Anything that conduces to this desirable development in the direction of actualization of the inner nature of man. What is psychopathological? Anything that disturbs or frustrates or twists the course of self-actualization. What is psychotherapy, or for that matter any therapy of any kind? Any means of any kind that helps to restore the person to the path of self-actualization and of development along the lines that his inner nature dictates (pp. 340–341).

This philosophy has been linked to biological concepts, at least on a analogical basis, for a long period. As an example, G. Stanley Hall, who was prominent in the development of American psychology, adopted a stage theory of child development, in conjunction with evolutionary concepts. In his view the developing child recapitulated the phylogenetic evolution of the species.

There have been many studies whose aim has been to show the importance of biological factors in the behavioral (personality) development of the child. For example, Carmichael (1926) showed that tadpoles prevented from gaining any experience in swimming, through the use of drugs, nevertheless immediately swam as well as their more experienced brethren as soon as the drugs had worn off. This study and its implications have been transposed to the human level many times. For example, Gesell and Thompson (1929) gave only one of two identical twin children training in stair climbing over a period of six weeks. At the end of training this twin could climb stairs better than the nontrained twin could. However, within two weeks after the training had ceased, the nontrained twin had caught up to the trained twin in skill in this task. Again, the conclusion was that learning was relatively insignificant, in comparison to the internal biological maturation factors regulating the twins' development. A limitation of such studies is that short-term learning is involved instead of the cumulative-hierarchical learning that produces personality repertoires and complex skills, where large and permanent differences can occur.

There also have been a number of studies that have measured the correlation of familial relationship, considered as the causative variable, with such personality traits as intelligence. The findings have been positive—identical twins showing more intelligence similarity than nonidentical twins, and so on—even when the twins have been reared in foster homes (see Jensen, 1969). Such studies assume that the foster homes in which children are placed are unrelated to the closeness of familial relationship of the children, an assumption that cannot always be met (Staats, 1971a).

On the other hand, there has also been an environmentalist orientation in developmental psychology. This philosophy suggests that the child's environment has causative effects upon the development of his personality. Many studies have correlated environmental factors—such as amount of education; parents' educational, occupational, and social class level; urban-rural differences; and so on—with intelligence. For example, in studies of identical twins reared apart, Newman, Freeman, and Holzinger (1937) found IQ scores among sets of twins to differ by as much as 24 points. Woodworth (1941) tabulated data from different environmentalist studies and found a correlation of .79 between differences in amount of education and differences in IQ scores. There have also been various studies that have enriched children's environments, usually with preschool experience, and have obtained increases in intelligence scores (Dawe, 1942; McCandless, 1940; Peters and McElwee, 1944). Fowler (1962) has provided an excellent critical review of these and other studies.

In the case of both the biologically and environmentally (nature and nurture) orientations, the studies are suggestive (Staats, 1971a) but not yet definitive. It has been suggested that in both cases specific and direct data are needed before general conclusions can be made and social implications can be drawn. In the biological realm, for example, it will be necessary to isolate actual biological mechanisms that affect human behavior development. The following quotation of an eminent neurosurgical researcher suggests that there is not yet substantial evidence of this direct type.

> You study the brain of a genius, and it doesn't show anything different from the brain of an idiot. Their tissue is the same, their brain waves travel in the same way. No chemical analysis, no electrical presence separates those two individuals. In a . . . [biological science] laboratory, you'll never discover why one person can write so well or paint so well or do mathematics so well, and another cannot. (R. White, 1967, pp. 112–13)

In the case of environmental studies, specific analysis of what constitutes intelligence and specific demonstration of the learning of such constituents are needed, as will be indicated further on. Ultimately, the environmentalists' position must rest upon specific proof that important aspects of human behavior, and human personality, are learned. To continue, however, the *various* levels of theory that have been included in describing the social behaviorism paradigm should have significance in contributing to this body of proof. Illustration of these possibilities will be the major function of the present chapter.

## The elementary principle level

The basic level of the paradigm in child psychology calls for the demonstration that the principles of conditioning found in the animal laboratory apply to children. If behavior development in children is said to be learned through the principles of conditioning, then it must be shown that children learn according to these principles.

### CLASSICAL CONDITIONING WITH CHILDREN

Hundreds of studies of classical conditioning have been conducted with children (Offenbach, 1966; S. H. White, 1962). An excellent summary has been given by Siqueland (1970). The studies range upward from neonates and include various responses. Krasnogorski (1907) conducted the first classical conditioning study, employing salivation as the response (measured by the child's rate of swallowing). He found conditioning in four- to five-month-old infants.

Lipsitt and Kaye have produced clear evidence of classical conditioning with infants (Kaye, 1967; Lipsitt and Kaye, 1964; Lipsitt, Kaye, and Bosack, 1966). For example, Lipsitt and Kaye (1964) produced a conditioned sucking response to a tone in three- to four-day-old infants. The unconditioned stimulus was a pacifier which, when placed in the infant's mouth, elicited the sucking response. Conditioning was shown as a function of the pairing of the $C_S$ and the $UC_S$, and extinction was demonstrated.

The eye-blink response also has been classically conditioned (Brackbill and Koltsova, 1967; Janos, 1965), as has the pupillary response (Brackbill and Koltsova, 1967), and the galvanic skin response (H. E. Jones, 1930). Possibly the best known study of classical conditioning is that of Watson and Raynor (1920), in which the unconditioned stimulus was a loud noise that elicited an emotional (crying) response, the conditioned stimulus being a white rabbit.

It may be concluded that there is very good evidence that the principles of classical conditioning apply to children down to the moment of birth and possibly before (Spelt, 1948). It may be concluded that the child is apparently ready as a newborn for the emotional conditioning involved in the formation of his A-R-D system.

## INSTRUMENTAL CONDITIONING

There is a plethora of evidence of instrumental conditioning in children, from neonatal stages onward. For example, Papousek (1959, 1967a, 1967b), using infants from birth to eight months of age, has demonstrated conditioning, extinction, and reconditioning. The instrumental response was simple head turning. The directive stimulus employed was a bell, rung for ten seconds. A response in the presence of the bell was reinforced with milk from a nursing bottle.

Siqueland and Lipsitt (1966) have shown instrumental conditioning with a sucking response, using newborn infants. This study is especially interesting in demonstrating that a response that is typically considered to be a biologically determined reflex is subject to learning according to instrumental conditioning principles.

While the first study employing operant apparatus and procedures was conducted by Warren and Brown (1943), Bijou's work (1955, 1957) gave impetus to the studies of reinforcement schedules with children. That the principles of reinforcement schedules apply to children has been clearly demonstrated (Long, Hammack, May, and Campbell, 1958; Orlando and Bijou, 1960).

Various reinforcers have been employed with children, ranging from food reinforcers such as candy (Bijou, 1957) to social reinforcers (Gewirtz and Baer, 1958). Negative reinforcers have been employed— such as electric shock (Lovaas, Schaeffer, and Simmons, 1965), and "time out" (Wolf, Risley, and Mees, 1964)—in which, in essence, positive reinforcement is taken away from the child by isolating him. Retarded as well as normal children have been employed as subjects, and various responses have been conditioned, ranging from vocalization (Rheingold, Gewirtz, and Ross, 1959) to knob pulling (Bijou, 1957). Again, the child appears to be ready from birth to acquire his vast repertoires of speech skills and other complex repertoires of sensorimotor instrumental skills. These direct findings must be considered basic to a conception that the child's behavior skill development results from learning.

## THREE-FUNCTION LEARNING

There has not yet been extensive study of A-R-D principles with young children, and it is suggested that this type of study would provide an important line of verification. One study dealing with the manner in which neutral stimuli come to have the reinforcing function can be summarized.

Silverstein (1972) used ten-month-old infants as subjects and reinforced them for making a response. Just prior to the delivery of the reinforcing (A-R-D) stimulus, a tone was presented. This is thus a classical conditioning procedure in which the tone is followed by a positive emotional $UCS$. Later it was shown that the tone in this process had also acquired reinforcing value. That is, the tone could be employed to train the infants to a new instrumental response by presenting the tone when that response was made.

This type of research should be elaborated to test the other types of A-R-D expectations. For example, it should be possible to demonstrate the three-function learning principles described in Chapter 2.

## S-R MECHANISM LEARNING

The study of basic learning has not included systematic investigation of the manner in which the child can learn complex combinations of stimuli and responses. There has been no model giving impetus to the pursuit of this type of investigation with children. However, many studies, although conducted under the aegis of different interests, can be seen to involve complex S-R mechanisms, especially in discrimination learning tasks where the child has to choose one of two stimuli that vary on several dimensions. For example, Zeaman and House (1963) have suggested that discrimination learning requires a chain of two responses—

first a central mediating response, which then elicits the instrumental response of making the choice.

Kendler and Kendler (1962) also attribute children's ability to shift responding in a discrimination task to mediating S-R mechanisms. They assume the better performance of older children is due to implicit responses, such as verbally responding to the stimuli involved in the discrimination, the verbal responses shifting readily and eliciting the correct instrumental response. This theoretical analysis has had a good deal of heuristic value.

It may be suggested, however, that these accounts have developed out of the need to explain particular experimental tasks. A systematic concern with investigating the manner in which complex S-R mechanisms can be learned by children, and the manner in which S-R mechanisms function in various tasks, it is suggested, would help produce the comprehensive data base that is needed.

## The behavior modification level

The preceding section refers to the first level in verifying a learning conception of child development, but it is only a beginning. The next level involved is that of the behavior modification studies. That is, as will be indicated further in Chapter 16, a dimension of progress in the movement from the laboratory study to the treatment of actual human behavior can be called the *representative sample progression*. After the verification of the elementary principles of conditioning with artificial (laboratory) samples of behavior, it is necessary to progress to the study of samples of behavior more representative of those that are significant in naturalistic circumstances. The development of behavior modification can be considered to constitute such an advance.

In the realm of child development, as well as that of clinical psychology, it is important to show that the elementary principles of conditioning apply to functional, albeit still simple, samples of behavior. Some of the studies that fall into this category have already been described, and additional examples will be given in discussing educational psychology. In suggesting a behavioral conception of child development in his 1963 book, however, the present author attempted to deal with certain areas of behavior and behavior problems in children that have traditionally been especially relevant to the field of child development. The analyses were based upon the author's own experiences, including studying and producing learning in his own daughter. Since that time behavior modification studies have appeared to support the analyses, and some of these areas of study will be indicated, as well as additional areas that are relevant to child development conceptions.

## SENSORIMOTOR SKILL DEVELOPMENT

As the child goes from infancy through childhood it can readily be observed that there is great advancement in his sensorimotor coordinations. Beginning with the infant stage, with a predominance of random and un-coordinated movements, the child progressively demonstrates more and more systematic, coordinated movements under the control of environmental events. His movements become aids to his increasingly complex adjustments. As an example, random movements of the hands and legs will give way to systematic skills of reaching for objects, conveying them to the mouth, and so on, to creeping and crawling, to pulling himself to a standing position. The child will be observed to gradually come to walk, and this skill will gain in sureness.

As will be indicated, child developmentalists originally considered this sensorimotor development to be primarily a function of maturation. The author's learning conception in the present case, however, led him to believe that sensorimotor development is largely a matter of learning. Behavior analysis of the skill of walking, for example, which has traditionally been thought to depend upon maturation, suggested that complex learning must be involved.

> Walking, as an example, appears to be a complex behavior consisting of sequences of different muscular responses that must be emitted in correct order. . . . [R]einforcing stimuli appear to have their characteristic function. . . . When the various responses in the chain occur in the proper order, the child attains the reinforcer, or he attains it more quickly than he does when some "incorrect" order of the responses is emitted. In addition, an "incorrect" variation in the order of responses may be followed by aversive stimulation—falling, bumping furniture.
>
> Stimulus control seems also to be another important variable in walking. . . . That is, [various] sensory stimuli must come to control responses in the skill of walking. Visual stimuli would be expected to control certain "walking" responses. . . . Appropriate "righting" responses to the "off-balance" visual stimuli should also be learned according to the principles of reinforcement. Thus, in the presence of visual cues of the room deviating too far from the vertical, if certain responses occur, aversive stimulation of falling is avoided. Without those responses the child falls. The visual cues, as directive stimuli, should in this way come to control the appropriate responses.
>
> The same is true of the stimuli of balance that arise in the semi-circular canals of the inner ear. . . . Since being off-balance (a certain pattern of stimulation from these sources) would frequently be paired with the aversive stimulation of knocks and bumps, it would be expected that the off-balance stimuli would become negative reinforcers. . . .
>
> In common-sense terms, it could be said that it should become "unpleasant" for the child to be off-balance, and it would be expected

that this source of negative reinforcement would be a factor in learning to make the correct walking responses. (Staats, 1963, pp. 370–371)

In addition, the analysis suggested that experience could be provided for the child that would hasten his acquisition of walking skills.

> Walking may also be produced in an accelerated manner through gradually providing the child with experiences that give him the constellation of skills involved. A child of only six months of age (or less) may be held so that most of his weight is supported, but he supports the rest of his weight by standing. If he is moved along as this is done, the friction of the floor will result in alternate leg movements. The parent may have to manipulate the child a little from side to side to ensure alternate leg movements. With such learning trials the child will begin to acquire the leg movement coordinations involved in walking. The leg musculature may also be gradually strengthened from this experience, as well as through gradually introduced standing experience. Later the child, when able to stand by himself, can be further trained to perform the walking movements when holding onto the parent's hands or, better, to the back of a stroller or some such device. When the child has learned to walk readily holding onto something, the experience can be extended where supports are gradually removed for short distances, and then longer distances. The attention and approval of the parents, which acts as a reward for the child, can be effectively employed throughout such training. (Staats, 1971a, pp. 57–58)

This general behavior analysis underlay (and then was influenced by) the systematic, albeit naturalistic, behavior modification investigation the author began conducting with his infant daughter in 1960. The various aspects of training were arranged and her manner of responding was noted. The author has photographs of this child walking alone at nine months of age—which is somewhat accelerated.

As with any specific behavior analysis, the above constitutes a theoretical structure that suggests experiments and may be subjected to experimental test. Only recently, however, has controlled experimental data become available that supports the behavior analysis. A newborn infant will make walking movements when supported so that his bare feet touch a flat surface. This reflex disappears after the infant reaches about eight weeks old. Zelazo, Zelazo, and Kolb (1972) tested the possibility that this reflex could be increased through training, thus preventing its decline during the eight-week period. An experimental group of children received, from parents, two and one-half minutes of specified "walking" training each day, from the second through the eighth week. The father supported the child's knees to stiffen them so they would support some of the child's weight and be strengthened. The mother supported the child in holding him erect. "A strong increase in walking was observed in infants who were allowed to use the walking reflex" (Zelazo, Zelazo, and Kolb, 1972, p. 314). Also, the responses were better executed by the experimental children than by infants in a control group. Follow-up records showed that the experimental children walked sooner than did the controls. In the author's opinion this was very likely due to the skills the parents acquired in training their children to walk, which they then applied, rather than to the truncated early training. However, the various results are very supportive.

The author's general analysis of walking learning and the several procedures indicated could be generally subjected to experimental test. For example, it is suggested that a group of parents instructed in the training could produce walking learning in children prior to the children of parents not so trained, and prior to the average of 10.12 months of the children trained in the Zelazo, Zelazo, and Kolb study. Evidence that the graded training described in the behavior analysis is important has been provided in a study by Loynd and Barclay (1970). The child was an eight-year-old microcephalic, profoundly retarded, who could stand but never had walked. The child was reinforced for walking while holding a broomstick. The therapist's hand was substituted for the stick, then a very small stick, a piece of tape, and a string, before the child would walk independently.

Analyses of other early sensorimotor skills have been described and tested with the author's children, including eye-hand manual skills (Staats, 1963, pp. 369–73; 1968c, pp. 422–25; 1971a, pp. 54–58). Bijou and Baer (1965, pp. 108–21) have also elaborated the behavior analysis of such sensorimotor development in some additional detail. In addition, the sensorimotor skills of swimming have been analyzed and dealt with in experimental-naturalistic research (Staats, 1971a, pp. 241–42). Chapter 6 also dealt with attention and imitation in writing.

## TOILET TRAINING

Toilet training has traditionally been a practical problem for parents, and this area of parent-child interaction was given special theoretical significance by Freud. Freud considered the child's elimination to concern one of his foremost biological needs. The manner in which training was handled was thought to impose lasting effects upon the personality structure of the child.

The behavioral analysis suggested how the parent could benignly train the child to desirable toilet habits, as Staats (1963) notes:

> On the behavioral level, a description of a "bowel-movement" may include the responses of going to a particular place, and in the presence of the appropriate situational, internal, and time cues, of assuming a particular postural response, and emitting certain other instrumental behaviors (such as straining and relaxing sphincters), which will expedite

the classically conditioned behaviors of the lower intestine that function in evacuation. On the basis of this behavioral description, the acquisition of toilet control may be considered as a complex learning task for the child and a formidable training task for the parent. . . . (Staats, 1963, p. 377)

A behaviorally oriented training program for the parent was then outlined. Indication that the primary behaviors involved were instrumental made the application of reinforcement principles relevant to this training problem. In original training it was suggested that the child be reinforced for sitting upon the toilet for a period proximate to the usual time of defecation. It was also suggested that successes should be reinforced. Through such systematic training, what was at first a random affair would become learned. Moreover, the responses of the lower intestine and sphincters would be classically conditioned in the process to the situational cues, the time cues, and so on. Another type of cue that was suggested as important to introduce was some word that, through association with the act of defecation, would come to elicit the responses and also could be employed by the child to indicate his internal state and need to defecate.

This method of toilet training was based upon procedures the author developed and tested with his infant daughter. "Although there is no formal research on the application of learning principles to toilet training, application of this method has yielded observations which conform to the analysis, and the tentative analysis itself should suggest further systematic investigation" (Staats, 1963, p. 377).

A similar analysis of toilet behavior was made by Neale (1963), based upon clinical cases involving treatment of children who suffered from encopresis (constipation and retention of feces) Neale employed positive reinforcement for successful bowel movements and was successful with four encopretic children.

Marshall (1966) has also employed a learning analysis to toilet train an autistic eight-year-old child:

> There are certain component behaviors to appropriate toilet behavior: approach to the toilet, removal of clothing, proper body position, straining, production, etc. It was decided to shape behavior by reinforcing each component of the chain. After 5 days of therapy only production in the toilet, either defecation or urination, needed to be positively reinforced . . . With an eye to the future, the counselor said, "Make poo" . . . at each session. . . . Jimmy has a high "pee-pee" sound as part of his repertoire of sounds, and we hoped after toilet training to shape this sound into a "poo" sound as a signal he could use to tell others that he had to use the toilet. (pp. 242–43)

A number of other studies have employed the learning analysis and behavior modification procedures to treat problems of toilet training (J. C. Conger, 1970; Edelman, 1971; Gelber and Meyer, 1965). (Kimmel

and Kimmel, 1970, have employed similar methods in treating enuresis.)
Kohlenberg (1973) conducted an experiment in which it was demonstrated that the anal sphincter response could be instrumentally conditioned, which establishes a basic support for the learning analysis.

As yet, however, there has been no work with infants to demonstrate
that successful toilet training may be conducted standardly with young
children by parents. In the author's experience such training may be
conducted easily before the child is 18 months of age, solely through the
use of positive reinforcement. It should be noted that it is the dissemination of a simple method for parents to employ that will eradicate the
problems of toilet behavior that result from being subjected to harmful
training experiences.

An example of these possibilities can be given here. Mowrer and
Mowrer (1938) have used a classical conditioning apparatus to treat
bedwetting. An electric sensor is built into the child's bedding so that a
buzzer sounds at the very beginning of micturation. This device will
prevent many children from bedwetting, although recent evidence
(Turner, Young, and Rachman, 1970) suggests that relapse of bedwetting
following this treatment is high.

The device is expensive, in any event, and not necessary. That is, there
is a training procedure by which parents can train their child so he
never develops enuresis. It is suggested that central to the child's not
wetting the bed is his awakening at night. Waking is a learned behavior
and can easily be acquired under the control of various stimuli. A simple
procedure for producing the toilet skill of awakening at night, under the
control of the stimulation of a distended bladder, and going to the bathroom can be readily instituted, as follows.

The young child is ordinarily put to bed much earlier than the parents'
time of retiring. The parent need only wake the child just prior to the
parents' bedtime. The child must be awakened and be directed (and
guided) to go to the bathroom and urinate. If the child has been in bed
from three to five hours, he will have a distended bladder then and will
be learning to awaken and to go to the bathroom under those stimulus
circumstances. At the beginning it will be aversive for the child to awaken
and to get up. More guidance will be required. This should be gradually
reduced. Toward the end of the training the parent will only have to tell
the child to get up and go to the bathroom. Then this directive stimulus
may be removed finally, and the child will perform the whole act without
guidance.

As with many of the other procedures the author has developed, this
one was employed with both of his children, with easy and lasting success.
If there is ever a relapse in bedwetting the training can be reinstituted.
The parent may himself have to get up in order to waken the child at a
sufficiently late point to "drain" the child for the night. As with many

other training tasks, however, it is far easier to inconvenience oneself for the brief initial training required than it is to attempt remedial action after the child has formed ingrained habits to the contrary. It is suggested that the present method can be employed standardly to initially train the child in such a way that a problem will never arise.

## EATING BEHAVIOR AND EATING PROBLEMS

This area of behavior was originally treated in a learning analysis (Staats, 1963, pp. 373–77) in part because of the significance given to the area by Freud's psychoanalytic theory of stage development, as will be described further on. The learning analysis of feeding and eating behavior suggests that important types of learning occur in the parent-child interactions involved, and clinical literature reveals that eating and related behaviors can constitute important problems.

The behavior analysis to be described indicated how the child ordinarily learns positive emotional responses to the parent in the process of being fed by the parent. It also indicated that feeding problems can arise because of the misconception of the parent that the infant's crying means he is hungry. A baby fed frequently because he cries will learn to ingest little food at a feeding and therefore will need feeding more frequently (see Marquis, 1941). Within the behavior analysis, the manner of devising a feeding schedule appropriate for parent and child was described and the topic of feeding problems was treated.

> As the child grows older, there continue to be important variables connected with the manner in which feeding is handled. One example is the child with feeding problems—the child who does not eat enough, refuses many foods, is overdemanding in terms of the kind of food, its preparation and its presentation, and so on. It is suggested that these problems may also be a product of the child's reinforcement history. It should be quite easy to condition a child *not* to eat at dinnertime. This could be done simply by reinforcing the child with an abundance of attention when he does not eat, and, on the other hand, by feeding the child snacks between meals, especially if they are delicacies. It would seem that in this manner very extreme difficulties with eating could be shaped which would continue into adulthood. (Staats, 1963, p. 375)

Ayllon and Michael (1959) very early showed that eating problems in psychotic patients could be maintained by social reinforcement of ward personnel, and the problems could be removed through behavior modification treatment (see also Ayllon and Haughton, 1962). Moreover, the behavior analysis of the development of feeding problems of children has received support from child studies manipulating reinforcement. For example, Berkowitz, Sherry, and Davis (1971) designed a set of reinforcement procedures for training profoundly retarded children to feed themselves. The child was reinforced, by receiving food, when he ate appropriately with utensils. When he did not the child was removed from the table and returned after a delay. The training included seven graded steps, beginning with very easy responses involved in the attendant spoon feeding the child.

The original behavior analysis had suggested that other problems of feeding involve inadvertent and inappropriate social reinforcement from the parents.

> Food preferences constitute another problem. . . . [E]ven though a particular food has no inherent aversive qualities, the child may be inadvertently trained to avoid it in a manner similar to that described above—coaxing and attention may serve to strengthen the behavior of rejecting the food. . . . On the other hand, manipulation of deprivation conditions so that edibles in general become strong reinforcers, and gradual presentation of the new food with scant competition from other foods, as well as social reinforcement contingent upon *eating* the food, would be expected to yield the conditioning desired, with no negative side effects. (Staats, 1963, p. 376)

Browning and Stover (1971, pp. 191–94) have conducted a study that elaborates and supports the behavior analysis. The child was autistic and would eat only hot dogs. In the treatment, hot dogs were completely eliminated from her diet. Social reinforcement for the child's eating or not-eating behaviors were withdrawn. "The program was simply to allow food to reinforce itself, which would be expected to occur if it were not allowed to be contaminated by the unknown effects of social reinforcement" (Browning and Stover, 1971, p. 191). After four days the child was trained to eat a variety of foods. She was also trained by means of reinforcement to eat with utensils, without temper tantrums, without throwing food, and without other undesirable behaviors. Berkowitz, Sherry, and Davis (1971) have also used reinforcement principles to train 14 severely retarded children to feed themselves.

## PROBLEM BEHAVIORS ASSUMED TO BE ORGANICALLY CAUSED

It is common to assume an organic cause of various behavior problems in children. This is done by the medical profession, as well as by lay people. There is widespread medical treatment of children with behavior problems, based upon this assumption, even when there is no other evidence of organic disorder. In recent times, for example, it has become standard procedure in many school districts and with many pediatricians to administer sedative-acting drugs to children who are behavior problems in school or who are hyperactive.

The author has described the case of an autistic or emotionally disturbed child who had no functional language and displayed hyperactivity (Staats, 1971a, pp. 309–11, 316–17). This child had been treated with drugs for hyperactivity and uncontrollable behavior, without solving his problems. It was possible to treat the problems and to control his hyperactivity through behavior modification, without the use of drugs, while also repairing some of the behavioral deficits that were the real cause of his major problems.

Many behaviors considered to be organically based may be more appropriately considered within the context of learning. The author had personal experience with an incipient case of seizures that appears to have such general significance. The case was with his daughter when she was about 20 months old. She had misbehaved and the irritated parent (the author) had angrily said "No," a stimulus previously associated with an aversive stimulus, and had given the child a sharp smack on the bottom. At that moment a gust of wind had slammed a nearby door, producing a very shocking sound. The several stimuli together resulted in a startle response that included the symptoms of shock, including a fainting type response. The child was at that time inadvertently rewarded with concerned solicitude.

Several additional times the child met with a falling accident and displayed unusual responses of shock again. Each time the behavior was rewarded with solicitous attention. Following this a pattern began to develop. When an accident would occur she would exhibit symptoms which appeared to include loss of consciousness. There was much inadvertent social reinforcement of this behavior by the three adults in the household. The behavior began to occur with increasing frequency, to increasingly mild circumstances. The symptoms resembled minor fainting-type seizures, which provided the context of anxiety giving impetus to the solicitous treatment of the child. The concern also interfered with the behavior analysis the author would otherwise have made. At one point a pediatrician was consulted on the problem; his advice was simply to wait and see what developed.

Finally, in spite of the obfuscation resulting from parental anxiety, the behavioral pattern became clear to the author. The child was very sensitive to pain, and falls elicited an anxiety response in her. In addition, however, her falls and consequent behaviors resulted in a great deal of social reinforcement. This reinforcement appeared to have strengthened the instrumental behaviors of taking falls and getting into accidents, as well as the instrumental behaviors of her consequent reaction.

The analysis suggested two modes of treatment, one to lessen the negative emotional response elicited by falls and the other to extinguish the learned "seizure" behaviors. To achieve the emotional change, the author played "London Bridge Is Falling Down," in which a fall could be paired with the enjoyment of the game. The falls were at first very mild and then were gradually increased in abruptness and roughness. In addition, the author began the general practice that when an accident occurred in which the child fell, she was to simply be left lying there (although the adult present would note surreptitiously that no damage had been done).

The intensity of the child's negative emotional response to falls decreased with the falling game treatment. Moreover, when the child's seizurelike response to the fall were no longer socially reinforced, they too began to disappear. The entire problem was eradicated in this way. It is interesting to speculate what undesirable behavioral consequences could have resulted if the social reinforcement of the behaviors had continued. It has been shown that reinforcement can be employed to strengthen emotional responses (N. E. Miller, 1969), and the instrumental behaviors might have been increased in strength by social reinforcement. It is possible that full-fledged epileptiform seizures could be produced in a child by such inadvertent training.

Such possibilities must be entertained and generalized in research to various types of behavior problems thought to be of organic origin. Neisworth and Moore (1972) describe the behavior modification treatment of a seven-year-old boy with chronic asthma. The boy had been repeatedly treated medically without improvement and had required frequent hospitalization of from one to seven days. "Specifically, it was hypothesized first, that asthmatic responding was being maintained or amplified by the presentation of verbal and tactile attention (as well as medicine) during or immediately after a seizure, and second, that behavior incompatible with coughing, wheezing, and generally 'being sick' was not being reinforced" (Neisworth and Moore, 1972, p. 96).

Creer (1970) describes the treatment of two children with severe asthma, using the principle of withdrawing reinforcement contingent upon having asthmatic attacks that required hospitalization. Each boy was isolated when hospitalized, with no visitors and no means of entertainment in his room except school books. Prior to this treatment one boy was hospitalized an average of 4.67 days a week. During treatment this dropped to 1.33 days per week. Similar results occurred with the other boy. In both cases the number and duration of hospitalizations decreased.

Additional studies must be conducted for specific purposes, as well as for the purpose of considering the generally accepted organic conception of such human behavior problems. Perhaps it is again the case that the conception can lead to treatments that exacerbate the problem, rather than treat it, as Neisworth and Moore's study suggests.

## DISABILITY AND REHABILITATION

The discussion in this section on disability and rehabilitation will refer to adults as well as children.

One of the areas in which the application of learning principles has appeared to me to be needed is in the field of rehabilitation. My discussions with therapists in this field have suggested that much of their work, especially with children, involves training new behaviors under circumstances in which the reinforcers are weak. For example, prior to developing skill with a prosthesis, the child secures reinforcement more easily for various already learned substitution movements. Thus, some way must be found at first to supply "extrinsic" reinforcement for the more difficult "prosthetic responses," since in a competitive sense these responses are not themselves "naturally" reinforcing. (Staats, 1964b, pp. 138–39)

[Other classes] of behavior problems associated with the reinforcement system may occur in cases of physical disability. Here, however, it is not that the individual has learned inadequate or unusual reinforcers, but rather that the customary reinforcers are not obtainable for some reason. Consider, for example, the young child with a hearing loss. [O]ne of his difficulties is related to the fact that language stimuli will not acquire reinforcing properties. The adult who suffers a severe physical handicap may also face many difficulties in obtaining his customary reinforcers. An athletic individual who finds physical exercise and prowess reinforcing may have these withdrawn because of disease or accident. The social reinforcers contingent upon this behavior may be eliminated. His friends, his wife and children, may have been acquired during the time when he was strong and healthy, and as a consequence of the injury he may lose these sources of reinforcement in part or even completely.

If his livelihood is connected to these activities, he may even suffer the loss or diminution of reinforcers connected with income. The result would be that in the face of the task of acquiring many new behaviors the individual might have available a very inadequate source of reinforcement with which to maintain the attentive, hardworking behavior required for the relearning (rehabilitation). Furthermore, responses when originally acquired may be provided with certain reinforcers that are not effective if the behavior must be relearned at a later time. The child when learning to walk has the reinforcement of attaining objects more easily and more quickly; in addition, social reinforcers may be contingent upon the first walking behaviors of the child. Thus, walking behavior is strengthened more than crawling behavior. On the other hand, when an adult has to learn to do something that is far inferior to his previous level of performance (such as learning to walk with a prosthesis), his behavior may be aversive to him, rather than positively reinforcing. (Staats, 1963, pp. 487–88)

This incipient behavior analysis of disability and rehabilitation has been expanded by Michael (1970). In addition to the analysis in terms of reinforcement, Michael indicates the importance of the negative emotions elicited by disability.

These . . . features [of the lack of positive reinforcers and the experience of punishers] of a serious disability often combine to produce an emotional catastrophe. The ordinary tasks of rehabilitation can be made very difficult by this overwhelming emotional condition. . . . It is not clear at the present time, and perhaps will always depend to some extent on the individual case, whether it is better to work directly with this emotional state by explicit manipulation of counteremotional variables of some sort (possibly even drugs), or whether one should simply use whatever reinforcers and punishers are effective and allow the emotional effects to subside with time. (Michael, 1970, pp. 59–60)

It is suggested here that the methods of behavior therapy and language behavior therapy may be employed to help allay the emotional responses of the disabled person to his condition. Further than that, the methods of positive reinforcement are still available for the rehabilitation task, even when there are inhibiting emotional conditions present. Meyerson, Kerr, and Michael (1967), at any rate, utilized the behavior principles summarized above (see also Meyerson, Michael, Mowrer, Osgood, and Staats, 1963) in directing a project on rehabilitation. In this project Karl Minke and William Heard employed the token-reinforcer system in training a boy to use a largely paralyzed left arm instead of using his right arm as a substitute. The token-reinforcer system was also employed by Edward Hanley and Albert Neal with a boy who would not walk alone. These behavior modifiers also gradually trained the boy to fall down, since his cerebral palsy made falling a probable eventuality, and his fear of falls was inhibiting his sensorimotor learning. This was done in the graduated way previously described in the use of the "London Bridge Is Falling Down" procedure. Social reinforcement was used by Brian Jacobson and Larry Sayre to improve the working behavior of an 18-year-old patient. The occupational therapist had chosen learning to typewrite by the hunt and peck method as a means of improving the patient's eye-hand coordination. However, the patient learned various ways of gaining the therapist's attention and avoiding work. This was changed so he received 5 minutes of social attention only after a 30-minute period of work which showed improved typing speed. Under this treatment the patient's work effort and learning increased. Further application of the simple principles of learning to cases of sensorimotor rehabilitation in the developing child may be expected to yield additional productive findings.

### ADDITIONAL STUDIES OF CHILD BEHAVIOR MODIFICATION

An exhaustive presentation of the various behavior modification studies relevant to developmental psychology cannot be made here. It is now recognized widely, however, that many types and problems of child development should be considered within the principles of conditioning. Additional examples of problems treated or analyzed through behavior modification are: (1) idiosyncratic speech characteristics (Staats, 1971a), including such disfluencies as stuttering (see Goldiamond, 1965, and Ingham and Gavin, 1973, for reinforcement studies on the treatment of well-established cases of stuttering, and Staats, 1971a, for behavior analyses of the onset of stuttering and its early treatment); (2) thumbsucking (Skiba, Pettigrew, and Alden, 1971); (3) inappropriate speech considered as neurologically caused (Lahey, McNees, and McNees, 1973; Rosen and Wesner, 1971; Rosen, Wesner, Richardson, and Clark, 1973); (4) preschool isolate behavior (Kirby and Toler, 1970); (5) the use of punishment in child training (Bostow and Bailey, 1969; Lovaas, Schaeffer, and Simmons, 1965; Staats, 1963, 1971a); (6) avoidance of outdoor play (Buell, Stoddard, Harris, and Baer, 1968); (7) hyperactive behavior (Patterson, Jones, Whittier, and Wright, 1965); (8) the lack of attention of conduct problem children (Quay, Sprague, Werry, and McQueen, 1967); (9) the lack of social reinforcers in schizophrenic children (Lovaas, Freitag, Kinder, Rubenstein, Schaeffer, and Simmons, 1966); (10) the lack of imitation in autistic children (Metz, 1965); (11) training in sexual behavior (Staats, 1963); (12) working habits (Staats, 1963); (13) parent-child love (Staats, 1963; 1971a); (14) undesirable crying (Etzel and Gewirtz, 1967; Hart, Allen, Buell, Harris, and Wolf, 1964; Staats, 1963; (15) the behavioral deficits of autistic children (Koegel and Rincover, 1974); (16) self-injurious behavior (Bachman, 1972); and so on. Ashem and Poser (1973) have compiled an excellent book of readings on behavior modification with children.

## The social behaviorism paradigm and child development

In the preceding sections of the present chapter several areas important in child development have been described in terms of learning principles. Some of these examples of behavior development were selected for treatment on the basis of interest that had been shown within the framework of other approaches, like Gesell's observations of sensorimotor development, Freud's theory of the importance of feeding and toilet training parent-child interactions, and so on. This was originally done to indicate that learning theory and the nonbehavioristic observations and concepts of the study of child development actually are not incompatible with each other—that the learning approach could be elaborated to include and extend such observations and concepts. In further developing this approach, it can be suggested that the social behaviorism paradigm has the potential for an even more extensive rapprochement than has been demonstrated thus far. In addition to the basic principles of learning and behavior modification, the paradigm includes additional levels of theory that should be employed in providing a general approach to child development. This suggestion will be exemplified first by reference to the principles of cumulative-hierarchical learning.

## CUMULATIVE-HIERARCHICAL LEARNING AND CHILD DEVELOPMENT

The traditional approach to child development has been very concerned with time as a variable. Observations that children can do something, or learn something, at one age that they cannot at an earlier age has been taken as evidence that there are indeed maturational stages. It has been felt that demonstration of an age-related break in behavior development or learning ability provides proof that rather than a continuous learning process a discontinuous maturational, or stage, process is involved.

It should be indicated, however, that there are learning conditions in which such breaks in behavioral or learning abilities can occur. For example, a propitious time for the child to learning speech occurs when he is still dependent upon the parent for all of his needs (reinforcers). The parent thus has powerful sources of reinforcement which he can manipulate informally in this training. An older child who gets his own food, and so on, is less susceptible to such training.

The cumulative-hierarchical learning conception suggests also that wherever there is a progression in learning—where one learned skill is basic to the learning of another—age-related limits to learning will occur. As an example, the child cannot learn to count until he has learned such language repertoires as imitating words. Since it takes a certain amount of training to acquire the language repertoires, the child must ordinarily be of an age that allows that training to have occurred. This says nothing, however, of the child's biological maturational stage, which may be the same at each age, in a learning sense.

There are also many social cases where the state of our cultural knowledge of child training is set to train a child to behavioral skills at one age but not at another. If the child does not acquire the behavioral skills at the stipulated time, he may not have the opportunity again. He may even be considered abnormal and from then on be subjected to conditions that not only prevent him from learning the necessary repertoires but also train him to undesirable repertoires, as has been indicated.

CUMULATIVE-HIERARCHICAL LEARNING AND LEARNING ACCELERATION. As will be indicated, the traditional developmentalist view of child development contained a distrust of introducing training procedures to the child too early. In contrast to the traditional view, it is suggested that there is an acceleration in learning ability that appears to result from early child training. The best example emerges from work conducted in training children to early cognitive skills. In Chapter 6 it was indicated that children who were trained to copy and write the letters of the alphabet required progressively fewer trials in learning new letters. This learning acceleration phenomenon has also been found in learning to read the letters of the alphabet. Each succeeding letter became easier to learn to read (Staats, Brewer, and Gross, 1970; Staats, 1968c). Moreover, this acceleration in learning contravened stage development and age-related expectations. For example, one finding was that a three-year-old child, when beginning training, learned at a slower rate than five-year-old children did. After only ten hours of training, however, he was learning at a faster rate than the five year olds had when they commenced training (Staats, 1968c). Suggestive evidence that the same learning acceleration occurs with number concept learning has also been reported (Staats, Brewer, and Gross, 1970).

Many more studies of learning acceleration in learning functional behavioral skills must be conducted. The findings suggest at this point, however, that far from damaging the child, properly conducted training may provide him with specific skills he needs in further learning, as well as skills that will generally accelerate his learning. The significance of this area of study can be dramatized by indicating that many people consider intelligence to concern learning ability. The learning acceleration phenomenon, in suggesting that the ability to learn is learned, is strong indication of a dynamic conception of intelligence as learned rather than biologically fixed. It is interesting to note that this phenomenon was originally isolated with subhuman primates in a discrimination task (Harlow, 1949). Learning to learn has also been found with chimpanzees in a type of language learning (Premack and Premack, 1974), as well as with adults in learning lists of nonsense syllables (Meyer and Miles, 1953). Accelerated learning in the child's cumulative-hierarchical learning sequences may be quite important and should be generally studied.

## CUMULATIVE-HIERARCHICAL LEARNING AND REGULARITIES AND UNIVERSALS IN CHILD DEVELOPMENT

[T]here is considerable stability of the individual patterns of development. . . . This conformity in the development of children is the most solid basis for the assumption of a maturational process. The growth of the abilities of the child seems—to some extent at least—an unfolding of innate potentialities. Such consistency of development is very difficult to explain in terms of the child's experience . . . (Baldwin, 1955, p. 368).

This has been a general interpretation in the field of child development. Children generally go through certain developmental sequences: They creep before they crawl, stand before they walk, and walk before they run. This sequence is not reversed. Moreover, children develop behaviors at roughly the same ages. Such observations lend themselves to an organic development explanation. The child who is atypical in these respects is considered to be atypical biologically (that is, abnormal).

The regularities in behavior development and the generality across children, however, can be considered within the set of learning principles employed in the social behavioral paradigm. First, sometimes regularities in development come about because of cumulative-hierarchical learning. One behavioral skill may be basic to another. The child, it is suggested, learns the sensorimotor skills of balancing himself in part while he practices standing (see Staats, 1963, pp. 369–73). The child can be seen to let go of the object he holds and fall backward, forward, or to the side, and he can be observed to learn in this experience movements to retain or regain his balance. Such skills are basic to free standing and walking. Moreover, there are skills involved in walking that are basic to learning to run, and so on.

But why do children generally develop the sensorimotor skills at roughtly the same time (although there are many exceptions)? It is suggested this occurs for two reasons. For one thing, the physical principles of the world are the same everywhere. A particular movement in the wrong direction results in an aversive loss of balance and the striking of a hard surface for all children. A move in the right direction prevents that occurrence. We learn balance according to the principles of reinforcement and other principles of conditioning, and the primary sources of reinforcement are in the physical environment. We also learn to walk straight toward objects, moving one foot after another, because we attain those objects when we do so—and we do not attain them with converse movements. Physical events, and physical laws of moving objects, in concert with the way we are structured, determine in large part how and when we learn the basic sensorimotor behaviors. These conditions are similar for most children.

In addition, however, there are other learning variables that make for generality in child development. It is suggested that there are practices of child training which are fairly common to a culture and which help transmit a homogeneity to development. To illustrate, it is not common in our culture for a parent to systematically attempt to train infants to sensorimotor skills. The principles and procedures have not been available for them to do this. Nor has there been a conceptual impetus to do so— the maturational conception cautions parents to wait for the child to develop at his own speed. We do not know how sensorimotor development could generally take place if all parents utilized systematic learning principles and procedures to train their children. The author's own experiences, analyses, and experiments suggest, however, that all kinds of child development—sensorimotor, intellectual, social, and emotional— could be moved up in time. The standards that now prevail are in good part a result of the limits of present cultural practices. It is suggested in general that regularities and universals in behavior are a result of regularities and universals in learning experiences.

It may be added that the learning conception can also take account of the fact that in spite of general similarities, there are also wide individual differences in children's behavioral development. The present approach does not lead to the inference of biological abnormality solely on the basis that behavior does not develop normally, for learning conditions can differ much more widely than a superficial analysis would suggest.

## SOCIAL INTERACTION LEARNING PRINCIPLES

It has also been suggested that when dealing with human behavior, the principles of social interaction must be added to the elementary principles of learning. The study of social interaction principles in developmental psychology is thus important. There are many studies in developmental psychology that may be considered in this light. An example can be seen in the operation of the principle described in Chapter 7 which suggests that the physical characteristics of the individual will elicit an attitude in others that helps determine how they respond to him. The example involves the fact that children reach maturity at different ages. Sexual maturity induces rapid growth in children, along with a change in physical characteristics toward resemblance to adults. These are aspects of the individual's physical self that elicit attitudes in his peers as well as in adults. It has been observed that early and late maturing youngsters are shown by tests to differ in personality (Jones and Mussen, 1958; Mussen and Jones, 1957). The learning and social interaction principles that have been presented can provide a conceptual context for such observations, as can be seen by example in the following description by Longstreth (1968):

> Heterosexual popularity in both sexes is more affected by physical maturity than same-sexed popularity. . . . The early maturing girl is not only physiologically a year or two out of step with her female classmates, being larger, taller, and heavier—decidedly unfeminine characteristics— but is three or four years out of step with the boys in her class. . . . Thus, the early maturing girl is approaching physiological maturity by the age of twelve, but not psychological maturity. Consequently, she may find herself overextended and at a disadvantage in dating behavior, associating with older boys with more experience and sophistication. The psychological discrepancy may well result in conflicts, fears, and confusions that contribute to the less acceptable personality of some early maturing girls.
> The early maturing boy, on the other hand, is at an obvious advantage on two counts. First, he is as large and as interested in girls as the average twelve-year-old girls are in boys. Since most of his male classmates are not yet interested in girls, he has little competition; he is in

demand. Second, his rapid development has also endowed him with characteristics admired by his own sex: good build, strength, height, a beard, and relatively large genitals. Thus, he enjoys increased popularity with boys as well as with girls. (Longstreth, 1968)

This example appears to involve the case where the time of puberty for the child, in comparison to other children, affects his physical appearance and sexual and other sensorimotor skills. Others' response to these physical and behavioral characteristics act according to social interaction principles in a manner that affects the further personality development of the child. It is suggested that combining learning and social interaction principles with the observations of developmental psychology would yield a richer fund of knowledge than either approach provides by itself.

## THE PERSONALITY LEVEL

The personality level of the social behaviorism paradigm also is a significant development for the present considerations. In specifying the several personality repertoires, for example, the paradigm provides a basis for extensive study of the development of personality. The study of personality development has been a central interest of traditional approaches to child psychology. This interest is not satisfied by the analysis of separate aspects of the child's behavior or by the solution of separate problems of behavior development—the forms of behavior modification study. It is suggested that a more complete account will have to involve description of the learning of the more complex personality repertoires.

SENSORIMOTOR PERSONALITY DEVELOPMENT. The concept of the instrumental personality repertoire suggests that a large class of normal behaviors is involved. It is interesting that the traditional child developmentalists have been concerned with specifying the nature of normal sensorimotor development (in a manner that has not been characteristic of any other approach) and have produced the most complete descriptions of instrumental personality development that exist. For example, investigators like Gesell and Thompson (1938) placed children in standardized situations and observed their behavior. The following types of behavioral description have emerged from such study. *Twenty weeks:* The infant holds his head erect when sitting supported. When held in a standing position he extends his legs and will support part of his weight from time to time. He will grasp a ring that is dangled before him. He will bring the ring to his mouth. He will turn his head to the sound of a voice, or a ringing bell, but not necessarily in the correct direction (Gesell and Thompson, 1938). *Fifty-two weeks:* The child is able to walk with some support. He can hold a cube in one hand and grasp another cube with his other hand. He can turn a bottle so that a pellet falls out. He will look at his image in the mirror and say things to the image. He will use two word utterances. In four weeks he will be able to stand alone (Gesell and Thompson, 1938). *Four Years:* The child can do a running broad jump and a standing broad jump. He cannot skip. He can put a knitting needle through a little hole, and he can fold along a diagonal of a rectangular piece of paper. But he cannot copy a diamond-shaped figure. He will ask many questions. His thinking may follow free associations; in responding to a question he may also include further responses to his own answer. He may also say things that are not so, and may not be able to separate the actual from what he has said happened. He will dress and undress with very little help except for lacing his shoes (Gesell, Halverson, Thompson, Ilg, Castner, Ames, and Amatruda, 1940).

These examples of the work of the early child developmentalists are important methodologically as well as substantively. In the former area, the resemblance to intelligence testing can be seen. That is, the child is presented with standard situations, and it is then observed whether or not the child displays the behavior. The descriptions of typical behaviors at a particular age serve as norms against which the individual child can be compared in terms of acceleration or retardation. Piaget's methodology is also in the developmentalist tradition.

It may be added that various research methods have been devised in the study of child development. These constitute a fund of knowledge themselves (see Mussen, 1960). For example, the progressive accretion of behavioral skills by the child has been studied longitudinally, in which individual children are observed over lengthy periods of time. The behavioral development of children over time has also been studied by the cross-sectional method of observing different children at particular age levels.

As has been noted, however, the focus upon the biological concepts of maturation and readiness has not provided a foundation for a detailed interest in how the child learns his behavioral skills. In fact, the child developmentalists as a group have tended to reject learning interpretations of child development and to regard with distrust the use of training procedures to promote early child learning.

> Some child psychologists, such as Ilg and Ames [1964] seem to think of readiness as something that develops automatically with age, just as a flower unfolds from its bud, and object to early instruction. . . . , feeling that artificial attempts to hasten development simply distort the maturational process without genuinely hastening learning. (Mathis, Cotton, and Sechrest, 1970, p. 320)

It is suggested that the body of knowledge produced by the child developmentalists is a very important one. However, it needs to be joined to a learning conception of child development that includes principles, methods of study, and methods of behavior analysis. Through this rapprochement, the manner in which behavioral development can be generally affected by learning could be systematically studied. The following general recommendations can be made in the context of the child's learning of instrumental personality repertoires, a vast area for potential research.

1. The sensorimotor skills that have been developmentally described should be subjected to the types of behavior analyses that have been demonstrated in preceding sections.
2. This would serve the theoretical function of suggesting the manner in which the repertoires are normally learned, and the analyses would serve as experimental hypotheses that could be tested.
3. The analyses should also be tested with children who have deficits in the skills involved, to establish whether or not the repertoires can be learned by such children.
4. Studies should be conducted with normal children to establish the procedures for parents to apply to insure their children's normal behavior development.

As a fifth area of study, research should also be conducted to see if development of behavior can occur earlier than is ordinarily the case. It may not be warranted to distrust early learning in children based upon systematic procedures, as has already been implied. For example, children generally begin to learn to read at five or six. However, this may reflect cultural policies rather than maturational readiness. Prior to the initiation of a public school system, most children did not learn to read at all. If reading were not taught until a later age (as is the case in the Soviet Union), this behavioral repertoire would "develop" later. It is suggested that conclusions concerning when some instrumental personality repertoire can be appropriately learned by the child can only arise from explicit study.

Some research results can be used to illustrate the need for learning data before making conclusions concerning readiness. The example concerns what may be seen as a sensorimotor and verbal skill, counting, which has been referred to by Piaget in terms of maturational stage development. His conception suggests that the child can learn certain things only after having attained a certain level of mental maturity.

> A child of five or six may readily be taught by his parents to name the numbers from 1 to 10. If ten stones are laid in a row, he can count them correctly. But if the stones are rearranged in a more complex pattern or piled up, he can no longer count them. . . . (Piaget, 1953, p. 75)

By this example Piaget suggests that until the child has matured sufficiently, counting (and other) training will produce only mechanical skills, not abilities that generalize across variations in superficial appearances. It has been shown, however, that a child may be systematically trained to general counting skills at much earlier ages than has been suggested. Based upon a behavior analysis of what is to be learned, and using a token-reinforcer system, the training can be accomplished easily. For example, the child can first be trained to label small groups of one to four objects by number, then to point to and number objects in sequence (counting) where the objects are in a row. Then the child may be similarly trained to count randomly arranged piles of objects. This can be done easily when the child is two years of age (provided he has normal language) (Staats, 1968c), and has been done standardly with four-year-old culturally deprived children (Staats, 1968c; Staats, Brewer, and Gross, 1970). Schimmel has substantiated these findings with two- and three-year-old children (1971).

Moreover, once the basic counting repertoire has been acquired it can be extended simply by verbal rote learning in which the child is trained to say additional numbers in sequence. The various results suggest that number concept development is entirely learned and is not dependent on reaching a maturational stage.

Investigators are beginning to explore the possibility that behavioral skills thought to be characteristic of a particular age, as in Piaget's conception, can be affected by learning (Brainerd and Allen, 1971; Gelman, 1969; Sullivan, 1967). Brainerd and Allen have conducted an experiment showing that one of Piaget's more advanced mental skills, the formal operation of density conservation, is trainable. Density conservation refers to considering the density (floatability) of a substance to remain unchanged even though its size and shape are changed. Piaget considered the concept of density to emerge first in children as mature as 11 to 15 years of age. Although there are studies that have found training to be ineffective in producing such Piagetian-type behavioral development (Kingsley and Hall, 1967; Smith, 1968), this may be due to the type of training employed (Halford, 1970). The positive cases provide support that such mental abilities are learned in the cumulative manner suggested for language-cognitive repertoire skills in general. At any rate, the social behavioral paradigm calls for detailed study of the cumulative hierarchical learning of all of the child's developing skills. Only when the results of such study begin to accrue will we know when the child is capable of and will profit from training. Moreover, such study would reveal how the training should be conducted. It is important to observe and describe the typical types of behavior development children display. It is also necessary to study the conditions of learning that can produce the behavior development.

The above examples have referred to instrumental repertoire personality development, but also to language-cognitive skills, a topic that will be considered more systematically in a later section. In any event, it is suggested that study of these several types—which combine traditional child development observations and methods with learning analyses and learning procedures—will help provide the basis for a more complete conception of child development and thus of the nature of man. Philosophical arguments and studies of the effects of environment versus heredity that manipulate unspecific variables cannot resolve the issues. But studies based upon specific analyses of the type outlined do have that potentiality (See also Staats, 1971a.)

EMOTIONAL-MOTIVATIONAL PERSONALITY DEVELOPMENT. As has been implied, psychoanalytic theory includes a theory of emotional-motivational development, based upon a conception of biological development influenced by social environmental factors. According to Freud, the child possesses an instinctual (biological) fund of energy. At the beginning this energy force (the libido) is localized in the child's oral organs, and in this oral stage the child's main satisfactions are of an oral nature— those involved in sucking. If the child's instinctual needs are well fulfilled in appropriate sucking and weaning experiences at this stage, his personality will be influenced in a normal direction. In the next stage, the anal stage, the child's libidinal energies are supposed to focally involve eliminatory activities, and the manner in which the child is toilet trained is thought to influence critically his personality development. By the time the child is three he is considered to have arrived at the phallic stage, at which his libidinal energies shift to his genitals. The manner in which the child is treated at this stage, especially with respect to the opposite sexed parent, is considered to affect personality development. Finally, in the genital stage, according to this view, the child's libidinal energies become invested in others, and his sexual energies become more socialized.

A great deal of research on child development has received its impetus from Freud's theory. For example, Levy (1934) conducted a study to see whether varying satisfaction of oral needs (sucking) in puppies would affect the later behavior of the animals, with positive results. However, Sears and Wise (1950) found that infants who were cup-fed from birth and presumably did not have their oral sucking needs fulfilled did not become fixated at this level (as psychoanalytic theory would assume) and show more nonnutritive sucking than infants who had received more oral gratification through sucking. Other researchers have constructed personality tests of Freud's psychosexual needs to measure the child's motivational development (Blum, 1950), and these tests have been employed to test expectations from the psychoanalytic theory (Cava and Rausch, 1952).

Other motives have also been considered to be important in the study of child development, with varying sources for the motivational concepts.

For example, Murray's list of motives described in Chapter 4 has formed the basis for research with children. Frenkel-Brunswick (1942) used nine motives—autonomy, social ties, achievement, recognition, abasement, aggression, succorance, dominance, and escape—to rate a group of adolescents. Judges asked to rate underlying motives and social behaviors were fairly reliable. One finding that is interesting in the context of A-R-D theory was that the same motivational state could be expressed with various behaviors. For example, girls who were high on abasement demonstrated behaviors of altruism and insecurity.

D. R. Miller (1960) has presented a very good review of studies of motivation and affect in children that refers to various types of emotional-motivational states. As one example, there has been a good deal of interest in the achievement need in children (see McClelland, Atkinson, Clark, and Lowell, 1955). White (1959) has differentiated the achievement need from competence itself. He refers to an interest and satisfaction in doing something well—reading, making something, or whatever—separate from that involved in the product that ensues. Tests have been constructed for measuring achievement needs of older adolescents (McClelland, Atkinson, Clark, and Lowell, 1955). Karolchuck and Worrell (1956) have also shown correlations between measures of achievement need and some types of learning, including school grades. Fales (1944) noted the extent to which two- and three-year-old children requested to be allowed to put on and take off their wraps. Then some of the children were trained to skill in this task, and others were given social approval for helping themselves. In both cases there was an increase in achieving the task without help.

The use of the concept of needs in the context of child development has been criticized, however, in the terms used by behaviorists to criticize the concept in general.

> [Needs] *are usually defined in terms of the very behavior they are supposed to explain,* and thus they explain nothing. Suppose, for example, it is observed that two-year-olds have a greater tendency to cry when away from mother than children of any other age. On the basis of this fact, a "dependency motive" is defined: two-year-olds have a strong "need for mother." Mary Contrary is then observed to cry when her mother leaves the room, and an observer asks, "Why does Mary cry?" Our need-inventor answers, "Because two-year-olds have a strong dependency need." . . . Crying is used to explain crying. (Longstreth, 1968)

A-R-D theory suggests that the behavior of the individual may be an index of a motivational state. However, to remove the circularity Longstreth cites, a conception of human motivation must separate the attitudinal, reinforcing, and behavior-eliciting functions of motivational stimuli and provide a statement of the development of such stimuli.

Following his example, it would not be circular to indicate the manner in which the child has come to have a positive emotional response to the mother and a negative emotional response to strange situations without the mother—with the additional indication of the way a negative emotional simulus elicits crying in the child. Such distinctions as those contained in the A-R-D principles could in this and other cases serve as the basis for more explicit studies, it is suggested.

To conclude, the A-R-D theory indicates that the human emotional-motivational system is very complex. Many different stimuli have A-R-D properties, not just a few types, as some personality theories would suggest. It has been suggested that the A-R-D system is learned. The child may be considered to begin life with biologically based emotional responses to relatively few stimuli. This is elaborated through classical conditioning to include cumulatively a vast number of different stimuli as emotion eliciters. It is necessary to study the cumulative-hierarchical formation and action of these various emotional-motivational stimuli in the normal as well as the abnormal child. This should include descriptive studies that yield developmental knowledge of the formation of the emotional-motivational system. Studies should also be conducted at different age levels to test the expected effect of differences in the A-R-D system on the child's behavior and behavior development. A similar recognition of the need for general study can be seen in the following quotation from Yarrow (1960):

> Attitudes and values of children have, for a rather long period, occupied fringe areas in the systematic literature of child development. Their incorporation into more general developmental theory would be valuable in developmental and social psychology alike. Their usefulness has been demonstrated in investigating the social contexts of children. The intrafamilial contexts, the effects of parental handling upon children's attitudes and value systems, and the interactive effects of children's attitudes and values and parental handling upon other consequents in child development are areas in which these concepts can be given further exploration. (p. 684)

LANGUAGE-COGNITIVE PERSONALITY DEVELOPMENT. The manner in which the child develops some aspects of his language-cognitive personality repertoires has been described. Additional suggestions can be made here as to the ways in which this type of child development requires further study.

The development of the child's language has been of concern to several disciplines, as an excellent review by Palermo (1970) indicates. For example, the developmental psychologist Irwin and his colleagues originally contributed greatly to our knowledge of the first vocal development of the young child (Chen and Irwin, 1946; Irwin, 1947, 1948, 1951; Irwin and Chen, 1941). The data indicate that during the first ten days of life the child's vocalizations are mostly vowels. There is an increase from a mean of 4.5 to a mean of 11.4 in the different vowel sounds that the child will make during the period from 1 to 30 months. Adults use 14 different vowel sounds. With consonants the increase in the same period is from 2.7 to 15.8. Adults use about 25 different consonant sounds of the type Irwin specifies (Chen and Irwin, 1946). Additional specifications of this development have been made by these investigators. It has been suggested, however, that the full range of sounds was not noted in these studies, since they were restricted to sounds relevant to the English language. Ervin and Miller (1963) also noted that nearly all the sounds that occur in the various human languages are made by the infant (Jespersen, 1922).

Jakobson (1968), a linguist, has been concerned with specifying the features that make up the various elementary sounds children make. The sound systems of various languages can be described in terms of these features. Jakobson has suggested that the child develops his sound system gradually by discriminating sounds that differ on a simple dimension, for example, consonantal (with the vocal tract occluded) versus nonconsonantal (with the vocal tract open). The easier contrasts, the ones that, Jakobson suggests, occur in a large number of languages, are acquired by the child first.

A number of studies have been conducted within this framework (Anisfeld and Gordon, 1968; Messer, 1967). One recent area has investigated the infant's perception of such simple speech sounds (Butterfield and Cairns, 1974; Eimas, 1974; Morse, 1974). One of the interesting types of research results involves infants whose rate of sucking responses are recorded. It has been found that introduction of simple speech sounds will have a reinforcing effect and increase the rate of sucking, whereas nonspeech sounds will not (although the finding has been questioned by Butterfield and Cairns, 1974). This suggests the child very early has special abilities for reacting to speech. Another finding also suggests that the child is biologically structured to receive speech sounds. That is, it has been found that speech signals are better perceived by the right ear (and thus the left half of the brain, which is dominant) while nonspeech signals are better perceived by the left hemisphere (Studdert-Kennedy and Shankweiler, 1970). Wood, Goff, and Day (1971) and Molfese (1972), using electrophysiological techniques, have shown greater cortical activity in the left hemisphere for speech and greater activity on the right side for nonspeech sounds.

Developmental psychologists have also observed and described the language development of children in the same way that other types of behavioral development have been described. For example, Smith (1926) indicated that between the time of acquisition of the first word and the age of two the child would acquire between 200 and 300 words. McCarthy (1954) and Templin (1957) have also studied the development of language. McCarthy lists the following developmental sequences: coos and babbles at 2–3 months; vocalizes at sight of person at 4–6 months; imitates sounds at 6–8 months; repeats syllables, such as "ma-ma" and "da-da" at 7–9 months; says one word at 12 months; says two words separately without knowing their meaning at 12–14 months; says five words (not together) at 16–18 months; says two words together, knowing meaning, such as "Ma-ma go" at 20–24 months; and names familiar objects, like key, spoon, and so on, at 22–26 months.

In recent years the developmental study of language acquisition has included grammatical analyses of the type employed by linguists (Brown, Cazden, and Bellugi, 1968; McNeil, 1966; W. Miller and Ervin, 1964; Slobin, 1968), as was indicated to some extent in Chapter 5. In fact, this work, in its focus upon the linguistic theories of language, has made important contributions that need to be supported by investigations of additional items important for language development (Staats, 1971b).

> The writer would also like to make several suggestions with respect to further developing a learning approach. First, the same types of detailed naturalistic observations must be made of various aspects of language that have been made of formal features such as grammar and phonology. The *functions* of various types of language must be studied as well. *In addition, the same type of detail in naturalistic observation must be made of the stimulus aspects that affect language behavior as has been made of language behavior itself. This must include detailed observation of the language training environment of the child—which will involve both verbal and non-verbal stimulus events. In addition, detailed naturalistic observations must be made of the present stimulus situation in which language occurs.* (Staats, 1971b, p. 146)

There is a growing tendency to include such investigations in the developmental study of language (Bloom, 1974; R. Brown, 1973; E. V. Clark, 1974). It is suggested, however, that a great deal of additional progress could be made in the study of language development by combining the various approaches that have been described. Learning theories have provided accounts of the principles involved in language learning (Mowrer, 1960; Osgood, 1953; Skinner, 1957; Staats, 1968c, 1971a). One of these includes specification of the conditions of early language learning, including procedures that can be employed in producing language development (Staats, 1971a). When developmental studies have included the study of learning elements the principles involved have usually been of a commonsense sort. It would seem that any of the above systematic learning theories would provide a better foundation for the developmental study of language than commonsense.

To elaborate, it is generally suggested that the methods of developmental psychologists, and the observations and analyses of linguistics, should be joined with the procedures and methods of learning research with children and a learning theory of language in a truly multidisciplinary approach to the study of language development. Utilizing this approach, it would be possible to study additional features of language development. For example, developmental studies have tended to record the developing *speech* of the child. However, this type of study only deals with some of the language repertoires. What, one may ask, are the developmental characteristics of the attitudinal, reinforcing, and directive functions of language, and so on? Speech is no more important than the other language repertoires that together compose language.

> But, in addition to the naturalistic observations, experimental-naturalistic research must be conducted. By this is meant studies where manipulations are made of learning variables and observations made of the language behavior produced—but not necessarily in the controlled conditions of the basic laboratory.
>
> Moreover, there is also a wider opportunity to conduct more controlled laboratory experimentation on language learning. This can be done at various levels of learning beginning with the development of language in infancy—prior to the emergence of actual speech. It can also be done at points where the child is introduced to new types of language learning—for example, in learning to read (which includes various forms of language and various learning principles), in learning number language responses, writing and so on. There are opportunities for conducting manipulative (causative) studies with normal children and with a variety of children whose language development is absent or defective. Thus, for example, retarded children, autistic children, and so on, could only benefit from research which would attempt to train them to normal behaviors—and learning principles could at the same time be tested and extended. (Staats, 1971b, pp. 146–47)

It may also be suggested that the functional aspects of language should be developmentally studied as well. As has been suggested, for example, the detailed analysis of the language-cognitive repertoires and the other personality repertoires in its general statement constitutes a theory of intelligence. This is a cause and effect theory that states if such and such learning conditions are presented to the child he will develop the behavioral skill and will measure to that extent as intelligent, and he will to that extent *be* intelligent.

The behavioral theory of intelligence provides many experimental hypotheses. It indicates how the ability to answer various items on intelligence tests is learned. This theory of intelligence could, and should, be employed as the basis for extended empirical tests. Experiments should be conducted to test the possibility that children can be trained, according to straightforward learning principles and procedures, to respond correctly to items on intelligence tests. For example, it was indicated that an intelligence item at the two-year level was the naming of objects shown to the child. This repertoire of language was said to be learned according to the principles of instrumental conditioning. It should be possible to take children who are a year or two of age and proceed to instrumentally condition them to a more extended labeling repertoire—including any of the items that appear on an intelligence test for toddlers. (The author's unpublished experimentation has done this.) In this way the manner in which children learn this crucial aspect of their language-intelligence repertoires could be systematically studied. Moreover, the learning conception of intelligence would be directly tested. In addition, the procedures verified in the research would be available for use by professionals in treating problems of child development and by parents in raising their own children.

Another item, the concept of length, may be given as an example because it may be considered to illustrate early number concept development. The item on the Stanford-Binet intelligence test involves presenting the child with two sticks and saying, "Which stick is longer? Put your finger on the long one" (Terman and Merrill, 1937, p. 84). This may be analyzed in terms of instrumental conditioning. The stimulus in this case in the length of the stick. The child must learn to respond to the shorter of two sticks (any two sticks) with the verbal response "shorter" and to the other stick as longer. This "intelligence" skill can be easily taught to a normal two-year-old child. The author has produced this and more difficult training with children down to the age of 18 months, in both formal and informal research (see Staats, 1968c, 1971a, 1974; Staats, Brewer, and Gross, 1970).

The main point here, however, is that this analysis should be subject to systematic experimental study. Two-year-old children could be used as subjects, selecting a group that could not correctly respond to the item. Half of the children would serve in an experimental group and would be given the special training, the other group would not. Then both children would be given the intelligence item. The experimental-group children should show that, as a result of specific training, they can succeed on the intelligence item. This research should be conducted with various items. Such research should also include systematic treatment of the types of abilities that other developmentalists, such as Piaget, consider to index intelligence. As has been indicated, some studies have already shown some of Piaget's cognitive skills to be trainable.

Systematic research across various indices of intelligence would test the learning conception of the development of intelligence. This is the evidence that is needed to lay at rest the controversy between the nature and nurture conceptions. This controversy, with its poignant social implications (see "Is Intelligence Racial?," *Newsweek*, 1973), is largely conducted on a philosophical level, because there has been very incomplete evidence to firmly support either a biological or a learning view.

DEPENDENT PERSONALITY DEVELOPMENT. The three major basic behavioral (personality) repertoires have been described in the context of child development study. It may be added that there are other areas of personality that have traditionally been of interest in developmental psychology, some of which can be considered to be combinations of elements of the three major behavioral repertoires. Several examples will be given in these next two sections.

One aspect of Freud's psychosexual theory involved the development of a dependent personality. It was suggested that if the child is too strictly treated in the anal stage of development—in his toilet training and so on —his energies may be fixated at this level. One result would be that the child would remain excessively dependent, methodical, neat, and stingy. Or the child might express his resentment at the harsh treatment by later "blowups" in different destructive ways. Under the impetus provided by this theory, dependency has been a central concern in the field of child development.

The characteristics that are employed to index dependency are more complex than simple behaviors and are usually considered to have the generality of a personality trait. A behavior analysis of this personality trait has been outlined in part (Staats, 1963). For example, it may be observed that children differ in the extent to which they (1) cling to their mothers in the presence of strangers; (2) are upset when they are even briefly separated from their mothers; (3) insist on having only their mothers care for them in feeding, and so on; (4) display few skills for entering a situation on their own and behaving appropriately, and so on. Later in life it may be observed that people differ in the extent to which they "make decisions" by themselves, behave appropriately on their own, depend upon others for beginning and completing any action, or are too closely attached to relatives and friends. The manner in which such dependent personalities could be learned was also suggested.

That is, individuals differ in their repertoires of adjustive behaviors and the manner in which these are under the control of different physical and social situations. Individuals with wide experience, which yields a large repertoire of skills in many different situations, are never helpless. There are relatively few entirely novel situations for such individuals, since stimulus generalization from the wide range of previous experiences will cover most any eventuality. On the other hand, the individual with sparse experience in this respect will more frequently be at a loss. "It would be expected that all sorts of behavior would be important in this sense— motor and verbal behaviors, as well as both direct experience with stimulus situations and vicarious contact through verbal experience, as in reading" (Staats, 1963, p. 382).

In addition, individuals differ in the range of stimuli which for them are effective social A-R-D stimuli. To illustrate one extreme, a child could be raised by a solicitous and loving widow who was uneasy except with her own close relatives. The child would thus have positively reinforcing occasions with his close relatives but would have uncomfortable (aversive) experiences with other people. It would be expected from this differential training that for this child a few people would come to have strong reinforcing value, and other people would tend to be somewhat aversive. If the child also had little experience with other children, this condition might be aggravated. "The restricted nature of these social [A-R-D stimuli] would shape up 'dependent' behavior in the sense that the behavior of clinging to the mother or close relative would be strong" (Staats, 1963, p. 383).

In contrast, a child could be raised in a situation involving many different adults and children in positively reinforcing circumstances. This would be expected to produce a different set of social emotional-motivational stimuli, and consequently a different set of behaviors.

In addition, the behaviors acquired by the child may also determine later whether other social interactions are reinforcing. For example, a child might be treated in the family in such a way as to produce "spoiled," unreasonable, demanding behavior, including frequent crying and temper tantrums, as well as aggressive and destructive behaviors. With a repertoire like this, the experiences of the child with other children could hardly be anything but aversive for him. Such experiences would have the effect of producing a dependent child, since they would be expected to result in people other than the close family members becoming negative emotional stimuli and thus would prevent the acquisition of adjustive social behaviors and social A-R-D stimuli. (This early analysis thus demonstrated by example the behavioral interaction principles that have been explicated herein.)

> Dependence might also be produced by keeping the behavior under the exclusive control of "directions" by the parent. If in a wide variety of situations the child is punished unless he does just what the parent does, or what the parent instructs, then his behavior will be largely under the control of these stimuli. In a new situation, this child will look to the parent for his [directive stimuli]—his behavior will be "dependent" upon cues produced by the adult. Parents who do everything for the child, or who instruct the child how to do everything, might be expected to set up this type of dependent behavior. (Staats, 1963, p. 383)

Longstreth (1968) has presented a view of dependency that coincides with this analysis and has organized research results to support this position. The research was that of the Fels study (Kagan and Moss, 1962), which involved 89 children and their mothers who were observed and studied from 1929 to 1939. For example, ratings of the extent of the mothers' protectiveness were correlated with the extent of the children's dependence. It was found that protectiveness—"the extent to which the mother encouraged and rewarded dependency behavior, by always helping the child when he (she) requested help" (Longstreth, 1968, p. 382)—was positively correlated with the child's dependent behavior. Moreover, "maternal protectiveness during the first 3 years of the child's life correlated highly with his dependency in early adolescence" (Longstreth, 1968, p. 384).

Longstreth also indicated that the Fels data shows the way that parental restrictiveness produces dependency.

> The restrictiveness-dependency relationship is easily understandable in terms of learning theory. A mother who demands conformity and obedience, who has lots of rules, and who threatens and punishes deviations from her demands (recall that in the Fels study *restrictiveness* includes a punishment component) is not likely to foster independent habits in her offspring. Her children have few opportunities for such behavior and are punished when they try it. They are rewarded for obedience, for adherence to her demands, i.e., for dependency and passivity. (pp. 384–85)

It may be noted that various personality repertoires can be included in the personality characteristic of dependency. Instrumental behaviors have been mentioned, as well as the emotional-motivational system. Language-cognitive characteristics may also be a part of the dependent personality.

> Thus, independent behavior would seem to depend in part upon having certain skills and upon the appropriate stimulus control of these skills. Even in such intellectual skills as independent "decision making," these variables seem to be involved. Decision making may be considered to consist of complex sequences of reasoning responses that culminate in some overt instrumental behavior. In order, then, for decision making to occur, the sequences must be established under the control of various situations. A child who has not been trained in the constitutent behaviors involved, as well as in the emission of these behaviors in actual problems, but who has, rather, been reinforced in such situations for asking instruction from parents, will be unable to make decisions for himself, or when he does make them, the decisions may be poor and turn out badly.
>
> The counteracting experience for such dependence in decision making would thus seem to involve social and physical reasoning sequences, as well as opportunities for these responses to be emitted and reinforced in problem situations. (Staats, 1963, pp. 383–84)

Additional studies in this area of behavior are needed. Harris, Johnston, Kelley, and Wolf (1964) showed that the regressed crawling behavior of a nursery school child could be changed by having teachers eliminate social reinforcement of this behavior, while reinforcing standing. This study has implications for considering dependency, as does any procedure that produces adjustive, independent behavior in the young child through learning.

More pointedly, Rosen, Zisfein, and Hardy (1972) have treated a child for immature and dependent behavior. The child was described as over-protected, with parents unable to be firm or to punish him. The child was said to consistently complain and refuse responsibility. "Lester does not like to *have* to do anything, responding to demands with anger and tears. Because of his belligerent attitude and poor response to peers, he has only one friend and tends to remain a loner" (Rosen, Zisfein, and Hardy, 1972, p. 5). A remedial program based upon a token-reinforcement system ad-ministered by the parents was employed to change his behavior. The par-ents were reported to have been pleased by the changes, and the child to have been gratified by the new privileges he earned.

SOCIAL BEHAVIOR PERSONALITY DEVELOPMENT. The manner in which people learn to respond to others was described in Chapter 7. The acquisi-tion of social behavior in children constitutes an important realm of in-quiry in considering personality development. The child may learn to be aggressive towards others, gentle and nonassertive towards others, coopera-tive or selfish, and so on. These social behavior characteristics, which are ordinarily considered to be due to biologically determined personality de-velopment, can also be considered to be learned. An example of the learn-ing of certain aspects of nonselfish or nonaggressive behavior can be taken from an earlier analysis of the acquisition of social control.

> It is suggested that . . . through . . . training many social stimuli, verbal and otherwise, come to control "cessatory" responses so that behavior which would otherwise be inappropriate does not occur. In addition to the verbal stimuli of social control already described [that is, learning to respond to requests to stop doing something], some others that must gain the same kind of control may be briefly mentioned. For example, the visual stimuli of another child playing with a particular toy will come to be [noneliciting stimuli] if the behavior of seeking that toy in the presence of those stimuli is not reinforced, while the toy *alone* can still be a stimulus in the presence of which seeking behavior is re-inforced. More broadly, the child must receive analogous, nonreinforced training in many situations in which a [directive-reinforcing] stimulus occurs in the presence of additional stimuli which indicate that the reinforcer "belongs" to someone else.
>
> As another example, the child must come to respond with "cessatory" behavior to social stimuli such as frowns and other facial and verbal expressions which indicate that another person is suffering aversive stimulation. In other words, the child who has not been so trained will be *solely* under the control of the cues of the reinforcing object or event ["what he wants"] rather than the social stimuli, and may there-fore persist in behavior that is aversive to others. People who have had the more appropriate training will find it difficult to accept such "un-restrained" behavior. (Staats, 1963, p. 385)

Azrin and Lindsley (1956) conducted an experiment, ostensibly on the manner in which two children can be conditioned to cooperate, that illustrates training that would help produce "cooperative" and "sharing" social behavior. In the experiment, two children were placed at opposite sides of a table with three holes and a stylus in front of each one. A cooperative "response" was defined as the two children placing their styli in opposite holes at the same time. Whenever this occurred, a single reinforcing stimulus, a jellybean, fell into a cup that was accessible to both children. Although no specific instructions were given, the children did learn to make coinciding responses and to play together in this manner.

It may be suggested, however, that another occurrence in this study actually had greater significance than the cooperative behavior as defined. That is, only one jellybean reinforcer was delivered on each trial. The situation was thus one in which the rules of reinforcement were such that "sharing" behaviors between the children were learned. This can be seen by referring to the cases (two out of the ten pairs involved) where one of the children at first took all the candy, a good example of selfish behavior. Such a circumstance meant that the other child's behavior was not reinforced. Extinction of this child's behavior would be expected, and this was the result. Since reinforcement demanded the participation of both, this result meant that the "selfish" child obtained no reinforcement. In order for either child to receive reinforcement the behavior of the other child had to be maintained, which meant sharing the reinforcement. This is what finally occurred with these pairs of children.

This experiment lends support to the analysis that social behavior, in this case selfishness-unselfishness, is learned. If children who begin by responding selfishly can be changed in a very brief experimental participation, then it can be readily seen how the behavior of the child could be affected over a long period of time which involved multifarious learning trials. It would be expected, for example, that a child who had been reinforced for selfish behavior in many situations with various people would come to display this type of behavior generally—to develop a selfish personality.

Additional examples can be cited. Aggression was described in Chapter 6 as an aspect of the instrumental repertoire and in Chapter 7 as a social behavior under the control of negative attitudes. The manner in which aggressive social behaviors develop in the child is also a topic of interest in developmental psychology. An early study that shows the effects of reinforcement on aggressive behavior was conducted by Brown and Elliott (1965). In the study teachers were instructed to ignore aggressive responses of either a physical or verbal nature with 27 three- and four-year-old nursery school children and were told rather to give attention to the children when they were interacting in nonaggressive ways. This treatment continued for two weeks, following a prior one-week preexperimental period during which the frequency of the aggressive behavior was tabulated. After another three weeks there was a two-week period of observation, and then another two weeks of the reinforcement treatment. It was found that, under the natural reinforcement rules the teachers followed, the children showed significantly more aggressive behavior than they did when the teachers began to ignore aggressive behavior and to attend to (reinforce) nonaggressive behavior. Three weeks later, when the teachers had reverted to their old ways, the children again became more aggressive. However, when the teachers were again instructed in the reinforcement practices, there was a significant drop in aggressive behavior. A study by Horton (1970) with adolescent delinquent boys also has shown that such behavior modification treatment of aggressive behavior in one situation will generalize to other situations.

A cross-cultural study in an Okinawan village conducted by Maretzki and Maretzki (1963) included a naturalistic description that shows one way in which a child may develop aggressive behavior through learning. The example indicates that a sex difference in training is involved.

> Boys are expected to be more aggressive than girls, who are, ideally, docile, gentle, mannerly, kind, reserved, and considerate. In some ways, adults seem to train little boys to be aggressive. Beginning in infancy, they tease and bully them, holding their arms and restricting them bodily, withdrawing desired objects and pretending to scold and hit. They glean great amusement at the expense of the child if he screams and strikes out in anger (Maretzki and Maretzki, 1963, pp. 369).

It is suggested that the development of aggressive personalities by children through learning may be studied experimentally and through descriptive developmental research. This suggestion may be extended to other personality characteristics.

Aronfreed (1968) has been the most explicit of the social learning theorists in the analysis of the development of aspects of the child's social behavior. He has dealt in detail with the development of the child's conscience and the manner in which this development affects the child's conduct. In this account he has taken a two-factor learning theory approach, specifying his analysis at the basic level more than is usual in social learning theory. His general definition is that conscience involves "those areas of conduct where social experience has attached substantial affective value to the child's cognitive representation and evaluation of its own behavior" (p. 6). Although Aronfreed does not specify the original acquisition of the basic behavioral repertoires of language-cognition, imitation, and so on, he does indicate how conscience involves observational learning, cognitive and verbal control, empathic and vicarious experience, as well as direct reinforcement. Aronfreed's analysis is a valuable contribution to the understanding of this area of personality development.

This work can be related to some of Piaget's ideas concerning moral development and to research in this area of study. For example, one Piagetian interview involves telling the child about two other children, one of whom has broken a number of objects while doing something with positive intentions, whereas another child has broken one object but with negative intentions. A number of studies have shown that younger children choose on the basis of outcome, while older children consider intentions (Bandura and McDonald, 1963; Cowan, Langer, Heavenrich, and Nathanson, 1969; Grinder, 1964; Johnson, 1962; Lerner, 1937, MacRae, 1954). A recent study has shown that whether the outcome is positive or negative interacts with how strongly intention is considered (Costanzo, Coie, Grumet, and Farnhill, 1973). In any event, this supports the view that moral judgment undergoes developmental changes in childhood.

One more example will be given to illustrate more pointedly the behavioral interaction principles involved. The example is taken from an analysis of the parent-child relation and shows that the behaviors the child acquires very early begin to affect the manner in which others respond to him.

> . . . [T]here are aspects of the parents' love for the child that would seem to be affected by their ability as trainers. Much has been said of the cold, loveless, rejecting mother. Little has been said of the fact that there are great differences in the "attractiveness," or positive reinforcing qualities, of the behavior of their children. Some children are "lovable," their behavior has many positive features and few aversive ones. The child whose behavior has aversive aspects . . . will not acquire positive [A-R-D] value (be lovable) as a function of this type of behavior. . . . It would be expected that the degree to which the parent is loving will depend to some extent upon the behavior of the child.
>
> Thus, the fact that a mother cares for a child as a duty without enthusiasm, or is cold, loveless, and rejecting, may be secondary. That is, her behavior may be a result of the fact that she does not have the skills as a trainer with which to shape behavior in her child which is reinforcing to her and to others. (Staats, 1963, p. 412)

A recent study provides evidence that is very relevant to this point and to the general suggestion that there is a behavioral interaction relationship between parent and child. Clarke-Stewart (1973) has conducted an extensive study of mother-child interactions. One finding was that the child's social behaviors (such as looking at the mother, smiling, and vocalizing) has the effect of increasing the amount of time the mother will spend with the child.

> . . . [I]t was concluded that the child's expression of positive emotion to his mother caused her to express more positive emotion to him later. Looking at and vocalizing to the mother also caused the mother to be less rejecting and more responsive to the child's distress and demands. Obviously, the infant's social behaviors are an influential force in mother-child interaction. The reciprocal nature of mother-child relations is demonstrated by these findings which illustrate that, over time, both

mothers and children affect each other's behavior (Clarke-Stewart, 1973, p. 88).

It has been suggested in an earlier section that naturalistic observations be made of the child's language learning involved in the parent-child interaction. The same type of study is important for all aspects of the child's behavioral development, including his social behavioral development—as the above findings suggest.

## PARENTS AS TRAINERS OF THEIR CHILDREN

Different conceptions of child development have different implications for child raising, which should be made explicit. For example, some of the early child developmentalists, as has been indicated, considered the child's behavioral development to take place largely through the child's biological unfolding (maturation). This conception had an unfortunate aspect. For one thing it led attention away from considering the role of learning in the child's development and delayed the investigation of this important realm of study. The conception has also influenced strategies of medical and psychological treatment. For example, many parents whose children are developing behavior normally have been advised to wait for hypothetical maturational processes to occur when other measures were actually more appropriate, as the following example indicates.

> *Question:* My 2½ year-old grandson, who has been talking distinctly for some months, has developed a stuttering problem. Is there a reason for this?
>
> *Answer:* . . . [I]t is a phase he's going through—a normal, developmental phase. . . . The important thing is to ignore the speech repetition. . . . By so doing, as the child's nervous and speech mechanisms mature, he will overcome the tendency to hesitate or repeat syllables and words. (Crook, 1969, p. C–3)

In analyzing speech hesitancies like stuttering, it was suggested that learning conditions are involved (Staats, 1971a, pp. 164–71). Frequently, for example, parents will be aversive to the child while he is speaking, to stop him from unreasonable requests, and so on. This acts to decrease the fluency of speech in a manner that can develop into stuttering. In such cases the parent needs to know the principles involved and the specific procedures that can be employed to discontinue the improper learning, while introducing conditions to remediate the child's speech disfluencies. As was indicated, such remediation can take place readily and in the home when remediation is begun early, before the undesirable behavior has become habitual. The advantage of early treatment is one reason why advice for the parent to be passive may be misdirected.

The traditional developmentalist ideal tended to be of parents who were warm and loving and who provided a generally good atmosphere within which the child could develop from within. The social behaviorism paradigm, however, provides a basis for a conception of the parent as actively involved in his child's development.

In summary, a learning analysis of the acquisition of behavior leads to a focus upon the parent as a *trainer*. Whether the parent intends to or not, he manipulates many conditions of learning that will determine to a large extent the behaviors the child will acquire. As long as the child's behavioral development consists of innumerable training experiences, many of which occur in the home, then the parent has many of the controlling variables in his hands and cannot relinquish them regardless of his philosophy of child development.

This suggests that the parent could be an active participant in arranging circumstances to most efficaciously produce an abundant, rich, adjustive, behavioral repertoire using a minimum of aversive stimulation and a maximum of positive reinforcement. Good working behaviors, good studying behaviors, the ability to work without immediate reinforcement; reasonable, cooperative, not overselfish behavior; a good language system about the world, his own behavior, and that of others; and a good system of reinforcers, including words of positive and negative reinforcement value; social stimuli that appropriately control striving and nonstriving behavior; social behaviors that reinforce other people as well as oneself; these seem to be some of the behaviors that the parents help determine by the conditions they present to the child. Thus, to a large extent, the learning conditions that occur in the home would seem to determine whether the child will grow into a "well-adjusted," "happy," "productive" individual.

Faced with a training task of such imposing responsibilities, it would seem that the parent would need an understanding of the principles of behavior by which children learn. In addition, it would seem that the parent would require an analysis in terms of those principles of the various specific training problems he faces. The parent needs to know how not to [train] undesirable behaviors, or when they have developed, how to decrease them benignly; and he needs to know how to [train] the many adjustive behaviors the child will require.

. . . It is thus evident that much research is necessary to clarify the problems of training in order to provide the knowledge needed by the parent with which to deal with those problems. (Staats, 1963, pp. 412–13)

This constitutes a rationale for behavior modification work with children. Since this was written there have been a number of behavior modification analyses and studies that support the conception of the parent's role as the trainer of various adjustive behaviors in the child. For example, Wahler, Winkel, Peterson, and Morrison (1965) pointed out that it was typical to attempt behavior modification work in artificial (laboratory-like) circumstances. They concluded the following, however:

> A logical procedure for the modification of the child's deviant behavior would involve changing the parents' behavior. These changes would be aimed at training them both to eliminate the contingencies which currently support their child's deviant behavior and to provide new contingencies to produce and maintain more normal behaviors. (p. 114)

These investigators selected specific problems of behavior in several children and had the mothers change the manner in which they responded to the behaviors, eliminating reinforcement for the problem behaviors and instead reinforcing more desirable behaviors. One child's problem "commanding" behavior was changed to cooperative behavior. Another child's dependent behavior was changed to independent behavior.

Similar behavior modification procedures were employed by Hawkins, Peterson, Schweid, and Bijou (1966) to modify tantrums, disobedience, excessive attention-demanding behavior, rudeness, and other objectionable behaviors in a four-year-old child. Walder (1966) and Wetzel, Baker, Roney, and Martin (1966) have used parents to treat their autistic children. Other behavior problems modified by parents as trainers have included emotional disturbances (Andronico and Guerney, 1967), destructive behaviors (O'Leary, O'Leary, and Becker, 1967), aggressive behavior (Patterson, 1973; Zeilberger, Sampen, and Sloane, 1968), excessive scratching (Allen and Harris, 1966), behavior problems in brain-injured children (Salzinger, Feldman, and Portnoy, 1970), chronic constipation in a three year old (Lal and Lindsley, 1968), and oppositional behavior (Wahler, 1969). Bernal (1969, 1971) used reinforcement with videotape replays to train parents how to train their children.

As has been indicated, these studies concern relatively simple behavior that can be treated in relatively brief behavior modification procedures. Ryback and Staats (1970) have shown, however, that complex behavior repertoires such as reading may be dealt with by parents over extended periods when the behavior modification procedures have been developed to be simple and easy to apply. Moreover, these procedures can be applied by parents with children having different types of problems (retardation, brain damage, emotional disturbance) in addition to their reading deficits.

More recently, Hawkins (1972) has suggested that young people should be taught how to be good parents by providing them with knowledge of reinforcement principles. McIntire (1973) has taken the notion of the importance of parent training to the extreme of suggesting legal enforcement of mandatory birth control unless such parental training is completed. A more moderate suggestion would be to provide incentives for parents to increase their skills for raising children. For example, socioeconomically depressed parents—whose children have a high frequency of adjustmental problems—could be given salaries (in lieu of welfare payments) to learn child training principles as well as to engage in educational activities to improve their intellectual skills in general, since the intellectual skills the child gains will depend to a large extent upon those of his parents (Staats, 1971a). This has been suggested as a means of breaking the cycle where deficient parents, subjected to life harassments because of deficiency, in turn raise deficient children.

Only a few of the studies cited have been examples of how parents are trained to appropriately train their own children. A number of other studies may be found in Ashem and Poser (1973), Fargo, Behrns, and Nolan (1970), and O'Leary and O'Leary (1972). Much additional work in studying various areas of child development can be expected. Some of this work will require more advanced analyses of human learning, as will be suggested. However, the potentialities of the approach have been demonstrated.

*11*

# *Social behavioral educational psychology*

In the author's early general learning analysis of human behavior (Staats, 1963), a chapter on behavioral educational psychology was included.. At that time there was little in the way of application of behavior modification principles or procedures to the educational field, other than the author's own work. Programmed instruction utilized the philosophy of Skinner's principle of breaking instructional materials into small, easily learned steps. His concern with precision was also displayed in the use of teaching machines, as was described in this account. Valuable as its products have been, programmed instruction relies upon the same reinforcers—achievement, moving ahead—as did traditional education. It was suggested that a more extensive use of learning principles, including reinforcement, is needed in the context of educational concerns (Staats, 1963).

The manner in which the behavior modification principles were developed and applied in clinical psychology has been described. These principles and procedures had similar implications for educational psychology, and in fact some of the early reinforcement studies involved educationally relevant behavior. Since the first studies, there has been a large number of studies that have provided additional support. The significance of this development has been shown with the publication of the *72nd Yearbook of the National Society for the Study of Education,* entitled *Behavior Modification in Education* (Thoresen, 1973a).

As in the other areas that have been addressed, there has been a

382

separatism between behavioristic and traditional educational and educational psychology approaches. The Yearbook is a step toward recognizing behavior modification. However, it is notable that most of the contributors to the Yearbook—such as Becker, Bijou, Kanfer, Krasner, Lovaas, Patterson, and the present author—have been identified more with the fields of learning and behavior modification than with educational psychology. It is suggested that a more general rapprochement is called for. The learning approach has value, as the behavior modification studies show. Moreover, the other levels of social behaviorism also have important elements for extension to education. A number of the preceding chapters have indicated this. Additional suggestions will be made in the present chapter.

## Behavior modification in education

The work that launched behavior modification in clinical psychology and child psychology also provided the foundation for behavior modification in educational psychology. The original statement of the behavior modification principles (Staats, 1957a) had several aspects that suggested different lines of development. For example, the analysis suggested that problems of behavior could be treated (1) by decreasing the undesirable behavior involved, or (2) by increasing a desirable behavior that could displace the undesirable behavior, or both.

Ayllon and Michael's (1959) study concerned the solution of simple problems employing these principles, as did some of the early behavior modification studies in education. Thus, as has been indicated, young children were treated in classrooms for inappropriate crying (Hart, Allen, Buell, Harris, and Wolf, 1964), for crawling and not standing (Harris, Johnston, Kelley, and Wolf, 1964), and for not playing with other children (Allen, Hart, Buell, Harris, and Wolf, 1964). These behavior modification studies were based upon the use of attention and social approval as the source of reinforcement.

Following these early demonstrations there have been a number of studies that have focused upon the manner in which teachers apply their own social reinforcement in the classroom. Hart, Reynolds, Baer, Brawley, and Harris (1968) were successful in changing the behavior of an aggressive girl, training her to be cooperative. Becker, Madsen, Arnold, and Thomas (1967) used the combined effects of ignoring and praising the child, according to reinforcement principles, to reduce classroom behavior problems with two disruptive children. Hall, Lund, and Jackson (1968) showed that a disruptive child—who spent only 25 percent of his time studying—could be changed through the systematic employment of teacher attention. When the teacher reinforced study behaviors and no longer attended to disruptive behaviors, the time spent studying increased to between 75 and 80 percent of total time. Hall, Fox, Willard, Goldsmith, Emerson, Owen, Davis, and Garcia (1971) showed, with a variety of problem children, that teacher attention could be changed to treat undesirable behaviors and that teachers could record data reliably while using such procedures.

The principles of behavior modification for dealing with problems of behavior in education were elaborated greatly through use of the token-reinforcer system (Staats, Staats, Schutz, and Wolf, 1962; Staats, Finley, Minke, and Wolf, 1964; Staats and Butterfield, 1965). The first use of the token-reinforcer system in a group procedure to affect misbehavior was made by O'Leary and Becker (1967).

> A pad was put on each child's desk in which the teacher placed a rating every 20 minutes. The ratings were exchangeable for backup reinforcers, which were initially available every day. The introduction of the token program resulted in a decrease in average disruptive behavior (talking, noise, pushing, eating) from 76% in the baseline period to an average of 10% during the 2-month token period. Delay of backup reinforcement was gradually increased to 4 days without an increase in disruptive behavior. The program was equally successful for all children observed, and anecdotal evidence *suggested* that the children's appropriate behavior generalized to other school situations. (O'Leary and Drabman, 1971)

In a later study, O'Leary, Becker, Evans, and Saudargas (1969) compared the effect of (1) stating rules of conduct to a classroom of disturbed children and structuring the classroom organization, (2) stating the rules and structuring the classroom, and providing social reinforcement for desirable behaviors while extinguishing undesirable behaviors, and (3) these conditions plus the use of a token-reinforcer system. The first two conditions were not effective in reducing disruptive behaviors, but the addition of the token-reinforcer system was effective in treating this problem of education. Meichenbaum, Bowers, and Ross (1968) showed the same effect of the token-reinforcer system with female adolescent offenders.

Increases in study behavior have been systematically reported as a function of introduction of a token-reinforcer system. Bushell, Wrobel, and Michaelis (1968) set up a classroom for 12 preschool children employing token-reinforcer procedures. They found that noncontingent reinforcement did not have the same effect in sustaining study behavior as did contingent reinforcement. This supports the contention (Staats, 1963) that reinforcers can be available in a school but not be employed to strengthen the appropriate behaviors. Broden, Hall, Dunlap, and Clark (1970) employed the token economy idea with a class of 13 academically retarded children in the seventh and eighth grades. Under ordinary circumstances the children were attentive 29 percent of the time. This was increased through the teacher's social reinforcement for studying to 57 percent. When the token system, which involved receiving points which were exchangeable later for extra time at lunch period, was used, the study behavior increased to 74 percent. When this procedure was extended to all the child's class periods, the average study time was 80 percent.

In a study by Walker, Mattson, and Buckley (1969), procedures that included token reinforcement increased study behavior of six disruptive children from 39 to 90 percent. This was accomplished in a special setting, but when the children were reintroduced to their regular classes they demonstrated improved behavior. At the end of three months their study behavior in their regular classes remained an average of 72 percent of what it had been in the experimental class.

A number of other studies have been conducted in which the concern was lessening the undesirable behavior of children in the classroom and increasing desirable behavior. The behavior modification principle that behavior problems can be treated by the production of desirable behavior, however, has additional implications that underlie another line of behavior modification work.

## Producing adjustive behavioral skills

The principle of behavior competition was outlined in describing principles of human learning relevant to abnormal behavior. It was said that the "acquisition of one behavioral skill can militate against acquiring another behavioral repertoire." Abnormal behavioral "skills," thus, may prevent the acquisition of normal behavioral skills.

The same is true in the opposite direction, however. Acquisition of normal behavioral skills in the individual can prevent the acquisition of abnormal, undesirable behavior. The present author, on the basis of this realization, focused upon the manner in which the behavior modification principles could be employed in the creation of normal, adjustive behavioral skills in children—a central goal of education. An early study that demonstrates clearly these principles of behavioral interaction may be summarized (Staats and Butterfield, 1965).

The subject was a 14-year-old juvenile delinquent. He disrupted classroom activities, cursed in class, was disrespectful to the teacher, fought with other students in and out of class, and had never passed a course in his school history. He was considered retarded and incorrigible, and he had been apprehended vandalizing a school. In addition, the boy had severe academic skill deficits. He read, for example, at the second-grade level—a severe retardation.

One could treat such problems by attempting to decrease the undesirable instrumental behaviors. A behavior analysis reveals in many cases, however, that it is the *deficit* in the normal repertoires that is central. The child who goes to school with cognitive and motivational deficits, as in this child's case, will cumulatively acquire additional deficits—as his academic achievement tests indicated. This creates problems for the child in school, ranging from the aversiveness of failing, to teacher rejection, to derision from other, more successful children. Such treatment, of course, will produce negative attitudes in the failing child, which then elicit the class of "striving against" instrumental behaviors he has learned.

Such an analysis calls for the reversal of these conditions rather than treatment of the child's undesirable "striving against" behaviors. Thus, the treatment employed consisted of special training conducted by a probation officer. The token-reinforcer system and the reading materials previously described were utilized with this boy for a period of four and one-half months of daily half-hour training sessions. With the reinforcement for reading this recalcitrant student became a vigorous worker. He made over 64,000 word reading responses, with an accelerating rate, even though the amount of token-reinforcers given per reading response was cut by three-fourths. He passed his courses for the first time ever. His receipt of demerits for misbehaviors in school decreased from ten the first month to none the last month and a half. Moreover, his reading achievement advanced to the 4.3 grade level. His attitude change was evident in his statements as well as in his behavior.

The results indicate the principles involved here. By treating the boy's behavioral deficits—and in so doing changing negative attitudes toward the positive direction—there were benign effects upon the undesirable behaviors themselves. More will be said of the implications involved in this study.

The next step in the advancing series of the behavior modification study of academic behavior and treatment of problems of such learning was the extension to groups of subjects. This was done in advancing the early child behavior modification procedures, as will be indicated, as well as in dealing with children requiring the remedial treatment described above. Staats, Minke, Goodwin, and Landeen (1967) utilized the token-reinforcer system effectively in reading training with 18 retarded readers, using high school students and volunteer housewives as the behavior modification technicians.

The next study with 32 culturally deprived Negro junior high school children living in the ghetto area of Milwaukee yielded additional findings (Staats, Minke, and Butts, 1970). With the reinforcement system and simple training materials it was not necessary to have a trained teacher perform the instruction. The technicians were literate, unemployed black adults or black high school students who were given three hours of special training. An average of $21.34 worth of reinforcers per child was allotted. The experimental subjects attended school better, learned to read more words, and scored better on reading and intelligence tests than did the 32 subjects in the control group. .

Camp (1971) has further shown the token-reinforcer system and reading materials to be effective in training problem children in a pediatric clinic. Camp and van Doorninck (1971) found that children given the reinforced reading training did better on sight vocabulary reading tests, in comparison to a control group, but not on the *Wide Range Achievement Tests* (Jastak, Bijou, and Jastak, 1965). However, the reading training was brief, and results in the study indicated that children with longer training did improve on the WRAT, findings that are supported by other studies (Harris, 1974; Ryback and Staats, 1970).

Wolf, Giles, and Hall (1968) also conducted a study in which the token-reinforcer system was employed. The children received points for excellence of work done, and the points were exchangeable for tokens. This allowed the experimenters to change the number of points they would award the correct completion of a task. Individual records were kept of the child's rate of completion of academic tasks, while the amount of reinforcement (points) was varied. The records for the individual children showed that the achievement of the child depended upon the amount of reinforcement he received for his completed studying behavior. It should be noted that the children were in the fifth and sixth grades in an urban poverty area. Standard achievement tests were given before and after the year of reinforcement training, as they were to a control group of children who did not receive the reinforcement training. The results showed the children trained under the token-reinforcer procedures advanced an average of 1.5 years in achievement in comparison to an average advance of 0.8 years for the control group children. This difference was statistically significant. Clark, Lachowicz, and Wolf (1968), Tyler and Brown (1968), and Hewett, Taylor, and Artuso (1969) have also successfully employed the token-reinforcer system in the schools to improve academic achievement. At the present time there is a growing number of educational applications of the token-reinforcer system. The terms token economy, contingency management, performance contracting, and so on, may be employed, each of which involves variations of the token-reinforcer system. Excellent reviews of this work have been provided by Hanley (1971) and O'Leary and Drabman (1971).

## EARLY CHILDHOOD EDUCATIONAL BEHAVIOR MODIFICATION

The competitive behavior principle has additional implications other than those already mentioned. If inappropriate behavior can be treated by helping the individual acquire normal behavioral skills, for example, it should be possible to *prevent* the original development of the inappropriate behavior by insuring that the normal behaviors are acquired first. Clinical psychology has had a focal concern with the treatment of behavior problems after they have arisen. Normal behaviors, because of the traditional conceptual background of clinical psychology, have been given less importance. The learning analysis shifts the emphasis to consideration of the learning of appropriate behavior (Staats, 1970b), rather than the treatment of behavior problems. This shift suggests a focus on early child learning, for this is the propitious time for insuring that the child will begin to learn his normal, adjustive behavioral skills.

Thus, the author very early in the development of the behavior modification principles and the token-reinforcer system began the study of the learning of the child's first academic skills. The studies of reading learning with four-year-old children under different reinforcement conditions was described in Chapter 9. Whitlock (1966) used these token-reinforcer procedures and the types of reading materials described in the earlier studies to train a six-year-old child who was retarded in reading. The child's reading skills improved as a consequence of the training. The importance of reinforcement in this training was shown with another six-year-old child (Whitlock and Bushell, 1967). Reading in the training was reinforced or not reinforced in various sessions. The child's reading rate increased in the reinforced sessions.

Additional behavior modification work in early childhood education will be described in the next section, along with the manner in which the token-reinforcer system was additionally extended.

## TOKEN-REINFORCER SYSTEM DEVELOPMENT

As has been indicated, the general implications for the extensions of the token-reinforcer system to human behavior were evident to the author after its first successful use in 1959. These implications were described by the author informally and formally, as has been indicated. This included extensions to the school situation.

> . . . However, there has been little systematic study of the possible sources of reinforcers in school training or of the potential ways in which these reinforcers may be made contingent upon the learning behavior of the child.
>
> As an example, consider the nursery school or kindergarten situation. Some of the reinforcers present in these school situations have been described—games, recesses, toys, snacks, rest periods, television, desirable activities of various kinds, and social approval. These are potent reinforcers but for the most part they are not made contingent upon the individual behaviors to be strengthened. Since most of these reinforcers occur only infrequently in the school day, however, even if they were made contingent upon a specific learning behavior, very few could be provided and thus only a few behaviors reinforced.

These reinforcers might prove to be very effective, however, if they were incorporated into a procedure involving a token system. For example, a recess could be exchanged for 100 tokens, and the tokens could be made contingent upon 100 appropriate responses. Through employment of such a token system effective use could be made of many reinforcers that are a natural part of school training.

The development of such a system of reinforcers might necessitate some changes in school organization. For example, one way to use a token system might be to have a "work" room for the learning activities and other rooms for dispensing the "primary" reinforcers. . . . The child would stay in the work room until he accrued a given number of tokens and then go to one of the "reinforcer" rooms to receive his primary reinforcer. . . . (Staats, 1963, pp. 457–58)

This was the first description of the token-reinforcer system for clinical or educational purposes. Moreover, the token-reinforcer system was described as applicable to groups of subjects. The monetary-like quality led to the use of the name *token economy* by Ayllon and Azrin, (1968) in their application of the token-reinforcer system to hospitalized wards of psychotic patients, and the term has also been used in education.

As will be indicated further, however, at the time the group token-reinforcer system was outlined, the author planned to advance his work from individual preschool children to children in the type of classroom described. This became possible in 1965 when a preschool classroom was set up by the author in a public school in Madison, Wisconsin. The procedures were a step on the way toward the more complete implementation of the token-reinforcer system described in the above quotation. The facility included a classroom in which the four-year-old, predominantly black, culturally deprived children engaged in preschool activities. Behavior modification principles were employed in this classroom to control such problems as temper tantrums in one of the youngsters who was emotionally disturbed.

The children each had three brief (about five-minute) academic training sessions per day in the "work" room, which used the apparatus depicted in Figure 2.5, (Chapter 2). One of the training sessions was devoted to teaching the children to read the letters of the alphabet, another to writing the letters, and a third to learning to discriminate the number of objects, to count objects, and to read the numbers. The children were reinforced with marbles as the tokens and could exchange the marbles for backup reinforcers. The children learned these skills, to a level in advance of children of their age who did not have socioeconomic disadvantages. During the seven and one-half months of the project the children advanced on the Metropolitan Readiness Test scores from the 2.3 percentile to the 23.8 percentile—when compared to a norm group almost two years older than the children in the project. Their Stanford-Binet (Terman and Merrill, 1937) scores advanced from 100.9 to 112.5, whereas expectations are that culturally deprived children will progressively fall behind other children in IQ. See Staats (1968c) for a more detailed description of the procedures and findings.

Hamblin, Buckholdt, Ferritor, Kozloff, and Blackwell (1971) have also applied the same types of procedures in a series of studies. For example, one study involved a class of 33 culturally deprived black children. The children had completed kindergarten but were judged not advanced enough to enter the first grade. This study and others conducted by these authors verified and elaborated the previous findings. For example, eight of the children judged to be lowest in academic ability were given training in reading for 20 days, 20 minutes per day. These children, who just previous to the training had failed the Metropolitan Reading Readiness Test, all passed the test after the training. The overall intelligence of the class, as measured by the California Mental Maturity Test, increased 18 points over the school year.

Additional advancements in the token-reinforcer system are possible, however, as will be indicated.

## TOKEN-REINFORCER MOTIVATIONAL SYSTEMS AND SCHOOLS OF THE FUTURE

In the author's 1963 passage outlining the token-reinforcer system (token economy) it was suggested that reinforcers could be activities and events as well as material articles (toys, and so on) and edibles (such as candy). Some of the behavior modification studies began to include such backup reinforcers for their token systems. For example, Bushell, Wrobel, and Michaelis (1968) included a special-event ticket as backup reinforcer. This ticket allowed the children to attend such events as a brief movie, a theater group rehearsal, a story-telling episode, an art or gym class, a trip to a park, and so on.

As another example, one of the important principles in the outline of a token-reinforcer system for employment in the school situation was that sources of reinforcement natural to the school situation could be used. It was suggested that games could be used as reinforcers, as could rest periods, recesses, snacks, television, and desirable activities of all kinds, as well as social approval. It was also suggested that there could be a work room and a "reinforcer" room where the child would receive his backup reinforcers in exchange for the tokens he had accrued. Hopkins, Schutte, and Garton (1971) have substantiated the principles and procedures suggested. Access to a playroom was used as the backup reinforcer for a token-reinforcer system. The rate and quality of printing and writing were reliably increased with children in first- and second-grade classrooms. As another example, Osborne (1969) employed free time as a reinforcer.

This acted as a reinforcer even though the children were restricted to the classroom, which is less reinforcing than such things as free time in a playground. Lovitt, Guppy, and Blattner (1969) also employed free time as the reinforcer with fourth-grade children to increase their spelling accuracy.

Wolf, Giles, and Hall (1968), in the study already described, employed backup reinforcers that were similar to those originally outlined for classroom application. Green stamps served as the tokens, and the backup reinforcers could be snacks and material items (candy, novelties, clothing, watches, and so on), or the children could exchange their tokens for desired activities such as going to a circus or a zoo, swimming, having a picnic, seeing a movie, and so on.

Thus, test of the token-reinforcer system had proceeded so widely and with such positive results that by 1969 it was possible to elaborate more fully the manner in which the system could be extended to education (Staats, 1969a; see also Staats, 1970b, 1972, 1973). It was suggested that *whole schools* be designed on the basis of reinforcement principles and the concept of the token-reinforcer system. The suggestions of the 1969 article are summarized below.

The fact is that a public school has resources that are very strong sources of reinforcement for all students, including those students who are now considered to be learning problems because of retardation or emotional or behavioral disturbances, cultural deprivation, and so on. Schools have special equipment for physical exercise, sports, games, dancing, painting, and so on. Most children do not have access to these sources of reinforcement at home, and the activities that can occur in such facilities are very powerful reinforcers for most children. In addition, schools have some facilities for films and television, which could be expanded, as well as for activities such as reading, special-interest clubs, free time, socializing, and so on.

When one examines the school situation, however, it becomes apparent that these strong reinforcers are largely wasted. The child has relatively long periods of restriction to the classroom and relatively brief access to the reinforcers. This system is ineffective for many reasons. First, the child's classroom behavior has no effect upon his receipt of the school reinforcers of play. Second, the time spent in the classroom by problem children is almost entirely a waste. They make a very low rate of learning responses. They do not attend to what is being said in class, they do not read, they do not work problems, and so on. *The reason that they do not learn is that they have very few learning trials.* Third, the length of the classroom period has been inviolate. This and other traditional procedures must be examined in terms of what is now known about the learning process. For example, when one works with an effective reinforcer system, it soon becomes apparent that one can get many more learning trials out of the subject in a given period of time than under conditions of no reinforcement (see Staats, 1968c; Staats, Brewer, and Gross, 1970). If the child acquired tokens for performing learning trials, and if after enough had been acquired he could then go out and play, learning trials would occur more rapidly. The child would then have to remain in the classroom for a much shorter length of time, so he would be spending a greater amount of time in reinforcing activities in school. We would then see fewer problems in the child's attendance in school.

To repeat, it is proposed that we develop a whole school designed on positive reinforcement principles.[1] The idea would be to provide reinforcing activities in the school which would maintain the child's coming to school. Within the school the child would first report to a work-learning situation where he would perform work to a relatively easy criterion and for which he would receive tokens.

After earning a certain number of tokens for work completed, children would then be able to go on to a reinforcement period of their choice. This would be followed by another work-learning period, and so forth. It would be expected that for young children, and for children who had not already learned better work-study skills, the work-learning periods would have to be relatively brief. As the child progressed, the ratio of work-learning time to reinforcement time would become greater. It would be expected that effective training of this type, even with problem children, would produce students who would spend as much time in work-learning activities in school as the better students now do.

This plan suggests a great deal of effort in the design of reinforcing activities within the school. In actuality, schools have only groped toward appropriate employment of their reinforcers and in making them contingent upon learning behaviors. The same is true of work to improve the reinforcers available. Lack of knowledge of reinforcement principles, and the misconception in education that learning should be for its own reward, have held back progress. With systematic development and engineering, and without the handicaps of prejudice, there could easily be a very abundant source of reinforcement to maintain the effective learning of all children.

It should be noted that various types of training could be introduced into a school with such a reinforcement system, after suitable research had been conducted. For children needing social learning, emotional learning, sensorimotor learning, and so on, in addition to intellectual learning, periods dealing with various subject matters could be designed. It would thus be possible to provide clinical treatment where it should occur—that is, in the situation where the child has to make his life adjustment. We could design a school to treat a wide spectrum of problems such as dropouts, incipient delinquents, children with school phobias, educational failures because of personal problems, cultural deprivation, motivational

---

[1] James Breiling has begun a project to apply this plan in the St. Bride School on Chicago's South Shore.

deficits, early cognitive deficits, deficits in social behavior, and the like. Possibilities and potentialities that cannot be achieved with a single iso-lated class, or a partial program for some students, could be realized if a whole school was developed along these lines.

Projects now accumulating do provide partial support for these pro-jections. The most elaborate has been conducted by Cohen and Filipczak (1971). The institution was not a public school but an institution for juvenile delinquents, the National Training School for Boys. The boys treated were all school failures, 85 percent dropouts. The tokens the boys earned through academic work could be exchanged for money, private bedrooms, gifts, and so on. In 90 hours of academic work, for example, the average gain of 28 boys was 1.9 grade levels on the Stanford Achieve-ment Test and 2.7 grade levels on the Gates Reading Survey.

## REINFORCEMENT IN HIGHER EDUCATION

Staats (1968c; 1971a, p. 229) has suggested several reinforcement corol-laries for use in training children: (1) the greater the number of responses, (2) the more effortful the responses, and (3) the greater the massing of learning trials, the more reward is necessary to maintain the child's learn-ing behaviors. These principles have not been sufficiently studied with young children in formal research. However, K. A. Minke and J. G. Carl-son have conducted a study at the University of Hawaii in which they ma-nipulated the effortfulness of the task of freshman students taking a "unit mastery" class. Students have to pass an exam at the end of each study unit before going on. The researchers found that lowering the criterion of passing from nine to eight of ten correct items improved performance in terms of units passed, rate of progress, and course grades. Minke and Carlson's unit mastery method, while dependent upon achievement rein-forcement, has also used token reinforcers. Bonus points given for perfect exams can be used by students to pass a unit on which their score is not high enough to attain criterion. These investigators are showing that ex-plicit data can be obtained in research in higher education utilizing rein-forcement principles.

## BEHAVIORAL COUNSELING

Krumboltz (1965, 1966), Krumboltz and Thoresen (1969), and Hos-ford (1969) have introduced the behavior modification principles into educational counseling. Krumboltz developed the concept in counseling that individualized goals should be arrived at between the counselor and client. According to Thoresen and Hosford (1973), "A major task of the counselor, therefore, is to work with each individual to clarify his concerns into specific action terms. In effect, the client is asked, 'What actions would you like to change?'" (p. 110).

One of the characteristics of behavioral counseling is an emphasis upon the environment. "For example, a counselor may spend relatively little time talking with the client; instead, he may work with the significant others in the client's life, such as parents, teachers, siblings, or mate to change the client's environment" (Thoresen and Hosford, 1973, p. 119). Following this view, Hosford (1969) trained teachers to reinforce shy students for participating in oral discussions in class. This had the effect of increasing the frequency of assertive verbal behavior in the students, and it also had positive effects upon the teacher's behavior.

There have been other uses of reinforcement by counselors. Liberman (1970) found the use of reinforcement to help develop counselor-client rapport. He also found that reinforcement produced positive group attitudes and more group cohesiveness than did intuitive, group-centered procedures. Furthermore, group members began using his reinforcement procedures on each other—with productive outcomes.

Warner and Hansen (1970) and R. W. Warner (1971) employed social reinforcement in the group to reduce alienation in students. Subjects were selected who were alienated. The counselor focused on the students' feelings of alienation and gave positive verbal reinforcement to statements made by students which indicated positive attitudes concerning their positions in the social structure. This treatment was more effective than the procedure used with a control group. Kramer (1968) also used reinforcement with students enrolled in a study-skills course to increase the amount of their questioning, responsibility, and positive verbal statements. This produced a greater effect than did the procedures used with a "traditional group" which employed non-directive techniques.

It may be added, however, that behavioral counseling has incorporated the rejection of traditional methods that are based upon the interview or therapy session. Thoresen and Hosford (1973), for example, contrast behavioral counseling to traditional counseling in those terms:

> Traditionally, counseling was neatly conceptualized as a unique kind of relationship wherein the individual called a client or patient sought the assistance of a professionally trained counselor or therapist. The counseling took place almost entirely in an office setting, often over many sessions of one hour or more in length. The change process involved primarily verbal interaction between two persons or within a small group.
>     In contrast, behavioral approaches often "treat" the person in his every day environment rather than in the counselor's office. (p. 112)

As was suggested in the context of clinical psychology, however, there is not necessarily acceptance of the one method at the expense of the other. Verbal methods of counseling and psychotherapy can be just as behaviorally justified as the behavioral methods employing reinforcement. As has been indicated, when the personality levels of theory are recognized, and when the importance of language as a means of changing behavior according to behavior principles is recognized, language behavior therapy can be expected to contribute to dealing with various problems in the educational situation.

One example that has been employed in the context of behavioral counseling that demonstrates the use of language behavior therapy methods concerns language induced aversion. In this procedure words are employed to elicit negative emotional responses. The method is arranged so that the negative emotional responses are paired with the stimulus that should be changed in a negative direction. This stimulus may be presented on a verbal level or in its concrete form, for example, alcohol versus the word *alcohol*. As an example, the client might be told to imagine himself dying of lung cancer. This would elicit negative emotional responses which would be paired with presentation of a cigarette. Such methods have been used to treat sexual problem behaviors (Barlow, Leitenberg, and Agras, 1969), smoking (Tooley and Pratt, 1967), and weight problems (Kennedy and Foreyt, 1968).

Recently, language methods have been employed in what have been called self-control techniques. For example, Kolb, Winter, and Berlew (1968), through verbal interaction, helped a group of graduate students to specify goals for themselves and to strive for changes in behavior, thoughts, and feelings in line with those goals. The students then participated in group therapy sessions. They were found to demonstrate greater change in behavior than other students who had not experienced the guided language procedures.

It may be added that a suggestion for the use of behavior modification principles in teacher training has been outlined (McDonald, 1973). As McDonald indicates, however, the basis for the use of reinforcement principles to strengthen certain behaviors rests upon ability to identify the behaviors. What makes a good teacher has not yet been stipulated by behavior analysis, except in terms of increasing the teacher's use of reinforcement principles with her pupils. Perhaps McDonald's suggestion will give impetus to further interest in this area.

## The personality repertoire level in education

It is suggested that the behavior modification studies, besides their specific significance, have general import as a body. The demonstration that learning principles extend to functional behaviors and can be employed to treat problems of behavior directly supports the learning conception of educational development.

However, important as our behavior modification findings are, they do not answer some of the questions central to educational and educational psychology. For example, the demonstration that reinforcement can be employed to teach problem children in reading has important uses and implications. However, such studies may not indicate what reading is or the myriad things we need to know about reading. The same is true of traditional educational research. Reading method A may be compared to method B with different groups of children who are given an achievement test at the end of a period of training. Such a study can indicate that method A is better than method B without indicating what reading consists of or what the characteristics of the learning process are.

It is suggested that the major cognitive repertoires that are (or are not) acquired in education are personality repertoires. They are as important to human adjustment as any of the other personality repertoires and require the same analysis and understanding. Reading may be considered a personality repertoire that some people have and some do not and that can be acquired to different levels of development. Mathematics is an exceedingly complex personality repertoire with various subrepertoires. Chemistry, history, physics, and so on may be considered as (mainly) language-cognitive personality repertoires composed of various interrelated skills and founded on the more basic personality repertoires.

Understanding man's cognitive characteristics depends upon understanding such cognitive repertoires. It is suggested that it is not possible to understand originality and creativity in any cognitive realm without a knowledge of the cognitive repertoires involved (see Staats, 1971a). It is important that educational psychology begin systematic, analytic research on its basic personality repertoires, as a foundation for more advanced concerns. It is suggested that behavioral educational psychology must involve directing itself to search for an understanding of the major intellectual tasks it sets for its subjects. Only on the basis of specific detailed analyses of these intellectual repertoires will principles and procedures (and a general concept of education) be provided for producing the repertoires efficaciously, for the individual and the society.

Other personality repertoires are important in education, for example the A-R-D system, sensorimotor repertoires, and so on. It is evident that comprehensive coverage is not possible here, however. Rather than attempt to describe a little about several personality repertoires, the focus will be on the description of one language-cognitive repertoire that is involved in all of education—reading.

## A THEORY OF READING

A word should be said about methodology. Traditional educational research that compares teaching materials provides some important information but it cannot provide certain types, including detail concerning the learning process. For this purpose detailed observations of the learning process with individual children are necessary. Behavior modification (or the social learning approach), it may be added, shows the importance of reinforcement in learning educational materials but also does not provide that necessary detail. Neither type of research has indicated what the personality repertoire of reading is or how it is learned in specific terms. Neither type of research has dealt with the beginning stages of reading, the cognitive skills on which reading is based, or what constitutes an accomplished reader, and the like.

The theory of reading to be presented includes behavior modification research and traditional educational data. In addition, however, it is heavily contributed to by what has been called experimental-longitudinal research findings (Staats, 1968b; Staats, Brewer, and Gross, 1970). This research has involved detailed observations of the learning process with individual children over extended periods of time. In the formal research every stimulus presented to the child was recorded, as was every response, every reinforcer, and so on. With the detailed observation of an individual child learning over a long period, the process can then be seen in intimate detail. For example, it is commonly said by professionals in education and psychology as well as parents that at least some children can acquire reading skills, such as alphabet reading, on their own, without training. When the process is made evident by experimental-longitudinal research, the patent impossibility of the common conception is shown very clearly. A long learning process is involved which must include many learning trials for every child. These may take place informally, perhaps where the child requests the information, and in this way be unnoticed. But the learning trials are necessary.

A good deal of the present theory of reading, it may be added, originated with experimental-longitudinal research the author made upon his own children (see Staats, 1968c). This made it possible to follow the reading learning process from the beginning until the complete repertoire had been acquired. The early aspects of the learning with the author's children led to more formal tests of certain aspects of the analysis with additional children. The entire reading repertoire should be studied further in this manner, however.

BASIC READING REPERTOIRE. In contrast to most approaches, it is suggested that reading is not a single process, to be explained by a single concept or principle. Reading is a complex repertoire, composed of subrepertoires. Different learning conditions and principles are involved at different stages of the learning process. Reading has a cumulative-hierarchical learning history, and the following account illustrates these principles. Reading also serves different functions in the individual's adjustment, involving different principles, and this is as important as understanding the acquisition of reading. The several repertoires that compose the reading repertoire will be described.

*The letter discrimination repertoire.* To the expert reader it seems easy to differentiate the letters of the alphabet. The letters appear to be quite different from one another. It is not realized that this is the case because of the expert reader's extensive training. Actually, in the world of visual stimuli the letters of the alphabet are quite similar to one another. If one mixes the letter A with a group of pictures the child can name and presents them one at a time to a young child, telling him the name of A when it appears, the child can soon learn to say the name of the letter. The child can quickly be trained to discriminate the A from the other pictures, and it may appear as though he can read the letter. However, if any other letter of the alphabet is shown to the child it will be seen that he will also call it an A. His discrimination learning of the letter A from the other letters is very incomplete.

As stimuli in the world, all the letters are highly similar. Yet, in English, the child has to learn a different response to each of the 26 letters of the alphabet, in both upper- and lower-case letters. By itself this is an imposing task, the difficulty of which can only be appreciated when a naive young child is taught the repertoire. Many learning trials are necessary before the child learns to name the letters of the alphabet (Staats, 1968c; Staats, Brewer, and Gross, 1970).

*The attention-discrimination repertoire.* The letters of the alphabet are not stimuli that in themselves have any significance. They are not animate and do not produce events of consequence. They have no intrinsic rewarding properties, and they are not cues that lead to other rewards or allow one to avoid aversive circumstances. They are thus not stimuli that the child will learn to attend to and discriminate on his own, without extensive learning experiences. It has been suggested, however, that discriminating the letters is basic to other aspects of reading acquisition (Staats, 1968b).

> [T]he fully functional skills of an expert reader . . . begin with the acquisition of the necessary attentional and discrimination skills. (Staats, Brewer, and Gross, 1970, p. 76)

> To discriminate the first few letters of the alphabet requires in many cases a set of attentional and scrutiny skills the child has not yet acquired. . . . [O]ne of the most difficult parts of the reading learning task—complex as it is—is that of learning to read the first few letters of the alphabet. After the child acquires this small reading repertoire, and the attentional and discrimination skills involved, he is able to learn new vocal responses more rapidly under the control of other highly similar, highly abstract, visual stimuli in the process of acquiring a full reading repertoire. It is suggested that this is what underlies (explains) the fact [Bond and Dykstra, 1967; Chall, 1967] that having an alphabet-reading repertoire is the best predictor of reading success when the child enters school. (Staats, Brewer, and Gross, 1970, pp. 26–27)

Samuels (1971a, 1971b) has a similar analysis as the basis for a theory of reading, indicating that various studies support this view. For example, Lahaderne (1968) found a significant correlation between attention and reading achievement with fifth-grade children. Noting that whether attention skills give rise to reading achievement or the converse cannot be ascertained from this study, Samuels describes an experiment by Turnure and himself in which children were employed as subjects before they had learned to read. Again, the child's attentional skills were predictive of how well he could learn to read.

Furthermore, training to attend to the distinctive features of letters appears to aid in alphabet learning (Jeffrey, 1958; Muehl, 1960). Pick (1965) trained kindergarten children in noting the distinctive features of letterlike forms. These children made fewer errors on a transfer task in which the stimuli contained the distinctive features. One study trained one group of kindergarten children to discriminate words and another group to discriminate the letters in the words (C. K. Staats, A. W. Staats, and Schutz, 1962). The children then had to learn the names of the words, and the first group did so more readily than the second. This suggests that discrimination of words, versus discrimination of the letters of which the words are composed, does not produce attention to different distinctive features. In addition, a study by Heard and Staats (see Staats, 1968c, pp. 241–262) showed that attention to the task of learning to discriminate letters improved with retarded children when this behavior was reinforced. Two studies (Staats, Finley, Minke, and Wolf, 1964; Staats, Minke, Finley, Wolf, and Brooks, 1964) employed a matching task to train children to discriminate similar letters from each other and to learn to read (name) the letters. In this procedure the child was shown a letter stimulus and then had to select from several alternatives the one that matched it. Samuels (1973) used this matching procedure to train kindergartners on the distinctive features of the difficult-to-discriminate letters *b, d, p, and q.* The children were then trained to name the letters and did so more readily than children who had not received the distinctive-features discrimination training.

Both the attention-discrimination skills and the ability to name the letters of the alphabet are basic behavioral repertoires. But the reading learning task involves much more.

*The unit reading (grapheme-phoneme) repertoire.* A child cannot be presented in his training with every word he will later be expected to read. It would be a virtually impossible learning task. This is not necessary, however. If the child learns to sound out letters and syllables, the learning task is immeasurably simplified. Then, when the child later encounters a new printed word, he can sound it out syllable by syllable and learn to read the whole word on his own.

Learning the letter sounds represents a very complex task, in the

English language especially. The vowels, and some consonants and consonant combinations, may require different responses in different words. For example, the *a* as a stimulus must elicit a different response in the word *fat* than in *fate* or *father*. This is thus a case where a *single-stimulus–multiple-response* S-R mechanism must be learned—the letter *a* being the stimulus which must come to elicit several different vocal responses, depending on the context. Moreover, the same vocal response in other cases must be elicited by different stimuli, which results in the *multiple-stimulus—single-response* S-R mechanism. For example, Soffieti (1955) has indicated that the vocal response "oo" must be elicited by nine different letters or letter combinations, for example, as in *rule*, *school*, and *group*.

Because the acquisition of so many complex S-R mechanisms represents such a difficult learning task, there have been various attempts to regularize English pronunciation with special alphabets. The most widely known system is called the *Initial Teaching Alphabet*, or i.t.a. (Downing, 1964, p. 15). Research suggests that this alphabet is learned more readily, although there are some problems with transfer to the regular alphabet (Downing, 1964). It has been suggested that a behavioral analysis of this aspect of learning to read, along with systematic research, could provide a basis for definitively composing a teaching alphabet (Staats, 1968c).

Using the standard English alphabet, however, there are various strategies of teaching. Some methods are based upon teaching the child to read meaningfully from the beginning, with less emphasis upon teaching the child grapheme-phoneme (or letter-sound) units. These are called whole-word methods. Detailed observations in teaching a single child have shown that "Although the child can easily learn a limited single word reading repertoire, it becomes evident very quickly that it is not possible to increase the repertoire beyond a primitive level in this manner (Staats, 1968c, p. 476). Samuels adds similarly that, using the whole-word method, classroom teachers "report rapid learning of a small sight vocabulary followed by a plateau" (Samuels, 1971a, p. 10).

Another method for training children to read that involves direct training is saying the sounds of the letters (Dechant, 1964). Sometimes this method involves the child learning to sound single letters, in which case problems may be produced. For example, the child may be trained to respond with the "p" sound when a *p* is presented, to the *a* and *t* with the appropriate individual sounds. However, on the basis of this training alone, the word *pat* will not control the word response of saying "pat." The author has observed this in individual work with a child. As a matter of. fact, the sounding out sequence is a very difficult skill to train using these particular types of grapheme-phoneme units, since the three phonemes vocalized discretely in a series do not produce a sound much like that of the word *pat*.

This difficulty of learning to sound individual letters that do not have the same sound as they do in a word may be overcome, however. The linguist Leonard Bloomfield suggested that the child can be presented with systematic series of words such that he learns letter-sound units (Bloomfield and Barnhart, 1961). This method has been employed and tested with individual children (Staats, 1968c; Staats, Brewer, and Gross, 1970). The child can learn units through suitably arranged whole-word training. He can, for example, learn the "puh" sounding out unit to the letter *p* through training in which he is presented with trials in reading the words *pat, sat, hat, fat, mat,* and *nat.*

Moreover, the research has shown that the repertoire learned through this type of training will generalize to new combinations of letters. For example, let us say the child had learned the "puh" response to *p* and had learned an appropriate response to the letters *an,* in entirely separate training. If the letters were then presented together, in the word *pan,* the child would read the word correctly on *first* presentation (see Staats, 1968c; Staats, Brewer, and Gross, 1970). This incidentally, is a very clear empirical demonstration of the originality mechanism described in Chapters 5 and 6, where new combinations of stimuli elicit novel response combinations. The learning principles and procedures involved in this crucial type of human learning could, and should, be studied with precision and detail. They are basic in learning to read as well as in other important personality repertoires.

*The sounding-out repertoire.* In addition to the pronunciation mechanisms described above, the child has to learn other skills to be able to sound out more complex words. As has been indicated, in English a particular letter in different word contexts must elicit different vocal responses (sounds). When the child is presented with an unfamiliar word which must be sounded out, it may not be apparent which vocal response will be correct. For example, the word *familiar* has four vowels in it, each of which comes to control several reading responses (a hierarchy of responses). The child who then sees this word for the first time and who has the necessary unit reading repertoire may still not be able to read the word. He may say, for example, "fay-my (rhymes with eye) -ly (rhymes with eye) -air." Or he may make other combinations of unit sounds that do not constitute the usual pronunciation of the word.

The word pronounced by "fay-my-ly-air" will not be meaningful or functional, nor will it match any known word in the child's repertoire. In order for the word to be successfully sounded out, he must repeat the process while varying the vowel sounds he makes to the letters. He must do this until he makes a combination that approximates the word sound, which is then known by him.

For the child, sounding out thus is composed of several skills: (1) first looking at the letters and syllables successively, (2) then emitting one of the responses from the hierarchy controlled by each letter or syllable, (3) listening to the word stimulus he has in this manner produced, (4) comparing this stimulus to those with which he has had experience, and (5) failing to achieve a match with a familiar word, repeating the process by emitting different responses to the letters and syllables until a match is produced. Research has yet to be conducted on this phase of reading learning or on the types of training that would expedite the learning involved.

*The whole-word reading repertoire.* It is quite apparent that the accomplished reader performs in the above manner relatively infrequently. After much reading experience, the accomplished reader advances to the stage where he has learned a repertoire of words directly. The whole words immediately elicit whole-word responses. It is not necessary to sound out the word. When this reading repertoire has become very large the individual will rarely encounter a word to which he does not have a well-learned reading response.

This might be employed as a rationale supporting the whole-word teaching method. It should be noted, however, that many reading responses will have been learned on the basis of the child's sounding-out repertoire. Using the previous example, the child may have laboriously sounded out the word *familiar*. When he has done so, however, he has provided himself with the learning trials necessary to learn a new whole-word reading response—that is, he now says the word while looking at its whole printed form. In the future the word will come to elicit the whole-word response more immediately, and on each future occasion the learning will be strengthened.

*The general configuration of words as controlling stimuli.* It has been indicated that the child's reading response comes under the control of the general configuration of a word (Dechant, 1964, p. 190; Staats, 1968c, p. 480; Samuels, 1971a).

> . . . [C]hildren often are taught to use tricks to identify the word, but these tricks betray them. For example, the word *purple* is identified by the one ascending and two descending letters. . . . Unfortunately, the word *people* has the same visual configuration and the ascending and descending letters fall in the same place. (Dechant, 1964, p. 191)

It is suggested that configuration is wholly functional for the accomplished reader, because he has already learned the specific letter discriminations very well and can utilize them when the need arises. With the child who is learning to read, however, configuration identification may serve as a crutch that prevents him from learning the attentional-discrimination skills and knowledge of individual letters and letter-sound units. Studies should be conducted comparing beginning and skilled readers in configuration identification.

*The phrase reading repertoire.* As was indicated in Chapter 2, traditional learning theories deal with simple stimulus-response events. This does not mean, however, that the stimulus or response events are always simple and elementary, in language or any other area. Larger responses can be composed of smaller units and, through training, come under the control of larger combinations of stimuli.

In the realm of reading, for example, the child may learn vocal responses to single letter stimuli, but later he may sound out single letters which produce a whole-word reading response. The combined letter stimuli (the word) can then come to control the combined response units (the vocalized word), as a unit.

The same is true on the word level. When combinations of word stimuli occur together a number of times, the combinations as a unit may come to elicit the combination of word responses. When this has occurred, one look at the phrase will control the unit response (composed of several word responses).

> The reading is then no longer dependent upon looking at each word stimulus. *Coca Cola* and *Pepsi Cola* may be used as examples of multiple word stimuli that come to be responded to as a unit by most Americans, as are the combinations *United States, General Motors, stormy weather,* and so on. It might be suggested in fact that there are many such phrase stimuli that control unit responses—and that there is probably a dimension of strength of control. Thus, the unit *Coca Cola* probably has more unit control than would *The blue sky.* It may be suggested that the extent to which a group of words had come as a unit to control a unit reading response could be tested by presenting the stimuli tachistoscopically—that is, for very brief intervals. Thus, it would be expected that it would take longer to recognize *shot tend* than it would to read *Coca Cola,* although both combinations of stimuli contain the same number of letters. (Staats, 1968, 482–83)

In the same way that the configurations of a single word come to control reading responses, the configuration of phrases in many cases is distinctive and can serve as a controlling stimulus for reading responses.

*Word associations and reading.* It has been suggested that verbal response associations or sequences are involved in reading.

> . . . [A]ssume a strong TABLE-CHAIR word association for children. In constructing a reading program, if CHAIR was to be introduced to the child and he already could read TABLE AND, it might be wise to introduce CHAIR in the phrase TABLE AND CHAIR. The two preceding words would contribute to the stimuli tending to elicit the vocal response CHAIR and make it a more probable response. (Staats, 1963, p. 463)

Rouse and Vernis (1952) demonstrated that when word associates are presented tachistoscopically (for a very brief period), recognition of the first of two words aids in recognition of the second word. Moreover, the stronger the word association between the two words, the stronger is this effect.

Samuels (1966) has directly tested the hypothesis that word associations can be involved in the child's learning to read:

> Pairs of words were learned by 44 1st graders during word-association training. They were then given reading training on the same or different pairs of words and tested for word recognition and speed of recognition on the 2nd word of each word pair. In the facilitation treatment, where word pairs were the same for word-association and reading training, the mean recognition score was 1.46. In the interference treatment and neutral treatment, where word pairs were different for word-association and reading training, the means were .84 and .82. In the control treatment, where no word-association training was given, the mean was .34. (p. 159)

The individual's extensive word associations may be considered to play a role in his reading facility. Reading material for which there are relatively strong associations is easier and faster than where this is not the case. For example, in reading such things as *Now is the time for all good men to come to the aid of the party*, or *I pledge allegiance to the flag of the United States of America*, each word does not have to be attended to. One look is enough to elicit the necessary reading responses. It is suggested in general that if the material that is read coincides with the individual's words associations, as formed through his past language experience, then it will be read relatively easily and rapidly. The reader will not be as dependent upon the word stimuli themselves. The reader's eye movements will cover more words per stop. When the material does not so readily elicit word responses of an associative nature, reading will be more laborious and slow.

READING FUNCTIONS. Words have various functions that can be referred to as word meaning, as was indicated in Chapter 5. These functions primarily concern the responses that words as stimuli elicit in the individual. Thus, some words elicit conditioned sensory responses, and some words elicit emotional responses. These words will also serve as reinforcing stimuli for the individual's other instrumental behaviors. Other words will themselves elicit instrumental responses from the individual. Some words elicit word associations, and so on.

These various functions of words can occur whether the words are presented by another person or the individual says the words to himself. These various functions can occur, also, when the individual reads the words; for example, the person may experience emotional responses or sensory responses (images). Some words will elicit instrumental responses —like a note left by a wife asking her husband to put something in the oven when he gets home from work. The various functions that words have in interpersonal interaction can also be produced through reading. The importance of reading lies in its functions, a topic that is generally overlooked in theories of reading.

It is suggested that at the beginning reading is a two-stage process. That is, the printed word stimuli elicit in the individual the reading of the words. Then the vocal word responses (which may be implicit) perform the "meaningful" functions that they have for the individual. The individual has to say the words to himself in order for the words to have their functions.

A learning analysis would suggest, however, that the functions of the vocally presented words should be conditioned directly to the printed words. When this has occurred the responses will be immediately elicited by the printed words. As an example, the written word *cancer* would be expected to elicit in the accomplished reader an immediate negative emotional response. This could occur at the same time as the pronunciation of the word, which would make meaningful reading occur more rapidly than it did when the word first had to be pronounced.

It would seem that the expert reader does not even make *all* the word responses to the printed stimuli he covers in reading. He makes the meaning responses to the printed stimuli so rapidly that he has moved on to the next word stimuli before having the time to vocalize some words. This stage of reading is even more pronounced when it is realized that the expert reader does not even have to look at all the word stimuli he covers in a passage, as the next section will indicate.

It may be added at this point that the above analyses should be tested by research. They stipulate cause and effect principles and conditions and can serve as the basis for experimental research, as has been demonstrated in a number of cases. However, much remains to be empirically tested, and such knowledge is requisite for basic and applied reasons.

REINFORCEMENT AND READING. The author has introduced the concept of extrinsic and intrinsic reinforcement into the context of child learning, including reading. By intrinsic reinforcement is meant the case where the behavior produces the reinforcement within the activity itself. The child who reads solely for the "pleasure" (A-R-D effect) of reading constitutes an illustration. Extrinsic reinforcement refers to the case where the behavior does not itself produce the reinforcement but where there are additional sources of reinforcement that maintain the behavior. A token-reinforcer system constitutes an extrinsic reinforcement source. It may be indicated further that in the case of extrinsic reinforcement, the reinforcers employed may be natural to the situation, or the reinforcers may be artificial in the sense that they are introduced specifically to maintain the individual's behavior. Thus, for example, use of recess as the back-up reinforcer in a school token-reinforcer system involves a natural reinforcer. The use of money, on the other hand, is not natural to education as we know it today.

However, the behavior involved in acquiring a complex repertoire such as reading is in the beginning not intrinsically reinforcing; it is work. Thus, some type of extrinsic reinforcement source must be introduced to maintain the behaviors required for learning. Schools rely upon coercive attendance for most children—which reduces the necessity for any reinforcement—in conjunction with such reinforcers as teacher and peer social approval, competition, learning for its own reward, parental approval, negative reinforcement (relief from disproval), and so on. These are all learned reinforcers and are ineffective for many children. They may be considered to be natural reinforcers, because of our traditional practices, but they are extrinsic reinforcers like any others.

The child's learning must be maintained for a great number of trials by such reinforcers. This is a very shaky foundation for education, as has been indicated. The behavior analysis and the research that has been summarized herein show that this is a primary area of breakdown of our educational system, in areas other than reading as well. That is, for many children the reinforcers are not adequate, and their learning behaviors are not maintained. From that point on they cease to learn. In fact, they will usually learn behaviors that are antilearning skills (see Staats, 1971a).

*Extrinsic reinforcement in reading.* To continue, however, the first stage of reading learning involves extrinsic reinforcement, as does the second stage. After the child has acquired some of his reading repertoires, not even to their fullest development, he has a behavior that will be maintained in brief activities. For example, the individual through reading may perform behaviors that allow him to avoid aversive circumstances and gain positive reinforcers. The child who can read the instructions on putting together a model car or airplane receives reinforcement—not from the reading itself, but from the behaviors the reading elicits. The individual may also read because he acquires new verbal response sequences (knowledge) for which he is reinforced. Many students read, not for the reinforcement involved, but because the behavior is maintained by success in taking course examinations. Some people read current events because this provides them with conversation pieces in social interactions, again not for the intrinsic reinforcement of the material.

*Intrinsic reinforcement for the expert reader.* The laborious, repetitive task of learning to read—involving many thousands of abstract learning trials—is not reinforcing in itself. This has been shown experimentally (Staats, Finley, Minke, and Brooks, 1964; Staats, Staats, Schutz, and Wolf, 1962). There is, however, an abundance of naturalistic evidence that indicates that the behavior of reading *becomes* intrinsically reinforcing for some individuals. For example, accomplished readers will perform the behavior in the absence of extrinsic reinforcers. Moreover, such a reader may perform arduous responses of various kinds, the behavior maintained by the activity of reading as a reinforcing event. He may walk a long way to a public library, move up and down rows of books, take out and examine various books, select several to check out, walk home with the books, and finally select and proceed to read one until forced to abandon this activity by some exigency of home life. The long sequence of behaviors is maintained in strength by the word stimuli in the book.

It is important that a theory of reading be able to indicate how it is and why it is that the word stimuli in a book become such potent reinforcers for some people. It is only when this stage of reading is reached that the individual will read voraciously. The importance of this stage cannot be indicated in full here. It will only be said that much of the child's and later the adult's facility with language, his reasoning ability, and so on, comes from the amount of reading he does. The child who reads a great deal, other things equal, will develop a greater general language facility—as indicated by intellective tests, writing ability, learning ability in various subjects involving language learning, and so on. And the amount that a child reads—outside of his school work—will be a direct function of how intrinsically reinforcing this activity becomes.

*Reinforcement through emotional meaning words.* In the present theory of reading (Staats, 1968c), it has been suggested that reading is intrinsically reinforced primarily by two types of meaning responses. As was indicated in Chapter 5, single words may be reinforcing because of the emotional responses they elicit. Finley and Staats (1967), for example, showed that positive emotional words presented after a motor response would increase the frequency of that response, while negative emotional words would decrease the frequency of that response. In addition, it must be added, words that in themselves may elicit no emotional response, and thus will not be reinforcing, may in combination have those functions. Thus, the words *received, the, you, place, contest, top,* and *in* as individual words do not elicit strong emotional responses. Together, in the sentence *You received top place in the contest,* the words may elicit a very positive emotional response. And reading the words will be seen to have strong reinforcing value. The individual who receives the notification in writing may read it over several times.

It may also be indicated that single words which themselves elicit negative emotional responses may in certain combinations elicit positive emotional responses. The word *murder* by itself will elicit a negative emotional response in most people. In the context of a detective story, however, it may elicit a positive emotional response. This is another example of the variation in units that are effective in eliciting a response in the individual. In linguistics it has been traditional to consider the morpheme to be the smallest language unit that is meaningful. A letter may be meaningful. For example, the *s* in *boys* conveys the meaning of plurality. A word may be meaningful, as has been described herein for emotional meaning. It should be noted, however, that a combination of words may be the unit that elicits an emotional response which the separate words themselves will not elicit. Again, units at one level may be combined to yield units at another level of complexity.

*Reinforcement through image-eliciting words.* The manner in which reading is reinforcing may be derived from another type of meaning that has been described. To elaborate, it has been suggested that printed words may elicit conditioned sensory responses (images). It has also been stated that several words, each of which elicits a conditioned sensory response, can when presented together elicit a composite conditioned sensory response or image (Staats, 1968c; see also Chapter 5). Thus, image-eliciting words can be brought together in different combinations to produce different composite conditioned sensory responses. Stories may be considered in terms of this analysis. Series of words that elicit conditioned sensory responses are presented, and what is elicited in the audience is a running sequence of conditioned sensory responses like those the actual objects and events would elicit. When these objects and events as actual stimuli are reinforcing, the story (the word stimuli) that elicits conditioned sensory responses like those the actual objects and events would elicit will also be reinforcing.

> For example, if the stimuli of a forest path leading to a grassy area by a brook elicits sensory responses that are reinforcing, then the words *forest path leading to a grassy area by a bubbling brook* will elicit a composite conditioned sensory response (image) in the expert reader that is also reinforcing. This is not to say that the reinforcement value will be large, certainly not necessarily as large as the actual stimulus objects. (Staats, 1968c, p. 532)

Thus, many stimulus objects and events are reinforcers for the individual. Words that have been paired with these stimulus objects and events will come to elicit conditioned sensory responses or conditioned emotional responses that have a reinforcing function. Storytelling is at least in part the art of putting together words which will elicit such conditioned responses and thus serve to reinforce the behavior of listening to or reading the story. This, it is suggested, involves the manner in which the words are strung together to elicit the most vivid and the most reinforcing responses. By way of example, *pornography* is a good illustration of word stimuli that elicit both conditioned sensory responses (images) and conditioned emotional responses that have observable physiological components.

*Reading for pleasure: Reinforcement in excess of effort.* It should be noted here that there is a corollary principle that is implicit in the basic principle of reinforcement. *An instrumental response will be maintained* when the aversiveness of the effort involved in the response is less than the reinforcement that the response produces. This is an important corollary in understanding the stage of pleasure reading, and another human learning principle.

Reading for pleasure does not take place until the act of reading reaches a certain (low) level of effortfulness and until it produces a relatively high level of reinforcement for the effort and time expended. The skilled reader does not need to stop to learn to sound out many new words, he does not need to consult the dictionary often, and he will not have to pronounce words out loud before they elicit meaning responses. He will read rapidly and effortlessly. This will enable him to glean a maximal amount of reinforcement per unit of time (activity), with little effort. The only thing necessary then to insure that the child reads a good deal for pleasure is to eliminate the competition of other more reinforcing activities when the child is supposed to be reading. The reading behavior will then be maintained by intrinsic reinforcement.

A frequent question that arises in the context of using extrinsic reinforcement in training children concerns the possibility the child will later depend upon such reinforcement. The present discussion has thus been introduced, in part, to indicate that a complete behavior analysis must in many cases *indicate the manner in which the behavior, once acquired, will be maintained by intrinsic or natural reinforcement.*

CUMULATIVE-HIERARCHICAL LEARNING, READING, AND LANGUAGE. Reading is learned only over a long period. When all the skills involved in reading are considered, the time required for reading acquisition, under very good circumstances, amounts to eight to ten years. Millions of different stimulus-response occurrences are involved. While this seems to be a very evident statement, ignoring the extensive learning involved in the acquisition of reading has led to a good deal of misunderstanding. To fully understand reading it is necessary to be concerned with learning throughout the period of acquisition. Moreover, this analysis indicates why long-term methods of experimentation are necessary to study this type of behavior (Staats, Brewer and Gross, 1970), because it is a very complex, extensive type of learning.

It may be added that reading acquisition is a good example of cumulative-hierarchical learning. For example, the attentional and discrimination skills acquired in the process of learning to read letters are basic to the later tasks of learning grapheme-phoneme units and in learning to read whole words and so on. The grapheme-phoneme units are basic to acquiring a sounding-out repertoire—although to some extent (as is the case with some of the other repertoires) these two repertoires are acquired concurrently. The grapheme-phoneme repertoire and the sounding-out repertoire are to a large extent basic to the whole-world reading repertoire, and this repertoire is in turn basic to the repertoire developments that follow. All of the repertoires are basic to the "reading for pleasure" stage of reading. Although it is not possible to cleanly separate the development of all the repertoires, a very extended cumulative-hierarchical learning process is involved.

Moreover, like the other personality repertoires, the reading repertoire has causal status as well as being a dependent variable. Much of the individual's later learning of all kinds depends upon his reading repertoire. The manner in which a variety of new cognitive, social-emotional, and sensorimotor skills are acquired through reading has been outlined (Staats, 1968c), the study of which should be a very large area of research in itself. Again, it is suggested that the behavior analysis of a personality repertoire can serve as the theory for research studies of both a basic and applied nature.

The specific point is, however, that reading is involved in a very extended cumulative-hierarchical learning process that leads to further learning and is dependent upon previous learning. In this context it is important to indicate that reading depends entirely upon the previous acquisition of the language repertoires. These are the basic behavioral repertoires upon which reading acquisition and reading function are based. Reading learning cannot proceed without the language repertoires, and it would be nonfunctional unless the child has the language repertoires. It is not possible to understand reading without understanding the language skills on which reading is based and of which reading is composed. The theory of reading, as this suggests, demands a full theory of language learning and function. The analysis also indicates that a child who does not have a normal language development will not profit from reading training, or will do so to a lesser degree depending on the extent of his language deficit. The analysis indicates that it is misguided to attempt to teach a child to read who does not have a functional language, as has been done with some mentally retarded children.

READING PROBLEMS: DYSLEXIA AND LEARNING DISABILITY. The value of the approach to education and educational psychology can be exemplified in the area of reading problems. For many years there has been great misunderstanding of what reading is, how it is acquired, and consequently to what to attribute cases of reading failure. The biological paradigm has stimulated the search for biological defect when the child has difficulty in learning to read. It has been suggested that there is congenital word blindness (Hinshelwood, 1900, 1917), poor associational learning ability (Otto, 1961), poor memory (Senf, 1969), organically determined brain defect (Critchley, 1970; Rabinovitch, Drew, DeJong, Ingram, and Withey, 1954). Various terms such as minimally brain damaged (MBD), neurologically impaired, learning disability, alexia, dyslexia, and so on all are concepts of brain defects that produce defects and deficits in learning to read.

The biological concepts of reading defects have attempted to account for the fact that the child can have normal language development and normal intelligence and yet fail to learn to read. This has led to the suggestion that there are very specific, selective brain defects. There are reasons, however, why almost all children learn language and a good percentage fail to learn to read, as the author has indicated (Staats and Staats, 1962). This occurs because of the differential learning conditions involved, not because there is a specific brain defect that prevents the child from learning to read. Analysis of language and reading reveals they involve the same things. It has been suggested definitively (see also Staats, 1968b, 1973) that *a child with normal language is perfectly capable of learning to read—under suitable learning conditions. The fact that he has learned language indicates he has all the abilities, and requisite skills, to be able to learn to read.* This is a definitive statement, with which many will disagree. But, as will be indicated, evidence to support the analysis has begun to be accumulated.

A major point of the present theory of reading is that thousands and thousands of learning trials are involved in reaching expertness. This can be seen clearly in the present exposition, abbreviated as it is. This fact in itself has many implications for understanding the acquisition of reading. It tells us, for one thing, that a central problem in any training program will involve the means of insuring that the child receives the multitudinous learning trials necessary. The procedures must insure that the child's attention and participation are maintained from moment to moment in the training which goes on over a period of years. Any breakdown in attention and participation, and slowing of the frequency of responding, will result in retarded learning. The analysis also tells us that there is no intrinsic reinforcement in the learning task itself. By itself, the task at the beginning is repetitive, effortful, and boring.

It is thus suggested that what has been referred to as learning disability or minimal brain dysfunction in reference to poor reading, or indeed alexia or dyslexia or any of the other terms implying internal weaknesses in the child, when the child's language is normal, involve a breakdown in the learning conditions for the child. Most usually, the child has simply not been presented with the conditions in which his participation has been insured in the innumerable learning trials he needs to acquire the various repertoires. When measures are taken to insure receipt of the learning trials, the most recalcitrant of learners make good progress (Staats, 1968c; Staats and Butterfield, 1965; Staats, 1967; Staats, Minke, and Butts, 1970; Ryback and Staats, 1970). Ryback and Staats showed, for example, that mothers could successfully train their own dyslexic children when the token reinforcement was employed in combination with an appropriately arranged set of reading materials. One of the children had been diagnosed as mentally retarded, one as an emotionally disturbed child, and two as learning disability cases.

Harris (1974) has tested both the behavior analysis of dyslexia and the previously described reading method using token reinforcement. Her subjects were six children specifically diagnosed as dyslexics, as well as six nondyslexic children with reading problems. Half of the children in each of the two groups were given traditional remedial reading training, and the other half received the motivation-activating methods using token reinforcement (Camp and Staats, 1970). The children who received the 40 hours of training with the reinforcement improved in their reading achievement a mean of 16.45 months. The children treated in traditional remedial methods advanced only 1.28 months in achievement. Moreover, there was very similar advancement between dyslexic and nondyslexic children.

> The fact that these procedures exerted equal control over dyslexic and non-dyslexic retarded readers suggests again that the diagnosis of dyslexia according to perceptual/IQ relationship, specificity, and the dyslexia syndrome is not meaningful in terms of reading remediation and underlines the importance of increased motivation, attentional behavior, and the immediate reinforcement of effortful reading behavior to produce reading achievement. The reading deficits traditionally viewed as due to genetic disposition, MBD and/or neurological, maturational lag are more productively viewed as learning history deficits. (Harris, 1974, p. 98)

These conclusions were backed up by other data. The reinforced subjects were given a perceptual skill test before and after training, and there was an increase in 15.84 months in perceptual skill for these subjects versus only a 2.20 month increase for the subjects trained in traditional ways. It was interesting that the dyslexics who were reinforced showed a greater increase in their perceptual skills than the non-dyslexics, which does not support the conception that the dyslexic's problem is a genetic perceptual defect, as has been suggested.

A great deal more research is needed in this area, based upon a specific behavior analysis and training methods that insure that the subjects actually receive training trials. Traditional classroom methods do not insure this. There are many ways the child can avoid learning trials— in fact it has been suggested that there are antilearning skills that children develop to avoid anxiety-producing and boring learning situations (Staats, 1971a). The findings thus far suggest that when these obstacles are removed, learning proceeds normally.

This discussion does not state that there are no biological conditions that hinder human learning like that involved in learning to read. Disease, injury, and genetic impairment that damages the brain can produce such conditions. It is suggested, however, that the learning of language involves the same principles and types of stimuli and responses as learning to read. A child who has learned normally the former has shown he is prepared to learn the latter. He has shown also that he has the necessary foundation in intellective basic behavioral repertoires to do so. This must be considered largely an hypothesis at this point, and one with wide implications. It requires much additional investigation, but there is theory and research methodology available with which to conduct such investigation.

## Conclusions

This chapter was entitled educational psychology because the material considered herein has special significance for this field—not as a suggestion that the topics of the present chapter delimit the field. Actually, the topics of all the preceding chapters are relevant to educational psychology.

As has been indicated, this is also true with respect to the personality repertoire of reading. This repertoire is only one of those that are important in the field of educational psychology. The various repertoires produced in educational institutions should be subjected to analysis. In addition to reading, cumulative-hierarchical analyses of number concepts, from original discrimination of numerosity through counting, addition, and multiplication, have been given (Staats, 1963, 1968c), and experimental work has commenced on the theory (Staats, 1968c; Staats, Brewer, and Gross, 1970). Similar analyses of counting have been made by Wang Resnick, and Boozer (1971). Schimmel (1971) has also employed the analysis of counting in working with two- and three-year-old children.

It is suggested that the paradigm provides a basis for a wide range of research, theory, and applied work in education and educational psychology. This is not limited to the language-cognitive repertoires; it should also include reference to the emotional-motivational system (of the individual and of the educational institution), as well as the other repertoires that have been mentioned.

# *Social behavioral*
# *psychometrics*

To A LARGE EXTENT the field of psychological measurement has been closely identified with the paradigm based upon the traditional concept of personality. In this view inner mental and personality structures or processes are considered to be the determinants of human behavior. Most personality tests were constructed as measuring instruments to index the internal traits of the individual. In the same way that the variation of the height of mercury in a thermometer is only an index of atmospheric temperature, not the temperature itself, so the psychological test, which makes observations of the individual's behavior, is not considered important for these observations. Rather, the measurement of behavior is important as it gives information about the underlying personality trait or process.

The field of psychological measurement is thus frequently not focally concerned with the behavior itself. The fact that on an intelligence test a child can or cannot give the names of several objects is not centrally important, according to this view. What is thought to be important is that the ability to name the objects reflects the quality of the child's internal intelligence. From this view, as a matter of fact, it is naive to be concerned about the behaviors the items of a test actually measure.

The field of psychological measurement is also based upon the biological paradigm. Thus, a strong assumption in most cases is that the internal states of personality are primarily or largely laid down in the individual's inherited makeup. It has been traditional, for example, to consider one's intelligence to be determined by biological quality. The same is thought to hold for various personality traits, abilities, motivational characteristics, and so on.

Along with, or perhaps because of, these points of orientation, the field of psychological measurement has had diagnosis as one of its central characteristics. Thus, tests classify and categorize individual in a static manner. If the child is judged to be intelligent or retarded, for example, he will be segregated with other children similarly categorized. An implicit assumption, ordinarily, is that individuals in a psychological category have the same underlying nature.

Following this orientation, the classification by intelligence test will not include specific directives concerning what to do to change any deficiencies of the child. The orientation toward measuring the unitary personality trait of intelligence does not provide a basis for consideration of the specific behaviors involved or methods of changing the behaviors. This is true, also, of personality tests used to clinically diagnose people for behavior disorders. Tests may be given to an individual that categorize him as a passive-aggressive personality, paranoid schizophrenic, sociopathic, and so on. However, these again are global classifications, which presumably index a unitary personality trait. They do not detail the behaviors that constitute that "personality," the actual behaviors that lead to the classification. As a consequence, such diagnoses do not yield specific prescriptions concerning what to do to modify the individual's behavior benignly in order to resolve his problems.

In sum, the field of psychometrics does not generally obtain knowledge concerning (1) what the various aspects of measured personality are *in terms of the behavioral repertoires that compose them*, (2) specifically how and by what principles these personality repertoires of behavior are acquired, (3) the functions of such personality repertoires in the general life adjustment of the individual, (4) the specific conditions involved in dealing with the original learning of the personality repertoires, or (5) the modification of them at a later date for treatment purposes.

## The elemental behavioristic paradigm and psychological measurement

As has been indicated, behaviorism and the field of learning have largely ignored psychological measurement. The search has been for general principles, uniform across individual organisms, by which the environment affects behavior. Individual variations in learning simple responses by laboratory animals can be overlooked or shrugged off as unsystematic variation. When the animal laboratory model is carried over into the human realm, the same lack of concern with psychological differences may result.

In eschewing the subjective concepts of personality, behavioristic approaches have generally rejected also the field of psychological measurement. Oddly enough, this has left elemental behavioral approaches—even those concerned with clinical psychology—in much the same position as the subjective approach of personality theory and psychological measurement. Thus, in rejecting the concept of personality, elemental behaviorism has not had a theoretical framework to provide a basis for the analysis of personality repertoires, the manner in which they are acquired, the functions the repertoires serve, or the ways in which such complex repertoires can be dealt with.

Elemental behaviorism does not provide a basis for the concern with clinical assessment related to clinical treatment. However, the foundation for behavior modification and behavior therapy has provided some interest in clinical assessment related to treatment. A few points will be made in describing the behavioral assessment procedures that have begun to develop in this field.

## Behavior assessment: Foundations and status

A prominent aspect of psychological measurement has involved concern with personality traits. To elaborate, if we observe a number of individuals' behavior we will see that while there are unique aspects of the behavior, there are also similarities among men. Some people, for example, will spend more time with other people than will others. They will do so consistently, and it can be observed that when in a situation where they have to select either a solitary activity or a social one they will more often choose the social one than will those who are more solitary.

### BEHAVIOR TRAITS OF PERSONALITY

The fact is that characteristics of behavior that are shared by large numbers of people can be described. In view of this it would seem inevitable that such descriptions would be made and that theories of why there are such behaviors would also be suggested. Hippocrates, for example, proposed that all human behavior resulted from a combination of four types, in various mixtures, each determined by the relative strength of its "causative" agent, a body humor (fluid).

In more contemporary times characteristics of behavior that are widely shared have been systematically studied toward the goal of measuring such characteristics. Thus, such investigators as Cattell (1946), Eysenck (1947), and Thurstone (1947) have been concerned with the measurement and description of traits of personality. One approach to the study of traits has been called the factor approach because the method used is that of factor analysis. In this method a large number of items are gathered together that are descriptive of human behaviors. The Guilford-Zimmerman Temperament Survey (Guilford and Zimmerman, 1949) is such a scale. Examples of such items are: "You give little thought to your failures after they are past" (No. 20), "You often feel grouchy" (No. 40), and so on. The individual is asked to answer the items yes or no, or with a question mark if he absolutely cannot decide.

The Guilford-Zimmerman was constructed by the methods of factor analysis. This means that a number of such items were given to a group of subjects, and the subjects' responses to the items were then related to one another. Groups of items that were highly related were selected for the test. In such a group of items, or factor, if one item is answered in a particular direction, then other items in that group tend to be answered in a like manner. The suggestion is that certain types of behaviors go together and thus reflect the operation of a personality trait. For example, one trait measured on the Guilford-Zimmerman is that of sociability. Items included on this factor are such as the following: "You find it easy to start conversation with strangers" (No. 44), "When you were a child, many of your playmates naturally expected you to be the leader" (No. 53), "You sometimes avoid social contacts for fear of doing or saying the wrong thing" (No. 54), and "You would be very unhappy if you were prevented from making numerous social contacts" (No. 59).

Such a cluster of related items is considered to reflect an underlying trait or characteristic. It is expected that the individual possesses this characteristic of sociability to the extent that he answers the items (30 of them) in the sociable direction. One individual might answer all 30 items in that direction, suggesting a very strong sociability trait. Another might mark *yes* to "You like to entertain guests" (No. 4) but mark *no* to "You find it easy to make new acquaintances" (No. 9). The more items answered in the sociable direction, the stronger the trait. Another trait, that of general activity, is represented by the following items: "You find yourself hurrying to places even when there is plenty of time" (No. 26), "You are the type of person who is on the go all the time" (No. 41), "You are quick in your actions" (No. 51), and so on.

There have been different interpretations of what traits of behavior are, and how they are to be explained. There is a general tendency to consider traits of behavior to be of organic, heritable, origin. For example, one of the foremost personality theorists, Gordon Allport (1937), spoke of traits in terms of mental structures and as springing from physiological processes or organic structures. Thus, as indicated above, one of the traits on the Guilford-Zimmerman is that of general activity. A common interpretation would be that some people are just naturally more active and energetic. Little consideration is usually given to the individual's past history and present circumstances in producing such behaviors, or to the motivational stimuli that maintain vigorous behavior.

Some of the investigators of measured personality traits have avoided attributing underlying explanatory mechanisms, preferring, at least on a formal level, to leave the questions open. Thurstone (1947), for example, has indicated that he prefers the approach where "No assumption is made about the nature of these functions, whether they are native or acquired or whether they have a cortical locus" (p. 57). The contemporary concern is with finding trait factors which can be employed to predict the individual's behavior for personnel selection purposes, for clinical treatment purposes, and so on.

As has been indicated, elemental behavioristic approaches have rejected such concepts as personality traits. This approach has been followed by social learning theorists who dealt with personality measurement. For example, Mischel (1968, 1971) has suggested that there are no broad behavioral traits that are general across different situations.

However, the present author was interested, from the beginning, in linking behavior analyses of personality with the psychometric approach, feeling that there were important constituents in each area. After summarizing several studies (for example, Nuthmann, 1957) demonstrating that human behavior could be affected according to reinforcement principles, he proposed the following extension to personality measurement in an early study.

> The foregoing results suggest that complex human behaviors, which we term personality, may develop according to the principles of [reinforcement]. . . . [I]t is reasonable that a life history of reinforcement for social responsiveness would culminate in a socially responsive person. In addition, the history of reinforcement for statements concerning oneself should also affect the behavior of an individual which we accept as indexing his "self-concept." . . .
>
> [S]tudies are needed to determine whether facets of personality behavior actually are acquired through [reinforcement]. Individuals behave in characteristic ways in many different situations. These characteristic ways of behaving are commonly spoken of as reflecting underlying personality traits. Tests have been constructed to measure these traits. . . . It is suggested that such measured personality traits have psychological status in that they are functionally unified classes of responses which have been . . . conditioned in the life history of the individual. Following from this, it is suggested that responses to trait items found through factor analysis are such classes of responses. . . . [S]trengthening the response to one item through reinforcement should strengthen this type of response to other items in the same trait. (Staats, Staats, Heard, and Finley, 1962, pp. 101–2)

The author's analysis constituted a beginning behavioral model of certain personality traits and provided hypotheses that could be experimentally tested employing learning principles and procedures. Two studies were thus conducted to test the possibility that subjects could be conditioned to respond to personality trait items. The personality traits employed were those of sociability and general activity, taken from the Guilford-Zimmerman Temperament Survey (Guilford and Zimmerman, 1949). Subjects were socially reinforced for answering the items in the direction of indicating sociability. When blocks of items were considered it was found that the subjects who were reinforced answered in the social direction progressively more items, in successive blocks of items, than did control group subjects. The same effect of reinforcement upon response to general activity items was found. The results thus showed that personality trait items, selected by statistical methods, did have a psychological reality. Strengthening the response to one item strengthened the individual's general tendency to respond to all stimulus items in that manner.

The results did not directly show that the behaviors of being sociable, or active, could be increased through learning procedures. They showed, however, that one of the types of behaviors employed to index personality traits—the individual's verbal descriptions of trait behaviors—was subject to conditioning principles. Moreover, as has been indicated herein, the individual's self-concept (self-description) can be considered to be an important aspect of personality itself. Such personality traits as those described refer largely to instrumental behaviors. Somewhat later the author applied behavioral analysis to another area of personality. After describing a preliminary conception of the A-R-D system (then called the reinforcer system) the following suggestion was made concerning measurement:

> Perhaps the function of certain types of "tests" used by the applied psychologist is at least in part to assess the reinforcers [A-R-D stimuli] that are effective for an individual or a group. Such tests might consist of items that control the appropriate verbal behaviors with respect to the reinforcing [A-R-D] value of stimulus objects and events, such as the behavior of others and various activities. For example, more than half the items on the Strong Vocational Interest Blank [Strong, 1952] ask the subject to state whether he likes, dislikes, or is indifferent to various occupations, school subjects, amusements, activities, and characteristics of people. This may be considered to involve a simple listing of [A-R-D stimuli] for the individual. A large part of the remainder of the test seems to concern the same thing. The subject is asked, for example, to choose from a group of ten factors influencing his work the three most important to him; from ten renowned individuals, the three he would most like to have been; from ten official positions in a club, the three he would like to hold, and so on.
>
> Items on the Study of Values questionnaire (Allport, Vernon, and Lindzey, 1951) may also be considered to be verbal questions that evoke responses under the control of the [A-R-D] properties of the described

stimulus objects and events. The following items illustrate the stimuli the subject is asked to evaluate. No. 9: "Which of these character traits do you consider the more desirable? (a) high ideals and reverence; (b) unselfishness and sympathy." No. 29: "In a paper, such as the New York Sunday Times, are you more likely to read (a) the real estate sections and the account of the stock market; (b) the section on picture galleries and exhibitions?" No. 4: "Assuming that you have sufficient ability, would you prefer to be: (a) a banker; (b) a politician." (Staats, 1963, pp. 305–6)

Kanfer and Phillips (1970) have also employed this concept of the reinforcer (A-R-D) system and the possibility that existing testing instruments measure the system: "[N]eed-oriented instruments, such as the Edwards Personal Preference Schedule, may be adaptable in yielding broad reinforcer hierarchies" (p. 516).

## ABNORMAL PSYCHOLOGY AND ASSESSMENT

The preliminary behavioral description of abnormal psychology was made in terms of deficits and inappropriacies in behavior and A-R-D systems and the conditions by which these were learned and maintained (Staats, 1963). It may be noted that traditional clinical test instruments were originally constructed to index the abnormal personality processes assumed to be the causes of abnormal behavior. However, consideration of different types of abnormal categories in terms of the behaviors of which they were composed, and the manner in which they were learned, suggests that such behaviors can be assessed for their own significance.

The growth of the literature showing that abnormal behaviors can be modified through reinforcement procedures provided a verified foundation for such interests. Moreover, as will be indicated, the need to specify in research that the behavior therapy methods of systematic desensitization could lessen phobias (anxieties) provided the need for assessment instruments to specify changes and to select subjects.

## REPRESENTATIVE SAMPLES AND BEHAVIOR ASSESSMENT

A concept of representative behavior sampling has been described in previous behavioral analyses (Staats, 1968a, 1968c). It has been suggested that the traditional view of human behavior as determined by internal mental and personality structures and processes does not provide the basis for an interest in the *content* of human behavior. "Stated simply, the experimental tasks we employ are traditionally considered to reflect an underlying mental process. In contrast to that approach, we need to accept our experimental tasks as a sample of some universe of learned behaviors (Staats, 1968a, p. 187). (See also Staats, 1968c, pp. 195–202.)

The concept of representative sampling of a behavior universe has focal implications also for behavior assessment (see Staats, 1971a). Other behaviorally oriented investigators have also considered assessment in these terms.

> [I]n the dynamic orientation the observed behaviors serve as highly indirect *signs* (symptoms) of the dispositions and motives that might underlie them. In contrast, in behavior assessments the observed behavior is treated as a *sample*, and interest is focused on how the specific sampled behavior is affected by alterations in conditions. (Mischel, 1971, p. 180)

> The sign approach assumes that the response may best be construed as an indirect manifestation of some underlying personality characteristic. The sample approach, on the other hand, assumes that the test behavior constitutes a subset of the actual behaviors of interest. (Goldfried and Kent, 1972, p. 413)

## BEHAVIOR ASSESSMENT

Behavior modification and behavior therapy studies, the concept of the reinforcer system, the analysis of complex human behavior, and the elemental behavioral conception have all been constituents in a foundation for productive developments in behavior assessment. Goldfried and Sprafkin (1974) have indicated, in addition, that the present author's taxonomy of abnormal behavior (Staats, 1963), as described by Bandura (1968a), provides one of the foundations for behavior assessment. These constituents have been creatively elaborated to form the area called behavior assessment. For example, Kanfer and Saslow (1965) have productively employed the conceptual framework in suggesting uses in clinical work. Their behavior assessment schema suggests seven categories to be explored for a patient: (1) analysis of the problem situation, (2) clarification of the problem situation in terms of who is maintaining the problem behavior, (3) motivational analysis, or in the present term, A-R-D analysis, (4) developmental analysis of biological equipment, (5) analysis of self-control displayed in life interactions, (6) analysis of social relationships to establish who the influential people are, (7) analysis of the social cultural-physical environment in terms of the agreements and disagreements the patient has with his social norms. Their development of the framework may be exemplified by elaborating one of the assessment areas, that of the reinforcement (A-R-D) system (Staats, 1963).

> Motivational Analysis: Since reinforcing stimuli are ideosyncratic and depend for their effect on a number of unique parameters for each person, a hierarchy of particular persons, events, and objects which serve as reinforcers is established for each patient. Included in this

hierarchy are those reinforcing events which facilitate approach behaviors as well as those which, because of their aversiveness, prompt avoidance responses. This information has as its purpose to lay plans for utilization of various reinforcers in prescription of a specific behavior therapy program for the patient, and to permit utilization of appropriate reinforcing behaviors by the therapist and significant others in the patient's social environment. (Kanfer and Saslow, 1965, pp. 534–35)

Cautela and Kastenbaum (1967) have also devised a self-report Reinforcement Survey Schedule which evaluates pleasurable feelings toward a large number of objects and events. Other psychologists have been interested in assessing the reinforcement value of stimuli as an aid to behavior modification treatment. Patterson (1967) used an operant conditioning type of task to assess the extent to which pictures of aggressive social interactions were aversive to children. Patterson (1965b) also employed a similar technology to test the relative reinforcing value of peers or parents (see also Weiss, 1968).

Behavior therapy, on the basis of its clinical concerns, also has involved the development of assessment procedures relevant to the A-R-D system. The primary interest of systematic desensitization has been with inappropriate anxiety. For example, rating procedures have been employed to identify fears (Geer, 1965; Wolpe and Lang, 1964). The instrument has been called the Fear Survey Schedule. Items include: sharp objects, failing a test, being a passenger in an airplane, worms, being alone, illness, driving a car, spiders, snakes, speaking before a group, and so on. The subject rates the extent of fear he feels toward the object or events, on a seven-point scale.

Goldfried and D'Zurilla (1969) have utilized the behavior sampling concept most systematically in behavior assessment. An analysis of the academic environment was made to obtain a sample of problems freshmen were likely to confront and the behaviors that would obtain reinforcement. Freshman subjects themselves, and faculty having close contact with them, provided the pool of data from which the assessment instrument was composed.

Direct behavior sampling has also been employed. For example, Lovaas, Freitag, Gold, and Kassorla (1965) employed a panel of buttons and an observer to record the behavior of disturbed children. Each button was employed for a category of behavior, for example, talking, running, self-hurtful actions, sitting alone, and so on. When the behavior occurred the observer could depress the appropriate button, and it would leave a timed record until released. Fears have been measured by direct behavior also, such as the proximity with which the feared object would be approached (Lang and Lazovik, 1963; Lang, Lazovik, and Reynolds, 1965; Bandura, Grusec, and Menlove, 1967). These various studies do not deal with the individual's A-R-D system as *a causative personality system that may lie at the root of the individual's problems, and in this sense require assessment as well as change through treatment.*

## THE BEHAVIOR ASSESSMENT CONCEPTUAL CONTEXT

In comparison to the traditional field of psychometrics, moreover, behavior assessment has produced relatively little in the way of psychological measurement instruments, especially those of a verbal (paper and pencil) variety. It could be said that this is solely a function of the recency of the development, but this actually seems due to characteristics of the approach.

THE OPERANT CONCEPTUAL INFLUENCE. Some of the conceptual elements for behavior assessment have been listed, such as the behavior modification and behavior therapy developments, the concept of behavior sampling, the concept of the reinforcer system, and the abnormal psychology which analyzes psychopathology into behavioral deficits and inappropriacies.

In addition, however, in recent times the conceptual foundation has come to include an operant conditioning orientation. This has had a considerable influence in suggesting rejection of the concepts, methods, and products of traditional personality and psychometric approaches. B. F. Skinner has provided the basis for this rejection in his general approach, as well as in many specific statements. In the passage from which the following was taken he eschews self-reports and verbal tests as a means of gaining information about the individual.

> Instead of observing behavior, the experimenter records and studies a subject's statement of what he would do under a given set of circumstances, or his estimate of his chances of success, or his impression of a prevailing set of contingencies of reinforcement, or his evaluation of the magnitude of current variables. The observation of behavior cannot be circumvented in this way, because a subject cannot correctly describe either the probability that he will respond or the variables affecting such a probability. If he could, he could draw a cumulative record appropriate to a given set of circumstances, but this appears to be out of the question. (Skinner, 1969, pp. 77–78)

As has been indicated, however, most real human behavior is not amenable to consideration in terms of operantly conditioning a simple response as measured on a cumulative record. Skinner thus illustrates the restriction imposed by elemental behaviorism. The fact is other concepts, principles, and procedures are relevant to the measurement of human behavior, as will be suggested. The potential contributions of other approaches need recognition. To continue, however, Lindsley (1956), in work that employed the term *behavior therapy*, followed Skinner's view and attempted to diagnose psychosis by placing patients in Skinner boxes and recording cumulative records of a simple motor response that was reinforced. This general conception of the use of operant conditioning methods for assessment purposes has been followed in other works (Patterson, 1967; Weiss, 1969).

Moreover, Skinner's rejection of the concepts, methods, and instruments associated with traditionally oriented psychometrics, in favor of direct behavior observation, has had a growing influence. For example, Kanfer and Phillips (1970) state that "Behavioral assessment in the future is likely to depend upon elaboration and extension of laboratory analogue or response sampling techniques" (p. 518). Mischel (1972) has suggested a general statement relating the ability of an assessment instrument to provide information about the individual's behavior to the extent to which it deals with that specific behavior:

> Predictive validity tends to decrease as the gap increases between the behavior sampled on the predictor measure and the behavior that is being predicted. On the whole, research regarding the relative specificity of behavior suggests that sampled predictor behavior should be as similar as possible to the behavior used on the criterion measure. (p. 323)

Acceptance of this principle would pretty well rule out the use of many of the verbal tests of traditional psychometrics. That is, frequently the behavior observed on such tests is verbal, whereas the behavior one is interested in predicting is nonverbal. Interest tests, for example, measure the individual's statements of likes and dislikes. However, interest tests are employed to predict the extent to which the individual will do well in a particular line of endeavor, which involves very different behaviors. As later analyses will indicate, there are very well-founded theoretical and experimental reasons—as well as traditional psychometric evidence—for the use of verbal test instruments, for measuring behaviors that are different from those to be predicted, and for not restricting oneself to laboratory analogues, which are laborious and time-consuming. The elemental behavioristic approach to psychological measurement would exclude or devalue very productive sources of information from the development of a behavioral conception of human behavior.

The central point here, however, is that behavior modification and social learning have not provided a conceptual foundation for describing and measuring personality traits (repertoires), as a general endeavor. While some social learning theorists (Mischel, 1973; Krasner and Ullmann, 1973) are coming to accept concepts analogous to personality traits, the approach has not yet recognized the need for indication of what constitutes personality in terms of the analysis of behavioral repertoires. For example, Mischel, writing in 1973, takes a step forward in indicating acceptance of the importance of certain personality traits. "[I]mpressive consistencies often have been found for intellective features of personality and for behavior patterns such as cognitive styles and problem-solving strategies that are strongly correlated with intelligence" (p. 253). In addition to this recognition, however, it is necessary to make analyses—theoretical and empirical—of what comprises intelligence, cognitive style, and so on. These analyses can then serve as a foundation for psychometric developments, as a later section will indicate.

THE TREATMENT ORIENTATION INFLUENCE. Behavior modification and behavior therapy have been restricted to concern with clinical problems of abnormal behavior. This has occurred partly because only patients with problems have been available as subjects of behavior modification treatments. Part of the restriction, however, stems from the clinical traditions of the orientations (Staats, 1970b). Abnormal behavior traditionally was seen to be more significant to understanding human personality than normal behaviors that are more commonplace. It has not been sufficiently realized that it is the learning of the normal repertoires that prevents the development of abnormalities.

> Actually, the abnormal behaviors of the problem child [or adult] are frequently simple behaviors—whose analysis and treatment may be handled relatively easily. Usually, on the other hand, the "abnormal" child's problems in normal repertoires may necessitate complex understanding in specific terms as well as long-term treatment procedures to rectify the deficits in skilled repertoires the child has "acquired." *Frequently, the primary problem of the abnormal behavior is the effect it has upon the child's normal learning.* Thus, the problem is not solved by attempting to reduce the abnormal behavior alone. In sum, what may seem to be problems of abnormal behavior many times have at a more basic level problems of deficits in normal behaviors. (Staats, 1970b. p. 27)

Behavior assessment, stemming from behavior modification and behavior therapy interests in clinical treatment of abnormal behavior (Staats, 1970b), focuses on the abnormal. The field of traditional psychometrics, with roots in education, educational psychology, child psychology, and personality has not had this limitation. Thus, there are tests of intelligence, aptitudes, achievement, interests, attitudes, and so on that are concerned with measuring normal and adjustive repertoires. Moreover, it is recognized that psychological assessment is relevant to purposes other than clinical treatment, as important as this purpose is. Thus, psychological tests are employed for personnel selection, for counseling in terms of career choice, for school placement, and so on. These interests are productive constituents of psychological measurement.

THE PERSONALITY THEORY LEVEL. The above limitation is related to other conceptual limitations prevalent in the incipient field of behavior assessment. As has been indicated, as valuable as behavior modification and behavior therapy work has been, the approach does not provide a conceptual basis for consideration of personality. It is suggested that a personality level is essential to a general approach to psychological measurement. Without the impetus provided by a general concept of personality, as well as specific personality analyses, measurement is likely to be restricted. Some of this limitation can be sensed in the following passage:

> There is little basis in current behavior theory or in a behavioral framework of personality development to point to historical or behavioral segments that are especially sensitive indicators of maladjustment or disturbance. Without such a guide for pinpointing primary response dimensions or behavior content, quick recognition of critical mechanisms or response patterns representing prime therapy targets is difficult. Thus, the behavior therapist often relies on both traditional personality theories to direct his probing and on adaptation of traditional measuring techniques for evaluation. (Kanfer and Phillips, 1970, p. 514)

The conditioning-oriented clinical psychologist may thus be placed in the position of theoretically rejecting psychological tests but needing them to perform research and clinical activities. For example, there are many studies that employ behavior modification procedures with children wherein the researchers employ intelligence tests to measure the changes induced but would reject the personality concept of intelligence and the psychometric methods of psychological test construction.

## Social behaviorism and psychometrics

It is suggested, again, that there is a separatism in the realm of psychological measurement between traditional approaches and contemporary elemental behavioristic approaches. Methods, concepts, and technology that are actually complementary remain separated. Again, a unification is necessary to achieve greater productivity, and such a unification is possible within the social behaviorism paradigm.

Direct observation of the individual in real-life situations may provide essential information in understanding and dealing with certain problems. For example, observing parent and child interacting can provide a great deal of information concerning the child's behavior problems. Laboratory methods can also provide useful information. The findings of Freund (1963) that sexual deviates show penile erectile responses to pictures of the inappropriate object, for example, could prove valuable when other means are not available for establishing the nature of A-R-D system deviation.

However, the rejection of traditional testing methods, or the restriction to operant conditioning or other experimental analogues, or the restriction to behavior sampling is not a necessary characteristic of an approach that is built on behavioral principles. It is quite clear, as will be further discussed, that verbal ratings and verbal tests of various kinds, along with verbal reports, can be employed to establish important conditions in the individual's life that help determine his behavior. The use of an operant conditioning apparatus to establish the reinforcing value of a stimulus (see Patterson, 1965b, 1967; Weiss, 1968) is cumbersome, costly, and time-consuming—when the person may simply indicate verbally the A-R-D value of objects and events. This is not to say that verbal tests do not have sources of invalidity—the person, for example, may not want to reveal the information. But the same conditions can pertain to the operant conditioning situation, which is not a direct measure of the emotional value of the stimulus—the basic condition that establishes reinforcing (and directive) value. The person could dissemble in the operant conditioning situation and not work for a stimulus that he actually would find to be gratifying.

In brief, it is suggested that the individual's verbal behavior is lawful —as lawful as any other type of behavior. Information concerning the individual's behavior and conditions that have affected or do affect the individual may be ascertained on the verbal level. Methods that have been developed by which to do this cannot be rejected. Verbal tests can play a productive role in providing information about individuals and groups. This is not to say that complete information can be acquired in this manner. There are certainly many limitations, which have been recognized within the field of psychometrics.

It is also suggested that personality is an important conceptual basis for psychometrics. However, the constituents of the personality repertoires require description. Thus, observation of the individual's life situation is important. It is also important, however, in many circumstances to ascertain what the individual's personality contribution is. If a child does not have the basic behavioral repertoires that constitute an adequate level of intelligence, it is important to know this. If a man does not have an A-R-D system of a certain type, as will be indicated, certain lines of work may be unrewarding for him.

## Need for unity

It has been suggested that elemental behaviorism has not included important contributions of traditional approaches. But the traditional psychometric approach, while containing valuable elements, also has not provided a conceptual basis for utilizing sources of knowledge in other areas of study.

> [T]he field of measurement has remained largely on an empirical level, concerned with the production of instruments for predictive purposes. There has been little articulation of the principles, concepts.

and procedures of psychological testing with those of general psychology. There is presently little effort within the field of psychological measurement to link test instruments with an underlying, more basic, psychological theory or with methods for clinically modifying behavior (Staats, Gross, Guay, and Carlson, 1973, p. 251).

The psychometric conception of intelligence, for example, does not provide the principles by which it is possible to understand how intelligence repertoires are developed or function. Neither does it provide the description of the repertoires involved, so that the behavioral meaning of intelligence can be made clear. The intelligence test, in providing an IQ number, leaves unanalyzed what the child does or does not have. Such a test, therefore, can provide little in the way of directives for treatment. This is true of personality tests in general.

It is important to indicate how learning and behavior modification principles, a behavioral conception of personality, and traditional psychometric methods can be unified. It is suggested that this unification should yield knowledge that is not provided now, either by the field of psychological testing or by the behavioral assessment efforts. This task will involve research to indicate how the learning principles and behavior analysis of personality are relevant to personality tests. Three general areas of personality have been treated in this book—the emotional-motivational system, the language-cognitive system, and the instrumental behavior system. Examples in each of these areas will be referred to.

## The A-R-D system and personality measurement

The A-R-D theory of the emotional-motivational aspects of personality provides an integration of classical and instrumental conditioning principles, and in doing so indicates how emotional and instrumental behaviors are intertwined. The theory states that the individual learns a class of instrumental responses under the control of positive emotional responses, and a class of instrumental responses under the control of negative emotional responses. Moreover, the individual learns positive or negative emotional responses to a large number of social and environmental stimuli, and this gives those stimuli either positive or negative reinforcement value for him. Anything that changes the emotional value of a stimulus will also change its reinforcing value for the individual, as well as the power of the stimulus to elicit "striving for" or "striving against" instrumental responses.

In suggesting this covariation of the attitude-reinforcer-directive stimulus functions, the personality theory has implications for psychological measurement. As has been indicated, the simple operant conditioning approach, with its focus on reinforcement, is led to the conclusion that the only way to measure the reinforcement value of a stimulus is by assessing its ability to reinforce instrumental behavior in an operant apparatus. But the A-R-D theory suggests that any of the three stimulus functions, since they covary, can be used as the measure.

This removes one basis for the schism between the traditional psychometrics and a behavioral assessment approach in this area. The clinician does not place a patient in an operant chamber or strap him into a physiological recording apparatus. He frequently judges by what the patient says in free interviews or in response to relatively unstructured tests like the Rorschach or Thematic Apperception Test. Or the clinician obtains the patient's verbal responses to verbal items on a personality questionnaire.

Such means of measuring the emotional-motivational aspects of human personality are not limited to the clinical psychologist. Social scientists frequently employ verbal responses to verbal items to estimate the individual's emotional-motivational characteristics, to obtain predictions concerning what the individual will do, or simply to characterize the person or group.

> An anthropologist in the field might simply observe stimulus objects and events for which members of the group "strive." . . . Or, they may possibly observe the verbal behavior of members of the group by asking the individual what he values, or what other members of the group value. (Staats, 1963, p. 305)

The A-R-D theory thus relates the operant conditioning assessment of the reinforcing value of stimuli to physiological measurement of the emotion-eliciting value of stimuli, as well as to the types of observation characteristic of traditional measuring procedures. Any of the three functions of A-R-D stimuli will provide information concerning the others. When physiological recording apparatus indicates that a stimulus elicits an emotional response, this also indicates that the stimulus can serve as a reinforcing stimulus and that it will elicit the appropriate class of instrumental approach or avoidance behaviors. When it is observed that the individual strives for or against a stimulus, this also indicates that the stimulus will elicit an emotional response in him and could serve as a reinforcing stimulus for him. When a stimulus that is delivered as a reinforcer is observed to strengthen or weaken a response, this also indicates that the stimulus will elicit an emotional response in the individual and could serve also as a directive stimulus for appropriate instrumental behaviors.

In view of the elemental behavioristic rejection of verbal methods of assessing the extent to which a subject can "describe either the probability that he will respond or the variables affecting such a probability" (Skinner, 1969, p. 78), it is necessary to theoretically and experimentally justify the use of verbal psychological test items. In the context of measuring the individual's A-R-D system, for example, it is necessary to indicate why verbal items can be used in the place of direct measurement.

These principles were presented in a general way in the analyses of language in Chapter 5 and can be elaborated with respect to the present problem. When the individual or someone else labels an object, event, or activity, this provides a classical conditioning occasion. Any of the responses elicited by the object, event, or activity will be conditioned to the word label. If a positive emotional response has been elicited, the word will, to the extent of the conditioning, come to elicit a positive emotional response for the individual. As a consequence, the word will have reinforcement value. Moreover, as a stimulus that elicits an emotional response, the word will acquire the tendency to elicit a whole class of "striving for" instrumental behaviors.

Of course these A-R-D functions are not acquired by the word with only one conditioning trial for the individual. It may take a number of conditioning trials. However, there are many words that come to acquire A-R-D properties for people in general, because they are systematically (if not always) paired with the objects, events, and activities that they label or with other words that perform the same functions as the actual objects, events, and activities.

At any rate, this analysis would indicate that by measuring the individual's response to the word it should be possible to assess his response to the actual stimulus. In addition, however, it is necessary to understand that the individual ordinarily learns various verbal ways of indicating the emotional value of a stimulus. For example, the child is asked if he wants more of a tasty food and learns to say "more" to desirable things (that is, stimuli that elicit a positive emotional response in him). He will learn verbally to choose, to indicate like for, to quantify, and to compare things on the basis of the emotional response the things elicit in him.

It is because we generally learn a repertoire of verbal behaviors that describe the emotional value that stimuli have for us that verbal items on personality tests can be effective. Thus, the Strong Vocational Interest Blank includes items the individual has to respond to as L (like), I (indifferent), or D (dislike). The Minnesota Multiphasic Personality Inventory includes items that measure aspects of the individual's A-R-D system, with the response being either "True" or "False."

There are also comparison items that resemble voting. On the Allport-Vernon Study of Values (Allport, Vernon, and Lindzey, 1951) there are items on which the individual must select which of two objects, people, events, or activities he considers the more desirable. The Strong Vocational Interest Blank includes several types of items like this, some involving selection among 10 listed objects, events, people, or activities. The Edwards Personal Preference Schedule (Edwards, 1953) also is composed of pairs of statements which the subject has to respond to by selecting the one he likes most. In attitude scaling, the task frequently is to rate the item on a scale of five or seven points, using terms like *pleasant* and *unpleasant* or *agree* and *disagree* to mark the ends of the scale. The individual may also be asked to rank order items.

Such items, it should be noted, *measure the directive value of the words in the item*. The respondent has to make a selection response, such as checking a rating scale, circling a word or statement, or making a pencil mark on a score sheet. These are instrumental responses, not emotional responses.

It is not this behavior that the test is interested in predicting, however; it is the behavior of the individual in his life situation that is of interest. It is suggested that such responses as rating a word can be predictive of other behavior because the rating response is under the control of the emotional value of the word, which is the same as the emotional value of the real-life situation. Thus, if the item is scored positively, the individual will have a positive emotional response to the real event. He will be reinforced by the real event in the sense that the real event will strengthen the learning of behaviors that obtain the event. He will be directed also to strive for the real event.

This is thus a case where the behavior in the life situation does not have to be sampled on the test instrument. It is suggested that the sampled predictor, unlike the principle posed by Mischel, does not have to be "as similar as possible to the behavior used on the criterion measure" (Mischel, 1972, p. 323). A central aspect of psychological testing lies in the ability to make general predictions from limited access to observations of the individual. It is thus important to provide empirical support for this theoretical analysis of measurement of the emotional-motivational system and for the predictive possibilities that can be obtained from traditional verbal personality tests. A first step in this direction is to show that the items on emotional-motivational personality tests do have the A-R-D functions that have been attributed to them.

## THE A-R-D PROPERTIES OF INTEREST TEST ITEMS

EMOTIONAL CONDITIONING THROUGH INTERESTS. Interest tests are examples of traditional paper and pencil personality tests. It has been suggested that the verbal items on interest tests, as one type of emotional-motivational personality test, actually measure the A-R-D values of the words in the item and hence of the actual life situations described. There are a number of hypotheses that can be derived from this theoretical model. It would be expected, for example, that interest inventory items should have the several functions of emotional-motivational stimuli. The items would be expected to elicit an emotional response in the individual, which would be conditioned to any other stimulus with which the item was paired. Let us say that there are a number of items on the interest test that elicit a positive emotional response in the individual. Let us say that he has a positive emotional responses to the verbal items *reading, swimming, playing golf, acting, going to parties, chemistry, biologist,* and so on. It would be expected that if these words were paired with some neutral stimulus, the neutral stimulus would come also to elicit the positive emotional response. The neutral stimulus would be the conditioned stimulus, and each interest item that elicited a positive emotional response would serve as the unconditioned stimulus. The same principle should hold with interest items that elicit a negative emotional response.

Whether or not interest inventory items can actually serve as emotion-eliciting stimuli that can impart emotional value to new stimuli has been tested by Staats, Gross, Guay, and Carlson (1973). They had 58 subjects take the Strong Vocational Interest Blank in their regularly scheduled class. The results were then tabulated so that 15 items could be selected for each subject that were scored "Like" and thus were expected to elicit a positive emotional response in the subject. Fifteen items were selected for each subject that were scored "Dislike" and which would thus elicit a negative emotional response, and 15 items were selected that the subject had scored "Indifferent and which presumably elicited neither positive nor negative emotional responses.

The hypothesis was that a neutral stimulus (nonsense syllable) that was paired once each with the positive interest items should come to elicit a positive emotional response, and a neutral stimulus paired with negative interest items should come to elicit a negative emotional response. The language conditioning method described in Chapter 4 was employed to test this possibility. Following the conditioning the subjects were presented with the nonsense syllables one at a time, ostensibly to see if they had been learned. The nonsense syllables were also rated by the subjects on a seven-point rating scale ranging from "pleasant" to "unpleasant." The results showed that the syllable paired with items from an interest inventory that had been scored as "Like," when the test had been taken two weeks before, come to elicit a positive emotional response as measured by the rating scale. The syllable that had been paired with "Dislike" items came to elicit a negative emotional response. The results indicated that interest items did indeed elicit emotional responses that could be classically conditioned.

INSTRUMENTAL CONDITIONING WITH INTEREST ITEMS AS REINFORCERS. If the interest items elicit an emotional response in the individual, then these stimuli should also serve as reinforcers in the learning of new instrumental behaviors. In an experiment conducted to test this hypothesis (Staats, Gross, Guay, and Carlson, 1973), subjects were first administered the Strong Vocational Interest Blank, and positive and negative interest items were selected for each individual. Later the subjects participated in an instrumental conditioning procedure in which the items were employed as the positive or negative reinforcers. The subjects were always reinforced for responding to the *size* of the stimulus—one half of the subjects were reinforced for choosing the larger stimulus, the other half for selecting the smaller stimulus. The reinforcement was the presentation of a card with an interest item printed on it. The subject was directed to read the item on the card and then rate it in the same manner as he had rated the items on the SVIB, that is, by circling the L, I, or D on the card. The subject was positively reinforced for selecting the "correct" stimulus with "Like" interest items. He was punished for selecting the "incorrect" stimulus with "Dislike" items.

Under these conditions the interest items functioned as reinforcers. The subjects increased their frequency in responding to the stimulus, for which they received positive interest items, while selecting less frequently the stimulus for which they received negative interest items. The analysis was supported by a similar study conducted by Reitz and Mac-Dougall (1969) which demonstrated separately that positive interest items would function as positive reinforcers and negative interest items would function as negative reinforcers.

THE DIRECTIVE VALUE OF INTERESTS. The theoretical model also suggests that interest labels should have a directive as well as reinforcing function, based on the emotional response elicited by the interest stimuli. The hypothesis derived from the model was that individuals with different interest systems should behave differently to the same stimuli. For this experiment a more naturalistic type of behavior was employed, that of selecting one of two materials to read on the basis of the label of the material.

Again, subjects were first given the Strong interest test. Groups of subjects were then selected who had interest systems like those of certain occupational groupings and unlike those of other occupational groupings. Ten subjects were selected to have interests highly similar to the interest systems of music teachers or music performers and interests dissimilar to the interests of chemists. Ten other subjects were selected to have interest systems that were just the reverse in similarity to the occupational categories. Under the guise of measuring physiological responses to reading materials, these 20 subjects were given the opportunity to choose, from one of two stacks of articles labeled MUSIC or CHEMISTRY, an article he was to read. As expected from the A-R-D theory of interests (and other aspects of the motivational system), subjects with strong music and weak chemistry interests chose to read from an article marked MUSIC. Subjects with the opposite interest systems responded in the opposite manner, selecting instead to read from an article marked CHEMISTRY. The determination of the subjects' behavior on the basis of their interests was so strong that almost every subject responded as predicted.

This experiment was then replicated with ten additional subjects who had high interests in medicine and low interests in business (either accounting or banking), and ten subjects who had the reverse interests. They had to choose from articles labeled BUSINESS or MEDICINE. Again, the nature of the individuals' interests determined how they behaved, and the experimental effect was exceedingly strong.

Actually, regardless of the stack the subjects chose from, half of the subjects received articles to read that were coincident with their interests, and half received the converse. It may be noted that the subjects rated articles coincident with their interests as more interesting, more informative, and more pleasant than did subjects reading noncoincident articles. Moreover, subjects gave better answers to questions concerning articles that were coincident with their interests (see Staats, Gross, Carlson, and Guay, 1973).

INTEREST THEORY, PERSONALITY, AND PREDICTION. As has been indicated, traditional psychometrics has not been strong on providing an explanation of personality or of the instruments constructed to measure personality. A number of investigators (for example, Clark, 1960; Loevinger, 1957; Shontz, 1965), while recognizing the predictive efficiency of the Strong test, have noted the lack of a theoretical foundation for interest tests.

> . . . [T]raditional statements concerning the origin, development, and definition of interests have often been characterized by a lack of specificity, and isolation from a comprehensive, rigorous personality theory (a theory that would provide a basis for deduction and the consequent extension of empirical work). Thus it has been suggested that interests develop as a result of the individual's practical adjustment to his environment (Carter, 1940); that interests are closely linked to personality development (Darley, 1960); and that "interests are the product of interaction between inherited neural and endocrine factors, on the one hand, and opportunity and social evaluation on the other" (Super and Crites, 1962, p. 410).
>
> A few psychologists, however, have discussed interests in somewhat more explicit terms and in a manner which may be related to the theory underlying the present investigation. Fryer (1931), for example, proposed an acceptance-rejection theory of interest measurement in which he suggested that likes represent a "turning toward stimulation" which may be correlated with pleasant experience, and dislikes a "turning away from stimulation" which may be correlated with unpleasant experience. Fryer stated that the like and dislike responses to interest test items are the result of learning but he did not specify the learning principles involved. Thorndike (1935), Tuttle (1940), and Strong (1943) emphasized the role of the law of effect in their theoretical formulations of interest development. Thus, according to these investigators, interests develop as a result of an activity being followed by agreeable or disagreeable consequences. This view, however, in addition to being inade-

quate because it was not part of a comprehensive learning theory of personality, was based upon an incomplete and ambiguous learning theory. The view is thus deficient in not considering the role of classical conditioning; it did not clearly indicate the principles in the formation of the emotional component of interest. Furthermore, the manner in which the classical conditioning of emotional responses determines the instrumental functions of the stimuli involved was not seen. (Gross and Staats, 1969, pp. 14–15)

*Interests are learned.* Unlike some conceptions of interests that assume a genetic basis for interests (D. P. Campbell, 1971), the present model suggests that interests are learned. Moreover, the manner in which interests are learned is specified. The basic principle involved is classical conditioning, some of which occurs on a primary level. For example, the child who has experience in which he is deftly led into athletic skills by a parent who insures positive emotional stimuli will be involved will help produce a child who has positive athletic interests. The child who experiences negative emotion-eliciting events in the context of athletic activities will be conditioned to negative interests.

In addition to primary classical conditioning experience, however, a great deal of higher-order conditioning, largely through language, must be considered to be involved in interest formation. Various experiments indicate that many words elicit emotional responses. Pairing such words with other stimuli will result in those stimuli also eliciting emotional responses. An interest in a moving picture, for example, frequently arises from reading a review in which positive emotion-eliciting words appear.

It should be noted that the classical conditioning model of interest formation solves one of the traditional puzzles concerning interests. That is, the theories of Thorndike, Tuttle, and Strong that interests developed as a result of activity being followed by rewards—an instrumental conditioning model—led to the prediction that abilities should be highly correlated with interests. This expectation was not supported. In puzzling over this fact, as well as other evidence contrary to his theory, Strong (1943) could only note that such interests might be the result of "social forces not yet recognized in this connection" (p. 13).

Classical conditioning, however, requires no skill development experience. The individual may be conditioned to positive or negative emotional responses through watching a movie, reading a book, conversation, and so on. We may come to "like" various people, activities, objects, and events solely on the basis of such classical conditioning, which involves none of the skill development that would be associated with instrumental learning.

*Interest function: Interests as a personality trait.* As has been indicated, when a stimulus comes to elicit an emotional response the stimulus acquires A-R-D functions. It is suggested that these functions give the individual's interest system the causative characteristics of a personality trait, in the manner described in Chapter 4.

The three interest experiments described above showed these personality trait functions clearly and are thus basic to the personality conception that has been elaborated herein. That is, the nature of the individual's A-R-D system, it has been suggested, helps determine how he responds, how he learns, whether he learns, what he experiences, his further personality development, and so on.

This principle was demonstrated in each study. In the first study the nature of the individual's interests determined how he would be classically conditioned to new stimuli. To elaborate, in the first experiment, interest items were selected to have positive or negative A-R-D value for the subjects, and the subjects learned emotional responses to new stimuli by pairing the items with those stimuli. Different items served as emotion-eliciting stimuli for different subjects, illustrating that the same stimulus situation may be experienced differently, depending on the individual's interest system. Subjects could have been selected to have opposite emotional responses to the same items. Then presentation of the *same* conditioning procedures to the two different groups would have resulted in just the opposite type of emotional conditioning. In general, based upon the nature of one's A-R-D system, one can learn vastly different new emotional aspects of personality than someone else will when subjected to the same experience.

The same indication that the nature of the individual's interest system could determine how instrumental behaviors are learned was shown in the second study. Interest stimuli which for someone else would be quite otherwise served as reinforcers for the individual in acquiring a new discriminative skill. This can be illustrated in its naturalistic implications. For one of two individuals, let us say, interacting with people in a persuasive role elicits a positive emotional response (interest), as does the successful accomplishment of a persuasive act. Because of this the individual would score relevant interest items as "Like." For the other of the two individuals these events elicit negative emotional responses, and he would score such items as "Dislike." Quiet, scholarly pursuits have more A-R-D value for the latter, as do the results of such activity (gaining scholarly information). Placed in a sales position, the first individual's appropriate behavior would be strengthened, and he would acquire additional skills in his job. The second individual's behavior would not be well maintained in this situation but, unlike the other, when placed in a scholarly research position his appropriate behavior would be strengthened. Each would acquire different skills, as determined by the reinforcing value of his interest system. It is suggested that these are bases for the predictive qualities of interest tests.

The ways in which different interest (A-R-D) systems can determine different behaviors, different experiences, and different further personality development were also shown in the third experiment. The subjects approached—that is, chose—to receive experience (reading an article) that was congruent with the emotional conditioning they already had. Individual differences in behavior to the same situation were determined by A-R-D personality differences. It should be noted that such instrumental "preference" behavior determines the type of later experiences (in this case reading) the individual will receive—a mechanism for insuring continuity of personality development. To illustrate, it would be expected that a person who had a positive emotional response to "liberal" people, actions, ideas, and causes would approach and be affected by such stimuli and would avoid those of an opposite, "conservative," type. Thus, the A-R-D personality mechanism would help insure maintenance of the individual's political personality characteristics. As another example, a male with feminine interests would approach different things and people than a male with masculine interests would. This would produce varying experience and varying additional personality development through learning.

*Interest measurement and prediction.* The A-R-D theory, elaborated in the area of interests, thus constitutes a model that provides a rationale for interest measurement. Assessment of individuals' interest systems appears to be feasible by verbal means, since the interest items demonstrate the functions expected of A-R-D stimuli. Moreover, because of the A-R-D functions of interests it would be expected that later behaviors could be predicted in part from a knowledge of the individual's interest system, as has been exemplified.

The model also suggests why the Strong interest method of test construction has produced an instrument with validity. To elaborate, it has been suggested that the individual's interest system is the result of a conditioning history that includes opportunities for millions of conditioning trials. Because of this complexity it must be expected that the individual's interests will be complex—that each individual's interest system will be uniquely different.

However, there are conditions that would be expected to make for similarity among individuals in aspects of their interest systems. The fact is that there are experiences that would be expected to be the same or similar for groups of individuals. A culture ordinarily provides for a finite category of activities in which an individual can engage, for example. This means that there will be a number of individuals who will experience relatively similar things. A sportsman, for example, may have experience in a finite category of sports activities. Those who participate in one sport will share similar experiences, and this experience will ordinarily result in a similarity in interest (A-R-D) conditioning.

There are also finite categories of types of books to read, subjects to study, social activities to participate in, periodicals to peruse, types of people to associate with, socioeconomic levels to be raised in, moving picture and television programs to watch, occupations to study and to engage in, and so on. Since the categories are finite and the number of individuals in the culture can be large in comparison, it is usual for many people to be in any particular category. This means that the individuals in that category will have conditioning experiences that are at least in part similar. As a result it may be expected that these individuals will share group characteristics in their A-R-D systems, in interests as in other aspects.

Because of these factors, it is suggested, it is possible to construct empirical tests of interests. Thus, as an example, it is possible to gather a group of items on which the individual indicates his liking or disliking and to give this inventory or test to different occupational groupings of people. It is then found that the occupational groupings of people indicate that they have different interest systems—they have learned different emotional responses to different objects, events, activities, people, and so on. Moreover, it is possible to characterize and categorize the interest system (or profiles) of different occupational groupings. Then the interest test can be given to an individual and his measured interest system can be compared to the characteristic interest systems of the various occupational groupings. If the individual's interest system is highly similar to that of some occupational grouping, it may be predicted that he will behave similarly to the people in that grouping, enjoy the same things, and so on.

It is suggested that this method of test construction should be relevant to construction of other tests of the motivation system. Thus, any grouping—national, cultural, economic, social, and so on—could be expected to have a commonality of A-R-D conditioning history that could have predictive possibilities. It should be noted that the present discussion is not meant as a blanket endorsement of all verbal tests. Mischel (1968, 1971) has performed a service in indicating that personality measurements frequently do not provide predictions across situations. The present position recognizes these limitations. The intent has been, however, to show in principle how verbal tests have the *potential* for providing predictive indices for behaviors different from those measured on the test, as well as to indicate mechanisms that underlie the test-behavior relationships. These mechanisms, as has been suggested, can operate over different situations and over long periods and can provide the basis for distal prediction in various ways. Much additional research is necessary to substantiate the functioning of these mechanisms, in the process of dealing with specific tests.

THE A-R-D SYSTEM AND OTHER PERSONALITY TESTS. The studies discussed above were conducted only with interest inventory items. It should be emphasized that motivational aspects of personality measured with various types of tests can be considered within the same concepts. A few additional examples will be given.

One of the most widely used instruments with which to assess "abnormal" personality characteristics is the Minnesota Multiphasic Personality Inventory. The instrument includes 561 items, among which there is a liberal sprinkling of items that assess the individual's motivational system. Examples may be seen in the following: "I enjoy many different kinds of play and recreation," "I enjoy detective stories," "I liked school," "I think I would like the work of a librarian," "I feel uneasy indoors," "I have several times given up doing a thing because I thought too little of my ability," "My sex life is satisfactory," "I am an important person," "I love to go to dances" (Dahlstrom and Welsh, 1960, pp. 55–78). These items were taken, respectively, from the following scales of the MMPI: Depression, Hysteria, Psychopathic Deviate, Masculinity-Femininity, Paranoia, Psychasthenia, Schizophrenia, Hypomania, and Social Introversion. The Masculinity-Femininity scale is largely composed of such motivational items. It has been suggested that deviations in the motivational system from "normal" are contributory to types of abnormal behavior (Staats, 1963, 1964, 1970c), and the inclusion of motivational items on the MMPI supports this hypothesis.

The Edwards Personal Preference Schedule is another test of personality (Edwards, 1953). The test primarily measures the individual's emotional response to various activities, people, and events. Most of the items involve selecting the one of two statements that the subject likes more than the other. The following examples illustrate the format of the personality test.

4    A    I like to tell amusing stories and jokes at parties.
       B    I would like to write a great novel or play.
15    A    I like to be independent of others in deciding what I want to do.
       B    I like to keep my things neat and orderly on my desk or workspace.
29    A    I like to tell amusing stories and jokes at parties.
       B    I like to write letters to my friends (Edwards, 1953, p. 2).

It is suggested that the items of various personality tests involve measurement of the individual's A-R-D system through verbal means. Such items and such tests should be considered within the systematic framework for experimental, theoretical, and applied purposes. It should be noted that this is not to suggest that all tests that measure aspects of the A-R-D system are valid or reliable. Investigations have indicated that different tests have differing abilities to predict. As has been indicated, many studies have questioned the ability of attitude measures to predict actual behavior, whereas, as one example, interest tests have been found to be valuable as predictors of learning and behavior.

## Intelligence (the language-cognitive system) and personality measurement

This section will be concerned with the manner in which intellective (language-cognitive) tests can be treated within the social behavioral framework to extend our understanding of this area of personality and personality measurement. As indicated in Chapter 5, the traditional concept of intelligence is that of an internal quality or structure or process, probably neurological in origin, and inherited. The elemental behavioristic approach has rejected such personality concepts, and behaviorists have generally ignored this area of concern. The social behavioral conception, in contrast, is that intelligence involves important personality repertoires which heavily involve language. Rather than denying the importance of this field of study, the conception coincides with the traditional view of the relevance of intelligence measurement and suggests additional avenues of study as well.

> To continue, however, it is suggested that the types of individual intelligence tests that have served as standards in the field (for example, the Stanford-Binet) are composed of items, each of which involves one or more behavioral skills. The items ask the child to label various objects, imitate certain actions, draw or complete pictures, and so on. . . . One may ask about the fact that there appears to be a general factor in intelligence tests. That is, the success of response to one item is correlated with success of response to other items. . . . [T]he relationship between items that seem to deal with different behavioral skills actually occurs, in part, because the items include overlapping or common behavioral skills. There are behavioral skills common to a number of items. If the child has the behavioral skill, he is in a position to respond to *various* specific items. If not, he will miss the various items. To give one example, a behavioral skill that has generality across various items involves the child's attention being under the control of the test examiner. In many items the examiner will verbally direct the child to look at something, do something, respond to certain words, and so on. If the words of the examiner do not control the child's attentional responses, the child will fail on each of the items. The attentional skills of the child thus are general. In addition, of course, each item would test for the presence of specific skills, the specific motor or verbal response involved—but the item would also test for the general skill. . . .
> One might parody what has been said so far by saying, "Well, if intelligence as tested consists only of specific behavioral skills, then let's make children intelligent by training them to respond to the in-

telligence test items." It should be noted, however, that generality is involved in the conception in various ways—including the type of general behavioral skills just mentioned that cut across various items. In addition, *the specific items on a test are actually samples of much broader classes of behavioral repertoires.* Thus, an item that tests whether or not a child can label a particular object involves more than this very specific response. The item may be considered to have sampled the child's repertoire of labeling responses—a class of behaviors that may have many, many members. Training the child on only the one item would leave him with glaring deficits in the other items in the repertoire.

The types of items that are included on intelligence tests have been arrived at empirically. That is, they have been selected because they correlate with school achievement, with performance on other intelligence tests, and so on. This means of selection of intelligence items, it is suggested, has guaranteed that the classes of behavioral skills on a valid intelligence test are important to the child's adjustment in school, his ability to learn in many situations, his ability to solve general problems relevant to his age group, and so on. It would be expected that intelligence tests would vary in their ability to predict children's future performance in their life situations to the extent to which the tests sampled the various behavioral skills that were relevant to success in those situations. The fact that intelligence tests are correlated with such things as school success attests to the efficacy of the tests and the relevance of the behavioral skills tested. (Staats, 1971a, pp. 41–42)

Examples of items on intelligence tests, analyzed in terms of the behaviors of which they are composed, were given in Chapter 5. It is suggested that the nature of intelligence (and other personality traits) and intelligence tests is to be understood by detailed analysis of the basic behavioral (personality) repertoires of which they are composed (Staats, 1971a). It is not possible to provide such an account here, but one area will be described in slightly greater detail to indicate some of the research possibilities that can stem from a behavior analysis.

## NUMBER-MATHEMATICAL CONCEPTS AND INTELLIGENCE

The ability to handle numbers and mathematical concepts has traditionally been considered to be a reflection of human intelligence. Numerical tasks were included on some of the earliest intelligence tests (see Goodenough, 1949). Piaget (1952, 1953) has considered the development of number-mathematical concepts to be central indicators of the child's intelligence development, although he emphasizes age as the independent variable rather than specific training.

A child of five or six may readily be taught by his parents to name the numbers from one to 10. If ten stones are laid in a row, he can count them correctly. But if the stones are rearranged in a more complex pattern or piled up, he can no longer count them. . . . [H]e has not yet grasped the essential idea of number. (1953, p. 75)

At any rate, the importance of number-mathematical concepts in the measurement of intelligence may be demonstrated by reference to intelligence tests. On the Stanford-Binet (Terman and Merrill, 1937), for example, aspects of the number-mathematical repertoire begin to be sampled at the level of two and one-half years. Item 5 at that age level consists of the examiner saying "Listen; say 2," "Now, say 4–7," (p. 79), and so on. This involves the verbal imitation repertoire, but the success of the child on this intelligence item is also a function of whether he previously has learned to say numbers. Similar items occur at the levels of 3, 4½, 7½, 9½, 10, and 12-years. Increasingly complex sequences of numbers are involved, sometimes to be repeated by the child in the order opposite to that given by the examiner.

At the five-year level, on item 6, the examiner presents the child at different times with four blocks, four square beads, and four pennies. The child must indicate how many objects are presented, that is, he must say "four" in each case. At the six-year level, on item 4, the child is presented each of four times with 12 blocks and successively asked to place on the table before the examiner 3, 9, 5, and 7 blocks. To succeed on these four problems the child must be able to count unarranged objects. It may be noted that items on the Stanford-Binet are located at the age level where 50 percent of the children will respond correctly. The use of these items at this age level provides support for Piaget's observations that this type of skill emerges at this time—albeit not necessarily so, as will be indicated.

To continue, at the age 14 level, item 4 consists of the following type of water jar problem.

> A mother sent her boy to the river to bring back exactly 3 pints of water. She gave him a 7-pint can and a 4-pint can.
> Show me how the boy can measure out exactly 3 pints of water using nothing but these two cans and not guessing at the amount. . . .
> (Terman and Merrill, 1937, p. 118)

At the average adult (15-year) level the person is given arithmetical reasoning problems. For example, one printed item states: "If a man's salary is $20 a week and he spends $14 a week, how long will it take him to save $300?" (Terman and Merrill, 1937, p. 122). These problems, as with the water jar problems, require the skills of arithmetic operations such as subtraction, multiplication, and division.

Other items relevant to quantity appear also. For example, at the level of three and one-half years, two match sticks of different lengths are presented to the child. He is instructed as follows: "Which stick is longer? Put your finger on the long one" (Terman and Merrill, 1937, p. 84).

NUMBER CONCEPT DEVELOPMENT THEORY. Without a theory to guide our interest, there is little basis for the analysis of such intelligence items. They are included on intelligence tests because they have predictive value. It is suggested, however, that the child will succeed on the intelligence items to the extent that he has had the learning history that provides him with the number concept skills involved. The items ask him to say numbers, to count, and later to add, subtract, multiply, and divide.

To simply say that these number concept skills are learned is not sufficient, either. For a theory of intelligence it is necessary that the number concept skills be specifically analyzed so that it is possible to see the stimulus and response components that compose the skills and the manner in which the skills are learned. The author has analyzed the early development of number concepts in terms of the repertoires of learned stimulus-response skills involved (Staats, 1963, 1968c, 1971a; Staats, Brewer, and Gross, 1970). As is necessary with any theoretical model, its statements must be empirically tested. Empirical verification was begun in experimental-naturalistic work with the author's own children and in the experimental classrooms with preschool children described in Chapter 11. The theory of number concept learning is of a cumulative-hierarchical nature. Several of the types of learning involved will be outlined.

*Numerosity discriminations.* One of the S-R mechanisms that was not described in Chapter 2 involves the manner in which a stimulus component of a complex of stimuli can come to elicit a response specifically. That is, a stimulus ordinarily is actually a complex of stimuli. An object may have color, but it also has shape and size, for example. When a response is learned in the presence of the stimulus object, each of the stimulus components comes to elicit that response to some extent. Thus, other stimulus objects that share one of the components will to some extent elicit the response.

It is important in human adjustment that the individual learn to respond to component stimuli such as color, shape, and quantity, apart from the specific stimulus objects involved. These are traditionally considered to be concepts. Some of them may be considered in terms of the S-R mechansim.

Take color, for example. A child will be said to have the concept of color if he can verbally label different color stimuli across various objects. This is learned in the following manner. Let us say that the child is led to say "orange" in the presence of an orange, and is reinforced. Through such training, he will learn to give the color name to the orange color of the fruit. But in addition, the spherical stimulus characteristics will also control the response. He may call a differently colored ball "orange." For the color component stimulus to elicit the labeling response, and no other stimulus, the child has to receive training in which he is reinforced for saying "orange" to various objects that have the color (and not to non-orange stimuli).

In a similar learning process, the first acquisition of number concepts involves the concept of numerosity—the child must learn labeling responses under the control of the *number* of objects, regardless of the other stimulus components of the objects. The author first tested the possibility that learning procedures and conditions could produce such early number concept skills with his 18-month-old daughter. Various objects were employed—such as raisins, pieces of popcorn, fingers, marbles, and so on. Some of them, such as the raisins, were used both as objects to be "counted" and as the reinforcers. For example, the author would hold one raisin in one hand and two in the other hand and ask, "Do you want one raisin (displaying the raisin), or two raisins (showing the other hand)?" His daughter soon learned the number discrimination of saying "Two" and pointing to the two raisins, being differentially reinforced by receiving the larger amount. As another example of progressively advancing training, later the author would hold up several fingers and ask "How many fingers are there?" Reinforcement for a correct response would be social approval, since the training trials were widely spaced and required only weak reinforcers (see Staats, 1968c, 1971a, for the principles of reward in child training).

These principles of numerosity concept learning were later studied with four-year-old culturally deprived children. Cards were employed with different numbers of objects pictured on them—for example, one, two, or three small pictures of a horse, or a ball, and so on. The children first learned to say "One horse," "Two horses," "One apple," "Two apples," and so on, to the appropriate picture. Then they learned simply to say the number. Each reinforcing stimulus delivered (the child learning apparatus depicted in Chapter 2 was employed), each stimulus presented, and each response the child made was recorded so the learning process could be described. To learn to respond appropriately to one or two objects required a mean of 253.6 training trials for the 11 children. The numerosity learning moved in a standard manner for all the children and added support to the learning analysis (Staats, 1968c, pp. 325–328).

Discriminating and labeling small numbers of objects can be learned by the child before he is two years old, although the item to test this repertoire does not appear on the Stanford-Binet until the five-year level. In any event, it is important to recognize that this intellectual skill is *learned* straightforwardly and is not limited to the age norms suggested by the maturational conception. It is also suggested that the ability to label numbers, an aspect of language learning, is important to the child's cognitive function. Blank (1974) has discussed evidence that the child's development of numerosity verbal labels and temporal labels aids in solving problems involving these results.

THE COUNTING REPERTOIRE. Although this number discrimination and labeling repertoire is basic to further number concept learning, it is a limited skill by itself. That is, this type of number labeling skill can only be developed to a relatively few objects. Above this number the discrimination of one group of objects from another becomes difficult. To advance much past the level of skill demanded by item 6 at the five-year level of the Stanford-Binet (where the child has to indicate there are four objects), the child must learn to count. This is a more complex repertoire which requires explicit analysis (see Staats, 1963, 1968c; Staats et al., 1970).

> It can be seen that counting consists of sensory-motor sequences which occur in conjunction with sequences of verbal responses. The child must look at [or point to, or touch, and so on] the objects in a systematic sequence, one by one, from left to right, when they are arranged in a series. At the same time the child must make a sequence of verbal responses, the series of vocal number responses of counting, beginning with saying "one." Moreover, these two sequences of skills must be coordinated, so that the stimuli produced by one elicit the appropriate response in the other. Thus, there are three types of learning that must take place in learning to count. . . .
>
> The child has to acquire the sensory-motor skill and the verbal response sequence, both in the appropriate order. And he has to have the pointing response in each case control the emission of the verbal response—and the verbal response, in turn, must control the next pointing response. In training the child, it will be seen that this coordination may be the last to be firmly learned. The child will first learn to make the verbal-number responses either faster or slower than the pointing responses. (Staats et al., 1970, pp. 46)

This analysis indicates the training that has to be conducted to produce the counting number concept skills in the child. Tests of this learning theory were done with the four-year-old children. The child was presented with a card with two dots on it, one of which was covered. The child, who had been trained to label by number, said "one." Then the other dot was uncovered and the child said "two." Demonstrating the procedure, the experimenter-trainer showed the child how to point to each dot in turn saying "One . . . two." This method was then extended to training the child with three- and four-dot cards. When a sequence of counting responses had been learned to the cards, objects (pennies, and so on) were arranged in a line, to be counted.

The results showed that the culturally deprived children could be standardly trained to these number concept skills. It took the children a mean of 284.3 training trials to acquire this repertoire (Staats, 1968c,

p. 327). Again, this aspect of number-mathematical skills could be acquired on the basis of the elementary principles of learning.

   *Counting conservation across object arrangement.* Piaget (1953), however, has suggested that there is a difference in mental maturity of the child who can count objects only when they are arranged in an ordered series and the child who can count randomly arranged objects. The latter he would regard as having the concept of number; that is, such children recognize number as invariant across changes in superficial aspects of the object display, such as change of arrangement of the objects. Children who could count objects arranged in a series, but not in other arrangements, would be considered by Piaget to have only a mechanical skill. This could be specifically taught to the child, but the real concept of number must await mental maturation, according to Piaget. As Staats, Brewer, and Gross (1970) have noted:

> [H]owever . . . there is no difference in principle between these two types of skills. The only difference, in fact, concerns the sensory-motor skill sequence. It is difficult to count a randomly arranged set of objects by pointing to the objects. Since there are no explicit reference points, for example, one may retrace oneself and point to and verbally count an object twice—or an object may be missed.
>
> A different sensory-motor sequence is required. A very simple one is to employ an appropriate set of objects like pennies, so the child can place a finger on each penny and remove it from the set. He must then continue to perform a sequence of such sensory-motor responses, moving each object from the original pile over to the newly formed pile—each time saying one number verbal response—until the objects in the original pile are exhausted.
>
> When the child has acquired the sequence of sensory-motor responses of withdrawing objects from the set, one at a time, in appropriate association with the number verbal-response, then he is able to count any random assortment of objects (up to the limit of his verbal-response chain). The totality of the skill as diagramed in S-R components is shown in [Figure 12.1]. The bottom row of responses and the stimuli they produce (the changed position of an object) represents the sensory-motor sequence. The top row of responses and the stimuli produced (the sound of saying a number) represents the vocal sequence of counting. The arrows indicate the various stimuli that come to elicit each response and, thus, how the coordination of both sequences is learned. The diagram indicates also how a complex skill can be seen more clearly when subjected to an S-R analysis (Staats, Brewer, and Gross 1970, p. 47).

This analysis receives support from the learning data produced by the 11 culturally deprived four-year-old children. After they were trained to count ten objects that had been arranged in a series, they were given the type of training described for counting unarranged objects. Ten pennies were placed in a pile, and the child was shown how to pull one out

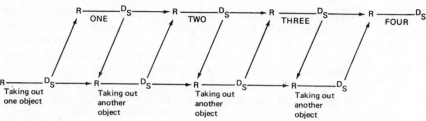

**FIGURE 12.1**

A *stimulus-response theoretical model of learning to count. The model depicts the two parallel and mutually controlling response sequences.*

at a time as he counted. It required a mean of only 63.0 learning trials for the children to learn this additional number concept skill. A mean of 6.3 training sessions was involved, with the training sessions lasting five to seven minutes (Staats, 1968c). Thus, an average of less than an hour was spent in taking the child from the one level of number concept development to the next—little evidence that mental maturity was involved. Again, the number concept skill appeared to be strictly learned. The fact that the children were four year olds indicates that the intellectual skills involved depend upon the child's learned repertoires, not on his age-determined maturation.

VERBAL EXTENSION OF COUNTING. One product of a behavior analysis in this area will be exemplified. The diagram in Figure 12.1 is a theoretical model of the learning of counting and contains certain implications.

> The theoretical model suggests, however, that in the specific S-R training the child also has learned *general* skills. Thus, for example, the sensory-motor response generally comes to elicit a number response. When the child pulls an object out of the pile, this provides the stimulus for the child to say a number response—*any* number response. The *particular* number response is dictated by the *previous* number response, that is, the word association between the two numbers. The child says "four" rather than some other number because he has just previously said "three" (and "one" and "two" before that), and the stimulus of having said "three" elicits the saying of "four." In the other direction, also, it is suggested that the child has learned general skills. The saying of a number response, any number response, provides the controlling stimulus for the motor response of removing the next object —until there are none left.
>
> It would be expected on the basis of the theoretical analysis that if a child with this "number concept" was now trained—purely as a verbal serial-learning task—to make additional number verbal responses, this verbal training *by itself* would extend his functional counting repertoire. Let us say, as an example, that the child with the counting repertoire shown in [Figure 12.1], where he has learned to count objects up to

four, was given verbal training in which he acquired the verbal-response chain of counting up to 10. That is, the upper sequence of *verbal* responses in [Figure 12.1] would be extended, with no extensions of the other S-R components that compose the number-concept skill of counting. The theoretical model would predict that if the child was then presented with from five to 10 randomly arranged objects, he would be able to count them—although this would represent an entirely novel behavior, since he would never have counted that many objects before. The rote verbal learning should result in a functional reasoning skill, when introduced at that point of training (Staats, Brewer, and Gross, 1970, p. 48).

This derivation from the theoretical model of this number concept skill was tested with nine children. After they had learned a basic counting repertoire, they were given *rote verbal* training in counting additional numbers. The extension of the verbal counting repertoires ranged from three additional numbers to eight. "In each of the nine cases, after the child's verbal counting repertoire had been extended, he was presented with the task of actually counting a set of objects of greater quantity than he had ever counted before, and did so immediately" (Staats, Brewer, and Gross, 1970, p. 51). The model of counting indicates that pure verbal learning in counting will be effective when the child has already acquired certain previous skills, and will be noneffective (rote-learning) when the other skills have not been acquired.

ARITHMETIC OPERATIONS. Addition and multiplication skills have also been subjected to a behavior analysis (Staats, 1963, 1968c). Furthermore, it has been shown that the beginning phases of these skills may be produced in four-year-old children according to stimulus-response principles and procedures (Staats, 1968c; Staats, Brewer, and Gross, 1970). The results suggest that under appropriate conditions of reinforcement this type of training could be extended in the cumulative-hierarchical learning process throughout the number-mathematical set of skills.

CONCLUSIONS. Number concept skills are important to intelligence. Whether the child can adjust to (solve) certain situations will depend upon the extent of his number concept skills. Whether he can learn new skills, when presented with training opportunities, will also depend upon the extent of his number concept skills, since arithmetic-mathematical learning develops in a cumulative-hierarchical manner, with one skill requiring the previous skill acquisitions. For example, the child had to have the verbal imitation skills to learn to say the numbers, as well as other imitation skills in performing the behaviors of pulling one penny at a time from the group as demonstrated by the trainer-experimenter. The counting repertoire was then basic to learning number operations such as addition and multiplication, and so on.

It is also suggested that these aspects of intelligence are learned. The number-mathematical skills that have been investigated thus far indicate that the early skills consist of specific stimulus-response mechanisms that are learned according to the basic learning principles. Moreover, more advanced skills have already been analyzed in these same terms (Staats, 1963, 1968c), and the analyses constitute theoretical frameworks from which additional experimental hypotheses can be directly derived.

Traditionally, theories of personality and child development have concentrated upon the observation of the types of behavior that develop, on the average, in children at a certain age. This assumes a development based upon maturation. Intelligence tests have followed this traditional approach. The studies of number-mathematical learning, however, show that such intelligence skills may be acquired by the child much before the usual time, when the training conditions are appropriate.

The general implication is that the various aspects of intelligence are learned in the same manner as the number-mathematical skills that have been described. It is suggested that these various aspects of intelligence could be subjected to theoretical and experimental analyses in a similar manner, to establish a direct, explicit, and differentiated definition of intelligence. More will be said of the need to bring the methods and principles of psychometrics into conjunction with the methods and principles of the social behavioristic theory.

## Instrumental behavior traits and verbal tests

Reference to sensorimotor skills was made in the above treatment of counting as an "intelligence" skill. The various performance items on intelligence tests also measure the individual's sensorimotor or instrumental repertoires. It would be quite feasible to analyze such skills into their stimulus-response components and to demonstrate the manner in which the skills are learned. The importance of such research will be indicated further on.

It has been suggested that there is good justification for the use of verbal tests in the area of emotional-motivational as well as cognitive psychological tests. Verbal tests also are relevant to the instrumental realms of personality. To elaborate, it has been suggested in Chapter 5, in discussing the labeling repertoire in langauge, that the individual learns to label himself. That is, the individual produces many stimuli. He has varied emotional responses which have stimulus characteristics. He has internal stimulation in his images and covert thoughts and plans. In addition, however, his own actions, and the effects of such actions on others and on the environment, produce stimuli. To the extent to which he has learned to label these types of events, he is potentially as capable of describing them as he would be in describing any experience he has had. It should be seen that it is this type of description that is involved in responding to the personality trait items on such tests as the Guilford-

Zimmerman. It is suggested that the validity of the test rests in part on the fact that when the individual describes himself as "You work more slowly and deliberately than most people of your sex and age" (No. 31), the individual has made veridical observations of the manner in which he works in comparison to others and can label those events. The same is true of an item like "People think you are a very energetic person" (No. 46). The latter involves observations of how other people respond to him, or inference of how they must respond.

Items such as "You daydream a great deal" (No. 50) require that the individual observe his own daydreaming behavior and have an idea of what a great deal is. To respond to item No. 136, "It irritates you to have to wait at a crossing for a long freight train to pass," the individual must have experienced the emotional response involved in a similar situation and have learned to label the emotional stimulation.

To say that the individual is an observer of his own behavior and can provide this information to others does not imply that this information is either complete or completely veridical. Item No. 66 on the Guilford-Zimmerman states, "You seem to lack the drive necessary to get as much done as other people do." A negative response to this item may occur because the individual has made veridical observations of his own behavior in comparison to that of others and has learned to veridically label such observations. Response to the item, however, may be negative because the individual has learned a deprecating set of self-descriptions and observational habits. Thus, the same item may contribute different types of information for different people.

Moreover, there may be important causal conditions in the individual's life, as well as behaviors characteristic of him, that he has not labeled veridically, or not at all, and which he is thus unable to report. There are also situations in which the individual will not report events that he could, or where he will dissemble. Determinants other than his previous observations will control what he says.

Thus, recognition of the potentiality of the person's verbal reports as a source of information does not suggest this as a complete or infallible source. It is important to recognize the contributions that the self-reports and various kinds of verbal tests can make, as well as to recognize the potential weaknesses. Direct observations as well as the reports of others also have weaknesses. The field of psychometrics must be prepared to utilize various sources of assessment to fulfill its goals.

## Behavior analysis, psychometrics, and behavior modification

The suggestion that there could be a productive rapproachement between a learning approach and a psychometric approach was made in the author's conclusions concerning. the experiment in which Guilford-Zimmerman personality trait items were shown to be capable of instrumental conditioning.

> Thus, it is found that the psychometric approach and the experimental S-R approach can be considered to have a complementary relationship. . . . These conclusions suggest, further, that S-R principles and methods can be extended to the psychometric study of [trait] factors in a way which may put psychological meat on psychometric's descriptive bones, i.e., discover the principles relating certain determinants to the development, consistency, and change of "traits" of behavior. It is suggested that such an extension of S-R principles will provide some of the theories of factors with the support of experimental evidence. In addition, the principles of trait development, once found, should introduce an element of control into the developmental situation not obtained from the study of S-R laws [alone]. (Staats, Staats, Heard, and Finley, 1962, pp. 112–13)

This remains a major suggestion of the present chapter.. There has been a rationale for test construction which includes the concept of the behavioral universe (Goodenough, 1949). Goodenough accepted Warren's definition of psychological trait as a "distinctive mode of behavior of a more or less permanent nature, arising from the individual's native endowments as modified by his experience" (1934). Although considering the internal trait process as the determinant of behavior, however, Goodenough indicated that the task of test construction consisted of obtaining a representative sample of the behaviors determined by the trait. This view coincides with the present view and indicates also that there is not an unbridgeable psychometric-behavioral chasm. It is suggested that a combination of the various elements would yield a rich harvest in personality theory, test construction, and experimentation on human behavior, as well as in treatment procedures for dealing with the problems of human behavior.

The field of psychometrics has not had a theoretical arm for systematic analysis of human personality, most of its products being based upon commonsense notions of personality. Psychometrics has been disconnected from general psychology and has not had avenues for utilization of knowledge in general psychology. Moreover, there has been no experimental arm for basic investigation of psychometric concepts and tests. Tests are validated by the extent to which they predict, that is, by the extent to which they are related to some criterion variable. For example, a test of interests will be validated by the extent to which it predicts success in a field of work. Why such tests do predict, the relationship of interest tests to other personality tests, the manner in which interests are acquired, the principles involved in the change of interests and in the function of interests in determining the individual's behavior—all these involve areas of knowledge with which psychometrics does not deal. The field of psychometrics has an advanced theory and technology for test construction, but additional elements could provide a means of dealing with the topics suggested above.

Elemental behaviorism is in a similar position. It has a rich fund of knowledge of experimental methods and many of the elementary principles of learning. But it has had no theory of human personality, or framework for attaining that knowledge. In terms of the present concerns, elemental behaviorism has not had psychometric's technology or theory for constructing tests to measure human behavior reliably. Behavior modification has had little measurement on which to base its clinical treatments. As has been indicated, traditional clinical psychology treatment is not intimately related to psychometrics or to the field of personality.

This is another example of the need for the parts of psychology to be brought together if the full value of the different fields and their products is to be realized. Intelligence testing was treated herein because it is such a good example of this combination. Presently, intelligence tests exist as useful instruments for characterizing the child for diagnostic purposes and in some tasks of prediction (such as selecting in training and vocational programs). But intelligence tests do not tell what intelligence is, how intelligence is acquired, how intelligence functions, or how to treat problems of intelligence. Intelligence tests and concepts of intelligence do not serve as springboards for conducting research relevant to the various fields of psychology.

The social behaviorism approach, it is suggested, provides a theoretical analysis of what intelligence is (Staats, 1971a) and what the functions of the intelligence repertoires are. The analyses suggest that the components of intelligence are learned and stipulate the learning conditions by which the repertoires are acquired. The approach includes, also, suggestions of methods by which to investigate the analyses and thus the general theory. Behaviors in children that are samples of intelligence repertoires have been subjected to studies in which the behaviors are produced in the children by learning principles and procedures. The findings thus far are positive and suggest that the various components of intelligence should be specified and subjected to experimental learning analyses. This would yield a specific, detailed theory of intelligence and its acquisition that would serve to guide practices of child training, education, and clinical treatment.

The theory and findings thus far suggest that intelligence acquisition involves a cumulative-hierarchical learning process. This is corroborated by the results from traditional intelligence tests such as the Stanford-Binet, which are composed of items that are age graded. It is suggested that a much more complete stipulation of the universe of behaviors that compose the intelligence repertoires at each age level for children is called for (see Staats, 1968c, 1971a; Staats, Brewer, and Gross, 1970). Stipulation of the cumulative-hierarchical behavioral advancement would serve as an improved basis for measurement (test construction) and prediction, and for clinical treatment as well. If, for example, the child is low in intelligence because he does not have certain number-mathematical skills, or language skills, or attentional skills, or whatever, measurement of such deficits would be the first step toward behavioral treatment. Intelligence testing, like psychological testing in general, presently does not provide such information. The behavior analyses could be employed to construct intelligence tests that do provide this information.

The rapprochement that is being suggested appears to have implications for various areas of personality measurement. Further analysis and experimentation are called for.

# Part four

## Biological science, social science, and humanistic levels

# 13

# *Social behavioral humanism*

THERE ARE various conceptions (theories, models, or philosophies) of man and his behavior. These conceptions may be part of a theology, or a world view, or of a scholarly or scientific tradition. Although it may not be considered in this sense, there is a conception of man that is commonly held in our culture. Most people, for example, consider human behavior to be free and spontaneous in a manner that they would not ascribe to physical objects in the world. As has been emphasized, the nature of the general conceptions concerning man are important, because in many cases they underlie our social actions and decisions (Staats, 1963, 1971a).

[T]he general system of explanation incorporated into one's thinking helps determine how one deals with practical problems. For example, at one time it was thought that an individual who behaved abnormally was possessed by the devil. This conception led to ways of treating abnormally behaving individuals that are now recognized as inefficient, if not detrimental to advantageous behavior change. At that time, the objective of treatment was to convince the devil that a given body was not a fit place for him to inhabit—that it was too uncomfortable. To prove the point to the devil, "therapy," involved whipping the person, burning him, immersing him in ice water, casting him into a snake-pit, and so on. The treatment was, of course, in line with the prevalent conception of the source of abnormal behavior; it was assumed that the devil would choose to vacate this particular body, and the individual would be "cured." . . .

> This example illustrates the intimate relationship between the system
> we use to explain behavior and our reactions to other people. The com-
> monsense conceptions we hold concerning behavior also determine how
> we respond to others. . . . (Staats, 1963, p. 33)

The matter of social philosophy and the effects of the philosophy were
treated more deeply in the context of presenting a social behavioristic
conception of intelligence and personality (Staats, 1971a). It was suggested that the common concept of intelligence, for example, which
permeates our culture, has roots in theology as well as science. It was indicated that the social philosophy that includes the concept of intelligence
as an inherited process or quality—sometimes with personal and moral
virtue attached—underlies various societal practices. The conception plays
a role in medical diagnosis, educational placement, in educational practices, in child-rearing, in racial conflict and racist views of man, and in
scientific research. It was suggested that the social behavioristic conception
of intelligence provided a more humanistic social philosophy—one that
gave impetus to more humanistic social actions—than did the traditional
conception.

It should be realized that theories of human behavior contain, at least
implicitly, a social philosophy, or parts of one. This may be seen very
clearly in Skinner's recent book *Beyond Freedom and Dignity* (1971) in
which he treats certain implications of the elemental behavioristic position
in opposing the traditional concept of man's self-direction, freedom,
purposiveness, and the like. The elemental behavioristic view as originally
espoused by Watson conflicts with traditional conceptions of man along
these various dimensions.

It has been suggested, in an address entitled "Social Behaviorism: A
Science of Man with Freedom and Dignity" (Staats, 1972), that elemental
behaviorism began as a conceptual revolution. As a revolution it strove for
identity and recognition and emphasized its differences from traditional
thought. This included a rejection of the various funds of knowledge that
existed in other approaches to the study of man. Thus, as an example, in
Skinner's work there is an absence of reference to, or use of, concepts,
principles, or findings from any of the social sciences, other areas or approaches in psychology, or even other behavioristic approaches.

It was also suggested (Staats, 1972) that the reasons for the elemental
behavioristic striving for recognition in the study of man are no longer
relevant. The importance of behavior and the study of behavior is now
recognized. Behavior principles have been shown to be valuable in various
areas of concern. Furthermore, it was said that there are conflicts still
given impetus by elemental behaviorism (as well as its opponents) which
are no longer necessary or advantageous—and in fact constitute obstacles
toward advancement in developing a scientific and humanistic conception
of man. A third generation behaviorism was called for.

> [T]oday the next level of the revolution, the new horizon, is the rapprochement of the new order with the old. This involves the development from the insular approach of elemental behaviorism to one that is general in the principles and concepts it incorporates, in the observations it makes, in the methodologies that it utilizes, and in the problems with which it attempts to deal. . . . The time has come to make the next advance, the incorporation of the various sources of knowledge of man. (Staats, 1972, p. 6)

It is productive to discuss, at least in part, some of the conflicts which elemental behaviorism has had with traditional conceptions of man—within the context of a social behaviorism that permits resolution of the issues. These issues may be organized by describing two traditional ways of looking at man, the subjective (humanistic) and the objective (behavioristic) views.

## Subjective and objective conceptions

Throughout the history of concern with understanding man's behavior there have been *objective* versus *subjective* approaches. Boring (1950) has indicated that the positions can be differentiated in ancient Greek thought. The continuing history of psychology involves the dichotomy of explanation of man's behavior by his internal subjective states, in contrast to the explanation of his behavior by some objectively specifiable realm of events.

> The main line of tradition from Locke and Berkeley to Wundt and Titchner, or, for that matter, to Brentano and Kulpe, was that psychology studies consciousness, even though you may decide to call it "physiological psychology" and attempt to specify what neural events underlie each conscious phenomenon. . . . There have been many more who have argued that it is unprofitable to study consciousness directly and that better data for the same problems are obtained by limiting research to the study of behavior. This last position has been occupied by the Russian school of Sechenov, Bekhterev and Pavlov, and by the American behaviorists, Watson, Weiss, Holt and the others who came after them. (Boring, 1950, p. 620)

The issues concerning the important elements of human behavior—whether internal, mental, subjective or external, behavioral, or objective—vary depending upon the area involved. For example, in the area of the study of language there has recently been a contest between a subjective approach (initiated by the linguist Chomsky) and the objective approach of learning theory. The former approach considers language to be largely given by internal subjective states—biological (Lenneberg, 1967) and mentalistic (Chomsky, 1968b). Chomsky, for example, suggests that man's mind, by virtue of its native (biological) nature structures or processes, predisposes him to language. He needs only a brief experience with the particular language for his mind to establish the particular categories and rules involved.

In developmental psychology, a dichotomous distribution of psychologists can be seen between those who look to learning experimentation as an explanation of child behavior and those who look to the internal, mental or cognitive development of the child as an explanation. Piaget may be seen as an example of the latter approach. He accepts that the child's experience plays a role in his cognitive development. However, his methodology—the observation of children's problem-solving behavior over different ages so that the nature of their mental development, and thus their behavioral abilities, can be formulated theoretically—indicates relative indifference to the learning conditions. The major concern is with the child's subjective states of development. The following quotation from Langer (1969) exemplifies a primarily subjective approach to the study of child development:

> Because humans spontaneously initiate their actions, they play a constructive role in their own psychological experience and development. Formally, they are organizations of organs or systems of action that operate (1) to interact with their environment and (2) to construct their own experience and knowledge of themselves and the world. Genetically, they are endowed with (a) the necessary systems for initial interaction with the milieu in personally meaningful ways and (b) the self-generative characteristics that ensure their own development and self-actualization. (p. 87)

The dichotomization in the study of man, however, is not limited to the field of psychology. The same opposing views have been noted in the other social sciences. As one example, one approach in sociology has stressed *objective* (or overt or motor) behavior relations, while the other stresses *subjective* (or covert or dispositional) behavior relations.

> Ideally, an objective behavior definition argues that the social is to be identified according to whether the *overt* walking, waving, talking, praying, singing, eating, fighting, writing, breathing, etc. of one participant is related to some similarly objective behavior of another participant, even when such relations are not accompanied by covert or subjective behavior relations. In contrast, an equally idealized subjective behavior definition identifies the social with the covert perceptions, goals, sentiments, opinions, habits, desires, needs (and it may be added, thoughts, attitudes, norms, values) etc. that participants hold in common, in complementarity or in conflict, even when this sharing remains latent so far as objective activity relations are concerned. (W. L. Wallace, 1969a, pp. 6–7)

Wallace (1969b) provides examples of the dichotomous theories by quoting statements of leading theorists. Among the objectivists he lists Engels, Merton, Blau, and Homans, citing the following quotations: "(S)ocial function refers to observable objective consequences, and not to subjective dispositions (aims, motives, purposes)" (Merton, 1957, pp. 385–386). "(P)reoccupation with value orientations has diverted theoretical attention from the study of the actual associations between people and the structures of their associations" (Blau, 1964, p. 13). Gibbs and Martin reject the work of human ecologists who "spend their research hours assaying their data hopefully for values, sentiments, motivations, and other elusive psychological elements" (Gibbs and Martin, 1959, p. 29).

Max Weber, Sorokin, and Parsons are listed among the subjectivists. "The component of 'meaning' is decisive in determining whether a phenomenon is sociocultural (superorganic). Without it there are no sociocultural phenomena . . ." (Sorokin, 1966, p. 13). Caplow, like Weber, states that "human interaction is social if it involves symbolic communication" (Caplow, 1964, p. 75). Parsons describes sociological theory as "concerned with the phenomena of the institutionalization of patterns of value-orientation in the social system" (Parsons, 1951, p. 552).

It is suggested that the division into the subjectivist and objectivist camps actually cloaks a number of individual issues which have not been resolved. Behaviorism was a revolt against the subjectivist view, primarily in Watson's rejection of the method of introspection and assertion of the importance of studying behavior. But elemental behaviorism, as it has continued this revolt in opposing the subjectivist position, has never examined the several issues involved. The issues require resolution if a comprehensive, unified conception is to result.

Some of the issues underlying the subjectivist (humanist) and objectivist (elemental behaviorist) camps are listed in Table 13.1. The aim of the present discussion is to fragment the "sides" that have been chosen in this conceptual contest and to include in the position elements from each of the traditional approaches. This can only be done by reconsidering some of the issues, which involve important questions requiring resolution. The problem in such cases is meeting the need within the framework of the scientific philosophy and the scientific method. Resolution of issues based upon realities cannot proceed by capitulation of one conception to the other. The present method is to approach each issue with the idea of finding productive as well as unproductive elements on each side. This does not mean the present approach represents an eclecticism. Rather, the intent is to build the social behaviorism paradigm by searching each issue for productive elements contributed by the various concerns.

**TABLE 13.1**

Characteristics and concerns dividing the humanistic and behavioristic approaches
to the study of man

| | Humanistic | | Behavioristic (elemental) |
|---|---|---|---|
| 1. | Subjective events | 1. | Objective events |
| 2. | Holistic (man as a whole) | 2. | Atomistic (elementary principle) |
| 3. | Naturalistic observation | 3. | Laboratory observation |
| 4. | Individual (Ideographic) | 4. | General (Nomethetic) |
| 5. | Qualitative description | 5. | Precision and measurement |
| 6. | Understanding | 6. | Prediction and control |
| 7. | Self-determination, freedom, spontaneity in causation | 7. | Scientific determinism, mechanistic in causation |
| 8. | Originality, creativity, and activity | 8. | Passive respondent, automatonism |
| 9. | Self-actualization, personal growth, personality development | 9. | Conditioning, behavior modification, and behavior therapy |
| 10. | Values in science | 10. | Valueless science |
| 11. | Applied, concern with human problems | 11. | Basic, pure, science; science for science's sake |
| 12. | Purpose and goals, future causation | 12. | Prior and present causation |
| 13. | Insight and awareness | 13. | Conditioning |
| 14. | Biological explanation of human behavior | 14. | Environmentalism |

## Feelings and sentiments and thoughts

As has been indicated, for a long time after its inception psychology
was considered to be the science of mental life. Its chief method of
observation consisted of presenting the subject with certain stimuli and
having him introspect. The individual reported his feelings, sensations,
perceptions, and so on. As it turned out, the method was not productive. The growth of the science based upon this method left much to be
desired, and there was great unreliability in such subjective observations,
with no way of validating by public appeal what the nature of the experiences was.

It was John Watson, the first self-conscious behaviorist, who indicated
the drawbacks to the subjective method of introspection as the basis for a
science of psychology. He rejected the method of introspection and indicated, rather, that the objective, public observation of behavior and the
stimuli that influenced behavior were the proper subject matter of psychology. He rejected concern with such matters as feelings, thoughts,
images, perceptions, sensations, and so on, unless they were specified by
direct observations of behavior. Thus, as an example, thought could be
investigated as subvocal speech, observed by special instruments.

This position was a useful corrective in the context of the problems
of the time. It has been continued, however, in present-day elemental

or radical behaviorism. Such concepts as attitudes, feelings, interests, purposes, goals, awareness, perceptions, communication, values, meaning, and so on are still not considered, even though theoretical and methodological developments, as has been indicated, have made this possible. Thus, elemental behaviorism has influenced a considerable number of contemporary psychologists to ignore internal responding, limiting themselves to concern with instrumental behaviors. This influence has led to inconsistencies in some positions, including internal response concepts in some analyses but not in others. For example, some social learning theories have considered attitudes to be inconsequential to the determination of the individual's instrumental behavior, while accepting other internal response concepts (Bandura, 1969; Bem, 1965; Mischel, 1968; Thoresen, 1973b).

As has been suggested, elemental behaviorism requires elaboration. It is not justified to rule out the study of subjective elements or to relegate such elements to the role of noncausal epiphenomena. On the other hand, the excesses of the subjective psychologist must also be avoided. "The humanist psychologist . . . gives primary concern to man's subjective experience and secondary concern to his actions, insisting that this primacy of the subjective is fundamental in any human endeavor" (Bugental, 1967, p. 9).

The fact is that it is unfortunate to reject or negate the one at the cost of the other. The important concerns of each can be unified, within a common framework. As has been indicated in the previous chapters, the basic principles can be elaborated to include such subjective states as feelings and attitudes and meaning, as well as cognitive states described under various terms. The same body of principles can account for overt, instrumental behavior. Moreover, the framework of principles indicates how the internal, "subjective" states and the overt behaviors interact. There are causal connections in both directions. The fact that there is objective research in each area indicates that the rapprochement is not merely a verbal convenience.

Salient features in the rapprochement that is being elaborated are: (1) emotional and cognitive events and overt instrumental behaviors can be dealt with in the framework of the same principles, (2) this can be done in an empirical manner, (3) verbal reports of subjective experiences can index subjective states, and (4) subjective states are causes, not epiphenomena. Other discussions in the present book have provided specifics in this rapprochement—for example, in giving a central role to the concept of attitude, in presenting a justification for "insight"-oriented psychotherapy and nondirective therapy, and in providing a basis for psychological measurement employing verbal tests, and so on. The methods of experimentation of behaviorism and the principles of conditioning, stated in the precise, causal language of S-R learning theory, can be developed with the personality level of the theory to deal with the events in which the subjectivists are interested, in a way that is productive conceptually, empirically, and in dealing with human problems. One aspect of producing a rapprochement of these two traditions involves confronting and resolving the specific issues that continue the division.

## Awareness

One of the issues emerging from the subjectivist-objectivist schism has concerned whether human subjects can be conditioned. For example, it has been shown that a person asked to say words, any words, in an experimental situation, and who was reinforced for saying a particular class of words by the experimenter saying "Good," or some such, would come progressively to say words from that class more frequently (Greenspoon, 1950). This follows the principle of reinforcement and was taken as proof that human verbal behavior could be learned in that manner— that the external environment determines human behavior. In opposition to this behaviorist interpretation, it has been said that the subjects of such experiments actually say the words for which they are reinforced because they are *aware* of the experimenter's purposes and the experimenter's "demands" upon their response (see Dulany, 1968).

A similar issue has arisen in the context of the classical conditioning of emotional (meaning) responses to words that has already been described. The cognitive (subjective) approach has been that what appears to be conditioning is only that the individual becomes aware of the situation and its demands and behaves accordingly (Page, 1969). The subject may be considered to test various hypotheses concerning what he is doing and what the experimenter is doing. When he becomes aware of the correct relationship—for example, that positive emotional words are paired with the conditioned stimulus—he responds the way he thinks the experimenter desires.

Although these cognitive interpretations of human conditioning have been answered in turn (Krasner, 1967; Kanfer, 1968; Staats, 1969a), it should be noted that awareness can be considered in behavioral terms. Experimental conditions, it is suggested, can indeed bring about awareness, and the awareness can bring about changes in behavior that appear to be like conditioning. That is, thoughts or hypotheses can serve as stimuli that direct the behavior. When the hypotheses are correct, correct behavior will tend to be mediated, and if reinforced it will be maintained. Incorrect hypotheses do not mediate behaviors that are reinforced, and the behavior and the hypotheses are not maintained. In addition, however, this position recognizes that reinforcement can directly produce instrumental conditioning (Farber, 1963; Postman and Sassenrath, 1961).

It is suggested that what is considered by cognitive psychologists as the mental event of awareness actually consists in most of these experiments as language behaviors the subjects have learned. Every articulate human adult has a vast history of the development of increasingly complex language behaviors, including those of labeling. The individual, because of the well-learned labeling repertoire, may label elements of situations he encounters. Where the situation has unknown elements, the individual may "label" possibilities, that is, he may hypothesize or guess. Since the individual also has a well-learned verbal-motor repertoire, in a problem-solving situation the hypotheses will elicit corresponding behaviors.

Through these basic behavioral repertoires (as well as others), the individual is thus prepared to make instantaneous adjustments. He will not require an extended learning experience to respond appropriately in some situations, as does the nonverbal lower animal or the preverbal or nonverbal child. To get a dog, or a preverbal child, to close the door would involve a time-consuming training effort. Once the individual has learned to respond to language stimuli, he can perform the action immediately, on his own or at someone else's request. The same is true in the experimental situation. If the individual labels the contingencies in the situation correctly, he is prepared to respond accordingly instantly, without long training (although his response will still follow A-R-D principles).

The cognitive theorists are correct in insisting that the human being has this very important feature, which elevates him in this and other ways high over the abilities of lower organisms. They are correct in objecting to the rejection of this human characteristic by elemental behaviorists. Human beings are capable of adjusting to situations immediately, making complex adjustments without requiring a learning process for the specific instrumental behavior involved. Many human experiments, because they include the operation of cognitive repertoires, show sudden changes in behavior that are not typical of the gradual increase of learning associated with animal learning.

This interpretation helps resolve the issue. It may be added that even in situations where the subject makes a correct hypothesis and thereafter makes a response to secure reinforcement, the A-R-D principles still are operating. A correct hypothesis concerning the fact that a particular behavior is followed by a particular reinforcer in an experiment does not by itself increase the frequency of the behavior. Negative A-R-D value will have the opposite result. It is still the A-R-D value that determines which behavior will occur. It should thus be possible to devise an experiment in which some aware subjects respond in one direction and some in another.

Accepting the hypothesis-making abilities of the human as givens, it may be added, is not productive. This leads one not to be concerned with how humans acquire the "hypothesis-making skills." To understand how such skills are acquired, to understand why there are individual differences in such skills, to be able to improve such skills, it is necessary to understand the acquisition process. The rapprochement called for, thus, is recognition of the importance of awareness as a causative event, but also recognition of the basic behavioral repertoires of which awareness is composed and the need for the study of the manner in which awareness repertoires are learned and function. This is the position of social behaviorism.

This issue also has implications for the issue of whether insight is necessary in psychotherapy. As has been indicated, the need for insight has been stressed by traditional psychotherapies and rejected by the conditioning therapies. With the inclusion of the personality level in the approach, and the elaboration of the language-cognitive basic behavioral repertoires, it is possible to resolve some of this conflict, as has been indicated. It is suggested that human behavior takes place to a very large extent on the basis of language. We affect people's behavior and are affected, more by language than by direct reinforcement, and this can be true of clinical treatment procedures.

The fact is—as the psychoanalytic approach has suggested within the confines of its concepts—the individual, in many of his problems, is not aware of the events or principles causing his own behavior. He is not aware of some of his behaviors and how they affect others and the situations he thus gets himself into. This is not surprising, since such knowledge has not been generally available. Perhaps it would be possible to directly condition the individual in such cases to display more appropriate behavior. But, getting the patient through verbal means—that is, insight —to recognize the conditions, principles, and events involved can lead to immediate behavior changes (see Staats, 1972).

Again, the importance of insight should not be rejected. It is necessary, in making a rapprochement with a humanistic approach, to be concerned with what insight is (in terms of the behaviors of which it is composed) and the principles involved in its acquisition, change, and function.

## The individual versus the general

Behaviorism, if not objective psychology in general, has been criticized by humanistic psychologists for being concerned with the nomothetic (the general) rather than the ideographic (the individual) (Bugental, 1967; Maslow, 1966). "First of all, we should be aware that [knowledge] . . . about *a* person is ruled out by many scientists as trivial or 'unscientific.' Practically all scientists (of the impersonal) proceed on the tacit or explicit assumption that one studies classes or things" (Maslow, 1966, p. 8).

The issue referred to above is of long standing. Concern with the individual has been in opposition to the study of general laws, and differing viewpoints have emerged based upon these realities. As has been indicated, search for general laws meant the study of artificially simple stimuli and responses. This meant that the elements studied were quite disparate from the conditions of human life. The clinical psychologist, on the other hand, deals with the individual case, based upon knowledge of the individual. The products of general psychology as a consequence seemed to offer very little to the solution of human problems.

As part of general psychology, behaviorism has been concerned with general laws. Tolman (1932), in his learning theory, did suggest that individual differences must be included as a variable. However, the suggestion was not developed and had little effect upon the development of the other learning theories. The field became concerned with the general laws of learning, although there was an exception to this in the influential work of Rotter (1954).

The extension of conditioning principles to the human level must first be concerned with whether or not the principles apply. Concern with generality was thus correct at the beginning, as in the first behavior modification studies (Ayllon and Michael, 1959; Staats, 1957a). But what was a necessity has been taken as a virtue. Behaviorists have sometimes assumed that individual conditions, individual history, and individual personality are irrelevant and that general treatment procedures applicable to everyone will suffice.

It is suggested, however, that the general by itself is inadequate. Adding a personality level to the theory recognizes the need for concern with the individual and his history. Again, a rapprochement is necessary. It is essential to find the general (elementary) laws of behavior, but these alone do not tell us about individual human behavior. The elementary principles tell us the general ways the environment can affect the individual's learned behavior. But what behavioral repertoires he learns will be a function of his individual experiences. By the *same* principles, two individuals can become quite the opposite in the personality repertoires they acquire and in the way they will respond to the same situation. At the point where such repertoires have been formed, understanding human behavior necessitates knowledge of the individual personality repertoires. These will be determinants of what the person does, of equal importance with the general laws, as examples already presented herein have suggested. When dealing with another person in the clinic, as indicated by humanistic psychologists, we must know the person's "situation, his past, and himself, as these objects appear to him" (Rogers, 1947, p. 360).

## *Self-direction versus passive responding: Humanistic versus mechanistic views*

"Mechanomorphic psychology [behaviorism] . . . views man as an object acted upon from the outside by various forces or driven from within by other forces which are to be characterized chiefly by their relation to the outside (e.g., thirst, hunger, sexual appetite)" (Bugental, 1967, p. 8). As another example, in the area of child development Langer (1969) has considered learning approaches as making man an automaton: "The view most closely associated with a passive model of man is that which we shall call the mechanical mirror theory . . . that *man grows to be what he is made to be by his environment*" (p. 4). As these quotations indicate, a primary dissatisfaction with elemental behaviorism has been that it characterizes man as a passive responder. There are several issues involved here (as the following quotation will indicate) which will be addressed in the next several sections.

> We have a deeply ingrained cultural conception that human behavior is spontaneous—it comes from within. It is not bound, determined, or set —it is free. Moreover, in our own experience we have innumerable cases where our behavior occurs under our own direction. We decide what we are going to do and then we do it. Any conception that suggests that we do not have a hand in determining what we do fails at the very beginning. Furthermore, there is something wrong with a conception that is unable to account for the originality we see so frequently in human behavior. People, with varying degrees of social significance, make original behaviors—not just behaviors that they have been trained to make. Humans are not just automatons and we resist a conception that suggest this, as traditional learning theories have.
>
> Do not these demands upon a conception of human behavior pose challenges to *any* learning conception? It is true that the demands have been obstacles to general acceptance of a traditional learning approach. . . . [I]t will be productive to indicate how the types of learning that have been described herein will produce a child who thinks, who directs his own behavior, and who will make original behaviors. Although we learn our basic behavioral repertoires, these repertoires then enable us to influence our own behavior and to be original, thus providing us with the means to rise above our specific experiences. The learning conception being proposed by no means suggests that man is an automaton. We see how behavior can be learned, yet also be self-directive and innovative. Self-direction and innovation are foundations for a belief in personal freedom. (Staats, 1971a, pp. 251–52)

To begin, as one example, it is suggested that the language repertoires described previously provide a basis for extensive self-direction. Ordinarily, as indicated, the language stimuli will come to elicit appropriate responses. At first this will occur to the words that someone else directs to the child;

when his appropriate response has occurred he will be reinforced (at least a certain percentage of times). But the child will also be reinforced for doing what he says. When the child says he will be home at a certain time, he will be aversively treated if he is not. In addition, however, the child learns to follow what he says because of other rewards. When the child says to himself, "If I have a dessert now I won't get one after dinner," and the language sequence elicits "waiting" behavior, the child is reinforced for responding to his own speech. His language sequences can mediate behaviors that avoid punishment and gain reward, provided the language sequences are appropriate for the child's circumstances and he also has the necessary overt behaviors.

Thus, the child's speech behavior is first learned under the reinforcement provided by other people, which can be considered passive responding. However, once learned, the speech can improve the child's adjustment by eliciting appropriate behavior to the nonsocial as well as the social world. At this point the child is not just the object of a training program administered by the parent. His self-directing behavior "finds" its own reinforcement. This independence from the reinforcement of another person can be seen clearly when the self-directing language is relevant to a nonsocial occurrence and mediates behavior that is reinforced by that occurrence. For example, the child who sees the sky darkening and says, "It looks like it is going to rain, so I'd better get some books to read because I won't be able to play outside" will be self-directing his behavior. The boredom he avoids will reinforce the behaviors. We continually employ our language repertoires for self-direction in a variety of social and physical situations. The language sequence is in such cases the primary determinant of what we do.

As this occurs the child emerges as an organism who is directing his own behavior. This is a new level of adjustment over the child who has only learned to use words for their effect upon others. It also frees the child, for many of his actions, from dependence upon the reinforcement provided by others. The principles of the language-cognitive repertoires indicate some of the ways that we can direct ourselves through our learned language repertoires. In so doing a view of *behavior* causation is provided that helps unify subjective and behavioristic demands.

## SELF-DIRECTION AND BEHAVIOR MODIFICATION

Interest in the possibility of self-direction in the treatment of problems has arisen in behavioral clinical psychology. Watson and Tharp (1972) and later Thoresen and Mahoney (1974; also see Thoresen, 1973b) have outlined programs for self-modification that are similar in their general features. Each considers identification of the problem behavior by the client; including observation and recording of the behavior, as central.

Thoresen (1973) has suggested that the act of observing and recording one's behavior, in making one more aware of one's behavior, also influences the behavior being observed. Research on self-observation suggests that behaviors desired by the individual will increase through the act of self-observation (Mahoney, 1972). A study by Bolstad and Johnson (1971) is cited by Thoresen as evidence that recording of their own disruptive behaviors reduces the incidence of such behaviors in young children. Broden, Hall, and Mitts (1971) had an adolescent student who wished to improve her use of study time in class record her incidence of study behavior. The behavior increased during the period she recorded her behavior and fell off when the recording was discontinued.

The second element in the self-directive behavior modification methods involves the use of learning procedures by the individual to affect his own behavior. Watson and Tharp (1972) demonstrate the simple use of self-reinforcement in the following case:

> One of our students decided to increase her study time. The reinforcement for studying was to have a carbonated drink. Before intervention, she had observed that she drank about five or six sodas each day. All she had to do was make them contingent upon studying the required amount. If she studied, she got the drink; if not, no drink. She would go to her study place, put in whatever amount of time was required by her schedule, and immediately go for a drink. Thus, she was providing reinforcement very quickly after performing the desired behavior. (p. 121)

Watson and Tharp also utilize token-reinforcer procedures in their self-directive behavior modification method:

> Many people, in modifying their own behaviors, choose a *point system* of token reinforcement, rather than using actual objects. . . . These points can then be "spent" for reinforcement. . . . For many people, the chosen reinforcer is something they can do at the day's end. They may use a particularly nice supper, or the opportunity to watch TV, or a talk with friends in the evening, as their reinforcer. . . . A man who wanted to substitute being-nice-to-friends for being-rude-to-friends selected as his reinforcer *watching TV in the evening.* . . . He carried a 3 × 5 card in his pocket and made a check on it when he performed the target behavior. Then, later in the evening, he would allow himself to watch TV if he had earned the number of points [required]. (p. 123)

Self-directive behavior modification also involves the individual manipulating environmental conditions to help induce the behavior he desires.

> [This] involves changing one's environment so that either the stimulus cues which precede the behavior or its immediate consequences are changed. This restructuring of the environment often involves the elimination or avoidance of daily life situations where a choice or de-

cision is necessary. Several studies have demonstrated that self-control is very effective through "stimulus control" in which a person alters the environment so that the problem behavior is associated with progressively fewer stimulus cues (Ferster, Nurnberger, and Levitt, 1962; Goldiamond, 1965b; Stuart, 1967). The overeater, or drug user, for example, may avoid those situations associated with the behavior that "stimulates" the problem behavior, or he may gradually narrow the situations in which he engages in the [response to be controlled]. Smoking cigarettes only in the basement by oneself after 10:00 P.M. represents a restricted stimulus situation. . . . Similarly, the obese eater may control eating by removing environmental cues, such as the television set, the cookie jar, and close friends, when eating. (Thoresen, 1973b, p. 407)

These two methods for self-directive behavior modification indicate the practical applications of the conception that human behavior is self-directive, as well as being a function of external environmental events.

## Purpose and scientific causation

Another expression of the humanistic-behavioristic antagonism revolves around the question of whether human behavior is purposive.

In dealing with persons, you must make your epistemological peace with the fact that people have purposes and goals of their own even though physical objects do not. . . . This simple fact, which is excluded systematically from the model of classical physical science [and thus behaviorism], automatically makes its methods less appropriate for studying most human behavior. . . . Because of this, it cannot discriminate between correct and incorrect instrumental behavior, between efficient and inefficient, right and wrong, sick and healthy, since all these adjectives refer to the suitability and efficacy of the means-behavior in actually attaining its goal. (Maslow, 1966, p. 18)

It is quite true. Man does not seem to be driven only by stimuli of the moment. Much of his behavior is relevant to events that may lie far away in the future. In many instances his behavior seems to be more influenced by these events than by those of the moment. Human behavior can have long-term, purposive characteristics.

On the other hand, scientific causation moves along a time dimension —earlier events can determine later events, but not the converse. Future events do not affect events that have already happened. When it is said that man's goals affect his present behavior, and the goals are conceived of as events that are still in the future, then this abrogates the "rules of causality" that have generally been found in science. Elemental behaviorism has not resolved this dilemma, restricting itself to the study of present events that affect the behavior of the organism.

How then is the dilemma to be resolved? It is suggested that future events, in a sense, can be conceived of as determining behaviors that occur before those events. This is "in a sense," however, for it is actually the reasoning and planning sequences of the individual that are relevant to those future events and determine the individual's action. As an example, take the previous case of the child who sees the sky darkening and says "It looks like it is going to rain, I'd better borrow a book to read because I won't be able to play outside." This is a reasoning sequence elicited by present circumstances, in correct causal order. However, the reasoning sequence can be emitted in a briefer period than it takes the climatic events to occur. Since the reasoning sequence can control the overt behavior—in the same manner as would the actual physical event—the child can behave in a manner that is relevant to the future event. To naturalistic observations this would appear to be goal-directed behavior—behavior under the control of the future events. It is actually behavior under the control of other *present* behaviors that are relevant to the future events. The behavior is independent of the future event, however, for it may not rain at all.

The example above was one of simple planning. There are in everyday life examples of much greater complexity. Sometimes when the complexity is great, when the reasoning sequences are long term, and when continuing aspects of the individual's behavior are directed by the sequences, this is spoken of as purpose, ambition, or goal. It may be said that the individual has the goal of becoming a great scientist. The manner in which he works hard in college, selects a top graduate school, reads widely over and above that which is necessary to be successful in his academic work, and later performs extended acts that result in approaching that "goal" lead one to describe his behavior as being determined by a goal—a future event that has not yet occurred. His behavior, as indicated, is actually influenced by a complex of behaviors, heavily involving language. His purpose consists of repertoires of behaviors that he has acquired through his past experience and that are maintained by present circumstances. Those organized complex reasoning sequences, which include potent and complex sources of emotional responses (which have reinforcing and directing qualities), allow for the organized, directed characteristic of human behavior that we can see. Such organized systems of response repertoires can result in organized human behavior that extends over a lifetime.

Thus, goals and purposes can be conceived of in a way that is coincident with natural science causality and yet can account for the purposiveness of human behavior. It is not the future events that determine the individual's present behavior. These future events are brought forward through the symbolism of the individual's language and other cognitive repertoires. What the future event will be cannot be a causative variable. But what the individual thinks and feels about those future events can be causative variables. These constitute his goals and purposes; they can play a decisive role in how he behaves and be considered to be a prominent cause of his behavior. His goals and purposes, in this way, can be considered to be causal.

And it may be necessary to understand those goals and purposes if one is to understand the individual's actions. It should be evident, however, that what one gains knowledge of, in assessing the individual's goals and purposes, is his conception of future events—not the future events themselves. His language sequences may parallel what later happens, or, on the other hand, be quite wide of the mark. In either case they can determine what the individual does. To understand the purposive behavior of the individual one must know of his complex symbolism of the future events, not the events themselves.

It is not necessary, it is emphasized, for the humanist-behaviorist schism to continue over this issue. It is suggested that the humanist is correct in criticizing the philosophy of the functional analysis of the external causes of human behavior, for example, as Locke (1969) has done. In contrast, in dealing with motivation in working situations, he hypothesizes that goals and intentions most immediately determine performance. In his analysis, incentives (A-R-D stimuli) affect performance through their impact on a person's goals and intentions (Locke and Bryan, 1969). When the language-cognitive repertoires are considered in conjunction with the emotional-motivational repertoires, it is possible to analyze such types of goals and intentions in terms that are appropriate for both the humanist and the behaviorist.

The humanist needs a conceptual framework within which he can be concerned with the individual's goals and purposes. Only in this way can the individual's behavior be understood. The scientist, on the other hand, needs a conceptual framework that allows him to study such events as human purpose in the search for cause and effect laws. He needs a framework within which he can study what purposes consist of, what the causes of purposes are, as well as how they have their effects. It is suggested that the conception of purpose in human behavior can embrace the concerns of both the humanist and scientist.

## Originality

A primary objection to behaviorism and learning theory is that it makes of man a "reactive," not a creative creature. It is said, and rightly so, that the simple conception that human behavior is learned does not provide for originality or creativity. The question is, how can behavior be novel and original if it first has to be learned? This is a paradox that requires resolution, for no conception of human behavior can be credible unless it accounts for originality. Some of the most important behaviors in man's continuing advancement are those that involve new behaviors hitherto not made by others. Such novel behaviors are ones the individual has not been specifically trained to make.

It is not possible here to make a statement of the various ways that the individual may arrive at behaviors that are novel to him, and sometimes to society. The type of process that produces originality in a wide number of behavioral areas has been outlined, and this may be elaborated in the present context. The essential aspect of the conception is that single responses may be learned to single stimuli. When new combinations of stimuli occur, however, new combinations of responses can be elicited.

A simple example was given in an earlier chapter of the child saying "running man" as an original combination. The components of one's language repertoires can combine and recombine in various ways that produce new combinations of behavior in multifarious ways. As another example, the preceding chapter stated that a child learns a repertoire of motor responses under the control of various verbs, as well as a repertoire of labeling responses for various objects that we call nouns. New combinations of components from these types of words will elicit new responses to the objects in the listener. Once verbs such as *push, pull, close, open, squeeze, touch, rub, lean,* and *kick* have come to control the appropriate motor responses, and once the child has come to respond to labels for objects such as *door, chair, dog, board, car,* and *ball,* the combinations of a verb with a noun can be applied to elicit a very large number of different responses. Thus, for example, the child may have learned the "pushing" motor response through being told, "Push the chair," "Push the button," "Push the carriage," and being shown how to push these objects. He may never have had experience with pushing doors, chairs, dogs, boards, cars, balls, and so on, through verbal instruction. If, however, he has learned the pushing response under the control of the word *push,* and he has learned to label doors, chairs, and so on and to look for them on verbal instructions, the very first time he is instructed to "push the door" he will behave correctly. When the combinations of verbal stimuli are novel—and there are an infinite number of possibilities—the verbal stimuli can elicit novel responses, ones the child has never before emitted.

As has also been indicated, a series of verbs (or verbs and nouns) can be put together in a way that an extended sequence of motor responses can be elicited that is original or novel. This happens whenever someone gives geographic directions to a stranger and the sequence of verbal stimuli elicit in the stranger a sequence of motor responses novel to him. The verbal stimuli will bring the stranger into contact with new geographic stimuli—the ones the stranger desires, if the directions are isomorphic with reality. This, of course, is the same function that a new theory can serve in science. It can lead investigators to look for things not observed before.

There are numerous illustrations in physical science—for example, astronomers have found and continue to find objects and events in space that the theory of relativity suggests. Mendeleev's periodic table of atomic weights yielded the search for new elements to fill lacunae in the systematic structure.

Since only one mechanism in the production of original behavior has been presented, the account is necessarily limited. But even with this one principle it is possible to consider more complex cases of creativity. Le us look at the periodic table of atomic weights in somewhat greater detai. to suggest some of the events involved in its production and extension. Early chemists were interested in finding the constituents of matter and in describing these constituents (elements). They found, for one thing, that different elements had different weights (mass). Looking for the elements involved skills the chemists had learned and elaborated. Recording the findings also involved learned repertoires. The listing of atomic weights, however, as the elements were discovered, constituted a new original stimulus circumstance. That is, the weights when listed had certain systematic features. Not only were there a number of different elements, but the weights of the elements increased in magnitude, with certain characteristics of order. This constituted a new stimulus. Mendeleev's response, let us say, was that the elements increased in a systematic manner, and a table was constructed to depict this order. As the table showed, the order was not perfect; there were gaps. That was also a new stimulus. The response was that perhaps there were elements not yet discovered that would fill the gaps and preserve the ordered system. This verbal conclusion then served as a stimulus to look further for new elements.

Furthermore, Mendeleev's periodic table, taken as a whole, served as a novel stimulus for speculation concerning what it was in the elements that made them increase in weight in an ordered manner. This led to the response, for example, that there must be commonality in all the elements, in all matter. Perhaps the elements themselves were composed of common constituents in varying amounts. This reasoning response sequence then elicited additional theoretical and observational behaviors.

This, of course, is only a loose and general description of the production of original behavior. However, it does suggest that individuals with certain repertoires of behavior, faced with certain novel stimulus situations, will come up with certain novel responses and products of those responses (discoveries, findings, and so on). These products will then serve as new stimuli which then produce additional novel responses, which then serve as new stimuli which then produce additional novel responses, which then serve to elicit yet other creative behaviors. It is suggested that a detailed history of such long-term acts of creativity, analyzed in terms of the learned skills of the men involved and the manner in which their findings served as stimuli to themselves and other individuals, would give us a more profound view of human creativity.

It should be realized that the various basic behavioral repertoires the individual learns consist of many, many units that may enter into novel combinations within the repertoire and across different repertoires. The central point is that (1) when the basic behavioral (personality) repertoires include many units, (2) when there are a number of repertoires, and (3) when the stimulus configurations impinging upon the individual are varied, there is a very large opportunity for creative acts of combination to occur.

Moreover, formerly novel units then can become available as constituents for other novel combinations. This is another type of cumulative-hierarchical learning—which operates within one individual over time, as well as from one individual to another. The author (1963) has given as an example Thales's original proof that the opposite angles formed by bisecting lines are equal. The several constituents of the solution were themselves at one time original mathematical response combinations, for example, the rule "Equals subtracted from equals leave equal remainders." This statement had to be discovered itself. Once elicited and learned, however, it could serve as a unit in another creative act. It is in this way that human creativity is cumulative and hierarchical in nature. These principles should be demonstrated experimentally, for they are human learning principles that do not derive from the animal laboratory.

Such cumulative-hierarchical learning provides a basis for the ever-increasing scope, abstractness, creativity, and power of the developing individual and, over time, of historically developing mankind. There is thus no reason for an incompatibility in this area between humanistic psychology and the social behaviorism theory of human behavior.

## Personal freedom and scientific determinism

The question of the determination of human behavior has been with us since antiquity. One early explanation was that there was a little man inside (a homunculus) who directed external behavior. But this concept simply defers the question, because one still has to account for what determines the homunculus.

The issue has continued between scientific determinism, on the one hand, and the very general experience of freedom of behavior. Our general study of physical events of our world has convinced us of determinism—events do not occur spontaneously, capriciously, from supernatural causes. The evidence is overwhelmingly in favor of natural, lawful causation.

In the realm of human behavior, however, there is still strong opposition to scientific determinism in causation. Contemporary humanists do not wish to treat man as an object of study—to assume his behavior to be caused by specifiable events. They hold out for spontaneity, for freedom, for self-direction, for activity. They reject a deterministic conception.

> The various behaviorisms all seem to generate inexorably such a passive image of a helpless man, one who (or should I say "which"?) has little to say about his (its?) own fate, who doesn't decide anything. . . . My crucially important experience of being an active subject is . . . either denied altogether . . . or is simply pushed aside as "unscientific," i.e., beyond respectable scientific treatment. (Maslow, 1966, p. 55)

What seems like a pair of mutually exclusive, antagonistic positions, however, need not be. Human behavior can be considered as caused by other material, natural events—not uncaused (spontaneous), capricious, or supernaturally caused events—without rejecting the originality, activity, and self-direction characteristic of man and the experience of freedom and spontaneity. It is suggested that the laws of learning, elementary and human, are causative laws and that the individual is what he has learned, as he is affected also by the present conditions to which he is subjected. The personal freedom, self-direction, creativity, and spontaneity that we experience can involve the way that individuals extend their past learning through reasoning, planning, purposeful goals, and so on, many times including original combinations of behavior that have not been learned.

In addition, however, freedom, self-direction, and spontaneity are *what we experience in our own behavior.* Our repertoires are *us.* They are our personalities, our "being." We do not and could not recall the infinitely complex set of learning experiences that molded our repertoires. What we experience are our own repertoires. For us our personality repertoires as they function are original causes—they are the givens. We experience our decisions, our plans, our reasoning—all complex repertoires of learned responses—as the causes of our behavior.

And these are also immediate causes. When the individual engages in reasoning or planning language sequences of behavior, and these sequences culminate in the elicitation of some overt action, the reasoning and planning are the causes of the actions taken. Other personality repertoires, not involving direct elicitation, can also result in the experience of personal freedom and choice. For example, two men may begin the same job in an organization. One may have a different A-R-D system than the other and thus find the work "uninteresting," "unmotivating," or "unrewarding." Later, when the two men have had different degrees of success, the one described may indicate his own personal responsibility and experience of freedom, with the following type of statement: "We both had the same opportunity. He worked at it and developed new skills. I did not. I am responsible for not continuing to advance." He sees himself, at a particular

time, as the "master of his fate." And he is right, for his personality repertoires have set the conditions for what he does and what he thereby becomes.

There is personal causation and there is the experience of freedom in
another way also, that which occurs through behavior-environment interactions. That is, the individual's personality repertoires will have effects
on other people that the individual may observe. He may see, for example,
that his behaviors (selfishness, demandingness, and untrustworthiness)
have alienated his associates. He may then describe his social environment
as having been a function of his own actions.

Thus, it is suggested that the phenomenological or existential experience of determining one's own behavior, of controlling one's destiny,
arises in the self-observation of the various causal processes. The individual
observes his decisions, wants, and interests, his effects upon others, upon
himself, and upon his environment, and so on. He sees that in these ways
he determines his behavior, what he becomes, what happens to him, how
people respond to him, and so on. He also observes similar events with
other people, sometimes more clearly than when he himself is the object
of observation.

He does not observe and is not aware of the learning causes of his own
behavior or that of other people. The determinants are too complex, and
many occurred long ago, even prior to the individual's verbalizing them.
Moreover, the external determinants have had such complex interactions
with the "internal determinants," like decision making, that the contributions of each may be difficult to conceptualize even in the general case,
let alone by the experiencing individual. For example, the individual does
things because he decides to do so. The way he decides will be a function
of his past experience, but this past experience will also have been affected
by many of his past decisions. By the time a person is an adult it would
be impossible for him to separate the two sources of causation (Staats,
1971a).

These various reasons underlie the pervasive acceptance of personal,
existential, phenomenological conceptions of human behavior causation.
Furthermore, it is suggested that a theory of man must have a place for
this pervasive experience of inner self-determination. Elemental behaviorism has not addressed itself to the resolution of this problem or provided
a conceptual framework by which to make the resolution. As part of the
social behavioral humanism, it is suggested that there be an existential
(phenomenological) level to the theory. That is, it is important to recognize that the individual experiences what he has done and the effects it
has upon his behavior, his behavior development, and his life circumstances. These experiences are labeled with words just as other experiences
are. These words will have an emotional content, as is also the general
case. The labels of one's experience of this type may be positive or nega-

tive. These descriptions of one's behaviors—decisions, plans, purposes, and so on—will help compose the individual's self-concept, or phenomenal self.

This awareness of one's self-determination may be considered an epiphenomenon in one sense. It arises from the actions of the individual's learned personality repertoires and the effects these actions have. It should be indicated, however, as outlined earlier, that the self-concept itself has determining potentialities. The self-awareness will enter into later decisions of the individual. At this point it is clear that this conception helps fulfill the need to understand our experience of personal causation. It also helps fulfill the need for understanding human behavior within the framework of scientific causation.

## ATTRIBUTION THEORY AND THE EXPERIENCE OF PERSONAL CAUSATION

It has been suggested that part of the conception of causality involves one's own experience of self-directive behaviors, as well as one's observations of other persons' behavior. In the field of social psychology, a set of concepts and findings has developed that is relevant to and extends this interest in the experience of personal causation and the general concept of behavioral causation.

Stimulated by Heider's work (1958), attribution theory has been concerned with the processes by which a perceiver infers the causal locus for a behavior. "In Heider's formulation, an action outcome is an additive function of the effective environmental force and the effective person force, the latter including both the person's ability and the effort he exerts" (Pines, 1973, p. 262). Rotter (1954) developed a related interest with his concept of the locus of control. This was defined as a generalized expectancy regarding the extent to which a person's own behavior or the external environment was perceived to be the controlling factor in obtaining reinforcement.

Kelley (1967) has elaborated on the variables that underlie the attribution of causality regarding another person's behavior in a situation. Thus, whether or not the behavior occurs when other stimulus situations are present, as well as the one stimulus situation, is a variable that influences one's conclusion that the stimulus situation is the cause. For example, if the person's behavior occurs in various situations, not only one, then it will more readily be assumed that determinants of the behavior reside within the individual. Also, whether or not the same behavior is produced by other persons in the presence of the same situation provides information concerning causality. For example, if various people in the presence of a particular situation behave in the same way, then it will more readily be assumed that the situation is the determinant of the behavior. Finally, whether or not the behavior occurs every time the specific situation is presented provides information concerning causation. If the behavior occurs when the situation is present and ceases when the situation ceases, then this is evidence for the causal nature of the situation.

Various studies have investigated the types of implications involved in attribution theory. For example, Himmelfarb (1972) presented information to subjects which was ostensibly based upon observations of an individual's behavior by one or more judges. He varied the information in terms of whether the individual's behavior was observed in only one situation or in several dissimilar situations. He concluded generally "variations across . . . different situations did produce increased weighting of trait attributions of a person" (p. 312).

It has also been proposed that similar attribution inferences are made by the person about his own behavior (Bem, 1965, 1967; Kelley, 1967). Bem and McConnell (1970) suggest, as an example, that the individual asks himself what his attitude must have been if he performed in a particular way in a situation—his behavior being the causal stimulus, rather than the attitude. Zanna (1973) has found evidence to support these hypotheses, although not unequivocably.

Schopler and Layton (1972) have extended these principles of attribution to consideration of the extent to which person A infers that he has influenced person B in an interaction. These authors propose that "Person A's attribution of having influenced Person B will be maximum if B's state after A's influence attempt is (a) not predictable from knowledge of B's previous state, (b) evaluatively positive, and (c) predictable from A's influence attempt" (p. 326). In the experiment, subjects were told what they thought was another person's score on the first half of a test. The score could be low or high. The subjects then helped that person on the second half of the test by telling him their own answers to test items. The results supported the expectation that high influence would be attributed when the person's performance changed, and the change was not expected from performance on the first half of the test. The results also showed that subjects attributed their own influence more when the person's performance changed for the better than when it changed negatively, as would be expected from A-R-D principles.

There are thus studies such as this one that attempt to isolate the principles and conditions involved in the way causes are attributed to behavior or changes in behavior. As has been indicated, the principles have generally been shown to be relevant also when the behavior is one's own. It has been noted, however, that different individuals often have different views of the causes of another person's behavior. Jones and Nisbett (1971), moreover, abstracted from Heider's writings a general hypothesis that there are different interpretations of causation by the actor—the person who performs a particular behavior—and the observer of the behavior. These authors proposed that actors attribute more causation to situational cues, while observers attribute more causation to the disposition (for example, attitude) or quality (for example, trait) of the actor.

Various studies support this hypothesis (Jones and Harris, 1967; Jones, Rock, Shaver, Goethals, and Ward, 1968). In a recent investigation derived from the hypothesis, a subject was presented with a situation which involved a request to volunteer service as a weekend hostess at a social function. This subject was the actor, and another subject observed the situation and the actor's decision. It was found that observers assumed that "actors would behave in the future in ways similar to those they had witnessed" (Nisbett, Caputo, Legant, and Marecek, 1973, p. 154). They thus saw the determinants of the behavior as residing within the individual, with generalization across situations. The actors themselves did not make this interpretation.

In another part of the same study other subjects were asked to write briefly why they liked the girl they dated most regularly and also to indicate why they had chosen their college major. They were then asked to write briefly, on the same topics, about their best friend's choices of girl friend and major. It was found that in describing their best friend's choices they referred to the dispositional qualities of the friend. For example, with respect to the friend's major a dispositional response might be "He wants to make a lot of money." In describing their own behavioral causation, however, the actors emphasized the properties of the girl friend or major. For example, "She is a relaxing person" (Nisbett, Caputo, Legant, and Marecek, 1973, p. 159).

These investigators also had subjects fill out questionnaires for themselves and four other persons, indicating on a number of items whether a trait term or its opposite fit, or whether the phrase "depends on the situation" fit. The subjects perceived their own behavior to be a response to the situation more than they did for the other persons, perceiving the behavior of others to be more trait determined.

Storms (1973) has suggested that actors and observers very literally have different points of view, with actors being unable to see much of their own behavior. Actors do, however, attend to the antecedents and consequences of their own behavior. He gives the following summary of a study testing the implications of his analysis.

> Two actor subjects at a time engaged in a brief, unstructured conversation while two observer subjects looked on. Later a questionnaire measured the actors' attributions of their own behavior in the conversation either to dispositional, internal causes or to situational, external causes. Similarly, each observer attributed his matched actor's behavior. Videotapes of the conversation, replayed to subjects before the attribution questionnaire, provided an experimental manipulation of visual orientation. Some actors and observers saw no videotape replay, while

other subjects saw a tape that merely repeated their original visual orientations. As predicted for both of these conditions, the actors attributed relatively more to the situation than the observers. A third set of subjects saw a videotape taken from a new perspective—some actors saw a tape of themselves, while some observers saw the other participant with whom their matched actor had been conversing. With this re-orientation, self-viewing actors attributed relatively more to their own dispositions than observers. The results indicated the importance of visual orientation in determining attributional differences between actors and observers. (Storms, 1973, p. 165)

It is especially interesting to bring this area of research and theory into conjunction with a behavioral analysis of the experience of personal causation that rests upon learning principles. Although attribution theory sprang independently from a cognitive conception, its present statement and the derived research deal with explicitly stated behaviors. As a result the findings and principles are very coincident with the social behavioristic view.

This coincidence is more general, it should be noted, than the specific topic of present concern. For example, attribution theory and the research findings provide support for the analysis of the self-concept that has been presented. This analysis suggested that the individual has a chance to observe his own actions and the outcomes of his actions, as well as his own subjective experiences. On this basis he acquires a verbal description of himself, and a set of related attitudes, and so on. This analysis provides behavioral specification of the term self-concept.

Part of this behavioral analysis also indicates how the self-concept has trait characteristics. That is, like other behavioral repertoires, the self-concept is a determinant of the individual's later behavior, learning, and adjustment. It is interesting to note that experiments in the context of attribution theory provide support for the manner in which self-attributions can affect later behavioral events. Rotter (1954) originally proposed the concept of the locus of control. Some individuals (internals) were seen to interpret the consequences of their behavior to be under personal control—to be determined by such factors as ability or motivation. Others (external controlled) tended to interpret their success or failure to factors not under personal control, such as chance, other persons, and so on. Various interesting findings have stemmed from this conception. For example, Strodtbeck (1958) found that middle- and upper-class individuals tend to feel more in control than lower classes. Gore and Rotter (1963) found that students in a Negro college differed in terms of interest in civil rights social action, depending on whether they were internals or externals (the former expressed more interest in social action). Other studies have shown that internals, who feel they personally control the outcomes of their experiences, make a greater use of materials given them in a task they have to perform (Crowne and Liverant, 1963; Lefcourt, 1966; Lefcourt and Ladwig, 1965; Rotter, 1954). Moreover, it has been shown that attribution of success or failure on an internal-external basis influences the degree to which one will predict future success or failure. When success or failure is attributed to ability there is a greater effect on future predictions than when the outcome is attributed to chance (Phares, 1957; Rotter, Liverant, and Crowne, 1961).

Several investigations (Lanzetta and Hannah, 1969; Leventhal and Michaels, 1971; Weiner and Kukla, 1970) have demonstrated that attribution of success to effort—an internal attribution—augments the rewards for achievement. Punishment for failure similarly is augmented when attribution is made to the lack of effort. Weiner, Heckhausen, Meyer, and Cook (1972) have shown that this occurs when the individual attributes cause in his own behaving. They found a relationship between self-attribution of effort and the extent that the individual would self-reinforce himself, using the self-reinforcement procedure of Kanfer and Marston (1963). Weiner, Heckhausen, Meyer, and Cook (1972) found that "Thus, the greater the tendency to attribute success, rather than failure, to effort, the greater the self-reward for success relative to self-punishment for failure" (p. 242). They concluded by saying "It is contended that attributions to effort play an important role in determining the direction, magnitude, and persistence of achievement-oriented activity" (p. 239).

> One of the controversies that divides psychology has been refusal of many cognitive psychologists to accept the [concept] that man does not contribute to his own behavior, but that his behavior is determined from external experience. Such psychologists insist, rather, that it is the individual's *awareness* that determines what he does; it is how the individual perceives the situation, and so on. There have been many unresolved experimental and theoretical controversies between such cognitive pyschologists and the traditional learning approaches. A rapprochement between these views is necessary, possible, and productive. A general conception of human behavior must thus show how human behavior is caused by the conditions the individual experiences, but at the same time also indicate how the "nature" of the individual contributes to his behavior. The manner in which self-determination or self-direction takes place is thus of central importance to a general conception of human behavior. (Staats, 1971a, p. 253)

It is suggested that the study of attribution is important in several ways. It contributes to our understanding of how we gain our conception of the causation of human behavior, and the experience of subjective causation. It contributes to the study of one realm of behaviors—attribution of causation—that has effects upon other behaviors. And it includes behavioral descriptions and research methodology that can be employed in affecting the type of rapprochement suggested above. It is suggested that additional research and theory on the basis of a rapprochement would be valuable in this important area.

## Personal responsibility

In rejecting the idea that man is free, elemental behaviorism also rejects the concept of responsibility. This is another area that has to be considered more deeply, however.

Originally, the concept of responsibility came from the belief that human behavior was divinely inspired. If behavior was good it was because the individual was righteous and divinely inspired to good behavior, and thus he was deserving of admiration. If his behavior was considered undesirable it was because of personal evil and the devil's influence, so it was deserving of punishment. As an example, abnormal behavior was once "treated" by subjecting patients to frightful and painful procedures intended to exorcise the devil.

Vestiges of this moral view of human responsibility have provided the basis for the use of punishment in penology. It was a step forward to consider man's behavior as determined. This view sees him as not morally responsible, and thus it provides a basis for a treatment orientation in penology rather than a punishment orientation.

But perhaps something is lost in the *simple* environmental conception. It seems appropriate to ask whether the belief in a religious personal responsibility has had a function. The fact is that the belief in moral responsibility has been a means of preventing certain types of behavior and insuring the occurrence of others. The religious person insists that he does not do certain things that are otherwise attractive because he considers them sinful and himself morally responsible. He may also do desirable things he does not wish to do, for the same reasons. Moreover, in addition to his personal experiences, he can point to the rise of self-indulgence in recent times in such things as drugs and sex that has taken place as the belief in religious morality and personal responsibility has declined.

Perhaps the concept of personal responsibility has had a function in our conception of man. Perhaps behavioristic analyses must examine the function of the concept. Perhaps, also, consideration must be given to ways of meeting this function in a scientific conception of man. While moral responsibility may not be supportable in a scientific view of human behavior, a *causal responsibility* can be. The concepts already developed are important here. That is, what happens to the individual at a later time is frequently the responsibility of what he has done at an earlier time. The individual who drops out of life's demands is *responsible* for what he experiences later in life. The youth who misses a college education may have to face a life with many fewer opportunities. The individual who is aversive to his associates, or his children, or whomever, and who later experiences aversive treatment in turn, is responsible for his unpleasant social-emotional relationships. The individual who loses his religious morality and engages in unrestrained sexual behavior may have a less pleasant relationship with his spouse. The criminal who injures some part of society and who is injured by society in return has contributed to these circumstances.

This does not imply the morally based eye-for-an-eye type of responsibility. It means that it is oversimplified to shrug off the notion of personal responsibility by a gesture that only suggests, in very general terms, that only the environment can be at fault, not the individual's ways of acting. A notion of causal responsibility, on the other hand, suggests that one's personality repertoires have a causal effect. The individual cannot dodge the responsibility for his behavior, for it produces systematic effects on other individuals and on society.

Moreover, it is suggested that man *needs* "restraining" skills (or beliefs). There are many things in life which are attractive but which will later prove undesirable to ourselves directly or because they are undesirable to others. We need behavioral mechanisms of restraint and mechanisms for doing unpleasant things today, because the outcomes of these things will reward us in the future. The conception of responsibility for one's actions in terms of the effects these actions will have on others, the effects they will in turn have on oneself, and the effects they will have on future conditions one will experience is an important one in personal decision making. Again, the simple rejection of a concept by elemental behaviorism does not appear to be sufficient. Another step is required, that of seeing the purpose of the concept and of providing one in its place to fulfill that purpose—one that is acceptable to science, it may be added.

## Reductionism in the study of man

It has been stated in many terms that man must be studied holistically, as a whole, not analyzed into more elementary constituents. Gestalt psychology was composed around this assumption—that there are emergent principles of the whole which knowledge of parts would not reveal. Humanistic psychology has accepted this position.

> The customary scientific technique of dissection and reductive analysis that has worked so well in the inorganic world and not too badly even in the infra-human world of living organisms, is just a nuisance when I seek knowledge of a person, and it has real deficiencies even for studying people in general. (Maslow, 1966, p. 11)

> The humanistic psychologist . . . disavows as inadequate and even
> misleading descriptions of human functioning and experience based
> wholly or in large part on subhuman species. (Bugental, 1967, p. 9)

Elementary, atomistic investigation of the principles of behavior—along
with attempts to explain human behavior causation by such principles—
is seen by the humanistic psychologist to be dehumanizing.

As will also be indicated later, the issue of reductionism versus holism,
or emergentism—the attempt to study man only in his naturalistic circumstances, without analysis—runs throughout the behavioral and social
sciences. Several points can be made here, however. First, it may be said
that the rigid statement that man can only be studied and dealt with as
a whole must surely be as extreme a position as a rigid injunction to deal
only with recognizably artificial samples of human behavior. It is suggested that neither is complete. The naturalistic descriptions are of central
importance to a comprehensive theory of human behavior. On the other
hand, if there are elementary principles of human behavior—as the present
approach suggests—they will not be discovered readily in the naturalistic,
wholistic situation. Analytic, albeit artificial, experimental research also
has its place.

## Understanding versus prediction and control

It is also contrary to the humanistic approach to consider the goals of
the study of man to be prediction and control, repeatedly stressed as the
goals of behaviorism. For one thing, such goals appear to deny the spontaneity, freedom, and self-determination of man. For this reason, also,
behaviorism is considered dehumanizing. These goals can also be considered to suggest an interest in manipulating man for ulterior motives, a
"1984" type of interest. There is also a feeling that if one person does
things purposely to change the behavior of someone else, this is undesirable. Rogers' client-centered psychotherapy (1951) is built around the
concept of providing conditions within which the individual can change
himself, rather than be changed from without. What is important to the
humanist is *understanding* the individual, rather than dealing with him to
effect some previously thought-out behavior modification.

The humanist also places great stress on accepting the individual as a
responsible, active participant in any treatment procedure. On the other
side, our extension of learning principles to human problems has given behavior modifiers an orientation to manipulate variables to change the individual's behavior according to the therapist's plans. In so doing the principles and procedures are taken directly from the animal laboratory. Thus,
as an example, the homosexual may be given aversive conditioning in
which pictures of males are paired with electric shock. The patient will

engage in the therapy under his own volition, of course, so this treatment is not the same as giving a psychotic a lobotomy or electroconvulsive treatment without his permission. Nevertheless, the conditioning methods are frequently practiced without the understanding of the patient. Although for benign purposes, the patient's behavior is manipulated.

Is it ever justified to arrange specific conditions to produce certain desired behaviors in another person? Are we limited to attempting to provide a generally warm atmosphere in which the individual can grow by himself? Such conclusions, as familiar as they are in humanistic approaches, overdraw the case. This would rule out, for example, efforts to obtain scientific information about child behavior development so that specific conditions could be applied by the parent to treat children's problems, or to prevent them. The fact is there are many individuals who do not have the basic behavioral repertoires in a sufficiently developed state to deal with them on an "insight" or "self-development" basis. There are autistic, emotionally disturbed, retarded, preverbal, and schizophrenic children, for example, whose treatment cannot be conducted on a language basis. There are many adults, also, whose behavior, if it is to be benignly changed, must be changed through more direct conditioning procedures. It would be unfortunate not to recognize then that the use of principles and procedures by which one can directly alter (control) behavior can be desirable. Moreover, such use can be conducted with very benign motives on the part of the therapist and with respect for the human involved.

On the other hand, to generally treat individuals, even those with problems they have not been able to handle themselves, as lower organisms not possessing personality repertoires that allow one to deal with them on a higher level is equally inappropriate. It may well be dehumanizing to subject the homosexual to conditioning treatment without providing understanding of the principles and the personal background involved. It may be that many personal problems can be dealt with through insight (language) procedures which involve the patient's growth of knowledge and decision-making powers. Considering psychotherapy in terms of language, along with the concept of the personality repertoires, suggests that such goals can be achieved without rejecting the methods and principles of scientific psychology (Staats, 1972).

The fact is, actually, the humanist does not mean by "understanding" only the knowledge of human behavior. "Understanding" in the humanist sense is heavy with acceptance of and sympathy for the individual and his problems. Again, however, understanding in this sense can only be improved by a theory that indicates the conditions by which the individual came to his state of affairs. This is especially the case where the individual demonstrates behaviors that are undesirable in one way or another. As the next section will indicate, an approach that indicates the conditions and principles that produce undesirable behavior removes the moral *blame* for the behavior from the individual. As a social conception for dealing with various types of undesirable behavior, such an approach is in this sense very humanistic.

## Values in science and humanism

There has been in the social and behavioral sciences a notion that scientific work should be academic and value free. At one extreme there are those who feel, for example, that psychological research is in the nature of a game—it does not have a significant role in understanding and solving important human problems. Another approach has been that psychology (or other social sciences) demeans itself by taking a position with respect to social problems and controversies—on the one hand, to do so would be to lose scientific objectivity, and on the other, to do so would be to enter an arena where psychological knowledge has no special merits.

There is a reason for scientific study unrestricted by social pressures of a religious, political, or economic nature. There are many examples from history where such pressures have worked to the disadvantage of science. As a matter of fact, it is in areas of high social concern that many times the society does not want an objective approach, whereas it might be in the interest of both the science and the society to conduct objective inquiry.

Academic psychology has had major traditions in natural science. For many experimental psychologists advancing psychology among the sciences is a central aim. This is a perfectly reasonable purpose in its general aspects, but it can lead to undesirable conclusions when the nature of science is seen in a restricted form. For some, the appearance of science— laboratory research, apparatus, the use of measurement and mathematics, and so on—are its essentials. Entry into less scientific-appearing activities has a lesser value. Such a rationale may put a low value upon concern with man in the naturalistic state and with his problems.

There has been contemporary dissatisfaction with the "ivory tower" nature of much of social science. Humanistic psychologists reflect this concern. There are very good points on the side of removing artificial barriers to the study of man's socially significant behavior. There should be nothing in one's explicit or implicit philosophy of science that would suggest that studies and treatment of complex human behavior are less central to the psychological and social sciences than studies involving apparatus, or whatever, that superficially have attributes more recognizable as "scientific."

It is one thing to grant that with some researchers there has been an overemphasis on the "objective," to the detriment of concern with the study of human problem areas. Some of the humanists' emphasis upon concern with the internal man, however, is itself at odds with interests of social relevance, and in this sense it also has an ivory tower aspect.

Humanists take their stand ostensibly to maintain the dignity of man against the "dehumanizing" effects of objective science. They refuse to accept man's behavior as caused, feeling that this takes away some of man's sublime nature. Man must remain spontaneous, inner determined, uncaused, free, and thereby noble. There are elements in this conception, it is suggested, that are antithetical to the present approach, because the conception is actually not "humanistic" in a humanitarian sense and is an obstacle to progress through science. That is, if man is inner determined and uncaused, then it is easy to conclude that there are evil as well as noble classes of people. There *are* people whose behavior is inferior, brutish, abnormal, defective, or evil. If man is inner determined and noble, he is also inner determined and evil, or abnormal, or whatever. Such a conception sends us looking for different things in science and provides a basis for different social solutions than does the present approach. Such a conception, it is suggested, allows us to accept others as less than men when they do not behave as we like.

The present approach, while recognizing modes of self-direction and freedom, sees the ability to direct oneself as arising in the individual's historical development. Recognition is given to the personal conditions and social conditions that give rise to the particular type of self-direction the individual displays. The individual in that sense is not abnormal or defective *personally*, or evil—nor for that matter, exalted, virtuous, or noble. His behavior may be any of these, but there is no personal, moral quality in the individual that makes him one or the other.

If we are interested in virtuous behavior and in the prevention of evil behavior, we must be concerned with the conditions that produce it. This is a humanistic position. It provides a framework for moralistic practices of human interaction, in cases of both desirable and undesirable behavior. There is no rationale in this conception for considering people as personally inferior, or for providing conditions for others that are inferior, or for exploitation of classes of men.

The study of man can be scientific—concerned with individual understanding and with principles that allow benign solution to individual and social problems through planned actions (prediction and control)—but be humanistic in its values concerning human welfare. It is central to utilize the values of humanistic and scientific psychology in forming such a conception.

## Conclusions

Not all of the points of issue between the subjective and objective approaches to man—as they occur in the various social and behavioral sciences—can be treated in a brief chapter. It is a goal of the present discussion, however, to indicate that there are no insuperable differences in the basics of the two approaches. There are essential aspects of both that can be combined to yield a framework for studying human behavior, for treating problems of human behavior, and for making social and personal decisions with respect to ourselves and others. Such a philosophy of human science, it is suggested, is more complete, less erroneous, and more productive than either approach alone. Establishing a unified general conception of man is an important part of the task of providing a paradigm that can deal with the various aspects of the study of man. Prejudices that are outgrowths of one or the other philosophical position can only continue to represent barriers to a unified approach. Finally, a primary purpose of the present chapter is to indicate that a social behaviorism also involves a social philosophy. This is an area of legitimate concern to any approach that intends to deal generally with man.

# Social behaviorism and the social sciences

IN THE 1960s there were several suggestions that learning principles should be employed as an integrated approach to at least some of the social sciences.

> It seems evident, however, that acceptance of a conception of man in terms of learning principles, as well as the utility of the conception in designing a culture, will require a more complete behavioral analysis of social and cultural practices, and of man's complex behavior in general, than has hitherto been available. The development and change of significant human behaviors will have to be treated systematically, and the science of behavior extended in greater detail to see if the statements that spring from a learning analysis check with the naturalistic observations collected throughout the history of man. . . . Such analysis might appear to be to a large extent the job of social theorists and other behavioral scientists—anthropologists, sociologists, historians, economists, and so on. The use of psychological conceptions of man by other behavioral sciences is not unusual today. Much of the work done in other behavioral sciences is based upon a "psychology," a conception of man's behavior. For example, much of anthropology in recent years has utilized psychoanalytic terms and principles in an approach to many problems of analysis. As yet, however, an integrated set of learning principles has not been comprehensively applied in the various fields of social science. It is suggested that a learning conception of human behavior is now available for this purpose. (Staats, 1963, pp. 353–54)

> . . . I now suspect that there are no general sociological propositions, propositions that hold good of all societies or social groups as such, and

that the only general propositions of sociology are in fact psychological.

What I do claim is that, no matter what we say our theories are, when we seriously try to explain social phenomena by constructing even the veriest sketches of deductive systems, we find ourselves in fact, and whether we admit it or not, using what I have called psychological explanations. I need hardly add that our actual explanations are our actual theories. (Homans, 1964, p. 818)

Homans employed Skinner's approach to deal with topics in sociology. (See also Homans, 1967.) In addition to these two statements, Kuhn (1963) attempted an integrational effort based on Thorndike's learning theory. He treated aspects of political science, sociology, and economics by "pulling each apart" and "throwing their components into a single pile, and then [attempting] to reconstruct them into a new single discipline" (p. vii). It is suggested, however, that the possibility of extending a learning theory to the social sciences is only in its incipient stages. Actually, there has not been a learning theory that has had the elements to perform the task to the satisfaction of many social scientists. As was indicated in the preceding chapter, for example, many social scientists are concerned about the subjective aspects of man—about values, attitudes, goals, ambitions, reasoning, and so on. In rejecting or ignoring these areas of human behavior, elemental behaviorism has not been an attractive basic conception to social scientists.

Perhaps most importantly, the traditional learning theories, in lacking a personality level, have not been appropriate. It is evident in the study of broader social events that men have personal characteristics that differ and personality must be considered as an independent variable in explanation. In addition, however, the conception must include principles of how individuals and groups interact and the mechanisms of interaction, such as communication, and so on.

It will not be possible here to extend broadly the social behaviorism paradigm to social science. This chapter, however, will attempt to demonstrate how the basic learning theory and the A-R-D and social interaction theoretical elaborations are relevant to various social sciences. This is meant to illustrate the suggestion that other aspects of the paradigm are also relevant. It will be suggested also that this exercise is not just interpretive, but that a basis for research may be projected from such analyses. Psychology has frequently withdrawn from treating broader social issues under the cloak of restricting itself to matters susceptible to "scientific" treatment. It is suggested, however, that a "psychology" should be *relevant* in this sense. It should indicate how it can be employed in dealing with broader social areas of study, hopefully to provide the student with a foundation in approaching these areas of study.

The three-function learning theory principles described in Chapter 2 are extensively supported by laboratory research with animals and humans. The principles would be considered scientifically objective. As has been indicated, however, the three-function learning theory can be elaborated to consider in an objective manner such subjective states as attitudes. When the theory has been elaborated in terms of theories of the emotional-motivational system and social interaction, the A-R-D principles appear to be relevant in all the areas of the study of man. The following sections will exemplify this and thereby the generality of the principles and the comprehensiveness (see Chapter 1) of the paradigm of which they are a part.

## A-R-D principles and the social sciences

The fact that three-function learning principles are important to human behavior is commonly recognized in some form. The principles involved, however, are generally very unclearly stated. Sometimes one principle will be referred to, and sometimes another, without making a distinction. Moreover, A-R-D events are referred to in different terms which are woven into different theories. Different aspects of the effects are stressed, and the A-R-D events themselves are defined in different ways. The basic premises and terms of the theories into which the commonsense statements of A-R-D principles are woven differ widely. Thus, the fact that the various areas of study deal with the same class of events and the same principles of behavior is presently disguised by the variegations in the theoretical language. Moreover, because the social sciences lack laboratory methods of study, the elementary nature and lawfulness of the principles are not easily seen.

### HEDONISM AND A-R-D PRINCIPLES

Let us begin with an old example, the philosophy of hedonism. The ancient doctrine was that the highest good to be attained was through pleasure. There were variations among the Greek philosophers in what objects or events, and under what circumstances, pleasure should be sought. Thus, Aristippus suggested that complete gratification of sensual desire should be striven for. Epicurus, on the other hand, taught that pleasure was best gained through the controlled gratification of one's desires and the avoidance of pain. It can be seen that these philosophies reflect or describe the philosopher's observations of the effects of "pleasure" upon the behavior of other men and upon themselves. As with the later British philosophers (such as Bentham), the observations of man's behavior and the pleasurable events that influence it led to proscriptive philosophies. The later hedonism was considered to be "the greatest happiness for the greatest number," an orientation that is labeled utilitarianism. There are contemporary researchers who refer their empirical results to the principles of hedonism (see Insko and Cialdini, 1969).

## THE PLEASURE PRINCIPLE AND A-R-D PRINCIPLES

It is interesting that Freud, in developing a theory of human behavior, included a central principle which he called the "pleasure principle." This principle can be seen as a variation of the hedonistic statements of philosophy. Freud first posited in his personality theory a fund of instinctual energy (the id) that was ruled by the pleasure principle. His account was cumbersome and complex, but it included indication that various objects and events can gain or lose what might be called pleasure value (cathexis). He also suggested that the individual's biological development, in conjunction with his experiences, had an effect upon what objects and events will come to be "pleasurable." The important thing here, however, is to note the central place in his theory given the principle concerned with the effects of pleasurable events upon human behavior. It is also true that Freud realized that attempts to explain all human behavior in terms of the pleasure principle failed—that behaviors occur which cannot readily be interpreted in terms of pleasure. People harm themselves, martyr themselves, and kill themselves. He thus also posited a "death instinct" as a principle to explain such behavior. Nevertheless, the important effects of pleasure stimuli on human behavior were seen by this most influential personality theorist.

## NEUROTIC GAIN AND A-R-D PRINCIPLES

In psychiatry today there is also wide reference to the phenomena called *primary* and *secondary gain*. That is, it is accepted that many symptoms of abnormal behavior are manifested by the individual because they are emotionally positive or rewarding. "The *primary gain* is the reduction in tension and anxiety which makes it possible for the patient to preserve the greater part of his integration" (Cameron, 1963, p. 273).

## VALUES AND A-R-D PRINCIPLES

In sociology and the other social sciences, there are many aspects of study that involve the same A-R-D principles, albeit stated in different terminology. The term *values*, for example, is prominent in sociological theory as well as in the theory of other social sciences.

> People cherish certain ideas or beliefs which are often called their "values." These ideas contain or express the judgments which people have of the relative worth or importance of things. . . . In America, for

example, we characteristically value highly such things as success, beauty, a high standard of living, and education. (Cuber, 1955, p. 42)

The concepts of values and of the value system are considered basic to understanding various types of social behavior. Thus, for example, it is widely suggested that there are differences in values across different groups in a stratified society, and the differences in the value systems result in different behaviors in the members of societal groups.

It should be noted that there is the same type of widespread interest in the differences in values *between* cultural groups in the field of anthropology. Anthropological descriptions of the value systems of cultures and cross-cultural differences in value systems are common. The excerpts of Goldman (1937a, 1937b) and R. J. Littman (1974) that were presented in Chapter 4 provide illustrations of such cross-cultural differences.

## THE LAWS OF MARGINAL UTILITY AND SUPPLY AND DEMAND, AND A-R-D PRINCIPLES

Economics is an area of social science that concentrates upon the study of certain classes of the objects and events that have A-R-D value, especially those types of objects and events apt to be considered under the terms *commodities, money, utility,* and so on. Economics has also been concerned with some of the principles that affect the A-R-D *value* of some object or event, depending upon its availability, the amount of labor that goes into its production, and the like.

To begin exemplifying this interest, Jeremy Bentham's 19th-century utilitarian hedonism of the greatest good for the greatest number also included more precise definitions and attempts to measure the utility or value of income. Some of his propositions, as summarized below, were important influences in the development of economics.

1st.  Each portion of wealth has a corresponding portion of happiness.
2nd.  Of two individuals with unequal fortunes, he who has the most wealth has the most happiness.
3rd.  The excess in happiness of the richer will not be so great as the excess of his wealth. (Bentham, 1931, p. 103 ff.)

As these propositions suggest, in economics there has been a high interest in mensuration of A-R-D value. This interest has been accompanied by an interest in the statement of the principles involved in the shifting values of objects and events, as a function, for example, of such variables as access to, and the effort (cost) in producing, the object or event (commodity).

To illustrate, the third statement of Bentham above suggested that there is a change in the happiness value of a unit of an object depending upon the amount the individual has of that object. This not yet precisely stated principle is of course in close agreement with the A-R-D principles that involve the effects of deprivation and satiation. The latter principles state that when the organism is deprived of the A-R-D stimulus, the A-R-D values of the stimulus are increased. An opposite effect follows satiation. The economic law of "diminishing (marginal or extra) utility" (Samuelson, 1958, p. 430) is closely analogous.

> According to this law, the *more* an individual has of some given com-
> modity, the less satisfaction (or utility) he would obtain from an addi-
> tional unit of it. . . . Illustrations of the law are available at almost
> every hand. . . . If pork were a rarity, the consumption of a small
> amount a week, say a quarter of a pound, might be the source of con-
> siderable satisfaction to a person with a taste for it. Now keep adding
> to his diet successive "doses" of a quarter of a pound of pork. Would
> not the added satisfaction from each new addition be less than that
> derived from the last? Assume that he finally obtains 10 pounds of pork
> per week so that he eats it at each of his three meals seven days a week.
> Now he is given an extra quarter pound. Is he overcome with tearful
> gratitude and unrestrained joy? Or with nausea? . . . It follows from
> this analysis that the more an individual has of a given commodity, the
> less he will be willing to *give* in exchange for an additional unit of it.
> (Ulmer, 1959, pp. 319–20)

This is thus a lively statement of the principle of the effect of satiation upon some of the values of a unit of an A-R-D stimulus. It is interesting to note, however, in this and the other cases that the behavioral functions of the commodity that are changed by deprivation-satiation are not speci-fied. Moreover, such accounts do not ordinarily refer to experimental verification of the basic principles of individual psychology that they employ. As will be indicated, the economic analyses could be bolstered and extended by such empirical support.

The law of supply and demand is another example of the way that economics has stated laws that involve A-R-D principles, including depriva-tion-satiation. John Stuart Mill stated the law originally as follows:

> Demand and supply, the quantity demanded and the quantity sup-
> plied, will be made equal. If unequal at any moment, competition
> equalizes them, and the manner in which this is done is by adjustment
> of the value. If the demand increases, the value rises; if the demand
> diminishes, the value falls: again, if the supply falls off, the value rises,
> and falls if the supply is increased. . . . [T]he value which a commodity
> will bring in any market is no other than the value, which, in that
> market, gives a demand just sufficient to carry off the existing or ex-
> pected supply. (Mill, 1969, p. 448)

The law of supply and demand can be seen in the light of a behavior analysis to refer to the effects upon the buyers' behavior of deprivation-satiation operations. When many individuals are relatively satiated (supply is high), the A-R-D value of the commodity is generally low. When the commodity becomes scarce, there are more individuals who will have been deprived of it. The increased A-R-D value of the commodity then more strongly controls buying behavior—which is to say that the A-R-D value of the commodity, *in comparison to money*, is increased. More individuals will pay more for the item. This law may be seen as a case of the principles concerning the relative strengths of items in the A-R-D system. These principles could be directly tested using some of the experimental methods described in earlier chapters. Integration of the laboratory studies of the A-R-D principles, with procedures such as the token-reinforcer system, and some of the economic principles would provide a number of hypotheses by which to extend knowledge of economic behavior, as will be suggested.

## CULTURAL EVOLUTION AND A-R-D PRINCIPLES

The concern of anthropology with cultural evolution (and of sociology with social change) may be analyzed also in terms of social behavior principles. The concept of cultural or social evolution has been derived from the concept of biological evolution. For example, it has been suggested that biological evolution occurs when "by adaptive modification the population is enabled to maintain or better itself in the face of a threat induced by a changing environment or that it is enabled to exploit the same environment more effectively than before" (Sahlins, 1968, p. 230). The principle is carried by analogy to the level of cultural evolution.

> Culture continues the evolutionary process by new means. . . . Culture diversifies by adaptive specialization until it successively produces over-all higher forms. Culture, like life, undergoes specific and general evolution. . . . [I]n a word, through adaptive modification, . . . cultures are organizations for doing something, for perpetuating human life and themselves. Logically as well as empirically, it follows that as the problems of survival vary, cultures accordingly change, that culture undergoes phylogenetic, adaptive development. (Sahlins, 1968, p. 233)

It is the thesis of the present analysis that the principles of cultural evolution and social change—along with the individual's development of behavior—are the principles of learning. It is suggested that the concept of cultural evolution arises because there is a close relationship between biological evolution and the laws of learning. For example, both the laws of biological evolution and the laws of learning produce effects that can be generally called "adaptive"—they enable the individual organism to continue its survival and to perpetuate its kind. What applies to individuals also applies to the individuals who comprise the group—the species or culture, and so on. Despite this compatibility in this superficial analysis, however, there is no utility in employing the biological model for consideration of human behavioral development, of either an individual or group kind. The laws of learning govern the acquisition, maintenance, and change of human behavior, and the laws of learning constitute a theoretical model that is sufficiently complex to really understand human behavior. The laws of biological evolution, as will be indicated later, do not appear appropriate for application to the phenomena of social behavior. One example can be referred to here which involves A-R-D principles.

> The present author . . . has suggested several mechanisms of a learning nature that could result in social evolution or change. For example, it was suggested that the various activities we call science are the behaviors of men subject to the principles of conditioning. These behaviors were seen as emerging into a more dominant position in our society because they result in greater reinforcement than other competing behaviors. Scientific behavior, as an example, has become more dominant because it results in avoiding aversive natural occurrences and obtaining positive reinforcers. The scientific behavior should consequently become more dominant in the society through several mechanisms. First, reinforcement will strengthen the behavior in the individual. In addition, however, others who are around this individual will experience positive reinforcement in his presence. This will give him greater social power and enable him to influence the behavior of a larger number of people. It will also put him in a position within the society where he will influence a much larger number of people through communication media, as occurs when the individual becomes a member of a university faculty. The individual whose behaviors do not result in positive reinforcement for himself and others will in competition find his influence weakening. In addition, social practices or social systems that differ in the extent to which they result in reinforcement (positive or negative) will control differing extents of "striving" behavior. Social practices or systems that result in stronger positive reinforcers will control stronger supportive behaviors than a competitive practice or system that results in little positive reinforcement or much negative reinforcement (Staats, 1964b, p. 336).

From this latter principle, for example, one could make predictions concerning the outcomes of cultural conflict and competition. Let us say that two societies are equally strong in terms of natural resources, population, technical advancement, and so on. Let us say, however, that one has social practices that result in more widespread reward for its members than has the other society. It would be expected that this society for this reason would win in the conflict. That is, its members would strive harder in support of the society than would the members of the society not so well rewarded. The same principles would be expected to function *within* a society where different institutions (religious, scientific, economic, and so on) may be in competition. The following quotation is interesting in this context.

> The tragedy of our role in Vietnam is but the current installment of an old story. Our commitment to "stop Communism" too often leads us to support corrupt and decadent regimes detested by the peoples of those countries. This is especially true in Vietnam where feudal authoritarianism, widespread corruption, and desperate poverty provide no incentive for the great mass of people to fight to preserve their way of life. (*The Progressive* [Madison, Wis.], October 1963, p. 5)

These, of course, are greatly oversimplified analyses, appropriate only for a very brief discussion. However, the analyses do indicate the possibilities for consideration of cultural evolution and change in terms of the effects of social behavior principles on individual behavior, group behavior, and cultural competition between groups. Kunkel (1970), a former colleague, has begun to elaborate the conception of social change as learning.

## HISTORY AND A-R-D PRINCIPLES

If the A-R-D principles are so widely demonstrated to affect man's behavior, it should follow that such principles will have a prominent place in any account of man's behavior. Thus, the principles should be frequently described in one form or another in historical writing. The following description may be used as a case in point.

> England was heartily sick of the War of the Roses. For thirty years it had gone on, its fires replenished by the fuel of hate which it created as it burned, but there was no great principle involved. It was merely the struggle of two noble factions for the crown, and by the year 1485 this had become obvious to all observers. (Mackie, 1952, pp. 8–9)

Such a statement is a summary description of a vast set of material and social reinforcing events that controlled to a very large extent the behaviors of individuals involved in the war—the behaviors of striving to attain the "crown." The War of the Roses was thus seen as a struggle of men whose behaviors were determined in large part by the A-R-D value of being on the side that had the crown. That man's behavior is in significant part determined by such principles is to be seen in various historical events, although it is not always in the interest of the historian to chronicle such determinants. Thus, in the example of history alluded to above—the War of the Roses and the central battle of Bosworth in which the army of Henry VII defeated the army of Richard III and killed the latter—the fact that no noble principles were involved and that the war concerned which faction would gain the economic and political rewards of the crown were apparently well known to contemporaries but were not given recognition by later historians.

> In support of this view it may be pointed out that although Bacon and later chroniclers lay stress on the fall of a tyrant, contemporary historians drew no such moral from the battle of Bosworth. The *Chronicle of London* refers to it somewhat casually between the mention of a sheriff's death and a notice of the appearance of the sweating sickness in the city. . . . For the author, plainly, the battle of Bosworth was not a historical landmark, the beginning of a new age. (Mackie, 1952, p. 9)

This is not to say that a complex social event which extended over 30 years in time and involved many people is simply to be tossed off as an example of A-R-D principles. If one wished to understand in detail the behavior of the individuals involved in that complex social and historical event, it would be necessary to make a detailed analysis of the determinants of their behavior, including the A-R-D conditions and their effects upon the actors' behaviors. The example is only employed to illustrate the fact that such conditions are important as determinants of the past behaviors of men which have historical significance. The result of the perusal of such historical events can only convince us that it is important to understand A-R-D principles if we wish to understand the manner in which human behavior significant to history is determined.

It may also be suggested that although the laboratory-specified A-R-D principles have not been available to historians in their historical theorizing, relevant commonsense principles that abided in the common language have been, and they have been employed widely in making historical analyses. Moreover, such concepts are employed not solely for cases of interpretation of individual historical events. The A-R-D principles—in one form of statement or another—have figured prominently in *general theories* of history. Perhaps the most famous of these is Marx's view of history. He stressed the role of economic factors (A-R-D stimuli) in determining man's behavior throughout history. In doing so he affected greatly the methods of analysis of other historians and social scientists of various kinds. Thus, Marx stated variously that man's behavior depends upon the consequences that behavior produces (Marx, 1959).

It is not being suggested that Marx's theory of history was correct. Actually, it follows from the present discussion that his understanding of the A-R-D principles and of the human motivation system was too undeveloped to have at that time produced a comprehensive theory of man's historical development. Marx's theory, however, is an important example of the recognition that A-R-D principles of human motivation are crucial to understanding man's behavior at any one time, as well as his historical development. Again, this is a topic of such breadth and importance that it can only be hinted at herein.

One other example of a Marxian interpretation of history will be given, because it again concerns a most illustrious general theory and because it specifies in closer detail the manner in which a particular historical event can be analyzed into its A-R-D determinants (in this case economic) to at least in part constitute an explanation of the complex event. The work referred to is Charles A. Beard's *An Economic Interpretation of the Constitution of the United States,* published in 1913. Beard's main thesis was that the framers of the Constitution consisted of a group of men with common economic interests, interests which largely determined the manner in which the document was written. The economic interests (A-R-D conditions) of these men were not general to the population, and in fact were in conflict with the interests of the small farmers and debtors. This was pointed out in the following quotations from Beard (1913):

> Large and important groups of economic interests were adversely affected by the system of government under the Articles of Confederation, namely, those of public securities, shipping and manufacturing, money at interest; in short, capital as opposed to land. (p. 63)
>
> Thus, the members of the Philadelphia Convention consisted of a small and active group of men immediately interested through their personal possessions in the outcomes of their labors [the Constitution]. . . . The propertyless masses were . . . excluded at the outset from participation (through representatives) in the work of framing the Constitution. The members of the Philadelphia Convention which drafted the Constitution were, with a few exceptions, immediately, directly, and personally interested in, and derived economic advantage from, the establishment of the new system. (p. 324)
>
> Inasmuch as so many leaders in the movement for ratification were large security holders, and inasmuch as securities constituted such a large proportion of personalty, this economic interest must have formed a very considerable dynamic element, if not the preponderating element, in bringing about the adoption of the new system . . . Some holders of public securities are found among the opponents of the Constitution, but they are not numerous. (pp. 290–91)

Beard's theory of this historical event (like Marx's general theory) rested upon a basic assumption that social progress in history occurs as the result of groups in society competing for economic interests. That this is a complete theory of history or theory of the particular historical event has been questioned (see McDonald, 1958). However, the fact that A-R-D principles have been seen by noted historians so strongly throughout history to be primary determinants of human behavior is an indication of the importance of the principles in constructing a general theory. Excellent examples in a recent analysis of Greek history may be found in an innovative history by Littman (1974).

But this is not the only area that has focused upon the manner in which the A-R-D consequences of human behavior are primary determinants of behavior—in the social sciences as well as the sciences concerned with individual human behavior. To enlarge this picture, the next two sections will deal respectively with social exchange and political power.

## SOCIAL EXCHANGE AND A-R-D PRINCIPLES

In the functional structuralism approach in sociology, which has been very influential, social events are explained by other social events that act as consequences for the first event (Wallace, 1969a). Within this general context, the concept of reciprocity has arisen. Gouldner (1960) defines his term *reciprocity* as "a mutually gratifying pattern of exchanging goods and services" (p. 170). As can be seen from this definition, the language (theory) employed is from the field of economics. Moreover, this has been the source for the principles employed by other social exchange theorists who have adopted economic principles with which to describe the behavioral interactions of men in groups (see also Blau, 1964; Homans, 1961; and Thibaut and Kelley, 1959).

It should be noted that the consideration of cases of human behavior that involve a mutuality or reciprocity is indeed central. As has been indicated, the basic principles of learning have been established in situations in which the concern is with the stimulus manipulations that affect the behavior of a single organism. The basic laboratory principles, however, can be elaborated to pertain to social interaction, as has been indicated. The social exchange theorists have employed some, but not all, of the possible behavior principles in dealing with social behavior. That is, the exchange theorists have indicated that social interaction between people can be considered in terms of the rewards that one person has for the other, and vice versa, as well as the costs of the behavior itself.

> The consequences of interaction can be described in many different terms, but we have found it useful to distinguish only between the rewards a person receives and the cost he incurs.
>
> By rewards, we refer to the pleasures, satisfactions, and gratifications the person enjoys. . . .
>
> By costs, we refer to any factors that operate to inhibit or deter the performance of a sequence of behavior. Thus cost is high when great physical or mental effort is required, when embarrassment or anxiety accompany the action, or when there are conflicting forces or competing response tendencies of any sort. . . .
>
> The consequences or *outcomes* for an individual participant of any interaction or series of interactions can be stated, then, in terms of the rewards received and the costs incurred, these values depending upon the behavioral items which the two persons produce in the course of their interactions. (Thibaut and Kelley, 1959, pp. 12–13)

An important matter to consider in the context of this reward-cost analysis of social interaction involves the manner in which the rewards that person A dispenses contingent upon certain behaviors of person B will be a determinant of B's behavior (and the converse is true, of course). The concept of power may be seen as an aspect of this behavioral influence.

> If two persons interact, the pattern of outcomes given in their interaction . . . indicates that each person has the possibility of affecting the other's reward-cost positions and, thereby, of influencing or controlling him. . . . Generally, we say that the power of A over B increases with A's ability to affect the quality of outcomes attained by B. . . . If, by varying his behavior, A can make it desirable for B to vary his behavior too, then A has control over B. (Thibaut and Kelley, 1959, pp. 100–101)

Current developments of "exchange theory" seem to be committing the approach more and more to dependence upon economic principles and language as the basic theory (Nord, 1969). The fact that the approach deals only with reward-cost principles may be seen as a limitation in the comprehensiveness of the approach for dealing with human behavior—as will be mentioned again further on. However, the importance of noting the effects one individual may have upon the behavior of another (and vice versa) cannot be overestimated. The laws of reinforcement as they function in this manner should be understood, as should the deterministic nature of such laws, as well as the vast support for the laws in basic research as well as social research and naturalistic observations.

## POLITICAL SCIENCE AND A-R-D PRINCIPLES

Similar principles can be discussed in the context of political science. It is clearly recognized in any number of examples that the simple principles of reinforcement—reward and punishment, in commonsense terms—are important determinants of politically relevant actions and events.

The description of several examples will begin with Froman and Ripley's article in the *American Political Science Review* (1965) which studied the differential strength of Democratic Party leadership in the U.S. House of Representatives and the conditions that led to differential strength in the leadership of the House. They allude at various places to the importance of reward-punishment principles in understanding the political leadership phenomena and also suggest more general principles for considering the strength of party leadership. These principles can be seen to involve A-R-D principles in situations of social interaction. For example, Froman and Ripley describe how the party leadership exerts certain pressures. This involves, in the present terms and at least in part, the manner in which the leadership serves to *direct* conforming behavior in addition to dispensing rewards and punishments for certain behaviors of the representative.

Froman and Ripley also indicate several other subprinciples. Stated in terms of the concepts employed herein, they suggest that where the representative's sources of reinforcement (the party and his constituency) are in opposition, the representative will be more likely to follow the party (1) if the issue is less public and will not get the attention of the constituency and its possible disapproval, or (2) if the representative's response (vote) is not public, which would also avoid the disapproval of the constituency. These principles of leadership in this political body are clear examples of effects of reinforcement and the other A-R-D principles.

Moreover, it is interesting to use the political science examples to elaborate a subprinciple of reinforcement—the case where there is more than one A-R-D condition involved and the conditions are in conflict. Frequently in life a behavior will have a positive reinforcing effect from one source but a punishing effect from another. The consequences may be reversed for the alternative behavior, as in the case if the representative votes in either of two ways on an issue. Which behavior will occur will be a function of the relative strengths of the various A-R-D consequences (both past and present). It may be suggested that other basic behavior principles, as well as social interaction principles, could be applied to the analysis of these types of behavior and thus produce additional understanding and additional research.

As another example of the relevance of A-R-D principles in political science, the concept of "power" may be treated briefly. In addition to the previously given statement of Thibaut and Kelley, in social psychology the central role of A-R-D principles in the analysis of power can be seen in the definition of the types of power given by French and Raven (1959). Several of the five types of power they name can be seen to involve A-R-D conditions: coerciveness (the control of punishing power), reward (the control of rewards), and attractiveness.

The concept of power has also been developed within the field of political science. In doing so, it is possible to see that A-R-D conditions maintain their central role, as exemplified in the following quotations:

> Interest is the primary propelling force and every action is based upon sharing of interest. Power configuration is basically the configuration of competing and struggling interests organized into groups. (Macridis and Brown, 1964, p. 139)

> When we use the word *power* . . . , it shall mean only, the numerical representation of rewards accruing to coalitions as evaluated by the members of these coalitions. (Luce and Rogow, 1956, p. 85)

> The community studies of who wields power are personal in the sense that power is associated with specific individuals. The estimation pro-

cedures are designed to determine the power of an individual. This power, in turn, is viewed as some function of the resources (economic, social, etc.), position (office, role, etc.), and skill (choice of behavior, choice of allies, etc.); but the study and the analysis assume that it is meaningful to aggregate resource power, position power, and skill power into a single variable associated with the individual. (March, 1966, p. 44)

March was commenting upon studies in which the power of individuals in a community is obtained by having other community members assess that power. (It may be noted that each of the types of power resources mentioned in the March's quotation may be considered as a condition of A-R-D value.) The results of such studies also show that there are certain types or classes of people who, through their possession of power (A-R-D) resources, have an inordinate influence upon the society.

> With respect to the distribution of power, most studies indicate that most people in most communities are essentially powerless. They neither participate in the making of decisions directly nor accumulate reputations for power. . . .
> With respect to the relation between power and other individual characteristics, rather sharp differences among communities have been observed. . . . First, in every study reported, the business and economic elite is overrepresented (in terms of chance expectations) among the high power holders. By any of these measures, the economic notable is more powerful in the community than the average man. (March, 1966, p. 45)

The concept of a power elite that has inordinate influence has been developed by such individuals as C. Wright Mills (1957). Mills suggests that there are three especially powerful groups in the United States: the military establishment, the political establishment, and the corporate (industrial) establishment. Although Mills does not focus upon the mechanisms and principles by which the power is exerted by the elite, he provides many passages which suggest the operation of A-R-D principles in determining men's political behavior.

> Money allows the economic power of its possessors to be translated directly into political party causes. In the eighteen-nineties, Mark Hanna raised money from among the rich for political use out of the fright caused by William Jennings Bryan and the Populist "nightmare"; and many of the very rich have been unofficial advisors to politicians. Mellons, Pews, and du Ponts have long been campaign contributors of note and, in the post-World War II period, the Texas millionaires have contributed sizable amounts of money in campaigns across the nation. They have helped McCarthy in Wisconsin, Jenner in Indiana, Butler and Beall in Maryland. . . .
> But it is not so much by direct campaign contributions that the

wealthy exert political power. And it is not so much the very rich as the corporate executives—the corporate reorganizers of the big propertied class—who have translated the power of property into political use. As the corporate world has become more intricately involved in the political order, these executives have become intimately associated with the politicians, and especially with key "politicians" who form the political directorate of the United States Government. (Mills, 1957, pp. 166–67)

As another example related to Mills's analysis, Robert Engler wrote a book published in 1961 entitled *The Politics of Oil* which describes the manner in which the oil industry has in the past heavily influenced political and economic decisions in the United States. One of the prominent mechanisms for producing this influence is through the use of A-R-D resources—with money, or influence and social and political position leading to money, as an important mechanism. The involvement of Texas oil multimillionaires in politics is considerably elaborated by Engler, who gives a number of individual cases of such influence by members of the corporate establishment of the oil industry. The following are examples:

> In 1957, William A. Dougherty, chief counsel and director of the Consolidated Natural Gas Company, whose $400 million empire includes the East Ohio Gas and the Hope and Peoples Natural Gas companies, gave a $5,000 "loan" to Orville Hodge, Republican state auditor of Illinois. . . . Dougherty later said he had heard Hodge had a good chance of becoming governor of Illinois. "I thought if I could do him the favor which he had asked me, that I would be in a much better position to get his help," . . . referring to a bill that "neither the company nor I wanted . . . passed." . . .
>
> Recalling his own experiences with the power forces of Texas that finally defeated him, as oil considerations gained political ascendancy over traditions of rural protest and depression factors in his district, a state legislator and then New Deal congressman with fifteen years of service made clear his painful conclusion that more than good citizenship is involved in this interest in elections. . . . "The big boys from the utilities, the banks, the railroads and oil and gas want to contribute to your campaign. They watch you to see if you are okay. Then they'll ask you how you stand on the key issues. If okay, they'll want to contribute. They want your name on the back of their checks." His record against tax loopholes incurred the wrath of the oil industry and he was replaced in 1939 by a congressman who went on to compile a sound oil and gas record. The chastised liberal found an obscure federal legal berth. His successor in Congress was active in steering offshore oil quitclaim legislation through the House. After his retirement, the latter was sought by the American Petroleum Institute to be its official Washington spokesman. (Engler, 1961, pp. 351–52)

These various examples may be seen to illustrate the manner in which the A-R-D conditions in a society help determine the behavior of the members of that society, including that of making social and political decisions. Interests (A-R-D conditions) help determine the individual's decisions or preferences, and the individual's resources (A-R-D capabilities) help determine the extent to which his decisions will be enforced through his ability to determine other people's actions. Again, his ability to influence others through employment of his resources depends upon the fact that their interests help determine their behaviors according to lawful principles. It may be suggested that a good deal of political analysis appears to involve the ways by which individuals affect one another through the A-R-D principles in a manner that is of importance to political activities. It would be expected, on this basis, that a detailed and precise understanding of the principles—as well as other principles of behavioral influence—would be of large importance in the foundations of political science.[1]

## EDUCATION AND A-R-D PRINCIPLES

The following statement succinctly describes the importance of motivation in the field of education.

> With respect to teaching, motivation may be defined as a conscious effort on the part of the teacher either to establish a motive, that is a drive, urge, or desire in his pupils so that learning goals will be attained or to link their already existing motives with the learning goals. It is vital to teaching since it determines whether a pupil learns at all. That is, just because a teacher has taught is no guarantee that learning has taken place, for a pupil could sit in a classroom all year without learning, simply because he had no reason—no motive—for learning. Indeed, were motivation not such an important factor in teaching, pupils of the same age, physical condition, and intelligence might reasonably be expected to achieve similar results in the same classroom situation. But, because motivation is so vital, some pupils—those with the stronger desires to learn—achieve better results than do other pupils of the same age, physical condition, and intelligence. Obviously, then, the first step in classroom teaching must be motivation. (Bortner, 1953, p. 5)

While recognizing the importance of motivation in educational learning, the field of education has not developed an objective, standardly accepted conception of motivation. Moreover, little progress has been made in understanding the principles by which the child displays "motivated" learning, or recognizing the conditions that can be provided for every child so that his learning is "motivated." One aspect of motivation that the field of education has traditionally and dogmatically eschewed is the one that is the focus of the field of economics. In education "Learning

---

[1] This section was written before the events referred to as "Watergate" occurred. These events, however, attest to the current relevance of the principles. It is thus suggested that these principles are relevant in generally considering national and international political processes.

is to be its own reward," and any material reward for learning is traditionally considered to be bad form and also harmful to the child.

The various motivations that education traditionally attributes to children can be straightforwardly seen as involving A-R-D principles. Bortner (1953), who was quoted above, lists several "drives" which children generally acquire and which can be manipulated in the teaching situation to provide rewards in learning tasks that would themselves provide inadequate sources of reward. He suggests that children have an "activity drive" and that this may provide rewards—as in planning things, building things, going to the blackboard, and so on—while the children are learning a subject matter. Other drives proposed are those for security, for mastery, for recognition, for belonging, and for new adventures. Bortner's account of motivation for classroom teachers shows a clear recognition that it is not the learning task itself that provides the rewards, but the reinforcing events that can be marshalled and presented to the child as he proceeds in the learning task. In this manner, the use of such rewards does not differ in principle in any way from the use of material rewards of the type that lawfully determine man's economic behavior, or other types of rewards. As has been indicated, the field of education could profitably lift its taboos in the study of the A-R-D principles and in use of various types of reinforcers. This study and the resulting practices should be dictated solely by the efficacy of the educational procedures, their economy, and so on (Staats, 1970b).

## ETHICS AND A-R-D PRINCIPLES

A-R-D principles are topics of concern in the humanities in the same manner as they enter into the social sciences. It is interesting to note, as a matter of fact, that the same concepts occur in the field of ethics as occur in the foundations of economics, political science, history, and sociology. Thus, the pleasure principle of Freud, the hedonism of Epicurus, the utilitarian hedonism of Bentham, the utilitarianism of J. S. Mill, the economic basis of value of Marx, and so on, are all prominent parts of the study of ethics. (See Jones, Sontag, Beckner, and Fogelin, 1969, as an example.)

Several brief examples will be given to indicate that study of what constitute A-R-D events for man is of central concern to considerations of ethics. To begin, the following definition of ethics indicates the close relation of the concerns of this field with the concepts, principles, and observations of the other areas already dealt with.

> Ethics may be defined as the philosophical study of morality. . . .
> Morality has to do with values, that is, with normative standards of
> evaluation and normative rules of conduct. By "normative standards of
> evaluation" are meant the criteria that an individual or a society uses

in judging things and persons as good or bad, desirable or undesirable, worthy or worthless. (Taylor, 1967, p. 3)

It is interesting to note that the ethical system of hedonism which began with the Greek ancients is based upon a psychology—which is true of other ethical systems as well. Thus, for example, the notion that good is to be considered as personal satisfaction has been termed an "egoistic" ethical approach, terms such as egoistic hedonism, ethical egoism, and so on being employed in this context. Taylor suggests that ethical egoism is based upon a psychological egoism, which may be considered a psychological theory of human behavior.

> The basic principle of psychological egoism may be stated in various ways, of which the following are typical:
> (1)  Every person acts always so as to promote his own self-interest.
> (2)  The sole end of every act is the agent's own good.
> (3)  All acts are really selfish, even if some of them appear to be un-selfish.
> (4)  Everyone always does that which he most wants to do, or that which he least dislikes to do.
> (5)  Concern for one's own welfare always outweighs, in motivational strength, concern for anyone else's welfare. (Taylor, 1967, p. 87)

The ethical system is that it is self-interest which is the foundation for the standards of moral behavior and ethical conduct. Another example is Thomas Hobbes, a classic philosopher who is considered to be an ethical egoist. He states the following:

> The moral philosophy is nothing else but the science of what is *good* and what is *evil* in the conversation and society of mankind. Good and evil are names that signify our appetites and aversions, which in different tempers, customs, and doctrines of men are different; and divers men differ not only in their judgment on the senses of what is pleasant and unpleasant to the taste, smell, hearing, touch, and sight but also of what is conformable or disagreeable to reason in the actions of common life. Nay, the same man in divers times differs from himself, and one time praises—that is, calls good—what another time he dispraises and calls evil; from whence arises disputes, controversies and at last war. (Hobbes, 1969, p. 225)

In the context of indicating the generality of the A-R-D principles that enter into the writings of various men who have been concerned with ethics, several additional statements will be sampled. Each is based upon naturalistic observations and commonsense language, but the pervasive importance of the A-R-D principles in the determination of human behavior is evident. Thus, for example, St. Thomas Aquinas, David Hume, John Stuart Mill, Arthur Schopenhauer, and John Dewey make the following statements:

Again, the end is that in which the appetitive inclination of an agent or mover, and of the thing moved, finds its rest. Now, the essential meaning of the good is that it provides a terminus for appetite, since "the good is that which all desire." Therefore, every action and motion are for the sake of a good. (Aquinas, 1969, p. 155)

. . . This is the second part of our argument; and if it can be made evident, we may conclude that morality is not an object of reason. . . . Take any action allowed to be vicious—willful murder, for instance. Examine it in all lights, and see if you can find that matter of fact or real existence which you call *vice*. In whichever way you take it, you find only certain passions, motives, volitions, and thoughts. There is no other matter of fact in the case. The vice entirely escapes you, as long as you consider the object. You never can find it till you turn your reflection into your own breast and find a sentiment of disapprobation which arises in you towards this action. Here is a matter of fact; but it is the object of feeling, not of reason. It lies in yourself, not in the object. So that when you pronounce any action or character to be vicious, you mean nothing, but that from the constitution of your nature you have a feeling or sentiment of blame from the contemplation of it. . . . Nothing can be more real, or concern us more, than our own sentiments of pleasure and uneasiness; and if these be favourable to virtue, and unfavourable to vice, no more can be requisite to the regulation of our conduct and behavior. (Hume, 1969, p. 263)

The only proof capable of being given that an object is visible is that people can actually see it. The only proof that a sound is audible is that people hear it; and so of the other sources of our experience. In like manner, I apprehend, the sole evidence it is possible to produce that anything is desirable is that people do actually desire it. . . . No reason can be given why the general happiness is desirable, except that each person, so far as he believes it to be attainable, desires his own happiness. This, however, being a fact, we have not only all the proof which the case admits of, but all which it is possible to require that happiness is a good; that each person's happiness is a good to that person, and the general happiness, therefore, a good to the aggregate of persons. Happiness has made out its title as *one* of the ends of conduct, and consequently one of the criteria of morality. (Mill, 1969b, p. 357)

The concept *good* . . . is essentially relative, and signifies *the conformity of an object to any definite effort of the will*. Accordingly everything that corresponds to the will in any of its expressions and fulfills its end is thought through the concept good, however different such things may be in other respects. Thus we speak of good eating, good roads, good weather, good weapons, good omens, and so on; in short, we call everything good that is just as we wish it to be; and therefore that may be good in the eyes of one man which is just the reverse in those of another. (Schopenhauer, 1969, p. 341)

*Judgments about values are judgments about the conditions and the results of experienced objects; judgments about that which should regulate the formation of our desires, affections, and enjoyments.* For whatever decides their formation will determine the main course of our conduct, personal and social. (Dewey, 1969, p. 459)

These statements should not be taken to mean that the various men quoted believed that the study of ethics concerned *only* the conditions of pleasure and pain and their effects upon behavior. Moreover, the several conceptions of man of these famous intellects would differ largely from a social behavior theory of man. It is also the case that each of the moralists' statements is couched within a general conception or theory of man which may be quite different from the others—with different terms and concepts. The examples are interesting, however, for the evidence provided that even with this diversity in view, and even in the case where propositions were not tested experimentally but were arrived at on the basis of naturalistic observations, there is a common theme that runs throughout. That is, there are events that men find rewarding and punishing, satisfying and dissatisfying, and these events have an important effect upon the man's behavior, including behavior relevant to ethical considerations.

## AESTHETICS AND A-R-D PRINCIPLES

It is finally suggested in this section that the field of aesthetics—the philosophy of art, or taste, or beauty—clearly involves psychological considerations, prominently included in which are A-R-D principles. For example, it has been suggested by Santayana that "The philosophy of beauty is a theory of values" (1965, p. 371). Santayana adds,

> . . . [W]e must widen our notion . . . to include those judgments of value which are instinctive and immediate, that is to include pleasures and pains; and at the same time we must narrow our notion of aesthetics as to exclude all perceptions which are not appreciations, which do not find a value in their objects. (p. 372)

The concepts of value, appreciation, and pleasures and pains, of course, cast the philosophy of beauty into the realm of the principles that deal with objects and events having A-R-D attributes and functions, in the same manner that has been illustrated in the other social sciences. It may be suggested that the objects and events of concern to this field are those that are labeled aesthetic. But from the present analysis it would be expected that the same principles and conditions would pertain in this field as in the others already illustrated. Only the class of objects and events considered to be aesthetic would be different from those considered to have value in other areas—for example, economic value, or social value. At this point, however, it will be helpful to present an additional excerpt that indicates the closeness of considerations of the philosophy of art to those that occur in other areas of study in the social sciences—and also indicates the need for a "psychology" in dealing with the questions involved. The following statements were taken from a list of ten categories of questions which F. J. Coleman (1968) indicated as important in the field of aesthetics.

> . . . What is an aesthetic experience? Or is there anything peculiarly aesthetic? And if there is, how can it be distinguished from a religious or a moral or a purely emotional experience?
> . . . What does it mean to say that an object has or does not have aesthetic value? Is aesthetic value a property of objects? Or is it a reaction that occurs in some or most persons when they observe certain objects? Or are there other possibilities?
> . . . What are aesthetic judgments? How are they to be verified? Can there be genuine disagreements about aesthetic worth? Is there a point at which the disagreement reaches an impasse? Are there any canons of taste, and if there are, what sorts of things are they? Empirical generalizations? Axioms? Proposals?
> . . . What sort of inquiry is aesthetics itself? Could the questions that it asks be more ably dealt with by psychology or some other science? (pp. 2–3)

These various questions concern principles and concepts that are explicit and detailed in the social behavior theory. Moreover, as will be indicated, there are experimental methods in the behavior theory with which to study some of the questions. It is interesting to note at this point that in the first category of questions Coleman sees the difficulty involved in distinguishing aesthetics from religious or moral experiences, or in fact from emotional experiences in general. This, of course, is the same dilemma faced by the various social scientists who have seen the goal of their science as the establishment of unique principles and concepts and events. The social behavior theory approach provides a framework within which to indicate the relationship of the various areas of interest, rather than to separate them uniquely in terms of principles.

Since others of the questions raised by Coleman are pertinent to later discussions, consideration of them will be postponed until then. It may only be added that the statements quoted herein exemplify the fact that aesthetics is concerned with the aspects of the A-R-D system usually labeled under the terms *art, beauty, aesthetic,* and so on. The field of aesthetics, moreover, is concerned with the principles involved in the origin of aesthetic value in objects and events considered aesthetic, with individual and cultural differences in such aesthetic value, with the measurement of such aesthetic value, and to some extent with the *effects* of aesthetic objects and events on human activity.

## CONCLUSIONS

This section has attempted to indicate by example the universality of some of the A-R-D principles and thus of the more general theoretical structure involved. While it is by no means possible to cover such a vast area of study in a brief discussion, the examples include diverse fields in the social sciences and humanities. Many additional areas could have been included. Moreover, in each area it could be shown how the A-R-D principles actually apply to a good proportion of the events in each area, not just to examples. The general suggestion is that the social behavioristic theory provides a helpful foundation for consideration of the phenomena in various areas of human behavior.

Even with the limited material covered, however, it was possible to see how some of the world's great thinkers of the past, as well as some of the important areas of study, had all in one way or another come to recognize the central nature of at least parts of the A-R-D principles in determining man's behavior. It is suggested that some of the success of the world's great thinkers with respect to considerations of human behavior may be attributed to the extent to which they recognized, dealt with, and conceptualized such emotional-motivational conditions as determinants of human behavior.

The suggestion that all of the disparate theories of these men, and all of the areas of study involved, could be unified through the use of the one theoretical language proposed—a language that explicitly and in detail includes basic principles and which has to support it the methods and findings of a laboratory science—would seem to be of some moment. More will be said of this in the following sections.

## Products of social behavioral analysis

In the process of indicating how the A-R-D principles apply to the various fields, examples of theorizing in these fields were excerpted for quotation. Perusal of these examples reveals that a bewildering number of terms were employed with reference to the three A-R-D functions of stimuli. A partial list would include *interests, gratification, satisfaction, pleasure-pain, happiness, primary* and *secondary gain, reward-punishment, values, norms, goals, aesthetic value, economic value, commodities, money, utility, adaptation, economic advantage, functional consequences* (for an institution), *rewards-costs, outcomes* (rewards minus costs), *power* (which was further separated into *resources, position,* and *skill powers*), *attitudes, emotions, motives, sentiments, needs, likes-dislikes, good-evil, desires, affections, enjoyments, appreciations,* and so on.

These examples are given by way of indicating that the analyses in these different areas rest upon a commonsense psychological theory. The terms are taken from the common language without systematic definition. The attributes of the terms are not specified. Even when the same referents are the concern of the different fields, different terms will be employed. In aesthetics the term *appreciation* may tend to be employed, in economics *satisfaction,* in ethics *good*—all referring to the same functions of stimuli.

The commonsense language is unsystematic and employs many overlapping terms that refer to the same types of A-R-D events. It also contains terms that are treated as if they refer to the same events but that do not do so. As will be indicated in further considering A-R-D events in social science, different functions of the stimuli are referred to in the same breath as if the same thing is being discussed, for example, *satisfaction* and *reward.* Because of these weaknesses in the commonsense language (theory) employed as the basic psychology for the social sciences, the unity in principle underlying the various areas cannot be seen, nor can analyses be made that maximally suggest further research. In the face of the confusion occasioned by the use of the commonsense "psychology," it is no wonder that understanding human behavior is considered such an impossible task.

The principles of the social behavior theory in general, and the A-R-D principles in particular, are explicit, systematically defined, and empirical in stating the antecedent conditions that produce the consequent outcomes. The terms are distinguished from one another, are nonoverlapping, and are relatively comprehensive. Various products should ensue with the substitution of a systematic language (theoretical structure) for the commonsense language.

## A-R-D PRINCIPLES IN UNIFYING SOCIAL CONCEPTS

It is clear that there are differences in the types of events of interest to economics, sociology, political science, education, and aesthetics, as examples. Aesthetics is concerned with works of art, economics with material goods and money, education with social and achievement rewards, sociology with social values in general. Different behaviors of man are also of concern to different areas. Political science concerns political organization behavior; economics concerns behaviors involved in production and consumption; ethics, aesthetics, and so on concern still other behaviors.

Such differences provide fertile grounds for the development of seperate conceptions in the various areas. In addition, however, this tendency toward separatism is enhanced by the fact that different areas are interested in events that involve different aspects of social behavior theory. In education, for example, the A-R-D stimuli involved may be of concern for their value as reinforcers. Thus, the extent to which attention and studying is *maintained* by achievement rewards, and so on, is of concern. In aesthetics the focus can be upon the emotional responses that works of art elicit. In political science, on the other hand, the interest may be in the directive value that party leadership has for members of the party.

The fact that different functions of the A-R-D stimuli involved are of concern also makes it difficult to see commonality.

When one looks through these several differences to the operation of the general principles, the relatedness of the areas can be seen. To see this unity, however, requires an appropriate theoretical structure, one that can refer to the different A-R-D events involved and their interrelationships, the different characteristics of A-R-D events, and the general effects they have on the various kinds of behavior of concern. It is suggested that the principles of the social behavior theory provide a potentially unifying structure. Several examples of the manner in which the A-R-D principles can be employed to clarify and extend social science concepts are relevant here.

## A-R-D PRINCIPLES IN CLARIFYING SOCIAL CONCEPTS

It has been suggested that the commonsense language of the social sciences does not provide a basis for clear consideration of the emotional-motivational events it studies. The commonsense psychology used by social science, for example, does not distinguish the emotional responses of pleasure-pain from the instrumental effects of reward-punishment, or from the third function, the manner in which emotional stimuli direct instrumental behavior. This has a confusing effect which is clear in the quotations taken from economics, for example, Ullmer's definition of marginal utility. Ullmer goes through an explanation of what the law of marginal utility is by describing how the more one has of something the less added satisfaction another unit of that commodity provides. But then his summary definition of the law states that the "more an individual has of a given commodity, the less he will be willing to *give* in exchange for an additional unit" (Ulmer, 1959, p. 230). There is no indication here that two different psychological processes—a satisfaction process and an instrumental process of exchange—have been given as the definition of the same law. Satisfaction refers to emotional-response elicitation. Giving in exchange refers to instrumental behavior.

This same confusion occurs in most of the examples that have been given herein. There is no clear social science statement that emotional-motivational stimuli have three distinct functions. There is also no statement of the principles by which these functions help determine human behavior, and no statement of the interrelationships of the three functions of emotional-motivational stimuli.

## A-R-D PRINCIPLES AND EXTENSIONS OF SOCIAL CONCEPTS

It is suggested that the A-R-D principles and other relevant aspects of the social behavior theory could be employed to considerably extend understanding of human behavior relevant to the social sciences and humanities.

As an example of the manner in which the several principles provide a productive theoretical base, let us consider how the concepts of norms and values—concepts that are widely employed in the social sciences—are distinguished and related and extended. Homans (1961) defines the term *norm* as follows: "A norm is a statement made by a number of members of a group, not necessarily all of them, that the members ought to behave in certain ways in certain circumstances" (p. 46). Bierstedt (1957) makes a more elaborate definition of societal norms:

> A norm, then, is a rule or a standard that governs our conduct in the social situations in which we participate. It is a societal expectation. It is a standard to which we are expected to conform whether we actually do so or not. It is a cultural specification that guides our conduct in society. It is also . . . the essential instrument of social control. (p. 175)

It may be added that a norm is also a description of the society's *values* in that area of behavior, although the relationship between norms and values has not been clearly seen. To be more explicit, it is suggested that certain A-R-D values with respect to behavior may be described. These descriptions may be considered to be statements of "rules" of reinforcement, of the types indicated in the later section of the present chapter entitled "The Group A-R-D System." That is, societies differ in the rules by which behaviors are reinforced generally, some behaviors being positively reinforced, some negatively, and in differing amounts. A norm may be considered to be a statement of one of society's rules of reinforcement for behavior.

The important thing here, however, is to indicate how the A-R-D theory may be employed to clarify the relationship of values and norms. The norm may be considered to be the rule or statement describing the type of reinforcer that will be given to a particular behavior generally within the society. In the terms of the A-R-D theory, it can be seen that the three functions of motivational stimuli should be involved. That is, the stated norm should have an emotion-eliciting function, a reinforcing function, and a directive stimulus function. To illustrate, the statement "Thou shalt not covet thy neighbor's wife" may be considered to involve a statement of the rules of reinforcement of the society (to be a norm) and also to involve the three A-R-D stimulus functions. The rule involved is that the behavior of coveting someone's wife, at least when detected, will be met with social disapproval (punishment). It may also be suggested that the usual person in the society with that norm will have been conditioned to a negative emotional response to the words "covet thy neighbor's wife." As a consequence the words would serve as a negative reinforcer (punishment). That is, a man who was told by someone "You are visiting your neighbor only to covet his wife" would find the statement a punishment and his visiting behavior, other things being equal, would be decreased in frequency. Furthermore, the norm should have directive stimulus value to the extent of its emotion-eliciting value. That is, the man who learns the norm, and in whom the norm elicits the appropriate emotional value, will be less likely to perform the behaviors of coveting wives of others. The point of the example is to indicate that the theory allows one to see the relationships of several functions of motivational stimuli and thereby to clarify what seem to be disparate events in social behavior. The relationship of norms and values can be seen, as well as the principles involved in their formation and function.

Consideration of the three A-R-D functions of emotional-motivational stimuli is relevant also to economic theory, to the concept of power, to exchange theory in general, and to the other areas that have been mentioned. For example, Jeremy Bentham described the acquisition of money in terms of happiness. A-R-D theory suggests that money will also have a reinforcing and a directive function, if it has a positive emotional value in eliciting happiness. A person with money should be capable of eliciting and conditioning positive attitudinal responses in others, reinforcing the behaviors of others, and directing their behavior to a greater extent than a person without money.

Exchange theory deals with the reinforcement function—with the effects of rewards and costs (analogous to punishment) on instrumental behavior. This account should be elaborated to include the other principles involved, and classical conditioning as well as instrumental conditioning. Thus, for example, when a person rewards an individual he elicits a positive emotional response in the individual. This has an effect of a more permanent nature than the immediate transaction. The experience will classically condition a positive emotional response in the individual for the other person. This will enhance the three-function A-R-D value of that person for that individual. As examples, the former will be better able to reinforce the individual in the future, as well as to serve as a directive stimulus for his approach behaviors of various kinds.

Besides the principles already mentioned, it should be noted that A-R-D theory also indicates the principles of the *formation* and *change* of values and other A-R-D stimuli. For example, in addition to the elementary principles of classical conditioning, the formation and change of the A-R-D conditions were seen to occur on the basis of language, which involves no material rewards. Language is a most central instrument of value formation and change, and language and the other personality repertoires need to be considered in the study of social behavior. It may be added that utilizing a set of empirical principles that stipulates the conditions that result in the formation and change of values should provide a foundation upon which empirical studies could be conducted, as will be indicated.

## A-R-D PRINCIPLES AND RESEARCH HYPOTHESES

AESTHETICS. The field of aesthetics is ordinarily considered to be one of the humanities, not a field that grows through empirical investigations. That is, there are divisions of study between the biological, psychological, and social sciences. Connected to these divisions are methods of study. The biological sciences employ biological science methods of exploration. Psychology employs some of these methods but deals more broadly with behavioral research methods, including laboratory and clinical methods, that approach the concerns of the social sciences. The social sciences depend to a larger extent upon systematic, naturalistic observations and research methods, such as surveys, rather than upon behavioral experiments that manipulate antecedent conditions that produce changes in the behavior of subjects. In the humanities there is less in the way of a research methodology involving observational procedures, and the areas generally are not considered to be sciences or concerned with scientific methods.

It is suggested that part of the importance of providing a unifying theory for the various areas of study concerned with man is to improve the extension of the methods of study that have been found productive, across the boundaries constituted by traditional separation of areas. It would also be expected that breaking down the barriers, so that problems of investigation were considered by individuals from various areas, would lead to the development of new methods of study. As has been suggested in some of the other chapters herein, a great deal of methodological development is necessary so that the study of complex, significant, human behaviors can be systematically advanced. While it will not be possible to provide an account that is in any way complete, the suggestion can be exemplified.

In the section on aesthetics herein a set of general questions concerning the field was quoted. One of the categories of questions listed contains some very central items for establishing the basis of the field. Thus, as noted above, F. J. Coleman (1968) asks ". . . What does it mean to say that an object has or does not have aesthetic value? Is aesthetic value a property of objects? Or is it a reaction that occurs in some or most persons when they observe certain objects?" (pp. 2–3). Research to answer such questions can be suggested. In the context of A-R-D theory, an aesthetic object would be said to elicit an emotional response in an individual. As a consequence, the object would have reinforcement value for him and would strengthen an instrumental response it followed. And, the object would have directive stimulus value for him in that it would elicit a class of instrumental behaviors that would "approach" the object. It would be expected that individuals or groups could differ with respect to any one aesthetic object, on the assumption that the emotional response to the object (and its other functions) had been learned. Besides the general classification of the type of A-R-D stimulus involved, it would thus be the individuals' response to the object that would define its aesthetic value, and with different conditions of learning there would be different aesthetic values. This analysis thus proposes an explicit definition of what is meant by aesthetic value.

The focus of the present discussion, however, is to show that this theoretical effort also implies empirical investigation. For if the theory is correct it should be possible to demonstrate empirically the experimental hypotheses drawn from the theory. For example, an object that elicited an "aesthetic" emotional response should be capable of transferring this emotional response to any new stimulus with which the aesthetic object was paired, according to the principles of classical conditioning. This type of experiment has not been done yet, although there are a number of experiments that suggest the principles would hold. For example, Geer (1968) recently conducted an experiment in which pictures of ghastly injuries were employed as the unconditioned stimuli to classically condition a negative emotional response to a previously neutral stimulus. This example is just the opposite of what would be expected if the unconditioned stimuli employed were aesthetic objects. There is evidence on the positive side also that has the same implications, although the research was not conducted with the present purposes in mind. Silverstein and his associates (see Silverstein, 1973) have conducted a series of experiments in which pleasant or neutral photographs were paired with nonsense syllables in a paired-associates learning task. The results indicate that syllables previously paired with pleasant photographs are learned more rapidly than syllables paired with neutral photographs. These results also have implications for the other aspects of research to be discussed.

It may be suggested that one of the important *functions* of aesthetic objects in a social sense is the fact that they can be employed to condition people to positive emotional responses to the owner of the object. Thus, in the same way that a politician who provides free food and drink at a political rally conditions positive emotional responses to himself, so does the religious order that assembles collections of art objects and displays them in their religious edifices and activities. The visiting dignitary who sees the religious or political leader in the context of arrays of highly "valued" art objects will thereby receive positive conditioning. In experimental terms, it would be hypothesized that neutral stimuli paired with emotion-eliciting aesthetic objects would come to elicit in subjects the emotional response involved.

The theory would also suggest the manner in which aesthetic values vary from individual to individual and from group to group—that is, the principles by which objects and events come to have aesthetic value. Again, the principles involved are those of classical conditioning, which could be of a primary or language conditioning or observational learning type. As one example, the individual should be capable of being conditioned to respond emotionally to objects and events on the basis of classical conditioning through language. If he experiences situations where certain objects and events are described by others as "beautiful," "lovely," "inspiring," and so on, those objects and events will come to elicit the "aesthetic" emotional responses that the words elicit. (Coleman, 1967, employing the present author's language conditioning methods, has shown how values can be classically conditioned.) Aesthetic emotional responses to objects and events could also arise in the individual's own instrumental activities. That is, if he is given piano lessons and receives positive reinforcement in this activity, the activity will come to elicit a positive emotional response. Also all the stimuli associated with the activity—most centrally the music itself—will come to elicit a positive "aesthetic" emotional response.

By such analyses (much more extensively conducted, of course), it should be possible to understand the differences between individual's in their aesthetic "preferences." Moreover, with supporting research, it should be possible to understand how societies differ in aesthetic preferences, as well as how changes take place in aesthetic preferences within one individual or society over time.

In addition to the emotion-eliciting properties of aesthetic objects, however, there are the other two functions of such stimuli that should help define what aesthetic value is. First, aesthetic objects and events should have reinforcement value and should affect human behavior in the manner that other reinforcing stimuli do. Thus, for example, an aesthetic object would be expected to serve as a reward (reinforcer) for any behavior it followed, and so on. It should certainly be possible to obtain aesthetic ratings on various objects from an individual and then show the behavioral implications of such aesthetic judgments. Let us say, for example, that two pieces of music were judged aesthetically by an individual (or a group) to be beautiful in the one case and lackluster in the other. It would be found that the former would act as a positive reinforcer to a greater extent than the latter. We could, for example, show this by putting the individual in a situation where the music played would fade in intensity, unless the individual pressed a key. The key-pressing response should be strengthened and maintained by the reinforcement of keeping the "beautiful" music clearly audible. The behavior should not be maintained, or not nearly as strongly, with the less aesthetically appreciated music.

This would of course provide an empirical demonstration of one of the behavioral effects of aesthetic stimuli. The same type of experimental study of the third, the directive, function of aesthetic stimuli could also be outlined. Some of the tangential results of Silverstein's research support this expectation. That is, pleasant (P) photographs appeared to control approach responses more than neutral (I) photographs.

> Highly correlated with the sharp separation of PL [pleasant] ratings of. the two subsets of pictures were differences in the amount of time that the subjects spent looking at the P and I pictures and in the percentage of the subjects who selected each as a keepsake of their participation in the rating experiment. (Silverstein, 1973, p. 193)

It is suggested that experimentation of the type proposed would indicate a general behavioral significance of aesthetic objects. In being demonstrated as one class of A-R-D stimuli, it would suggest that aesthetic objects are powerful stimuli that have affected, and presently affect, important human behaviors. Historical acts have been committed under the influence of aesthetic objects. This is true not only for present and past events of historical significance, however, but for the behavior of every man. The behaviors that are maintained by the A-R-D value of aesthetically rewarding cars, houses, clothes, husbands or wives, children, movie stars, and so on, as well as music, paintings, sculpture, literature, and the like, is certainly of significant measure. By the same token, the study of such stimuli and the effects of such stimuli on human behavior can be considered to have no small significance.

ECONOMIC PRINCIPLES. In developing the token-reinforcer system for use in behavior modification, the present author considered the token as analogous to money (Staats, Staats, Schutz, and Wolf, 1962; Staats, Minke, Finley, Wolf, and Brooks, 1964). As has been indicated here, it has been suggested that the principles of reinforcement coincide with certain economic principles (Staats, 1963, p. 309). The feasibility of directly testing this interrelationship has already been shown in several studies. Castro and Weingarten (1970) have proposed that some of the principles of economic behavior that are employed in economics could be tested with animals, employing reinforcement principles and instrumental conditioning apparatus. In the context of token reinforcement (token economy) conditions, Ayllon and Azrin (1965) found that wage rates affected the supply of labor. Winkler further studied, in the token economy, the economists' principle of a relationship between savings and the amount of work patients performed. He found that as savings go up, less work is done (Kagel and Winkler, 1972) and suggested further that this deterioration in performance could be forestalled by increasing the range of consumer goods or by an increase in prices (Winkler, 1971). Phillips, Phillips, Fixsen, and Wolf (1971) found that savings could be encouraged in a token-reinforcer study with adolescents if interest were given for the savings.

The possibilities for interdisciplinary research when the principles of economics and the technology and principles of reinforcement are joined has been further suggested in a paper by Kagel, who is an economist, and Winkler, who is a psychologist.

> A token system for ward populations, whatever else it might happen to be, closely approximates the economist's concept of a closed economic system where tokens are money, deliveries of tokens as conditioned reinforcers are wage payments, and exchange rates of tokens for primary reinforcers are prices of consumption goods. As such, cooperative research in token systems would provide economists with unique opportunities for controlled observation and experimental analysis of economic behavior. (Kagel and Winkler, 1972, p. 335)

This is a very promising beginning. It should be emphasized, however, that A-R-D theory suggests that attitude formation and change are important, as is the directive effect of money and economic products. The various principles and implications need study, along with the principles of reinforcement.

VALUES-NORMS.  It has been suggested that descriptive statements concerning valued behaviors constitute norms, as dealt with in sociology and anthropology. This analysis suggests that norms as verbal statements should possess A-R-D qualities. Stated in the positive case, for example, norm statements should elicit a positive emotional response in members of a society who accept the norm. It should be possible to classically condition attitudes toward a person (or name) by pairing the person with positive norm statements, using the language conditioning procedures that have been described. Negative norm statements, on the other hand, should elicit negative emotional responses. "He respects his parents" might exemplify a positive behavioral value, and "He does not respect his parents" might exemplify a negative one.

The A-R-D theory would also suggest that such value-norm statements would function as reinforcing stimuli, positive or negative, when presented following an instrumental response. Golightly and Byrne (1964) have used an instrumental conditioning procedure in which verbal statements serve as reinforcers. This procedure could be employed to test the hypothesis that norm statements have reinforcing properties for the members of the society who hold those values.

Thirdly, norm statements should also have directive value. Positive norm statements should control approach responses and negative norm statements should control avoidance responses in members of the society who hold relevant values. This possibility could be tested in the type of procedures employed by Solarz (1960), Staats and Warren (in press) or Staats, Gross, Guay, and Carlson (1973), which have previously been described herein.

It would also be expected that cross-cultural research could be conducted on the basis of this analysis. That is, members of societies with different values should respond differently to norm statements, in the several ways that have been described. Experiments that demonstrated this would provide behavioral specification to the traditional assumptions of social scientists that the values and norms of the society help determine the behaviors of the members of the society. Such theory and experimentation are necessary to indicate the basic principles that are involved in this general conception.

The analysis of values and norms suggests that social groups can also be described by their A-R-D systems, a concept that will be elaborated in the next section.

## The group A-R-D system

Chapter 4 outlined the concept of the individual's emotional-motivational personality system, and Chapter 7 showed how the A-R-D principles applied to social interaction. This conceptual basis has been employed in the present chapter to indicate how social behavioral principles are relevant to study in the social sciences and humanities. In this context it is appropriate to elaborate the concept of the individual's A-R-D system to the group level, since it is this area that can be of central concern in social science considerations.

To begin, it is suggested that a group may be described as having an A-R-D system, in the same way that an individual is so described. That is, the members of the group may have commonalities in their A-R-D systems in contrast to other groups of individuals. One family, for example, engaged in business occupations may have high A-R-D value for shrewd trading skills and the acquisition of material wealth. Another family, engaged in scholarly pursuits for generations, may be characterized as having high A-R-D value for skills in the acquisition and demonstration of knowledge. A third family, having been involved in physical work and military occupations, will have high A-R-D value for sensori-motor skill, strength, courage, and audacity. Each family will have much less A-R-D value for the characteristics that are so positive to the other families.

It is important to realize that groups of individuals who are subjected to common conditioning experiences will acquire A-R-D systems that have commonality. Moreover, different groups of individuals may be subjected to conditioning experiences that differ. In this case the common A-R-D systems that the members of each group display will differ.

The groups involved may be small personal groups such as a family or social group. They may be larger, such as a club or church or business organization. Members of socioeconomic groups may have some common conditioning experiences that produce commonalities in certain aspects of their A-R-D systems. The same is true for members of social institutions, societies, and cultures.

This topic is one of exceedingly great scope, entering into sociology, political science, economics, and anthropology, for example. A good deal of interest in each of these fields concerns description and comparison of group A-R-D systems and the effects of A-R-D systems on the character of the group.

RULES OF APPLICATION OF A-R-D STIMULI. In considering group A-R-D systems and their effects, the manner in which the group applies its A-R-D stimuli must be considered. To begin, in the laboratory there is a certain rule in existence when the principle of reinforcement is studied. The rule involves what behavior the investigator elects to reinforce and in what manner, depending on the principle studied. In the rat he will reinforce bar-pressing behavior, or running down a runway, or turning in one direction in a T maze; in the pigeon he will reinforce pecking a key, or a key of a certain color, and so on. The behavior is selected for various practical reasons relevant to laboratory work: to be specifiable objectively, naturally occurring, simple enough to be treated as a unit, of limited duration enabling repeated trials, and so on.

This aspect of the principle of reinforcement—the particular behavior reinforced—which has only practical importance in the basic study involves some of the most significant matters in the study of human organization. Independently of *what* the A-R-D stimuli are for a particular group or culture, there can be differences between groups in the *rules* by which these stimuli are applied. The ways that groups differ in this respect, the ways that these differences develop and change, and the effects that are produced on human behavior may be seen as primary topics of study for scientific and professional areas concerned with man. A few examples indicating the importance of the rules of applying A-R-D stimuli will be given.

As one illustration, in our society some of the stimuli that have a good deal of A-R-D value are titles, positions, status roles, social and personal attention, acclaim and respect, money, fine clothes, expensive cars and houses, and various honors and awards. There are also rules in the society (not necessarily formal or explicit) for the application of these stimuli. For one thing, these stimuli (or tokens which can be exchanged for them) are delivered contingent upon some kinds of behavior but not upon others. Thus, large amounts of these stimuli are delivered contingent upon exceptionally skilled baseball, football, acting, dancing, or comic behaviors, among others. Relatively small amounts are delivered contingent upon the behaviors of skilled manual work, studying, unskilled manual work, nursing, and many others.

These characteristics of our A-R-D system and its rules of application, to continue with the example, have an effect upon the manner in which behavior in our society is molded. Consider a boy who has two classes of skilled behaviors, one a set of intellectual skills consisting of knowledge and well-developed study and scholarly work habits, and the other consisting of some form of fine athletic prowess. Let us say that either behavior could be developed to "championship" caliber. Now, in a situation in which the societal rule is that the larger amount of reinforcement is made contingent upon the one behavior, this behavior will be strengthened, and as must be the case, at the expense of the other, to the extent that the behaviors are incompatible. In our society, of course, many of the strongest reinforcers are more apt to be more liberally applied to athletic rather than scholarly behavior.

When groups are considered, it would also be expected that the A-R-D system and its rules of application will determine the types of behaviors that are dominant. A society that has a differing set of stimuli and rules will evidence different behavior over the groups of people exposed to that set of conditions. For example, a society whose reinforcers are made contingent upon scholarly behaviors to a larger extent than in another society will create stronger behaviors of that type, in a greater number of people, than will the other society. In general, many of the different cultural, national, class, and familial behaviors that have been observed in sociology, anthropology, clinical and social psychology, and other behavioral sciences can be considered to involve this aspect of human motivation—the A-R-D system and its *rules of application*. Differences between the behavioral skills of blacks and whites in the United States (for example, in intellectual skills versus athletic skills) may be seen to involve these principles.

It should be indicated that sometimes the rule specifies a particular *behavior-social stimulus-reinforcement* relationship. Thus, sexual behavior and sexual reinforcement occur in all societies. However, the rules regulate the type of behavior and the type of social stimulus. Thus, in our culture, sexual behavior is associated with positive A-R-D stimuli, but only in certain situations with certain people. People such as siblings, parents, children, same-sex partners, unwilling partners, and so on are largely excluded. We also have many examples of cultures and subcultures with markedly different rules, for example, the ancient Greek, Roman, Hawaiian, and Egyptian cultures, homosexual groups, marital 'switching' clubs, and so on.

It may be added that there are rules for the application of negative A-R-D stimuli as well as positive ones, and these can differ for families, subcultures, and cultures. For example, a family has rules by which certain behaviors are punished. A group or a society also does. Many rules concerning the application of negative A-R-D stimuli when the behavior or social stimulus involved is inappropriate are made explicit in the form of legal or religious laws. Certain *behavior-social stimulus-reinforcement* relationships are relatively likely to be heavily controlled by laws, as occurs in the area of sex. These rules concerning negative and positive A-R-D stimuli, as has been indicated, may also be called mores or values or norms in the social sciences. Generally, studies that describe a group's value system, as in anthropological study, are describing the group's A-R-D system.

THE A-R-D SYSTEMS OF SOCIETIES AND SOCIAL INSTITUTIONS: SOCIOLOGICAL AND CROSS-CULTURAL RESEARCH. It may also be suggested that the concept of the group A-R-D system could serve as the basis for additional research in social science areas.

> For some time there has been a waning enthusiasm for the use of psychoanalytic theory in the study of differences between cultures. Cross-cultural studies have tended to become more strictly empirical because of the paucity of theoretical principles that appear to have generality across various peoples (Guthrie, 1966).
>
> However, basic psychological principles should be evident with different cultures and groups within a culture, although the particular stimulus and response events may vary . . .
>
> In the present case, the principles in the human motivation theory should be useful as a structure within which to study and compare peoples (cultures) as well as institutions and groups. It is suggested that a society or social institution may be described, and compared to others, in terms of its A-R-D system. This would involve observation of the stimulus objects, events, activities, behaviors, positions in society, and so on, which elicit positive attitudes, which thus serve as positive reinforcers and discriminative stimuli. The hierarchical aspects of the A-R-D system should also vary over groups and cultures. Moreover, the rules for application of the reinforcers to specific behaviors should also vary. Each of these variations would in terms of the present theory be expected to be a determinant of the behavior of the members of the group and thus of the character of the group. As already suggested, the conjunction or disjunction of the A-R-D systems of the social institution and the individual must also be studied for the social problems produced.
>
> It should be noted that in the present context of principles, experimental research should also be derivable from the analysis. That is, the description of the various aspects of the group's A-R-D system would constitute a set of hypotheses concerning the behaviors that the (conjunctive) members of the group should display. These hypotheses could then be tested by observing whether or not the behaviors occur as predicted. It is suggested that such studies could make objective and scientifically interesting cross-cultural comparisons and the study of national (or group) character (about which there has also been waning interest because the endeavor has not been tied into a meaningful

theoretical-research structure ([Hoebel, 1967]). (Staats, 1968d, pp. 63–64)

## INTERACTION OF INDIVIDUAL AND GROUP A-R-D SYSTEMS

There has been an interest in the relationships between the individual's personality and the social system, but there has been no set of principles within which to consider systematically the possibilities (Smelser and Smelser, 1964). A systematic framework, however, is provided by the conception that (1) the A-R-D system of the individual is a prominent aspect of his personality, (2) that social systems also have A-R-D characteristics, and (3) that the two interact. That is, in a complex society formed of different groups, the individual may learn an A-R-D system that is conjunctive with social institutions with which he relates, or he may learn a disjunctive A-R-D system. The nature of the behaviors produced in the individual as a result of this relationship may vary, depending upon whether his system is conjunctive or disjunctive.

As Chapter 8 indicated, a child has an A-R-D system, and the educational institution may also be considered to have an A-R-D system, including a set of rules for applying its A-R-D stimuli. When the child's A-R-D system is disjunctive with the A-R-D system of the school, the child's behavior is likely not to be adjustive for him or for the society. It is suggested that social problems for individuals and groups can result because the social institution has not been constructed to be appropriate for those particular individuals or groups.

Research concerning individual and institutional interaction can be designed when the principles have been understood, as has been indicated by Staats, Gross, Guay, and Carlson (1973). In this study three experiments suggested that, depending on the institution's A-R-D system and the individual's A-R-D system, the individual in interaction with the institution could (1) be classically conditioned to positive or negative attitudes, (2) be instrumentally conditioned to perform varying instrumental behaviors, and (3) have varying instrumental behaviors elicited that would result in different types of learning. The possibilities of elaborating research in this area should be given additional attention.

## Large-scale social interactions: Individual-group and group-group interactions

It is not infrequent in the social sciences for an elementary principle found in some other area to be applied to a social phenomenon, not always in an appropriate manner. Thus, the principle of biological evolution was applied to social evolution, in a manner that has been criticized in the social sciences.

It has been suggested herein that some learning-behavior principles apply to groups as well as to individuals. This is supported by studies conducted by Klaus and Glaser (1970) that showed that small groups (teams) functioned according to the principles of reinforcement, the reinforcement being experienced by each member of the team.

It should be noted, however, that whether or not the elementary principles can be directly applied to organizations of individuals can only be projected on the basis of analysis of the specifics involved. Organizations of individuals cannot always be considered to be analogous to an individual. For example, the fact that a commercial organization realizes a profit is not the same thing as an individual receiving reinforcement. When making basic explanations, groups must be considered to consist of individual actors rather than as organisms themselves—although, as will be seen, even large groups may follow the same principles of behavior as the individual. A statement of effect on a group may require indication of how individuals' behavior in that group has been affected if precise knowledge is desired.

Thus, for example, the principle of reinforcement may require qualification when applied to groups. Organizations that are rewarded may be strengthened in their organizational "behaviors" *if* increase in organizational rewards is reflected in some way in the rewards made contingent upon the individual member's pro-organizational behaviors. Organization reinforcement that does not contact the individual will not have such an effect. A beginning should be made in extending learning-behavior principles into functional social analyses. Suggestions can be made in the process of presenting additional points involved in extending learning-behavior principles to large-group situations.

## LANGUAGE AS A MECHANISM IN LARGE-GROUP INTERACTION

Language is an important mechanism by which experience received by certain members of a large group can be transmitted to affect the other members of the group. Thus, a stimulus applied to relatively few members can result in a general group response. The process can also follow learning-behavior principles.

As an example, let us take the group of black people in the Union of South Africa who share highly similar role stimuli in various respects. An experience, let us say the Sharpesville massacre of 1961, which personally affected relatively few individuals, can nevertheless have a large-group effect. To elaborate, the massacre occurred when a group of demonstrating blacks were repeatedly fired upon by soldiers, resulting in a number of killed and wounded. The demonstrating blacks, according to accounts, were not threatening the soldiers, and the firing continued while the blacks were running away. Most of the wounds were from the back.

In the unseemly neutralness of technical terms, it may be said that this experience would be expected to elicit a very intense negative emotional response in those members of the group who were present and who lived through the experience. It would be expected, further, that these intensely negative emotional responses (hate, fear) would be conditioned to the soldiers, and to all other whites who share role stimuli with the soldiers. These negative emotional responses would be expected to elicit the large class of "striving away from" or "striving against" behaviors.

Such experience and the resulting learning can be transmitted through the group on the basis of the simple language description of the happening. Elder (1971) has presented evidence that supports the suggested principles of language learning of group conflict. He showed that maternal explanations of racial conflict were related to awareness of the existence of attitudes toward Negro protest groups (like NAACP, CORE, SNCC, the Black Muslims) and the holding of such attitudes. Such awareness was also correlated with access to other types of language learning. Exposure to newspapers and magazines increased the recognition and holding of attitudes towards Negro protest groups. These results held for both black and white children.

Following the above example, sufficient negative conditioning—as has occurred to blacks in South Africa, as well as in the United States— would be expected to yield strong negative attitudes towards individuals with the same role stimuli as those associated with the negative experience. Thus, when the situation permitted, it would be expected that the negative emotional response elicited by a white man in a black man would elicit instrumental behaviors that would be harmful to the white man, at least in the black man who had also learned violent behaviors. One might see cases in the society of a black man stabbing a white in a dark street, of stealing from the homes of whites, of organizing in attempted opposition, of villifying whites, and so on—all actions that have been reported to be endemic in South Africa.

At any rate, it is suggested that where communication exists among the members of a class of people, the experience of one member of a group, if it occurs to him because of being a member of the group, may be transmitted to other members of the group, on the basis of language. This can have the effect of making the group appear as if it responded directly as a unit in following the learning-behavior principles—and in this sense it does, although not through the working of one elementary behavior principle.

## RECIPROCAL GROUP INTERACTIONS

As in small-group interactions, large groups can interact with each other over a period of time in which the actions of one group constitute conditioning experience for members of the other group. The latter, as a consequence of the conditioning experience, then performs actions that in turn affect the conditioning experience of the first group. This reciprocal interaction may continue over long periods. In this manner two or more groups can affect each other's "behaviors" in a manner analogous to the way that individuals may affect each other's behavior in social interaction. Such considerations could also serve as the basis for extended research.

## Conclusions

The last sections of this chapter have been added to provide additional indication that the elementary principles of reinforcement established in the animal laboratory cannot be extended to the social level without elaboration and modification, if social events are to be dealt with comprehensively. The A-R-D principles themselves have been elaborated from the separate elementary principles. And the consideration of group phenomena requires additional developments.

Moreover, as has been indicated, all of the other concepts and principles that apply to individual human behavior need to be extended to the social sciences. The various principles of language theory and the language system—involving concepts of belief, opinions, communication, propaganda, ideas, reasoning, and so on—need such elaboration.

It is suggested, focally, that the various fields of the social sciences and the humanities utilize a psychology in important areas. The psychology is ordinarily based upon commonsense usage. It would seem that a systematic psychology, based on various types of systematic evidence, would provide an advantageous foundation. It is thus suggested that the social behavior theory should be employed in a systematic manner to unify, clarify, and extend social concepts in general, and also to serve as a theoretical structure for indicating lines of research. When the social sciences and humanities have been tied into an empirically based psychology, the experimental methods of the psychology become available. It has been suggested that laboratory and other research methods of psychology could be utilized extensively in the study of hypotheses derived from a social behavioral approach to social science.

A further purpose of the present chapter has been to suggest that psychology should interest itself in areas that deal with broader issues in the study of man. It is important to study human behavior in ways that follow in the tradition of the laboratory. It must be remembered, however, that psychology is interested in the general behavior of man. In pursuing this general interest, psychology must link up with the various disciplines that deal with man, including the social sciences and humanities.

# 15

# *Social behaviorism and the biological sciences*

THE FIELD OF PSYCHOLOGY has always had a significant relationship to biological science. In fact, some of the roots of psychology came from areas in biological science. For example, Edwin G. Boring, in writing his classic A *History of Experimental Psychology* (1950), indicated that one of the centrally important origins of modern psychology is within physiology. Moreover, major aspects of present-day psychology involve such research.

As will be suggested, when this aspect of psychology is included, psychology is centrally located in the sciences that deal with man. Psychology is focally concerned with the elementary principles of behavior, which may be considered to be somewhere in the middle of the range of the sciences involved. Extending in one direction, psychology relates to, has roots in, and has an active branch concerned with, the biological foundations of behavior. In the other directions, similarly, the science extends to clinical, social, child, and educational psychology and thus to the concerns of social behavior. This tradition, at the center of the dimension—a dimension ranging from the biological foundations of behavior, to elementary behavior principles, through social events—constitutes invaluable potentiality for a unification of science of the type proposed herein.

The fact is, however, that this potentiality has not been entirely recognized. There has not yet been a successful wedding of the biological and behavioral (any more than there has been a successful wedding of the behavioral and the social), at least in terms of establishing a general

531

conception that can range from the biological to the social within a common context of principles and methodological and philosophical concerns. Some of the strongest antagonists of behaviorism (learning) have been biologically oriented psychologists. As another example, cognitive theorists traditionally employ inferred concepts relating to the mind— concepts that have a biological flavor—while behaviorism traditionally looks to external environmental conditions and overt behaviors in its research and theoretical efforts. B. F. Skinner, as an illustration of a traditional learning theorist, has been criticized for his "empty organism" approach, that is, for a lack of concern with anything internal to the organism. The fact is that traditional learning theories, the models for theory construction in psychology, have not generally provided guidelines to indicate how the biological study of man and the behavioral study of man are to be related. In the present view, a unified theory must indicate the relationships of the various realms of study—from the biological to the elementary behavior principles through the most complex social activities of man, as will be further indicated. It will also be suggested that a continuity exists between these realms.

What is proposed is that in the study of behavior various elements are involved. There are environmental events, or stimuli; these have to have some way of affecting the organism. There are also the observable responses the organism makes. Between these related events there are complex biological occurrences that allow the expression of the principles of behavior. The two realms of study—the external stimulus-response events and the internal physiological events—must be related in a way that is relevant to each. The purpose of this chapter is to provide a conceptual base within which to consider the continuity between the biological and behavioral realms in a way that enhances each and leads to the development of a larger, more unified theory.

It will not be possible to describe in detail the manner in which there is continuity between the behavioral level of study and the study of the biological structures and processes that are involved in behavior. The work done in these areas is too abundant to outline here. The following sections thus will seek only to provide a general framework within which to exemplify the actual and potential continuity that exists between the two levels of study.

For organizational purposes, this discussion will be broken down in the following manner. In the description of the learning-behavior theory principles, the classes of events dealt with at the most elementary level were the *stimulus* and *response*. In addition, principles of learning that described laws by which stimuli and responses become related were stated. Thus, in the following sections examples will be given to indicate how basic behavioral study—in the several areas of stimulus sensitivity, response ability, and learning—shades easily into the study of the biological processes involved. The counterpart of the behavioral term *stimulus* is the biological term *receptor*, which refers to the organs that receive stimuli. The counterpart of the behavioral term response is the biological term *effector*, the organs of response. And the counterpart of the behavioral term *learning* is the biological term *connector*, which refers to the brain mechanisms that connect receptors and effectors. As will be seen, no change in any constituent of behavioral science is involved in moving to the biological level, except that different events are dealt with.

## Stimulus-receptor continuity

The basic concern of a learning-behavior theory involves the manner in which environmental events affect behavior. The environmental events, in analytic terms, are called stimuli. If one had the task of discovering the stimulus-response behavioral laws for a new organism, one of the tasks would be to ascertain what energies of the environment would be effective stimuli for the organism. It is quite conceivable that there would be behavioral differences between organisms because different types of stimuli were effective for them. (We know, of course, that such is the case.)

It is a central aspect of science that it systematically and in detail explores the realm of events in which it becomes interested. Thus, when man's behavior (actually, to early investigators, his mental activity) was seen to be related to the stimulus conditions occurring, it became the task of some scientists to begin to study systematically the nature of effective stimuli for man and other organisms. For example, in the middle of the 1800s Gustav Fechner studied the magnitudes of stimuli necessary for subjects to respond to their presence (absolute threshold), as well as the amount of change in a stimulus necessary before the subject would report a difference (difference threshold). His interest led to specialized concepts and methods of study. Thus, for example, Fechner developed new methods of measurement which may be considered to be the beginnings of quantitative psychology. The methods included experimental procedures and mathematical treatments by which to establish when a stimulus is reliably sensed, ruling out such effects as the order of presentation of the stimulus in a series. Large areas of study—for example, psychophysics and signal detection—have sprung from these early interests.

This study of the stimuli to which an organism is sensitive did not depend upon physiological exploration, however. It essentially involved the presentation of a stimulus to a subject, with the sensitivity established by the subject's verbal response to the stimulus. The same methods can be employed with subhuman organisms. An animal can first be trained to respond to a stimulus, let us say a light. Then, as one example, the light can be varied in intensity to establish the range within which the stimulus will still elicit the response (Guttman and Kalish, 1958).

It is a close interest, however, to move from the behavioral facts of to what stimuli an organism can respond, to the study of the biological mechanisms that underlie the ability to sense stimuli as well as to the biological differences that determine differential sensitivity in organisms. As an example of the closeness of the behavioral and biological mechanism interest, it has been found that different parts of the retina are differentially sensitive to light. The most sensitive part of the retina in the detection of a faint stimulus is not the central part of the retina, the fovea, which is employed when we focus upon some stimulus. The eye is more sensitive when the light falls on the retina some 20 or so degrees away from the fovea (Woodworth and Schlossberg, 1954).

This type of sensitivity is observed on the behavioral level. However, the behavioral facts of sensitivity lead to the question of the mechanisms that account for the behavioral manifestations. It is thus interesting to note that there are two types of sensory cells on the retina, rods and cones. The cones are more profusely distributed around the fovea and occur more sparsely with distance away from the fovea. Rods, on the other hand, are not present in the fovea and occur more profusely with distance away form the fovea. The anatomical finding that the rods are distributed where the retina is most sensitive to faint light suggested that it is these structures that are responsible.

This example has been given to illustrate how isolation of elementary stimulus-response relationships could provide information to serve as a basis for investigating biological mechanisms. The converse is also true. It could have been established first that there are two types of cells on the retina of the eye which had different distributions. This finding could then have led one to search for differential behavioral events resulting from the two biological mechanisms.

This example thus suggests that movement from either the behavioral to the biological, or from the biological to the behavioral, can be just as productive. Nothing in principle would dictate that one direction would be favored, and in actual practice movement has occurred bidirectionally. This is not the place for detailed consideration of the study of human sensitivity to various stimuli and the many facts, concepts, and theories involved. Nor is it the place for consideration of the study of the biological structures and processes involved in this sensitivity. The illustration is presented, however, to see the continuity that is involved. There is an easy relationship between the study of what stimuli are effective for organisms and the study of the biological mechanisms that are responsible for the differential effectiveness of stimuli. No changes in scientific philosophy are necessary as one moves from one level of study to the next. There is continuity in the general principles and methods of scientific investigation. And, importantly, there may be possibilities of gaining direct suggestions from one level for study in the other level. As will be indicated, the possibility of moving directly and productively, in a reciprocal manner, from one level of study to the other provides a central definition of actual continuity between levels of study—versus the pseudo continuity to be described in the next chapter. It is important to indicate the continuity that exists in the present case, and thereby to characterize the present approach.

## ADDITIONAL CHARACTERISTICS OF BIOLOGICAL-BEHAVIORAL CONTINUITY

It should also be noted here, suggesting a point that will be touched upon again, that the continuity may be seen at its point of occurrence more easily than between areas of study not so closely related. For example, the anatomical study of the biological basis of stimulus sensitivity is close to the study of stimulus sensitivity in organisms, which is close to an interest in stimulus-response relationships. However, when one gets into the detailed study of the physiology of the eye (or other sense organ), the chemistry and physics involved, and so on, the relationship of the study to the behavioral level is not so easily seen by the usual specialized scientist.

It should be noted that one does not have to be interested in complete detail in any particular area of study. The extent to which one pursues, in his chosen area of investigation, the ramifications of related events into other levels of study is purely a matter of pragmatics. Although one can see continuity from the study of learning-behavior principles to the study of physiology of the sense organs (and it is important to see this continuity), this does not imply that when dealing with problems of sense physiology it will necessarily be productive to refer to the learning-behavior principles. Conversely, it must be indicated that if one can obtain reliable relationships between grossly stipulated stimuli and their effects upon responses, this level of relatively gross stipulation may be the one that is appropriate. One does not have to indicate the neurophysiology of gustatory (taste) sensation, as an illustration, to state the principle that there is a lawful relationship between the strength of response and the fact that the response has in the past been followed by the consumption of a food-reinforcer. To go into the detail of the gustatory sensation mechanisms could be unproductive in the statement of the learning-behavior theory laws. There is no special nobility in specifying to a lower level than necessary to secure the products of science for the purpose at hand, although it is important to indicate the continuity so that it may be drawn upon where it is productive to do so.

## REINFORCING STIMULI AND BRAIN STIMULATION

The physiology of the sense organs includes areas of extensive research and findings. The major structures of the sense organs are well known and are standardly taught to students of physiology and psychology (Morgan and Stellar, 1950). The continuity from concern with environmental stimuli on the behavioral level and the anatomy and physiology of the sense organs can readily be seen. It is not necessary to present this material here. However, the biological study of one important stimulus function will be additionally mentioned.

One of the important functions of stimuli that has been dealt with in the preceding chapters, beginning with the statement of the basic learning-behavior principles, is that of reinforcement. Some stimuli have the ability to reward in the sense that they strengthen future occurrences of the responses they follow; and other stimuli punish in the sense that they weaken responses they follow. The importance and universality of the reinforcing function of stimuli has been indicated, beginning with Thorndike's work. Perhaps all of the organisms in the animal kingdom are sensitive to reinforcing stimuli of one kind or another, and much of their variable behavior is determined by the action of those stimuli. In the realm of human behavior, the effects of reinforcement permeate all areas of significance. The principles involving reinforcement appear to be ubiquitous for human actions, from the simplest learning to the most complex social behaviors.

It is thus of interest to note that the physiological mechanisms involved in the action of reinforcing stimuli have begun to be explored. Although the work began in serendipity (Olds and Milner, 1954), the discovery of areas in the lower brain which, when stimulated, provided reinforcement for instrumental responses has led to a great deal of research on the phenomenon. Olds and Milner permanently implanted electrodes in the limbic systems of rats' brains. The rats were then allowed access to a chamber which included a lever. When depressed by the animal, the lever would produce a slight electric charge to the brain. The animals' bar-pressing response was increased in frequency and maintained by the reinforcing action of the light shock to that part of the brain.

Investigations have also found that stimulation of some areas of the brain will serve the same function as a negative reinforcing stimulus (Cohen, Brown, and Brown, 1957; Delgado, Roberts, and Miller, 1954). The animals would learn to perform responses that would terminate the brain stimulation. It has also been found that deprivation (for example, of food) can affect the extent to which brain stimulation will serve as a positive reinforcer (Brady, Boren, Conrad, and Sidman, 1957). Sex deprivation has also been studied for its effect on the reinforcing value of brain stimulation (Olds, 1958). The results follow the same laws as an external reinforcing stimulus.

Many questions concerning the brain stimulation phenomena remain to be explored. Furthermore, there is some evidence that brain stimulation does not have all the same characteristics as traditional reinforcing stimuli (Olds, 1955, 1956; Olds and Milner, 1954). It is pretty well accepted, however, if not precisely isolated and understood, that areas of the brain have been identified, at least in part, as the physiological mechanisms involved in reinforcement and motivation. This work must also be considered together with the physiological evidence on emotional-motivational responding described in Chapter 4.

It is very relevant, hence, in this central area of reinforcement to indicate there is no contradiction in moving from the behavioral level to the biological level. The terms employed on the behavioral level—positive reinforcing stimulus and negative reinforcing stimulus—display no discontinuity when transferred to the use of electric shock to the brain. The other terms and principles, for example, frequency of response as a measure of response strength, or deprivation of reinforcers, extinction, and so on, also enter into this biologically oriented study with no difficulty. No change in philosophy of science is necessary; no change in the standards of experimentation. Moreover, it is especially illuminating to indicate that the naturalistic observation of reward-punishment effects that has been evident in social philosophies for many years and the laboratory isolation of reinforcement by Thorndike were available prior to the biological discovery of reward-punishment centers. The behavioral knowledge formed a context for the biological finding, and the prior knowledge could have provided the rationale for search for the biological mechanisms. The continuity from the behavioral level to the biological level may be seen in this case again to be easy and potentially productive.

Perusal of a good textbook in physiological psychology (see Deutsch and Deutsch, 1973) will provide information on the anatomy and physiological functioning of the organs that receive auditory stimuli (pp. 123–69), visual stimuli (pp. 71–122), taste and smell (pp. 252–85), the skin senses (pp. 170–216), and so on. In each case there is continuity from the behavioral knowledge of stimuli to the physiological level of study.

## Response-effector continuity

On the psychological level of investigation it is the behavior of the organism that is to be accounted for. In the study of social events, for example, it is the complex behavior of people interacting that is of concern. In the study of the elementary laws of behavior, it is the behavior of the organism that is to be explained by previous experimential (stimulus) events, according to the laws of learning.

As in the preceding section, the question of the physiological mechanisms of behavior—the detailed investigation of the biological structures or processes involved—is closely related to the study of behavior. This study in traditional terminology concerns the effectors, the muscles and glands that can make responses to stimuli.

## LEARNING THEORIES INFLUENCED BY BIOLOGICAL FINDINGS

The muscles are divided into two major classes, striated or skeletal muscles and smooth muscles. The striated muscles are largely those that attach to and move the skeleton. These are the muscles that are involved in all bodily actions. Within the body there are also the smooth muscles, which, rather than connecting to the skeleton, ordinarily reside in the linings of the visceral organs—in the walls of the stomach, in the walls of the blood vessels, in the iris which controls pupillary size, and so on.

In addition, there is also another class of effectors, the glands. Thus, to the stimulus of food in the mouth the salivary glands respond by generating saliva and pouring it into the mouth through ducts, for example, those located where the front of the tongue attaches to the floor of the mouth. There are also ductless glands, the endocrines, that secrete the products of their response directly into the bloodstream, where additional effector responses may be generated as a consequence.

It is interesting to note the actual, if not always recognized, continuity that has occurred in this realm between the physiological study of the effectors and the behavioral study of the learning laws that involve the effectors. That is, as indicated in Chapter 2, two-factor theory has suggested that there are two types of conditioning or learning. The first type of conditioning is considered to involve classical conditioning of the glandular responses and the responses of smooth muscles. The other type of learning is instrumental conditioning, which involves the skeletal muscle responses. This principle is considered to be that of reinforcement, versus the contiguity principle that functions with classical conditioning.

Schlossberg (1937), following available physiological knowledge, suggested that different nervous systems are involved in the two types of conditioning. The autonomic nervous system was seen to be involved in classical conditioning, while the skeletal nervous system was seen to be involved in instrumental conditioning. As N. E. Miller (1969) has noted, Cannon (1932) has suggested that nerves in the sympathetic part of the autonomic nervous system all fired simultaneously and were not capable of discrete activity and thus did not have the detail and precision available to the skeletal nervous system. This observation was supported by the fact that visceral pain is poorly localized, as well as by the fact that there are gross, overall responses to stimulation of the autonomic nervous system. For example, when part of the body is subjected to cold, blood vessel constriction occurs over the whole body.

## BEHAVIORAL EXTENSIONS OF THE RELATIONSHIPS BETWEEN TYPE OF RESPONSE AND TYPE OF LEARNING

These interpretations, based upon both behavioral and biological findings, generally considered that classical conditioning involves the autonomic nervous system and is involuntary. Miller, whose research has helped challenge this position, characterizes the interpretations as follows: "[T]he inferior classical conditioning was seen as the only kind possible for the inferior, presumably involuntary, visceral and emotional responses mediated by the inferior autonomic nervous system" (1969, p. 434). On the other hand, "[T]he strong traditional belief [is] that the superior type of instrumental learning involved in the superior voluntary behavior is possible only for skeletal responses mediated by the superior cerebrospinal nervous sytem . . ." (Miller, 1969, p. 435). These traditional positions, it should be noted, had empirical support.

> They were heavily influenced by the fact that many experimenters had tried to condition autonomic responses instrumentally—and had failed. The evidence was so uniformly discouraging that Skinner declared in his influential treatise on instrumental learning in 1938 that the ANS [autonomic nervous system] could not respond to instrumental training techniques. Other prominent learning psychologists shared this conclusion, so that for two decades psychologists influenced by their views essentially ignored the problem, assuming that the issue had been closed (Katkin, 1971, p. 3).

In recent years there has been an upsurge in interest in the possibility that visceral (emotional) responses can be conditioned on the basis of reinforcement, according to the principles of instrumental conditioning. Fowler and Kimmel (1962) reported the first significant results in conditioning an autonomically mediated response—sweat gland activity—through the use of reinforcement. Engel and Hansen (1966) have showed that the rate of heart beat response, a visceral response, can be changed through reinforcement, thus having characteristics like an instrumental response. This finding has been repeated employing better control for muscular activity (Trowill, 1967). Miller and Banuazizi (1968) have also shown that rats could learn to contract or relax the large intestine when reinforced for either response. DiCara and Miller (1968a) demonstrated that rats could learn to either increase or decrease their blood pressure to escape from mild electric shock—and that the blood pressure response was unrelated to changes in heart rate.

There is still some doubt whether the autonomic responses are being learned directly through reinforcement. That is, as an example, perhaps the individual who learns to slow down the rate at which his heart beats is simply learning muscular relaxation responses. Attempts have been made to rule out this possibility by giving animals drugs (curare) that relax the muscles so this cannot be varied (Miller, 1969). However, it is granted that this is not a foolproof method. There have also been demonstrations that attempt to rule out the action of skeletal muscle involvement in other ways. For example, DiCara and Miller (1968b) showed that they could train rats to increase the flow of blood to one ear while decreasing the flow of blood to the other. It is felt that this result cannot be interpreted as due to skeletal muscle responses (Katkin, 1971). However, animals (and some humans) can move ears independently of each other, and the activation of the musculature that moves the ear may affect the blood flow to the ear.

An important study in this context, as well as in others, has recently been conducted by Harris, Gilliam, Findley, and Brady (1973). They used food and shock avoidance as reinforcers for increases in diastolic blood pressure, in daily 12-hour conditioning periods. In this manner it was possible to produce sustained and significant increases in both systolic and diastolic blood pressure, accompanied by elevated heart rates. This study may be interpreted to suggest that individuals can develop high blood pressure through learning. The study also showed "changes in blood pressure which cannot be accounted for on the basis of short-term 'voluntary mediators'" (Harris, Gilliam, Findley, and Brady, 1973, p. 177).

It should also be noted that, whether or not the autonomic trainability turns out to depend upon skeletal musculature learning, these various findings have important practical implications. There are many medical problems that involve the responses of the internal organs that are governed by the autonomic nervous system. For example, in part because of the response of the blood vessels, the individual may have high blood pressure. There are individuals who have medical problems because their hearts beat too fast, and so on. The suggestion that such internal responses can be learned through reinforcement has implications for the treatment of medical problems involving the responses. The suggestion is that these responses, thought to be uncontrollable, can actually be learned as voluntary responses. This possibility is being actively explored at the present time. Cheap and reliable apparatus is now being manufactured with which to measure autonomic responses, and these are being employed to treat individuals with problems ranging from migraine headaches to skin problems. The value of such treatment will of course require systematic experimental demonstration.

There is also evidence that there is overlap in principle in other ways between the conditioning of skeletal responses and autonomic responses. As was suggested in the second chapter, in the section on higher-order instrumental conditioning, instrumental conditioning can also occur through continuity. If a skeletal response has been learned through reinforcement to a stimulus, when a new stimulus is paired with the original stimulus the new stimulus should also come to elicit the response. The author has informally demonstrated that this occurs (Staats, 1964a, 1968c), and there has been a recent study to that effect (Birkimer, 1966). The area, however, needs focal attention in a series of studies.

The central point of this discussion, however, is to illustrate in this area the manner of interaction between the behavioral and biological levels of study. That is, there were a number of biological findings concerning the effectors and the nervous systems that activated the effectors. Learning theorists such as Schlossberg and Skinner were aware of the biological findings and concepts of classical and instrumental conditioning. With knowledge of two of the types of effectors (skeletal muscles versus glands and visceral muscles) and the two nervous systems (the cerebrospinal nervous system and the autonomic), it was natural that this knowledge should be influential in theoretical systematization of the evidence of two types of learning—even if the biological realm was not recognized in the philosophy of theory construction involved (Skinner, 1950).

It is interesting to note, however, that further behavioral study has suggested that the principles of learning may not be strictly dichotomized according to the types of responses (effectors) involved. Learning through reinforcement may occur with responses supposedly associated with the autonomic nervous system. And instrumental conditioning through contiguity suggests also that the supposed "inferior" principle of classical conditioning is relevant to skeletal responses.

Thus, although the evidence is not yet definitive or completely positive, the recent behavioral findings suggest changes in traditional learning theory and also implications for the biological understanding of learning. If it is definitively indicated that responses mediated by the two different nervous systems follow the same learning principles, this will suggest that there is a common mechanism involved, or at least mechanisms that obey the same laws of learning. This would support the speculation that the seat of learning for the various responses is in the central nervous system, the brain, and that the same laws pertain there regardless of the response involved. Such possibilities could provide a basis for additional research on the physiological and anatomical levels.

And, as has been indicated, the learning theory of effector action also feeds back to biological concerns in other ways. Thus, the behavioral study of the functioning of glandular and visceral effectors has significance in physiology and medicine. The laws of learning can be seen in this light to be relevant to the treatment of medical problems traditionally considered to be of biological origin and to concern biological treatment. The main purpose of the present discussion, again, has been to illustrate further the manner in which basic issues at the behavioral level can move to the physiological level of study for resolution. This again suggests the reciprocal relationships involved.

## IMAGES, SENSATIONS, AND RESPONSES

There are several other areas of overlap in behavioral principles that help show the continuity of principle from the behavioral to the biological levels of inquiry. Take, for example, the separation in physiological study of receptors and effectors, and in behavioral study of stimuli and responses. The fact is that consideration of certain human behaviors suggests a reconsideration of the two dichotomies in certain respects. In commonsense terms, we have traditionally spoken of sensations and also of images. A sensation occurs when a stimulus is present; an image may occur in the absence of the stimulus. Subjectively, people describe images as different from sensations. While some hints may be obtained from commonsense descriptions (introspections) of these processes, it has been found that not much progress can be made when limited to such descriptions.

However, when the analysis is made in terms of conditioning principles, a more penetrating understanding of images is possible. As already indicated, it has been suggested that sensations have not only stimulus characteristics but also response characteristics. There was suggestive evidence that sensations could be classically conditioned (Ellson, 1941; Leuba, 1940; Phillips, 1958). Test of the possibility in the realm of language conditioning (Staats, Staats, and Heard, 1961) gave further support to the analysis. Moreover, the conditioning characteristics of images have been shown in clinical treatment procedures (Cautela, 1967, 1969, 1970). The conception that has been presented is that a sensation is the individual's response to an actual stimulus. That is, an effective stimulus may be considered to be a $UCS$ for an unconditioned sensory response. As is the case for unconditioned responses in general, it would be expected that part of the sensory responses could be conditioned to a new stimulus, a $CS$. Such a conditioned sensory response would then constitute an image. The image is considered as part of the unconditioned sensory response.

The purpose of this recapitulation here, however, has been to indicate another example of productive continuity between an analysis on the behavioral level and investigation on the biological level. To begin, it may be suggested that the conditional part of the sensory response may on a speculative level be considered to occur in the central nervous system. For example, it is not the activity in the retina of the eye that constitutes the response on the behavioral level. It is only when the neural impulses from the retina have been carried to the brain (visual cortex) that the person experiences a sensory response (sensation). In this context it is interesting to note that Penfield (Penfield and Roberts, 1959) has shown that it is possible to produce sensory experiences in humans by directly stimulating certain areas of the brain. This work was done during brain surgery. Moreover, stimulation of the same area of the brain always yielded the same sensory response. This finding supports the conception of conditioned sensory responses as learned responses of the central nervous system.

The behavioral concept of the conditioned sensory response, it may be added, would suggest further hypotheses for direct biological study. Thus, for example, it would be expected that it would be possible to produce conditioned sensory responses (images) through the type of direct brain stimulation that Penfield has dealt with. To elaborate, one patient reported that he heard a certain song when a certain cerebral area was stimulated. If some other stimulus—let us say a light—had been paired with the cerebral stimulation (and the sensory response of the song), the new stimulus would be expected on the basis of the behavioral analysis to come to elicit an "image" of the song, a fainter hearing of the song. The principles of classical conditioning should apply.

At any rate, there are additional types of evidence that lend themselves to the interpretation that stimulus and response events occur in the brain, subject to the laws of conditioning. The studies help indicate further interactions of behavioral and biological concepts and methods of study. Thus, for example, an early experiment of Durup and Fessard (1935) involved the serendipitous finding that electrically recorded brain waves (EEG) could be conditioned. These investigators were studying the EEG response of a cat to a visual stimulus. In the procedure, for purposes of recording the EEG on an oscilloscope, a camera was activated just prior to the presentation of the stimulus. The apparatus was subject to failure. Sometimes the camera would be activated, but no visual stimulus would be presented. The investigators then noticed that the camera shutter click alone served to elicit the EEG. They astutely realized that they were observing the classical conditioning of a brain wave response. The camera click served as the $CS$; the visual stimulus served as the $UCS$ that elicited the EEG response. Because of the systematic pairing, the camera click had come to elicit the brain response, the EEG response.

This finding has been widely confirmed under precise laboratory conditions (see Morrell, 1961; Grossman, 1967, pp. 670–91). For example, Doty (1958) presented an auditory stimulus to a cat many times until the cortical (brain wave) it elicited had adapted out. Then the auditory stimulus, as the $CS$, was presented with a shock to the animal's foot, as the $UCS$. After a few pairings the auditory stimulus again came to elicit the EEG response, as well as a foot withdrawal motor response. It is interesting that when the $^{C}S$ continued to be presented by itself without pairing with the shock, the motor response extinguished with fewer trials than did the conditioned brain wave response. The same type of conditioning has also been shown where the $^{UC}S$ is a sensory stimulus that also elicits a brain wave response. For example, the unconditioned stimulus could be a light. In this case the conditioning is termed sensory-sensory conditioning. It would seem important to bring these studies into relationship with the possibility that images (sensory responses) elicited by direct brain stimulation in humans can be classically conditioned. Demonstration of the human learning of such typical cognitive events would have many implications. For one, the consideration of such traditional cognitive events as images, within a learning-behavior theory, helps remove the traditional learning theory and cognitive theory schism.

## RESPONSE-PRODUCED STIMULI

In the previous discussion it was suggested that receptor processes (sensory processes) have closely associated effector characteristics (as in the conditioned image). The general suggestion may be made also in the realm of emotions and with instrumental responses as well. Emotions are usually considered from the standpoint of effectors, as responses to stimuli. Emotional responses, in the behavior theory, however, may be considered in addition as receptor processes. That is, as was suggested in Chapter 2, when a conditioned stimulus has come to elicit an emotional response, the conditioned stimulus will also require a reinforcing function. The conditioned emotional response will confer a receptor function—the reinforcing function—on the stimulus. The behavior analysis will suggest that emotions have to be studied as acts themselves, but in addition emotions have to be studied for the way they serve a stimulus function in reinforcing instrumental responses. The stimulus aspects of emotions have been generally referred to as feelings and sensations (Pribram, 1970a; 1970b; Stanley-Jones, 1970) and even perceptions (Leeper, 1970).

It need only be added in this context that instrumental (as well as emotional) responses, which again are considered as effectors, also have receptor functions. There are nerve endings in the muscles and tendons. Thus, whenever, a muscle responds it also activates the nerve endings in it and in the tendons to which the muscle attaches. Knowledge of these processes has given rise to the term *response-produced stimulus* (Hull, 1943). The concept in part enables an understanding of the manner in which one response can come to elicit another response in the types of sequences described in previous chapters.

In this context there are several points to be noted. First, this indicates again that there is not a clear separation of receptor and effector processes; there is overlap, as there seems to be in the principles of classical and instrumental conditioning. Receptor processes (sensory organs) have, in their elaborated form in the central nervous system, response characteristics. It was suggested, thus, that sensations could be conditioned and become conditioned sensory responses or images. On the other hand, it appears that effector processes have stimulus characteristics and can serve the several functions that other stimuli serve.

It should also be noted that a great deal of the justification for the concept of the internal, response-produced stimulus comes from the biological level of study. Anatomical observation of free nerve endings in the muscles and tendons, for example, is basic to the conception. In order for the learning theory to make use of such evidence, however, it must have a philosophical methodology that recognizes the actual and potential importance of the biological level of observation in the development of behavioral theory.

ANATOMICAL LOCATION OF RECEPTOR-EFFECTOR MECHANISMS IN THE BRAIN. It is also relevant in this context to indicate that there are different areas of the brain that are specialized for receptor or effector functions. Physiologists have traced processes involved in reception. External visual stimulation, for example, creates electrochemical processes in the retinal cells of the eye. These activate nerve action in the optic nerve that leads to the brain. The result is activity in the back of the brain (the occipatal lobe) which can be measured in terms of electrical potential. The occipital lobe is thus an area for the receipt of visual receptor stimulation. Stimulation from the ears goes to the sides of the brain (the temporal lobes). For the receptor processes to occur, the sensory activation has to be carried to the appropriate part of the brain by electrochemical neural impulses.

It has also been found by stimulating the cortex of the brain with electric impulses that the brain contains areas that are the *sources* for the individual's responses. Fritsch and Hitzig in 1870 stimulated the brain directly with weak electrical shocks, using dogs as subjects. A certain area on the right side of the cortex when stimulated would produce a movement of the left leg. Stimulation of the cortex on the other side would produce a movement of the right leg. The movements were the result of the brain stimulation that in turn produced stimulation of nerves leading to the muscles attaching to the leg.

All muscle activity in humans appears to arise from or involve activity in particular areas of the brain cortex. This is shown by various types of research, including brain stimulation as well as the evaluation of the motor deficits that result from specific brain damage. There are certain parts of the brain that control the muscles of the tongue, lips, face, head, torso, arms, legs, hands, and feet. The areas of the body that are capable of the most complex muscle coordinations require the most abundant neural mechanisms to control them. Thus, the cortical area controlling the tongue and lip movements is relatively large, as is the area controlling finger movements.

Again, the introduction of the primary concepts of stimulus and response does not appear to be arbitrary in terms of reductionistic relationships to biological science. There are important anatomical and physiological counterparts of these behavioral terms. The behavioral concepts have relevance for biological science considerations, and the converse is also true.

## Learning-connector continuity

Behaviorally, the scientific study of learning began because it could be seen clearly, on the basis of naturalistic observation, that various types of change in behavior occur as a result of experience. The systematic study that has been made of the various learning processes has yielded very reliable and detailed information. It might be expected that there would be continuity here also from the behavioral level to the biological level. A natural area of interest in extension of the knowledge of the behavioral principles of learning is to establish the biological processes and structures underlying the principles involved in stimuli coming to elicit responses.

In physiology, as has been indicated, there are the terms *receptors* and *effectors* that are analogous to the behavioral terms *stimulus* and *response*. There is also the behavioral term *learning*, which concerns the principles by which stimuli and responses become functionally related through experience. Following the rationale that has been developed, it may be noted there is a corresponding term on the biological level, that of *connectors*. Thus, there are biological organs for the reception of stimuli, the receptors. There are biological organs for responding, the effectors. And there are biological organs which bring the two into relationship; these are called connectors. There are various aspects of the study of the connectors. As the next section will suggest, there is a special part of the brain whose major function appears to be that of a connector in complex learning skills.

### COMPARATIVE (EVOLUTIONARY) BRAIN-LEARNING RELATIONSHIPS

The fact is the continuity between the study of the behavioral principles of learning and study of the biological basis of learning has been a topic of interest since the turn of the century. For example, in one early experiment Lashley (1929) destroyed parts of the brains of rats. Later the animals were provided experience by which they learned a maze and then they were killed and their brains examined. The amount of brain destruction that had occurred could thus be ascertained. It was found that the degree of loss in learning ability corresponded with the amount of, rather than location of, cortical destruction that had occurred. This seemed to suggest that the various parts of the brain are of equal importance in learning.

More generally, however, it has been thought that various parts or areas of the brain have different functions in learning. As has been indicated, there are sensory areas of the brain, where stimulation from the sensory organs is carried. There are also motor areas of the brain, where cortical activity results in motor movements of particular parts of the body, depending upon the locus of the cortical activity. Neurology has revealed, moreover, that there are cortical regions intercalated between the sensory and motor zones, the so-called association area. Moreover, comparative neurology has indicated that in mammals, as contrasted with lower organisms, there is the development of a new part of the cortex, which is called the neocortex. In man, the greater portion of the neocortex is made up of the association cortex. It is interesting to note that "The primary locus of evolutionary advancement in mammals thus can be further narrowed from neocortex to a particular subdivision of the neocortex, the association cortex" (Diamond and Hall, 1969, p. 251).

As these authors indicate, the facts of the expansion of the association cortex coincided with the prevailing view of psychology in the early 1900s. That is, the association cortex was seen as the site of higher functions such as perception and learning, in contrast to invariant sensory response to external stimulation. The trend toward greater complexity in behavior with evolution was seen to result from the increase in the size of the association cortex and the number of connections possible between stimuli and responses. Evidence did not support this conclusion, however.

> The concept that the size of association cortex is correlated with performance on simple learning tasks was not supported by experimental studies, which showed that all species, regardless of the size of their association cortex, formed simple associations at about the same rate. . . . The net effect of these developments has been that comparative psychology, with some notable exceptions . . . has lost its close connection with neurology—a connection originally envisioned by anatomists and psychologists alike. . . . (Diamond and Hall, 1969, p. 252)

Using the considerable progress that has been made in anatomical, physiological, and behavioral ablation studies, however, Diamond and Hall have reasserted the unity of psychology and comparative neurology— that is, in the present terms, the continuity from the behavioral level to the biological level in the area of learning. Their work has had several focuses. First, they traced the evolution of certain brain structures. For example, they found that there is development of certain structures in the tree shrew that is absent in the hedgehog. Moreover, these structures appear to be precursors of advanced cortex areas in evolutionarily higher animals. The brain tissue of the tree shrew includes an evolutionary development that the hedgehog does not have. In view of this, their ablation findings were significant in showing that the new areas had functions in a learning sense.

Diamond and Hall also indicated that the cat shows marked development of the cortex in comparison to the hedgehog. Some of the changes parallel those found in the primates. However, there are large differences in the development of the cortex in cats and primates. They suggest that the different behavioral requirements of carnivores (night hunters) and primates (semiarboreal), each dependent upon different sensory cues, results in different evolutionary brain development. In the former the auditory centers are more predominant, in the latter the visual centers predominate. Moreover, the authors hypothesize that similar behavioral requirements can lead to similar evolutionary brain development in species that do not stem from the same biological family trees. This is exemplified by comparing the brains of the evolutionally unrelated tree shrew and the grey squirrel. The two arboreal animals exhibit very similar brain elaborations of the visual centers and, in general, show much similarity. Moreover, Diamond and Hall suggest that evidence supports the possibility that the similar structures have similar behavioral functions in the animals, although the animals are of different species.

The continuity between certain aspects of the behavioral level and certain aspects of the biological level—namely, those involved in evolutionary concepts and comparative neurology—is shown in this study.

THE "BRAIN SIZE AND NUMBER OF ASSOCIATIONS" HYPOTHESIS. As indicated by Diamond and Hall, the hypothesis of a relationship between size of the association area and learning has been rejected, largely because "all species regardless of their association cortex, formed simple associations at about the same rate" (p. 252). But, as has been suggested herein, it is oversimplified to consider human behavior only in terms of the elementary learning principles. Thus, from this view a simple learning task would not seem to be one with which to test differences in learning prowess between species. It is the ability to learn hierarchies of skills— more advanced skills built upon the prior acquisition of basic skills— that distinguishes man's learning ability. It is the number and complexity of repertoires that can be learned that is probably related to the size of association areas. Thus, as an example, language—man's most unique repertoire—is based upon simple conditioning principles. And lower animals can learn by those same principles; they can even be trained to various aspects of language repertoires (Hayes, 1951; Premack and

Schwartz, 1966) and number concepts (Ferster and Hammer, 1966). But, it is suggested, they cannot learn as readily and in as great number the fantastically complex repertoires of which a full language is composed, with new skills learned on the basis of previously learned repertoires. The hierarchical form of skill development, involving multitudes of individual stimulus-response learning, occurs in all of man's skill accomplishments. It is suggested that animals could be distinguished in terms of their brain development by the level of increasing complexity that they were able to attain in learning. Possibly such hierarchically complex types of learning procedures could be developed for differentiating along a scale of increasingly complex brain structures. Such a demonstration would help reestablish the relationship between neurology and behavioral study. It is thus suggested that the social behavioristic learning theory could lead to research at the biological level.

## EVOLUTION AND LEARNING RELATIONSHIPS

It has been indicated that the study of evolutionary biological development and evolutionary behavioral development includes biological-behavioral continuity. There are additional sources of continuity in this area. It has been suggested briefly that there is considerable similarity between the laws of learning and those of biological concepts of evolutionary adaptation. This continuity may be detailed a bit here in quoting from a passage the author wrote entitled "Adaptiveness and Reinforcement, Extinction, and Scheduling Principles."

> Organisms, including humans, function according to principles that allow for a great deal of modifiability and consequent adjustment to environmental occurrences. It was seen that behavior which is followed by certain consequences (positive reinforcers) increases in strength, that is, is more likely to reoccur. It is not surprising that organisms have evolved in this manner, since a species of organisms which did not function according to this principle would not be likely to survive. An organism that took one branch of a winding path and found water and became less likely to make that same response might not live to reproduce its kind. It seems that it is only because the consequences of an organism's behavior affect its later behavior in a certain way that an organism adapts to its environment and survives.
>
> Sometimes, however, the environment changes. Although a certain type of behavior may have been at one time, later this type of behavior may no longer be reinforced—the water hole may dry up. It is, therefore, also important for the adaptability of the organism that, even though a response has become strong, it is not immutable. When a response ceases to be reinforced, it becomes weaker. After the response occurs a number of times without reinforcement, it will return to its original level of strength.

Further, the adaptability of the organism to the conditions of re-inforcement in the environment can be seen perhaps from the effects of intermittent reinforcement. Different schedules of reinforcement have specifically different and lawful effects upon the behavior of the organism. It is important to see that behavior makes very sensitive adjustments to the reinforcing contingencies in the environment. As the reinforcing contingencies in the environment vary, distinct and lawful effects are imposed on the characteristics of the emission of a response and the extinction process.

It has been pointed out that the principles by which organisms behave are those of adaptability to environmental events, and that this adaptability must have been necessary for the survival of individual and species—thus, perhaps, the learning characteristics of living organisms arise in biological evolution. (Staats, 1963, p. 69)

A related thesis has been developed by Skinner, using the terms *phylogenetic* for species and *onogenetic* for individuals.

Another apparent characteristic in common between evolution and learning is "adaptation." Both kinds of contingencies change the organism so that it adjusts to its environment in the sense of behaving in it more effectively. With respect to phylogenetic contingencies, this is what is meant by natural selection. With respect to ontogeny, it is what is meant by operant conditioning. Successful responses are selected in both cases, and the result is adaptation. (1966, p. 74)

Actually, the principles of classical conditioning are as relevant as instrumental conditioning. That is, it is adaptive that the organism learns to respond emotionally to a stimulus that is associated with a stimulus that already elicits the emotional response in the organism. The young antelope that smells the spoor of a lion just preceding its aversive stimulation by its mother in propelling it into headlong flight will later on respond with fright, and running, to the lion smell itself. All of the learning principles, it is suggested, have come to have their particular natures because they enable the organism, in general, to better adapt to his environment. In doing so the principles insure the survival of the organism and the procreation of a line of organisms that operate according to the same principles.

In addition, however, biological concepts are involved in the present conception in other ways—and it may be stated in general that a learning conception of human behavior should be in agreement at major points with basic facts and principles of the biological sciences. Thus, one of the points underlying rejection of the present notion of the inheritance of specific behaviors or behavior potentialities is that the notion is not a good biological concept [for considering man]. While there is ample evidence to indicate that the biological structure is the mechanism by which the individual learns and retains complex responses to complex environmental (stimulus) configurations, it is quite a contrary thing to suggest that the fantastically complex stimulus-response coordinations man displays are predetermined biologically in any way. That would be a poor biological conception, as a matter of fact.

An important evolutionary finding is that specialization *hinders* adaptation to a changing environment. Extinction of species may occur when the species has biologically evolved in a manner specialized for one type of environment, and then the environment changes. Specialization of biological structure and function can also limit the species to a narrow environmental circumstance and prevent widespread proliferation of the species.

Thus, the conception that is rejected—which conflicts with basic biological concepts—is that specific *complex* behaviors (not simple reflexes) are laid down in the biological structure of man, as they are in lower animals. It is suggested that man does not inherit in his biological structure *any* complex human behavior. To inherit specific behavioral skills as a member of the human species would have been maladaptive. A Stone Age man who had a repertoire of higher mathematics, or chemistry, or courtly manners, perfect pitch, ethical behaviors, a pacifist conception of human interaction, or what have you, would have had a useless set of skills. Fortunately, such men had no such skills, because training that would have produced these skills was absent. Rather, such men *learned* to shape rocks, fight savagely, throw spears, club prey, make fire, plan group hunts, stitch furs, carve fishhooks, find and eat insects, discriminate subtle cues in tracking prey, communicate, and so on. The intelligent cave dweller was the man who had the advantages of learning in these areas of skill and profited from them. It is also fortunate that we do not inherit most such skills as he displayed, for they would be largely useless and interfering today.

These points are important in the context of a conception of man. *It may be suggested that the marvelous adaptive powers of man are due to his nonspecialization—to his generalized adjustmental (learning) capabilities.* His hand, for example, is not a special tool as is a hoof for running, a set of claws for predatory purposes, or a nonopposable thumb for swinging in trees. His hand is a *general* tool, capable of many different skills, if not so good for certain special tasks.

It is suggested that *behaviorally* specialization refers to a specific stimulus configuration that calls out specific responses, built into the organism by the biological structure. Thus, to illustrate, the blue crab responds in a specialized manner, a stereotyped dance, to the stimulus provided by another blue crab which is a potential mate. The stimulus complex is specific, as is the response. Generality, on the other hand, is given by having no, or few, stereotyped behaviors to stimuli—but, rather, in having the structure by which a stupendous number of responses can be learned to a stupendous number of stimuli. If man were specific like the blue crab in his sexual behavior, he would respond specifically to specific stimuli. This is hardly the case, however. Historical, anthropological, sociological, and psychological data reveal the differences that occur in man's sex behaviors as a function of experience. Man's variations in sexual responses are of course multitudinous. They are not performed via biological structure. . . .

Evolution of the species has been heavily documented; better biological specimens are selected by better adjustment to environmental conditions—including competition with members of the same species as well as members of other species. It is also quite evident that as one moves up the animal scale there is a progressive increase in the complexity of species' sensory input mechanisms and response mechanisms, and the integrating neural mechanisms that connect the two.

It is suggested that an improved biological conception is that as a species the evolution of generality reaches its zenith in man with the development of a central nervous system—the human brain—which is the *least specialized* of all the species. The generality of the human brain has such vast generalized complexity that is capable of "handling" input that for all intents and purposes is infinite in complexity.

With his biological structure, man inherits organs for responding in a fixed manner to certain stimuli in a sensory and emotional fashion, as well as a number of fixed reflex muscular connections of a simple sort. He inherits also neural pathways connected to muscle groups—and sources of sensation internal to the body in muscles, tendons, internal organs, and glands. In addition, he inherits vastly complex mechanisms for "associating" the stimulus and response processes according to learning laws. . . . But again the suggestion should be stressed that he likely does not inherit *any* specific complex human behavior or skill. Such behaviors are ordinarily composed of very complex sequences and combinations of responses under very complex and subtle sensory (stimulus) control involving sensory responses, muscle responses, emotional responses, and so on—grouped together through learning in exceedingly complex ways. Moreover, these complex skill constellations must vary widely for the individual and across individuals and groups.

Man's biological structure and biological nature may be considered that of a marvelous mechanism by which behaviors can be acquired through experiential conditions. The child, for example, does not inherit the skills we loosely call intelligence—again consisting of variegated responses under variegated stimulus controls. The child is a superb mechanism who can "receive" complex and subtle environmental stimuli, who can learn and retain marvelously coordinated responses, and in whom the stimuli and responses can come to be related by means of the intricate associations provided by the brain and central nervous system. It is the writer's conviction that the individual becomes what he is largely through learning. He would not be the magnificent biological organism that he is were this not the case. It is in large part because of his stupendous ability to learn, to acquire different complex repertoires, that man is set apart from lower organisms.

To continue, it is quite reasonable to speculate—within the context of biological concepts—that the conditions which, in an evolutionary sense, selected for man's biological superiority in brain structure occurred long ago. "[W]hile the psychobiological characteristics of the hominids would seem to be relevant to an explanation of the emergence of cultural traditions from their infrahuman background as well as the early development of these traditions, there does not appear to have been any significant change in the neural structure of the human species since the Upper Pleistocene, perhaps as much as 50,000 years ago" (Kaplan, 1968). This means that man's biological structure has remained unchanged during a period when his intelligence—his skills of various kinds—has changed fantastically. (Staats, 1971a, pp. 47–50)

This has been presented to suggest that behavioral concepts and biological concepts are relevant to each other in areas other than neurology, anatomy, and physiology. Moreover, the conception of man we draw from biological study can vary. The above stated view of evolution and man's nature is only one, in what is generally a controversial area. It is suggested, however, that well-established concepts at each level of study should not be in opposition to one another. When they are interpreted to be so, one possibility is that some aspect of the interpretation is at fault. Conflict between the behavioral and biological levels is an obstacle to unification, and it would seem worthwhile to be concerned with indicating avenues of general continuity.

## BASIC NEUROLOGICAL-LEARNING RELATIONSHIPS

Another type of research on the biological bases of learning involves attempts to establish the actual mechanisms by which elementary learning occurs. To begin, Pavlov hypothesized that classical conditioning depended upon cortical processes. Research has indicated, however, that the cortex is not essential to simple conditioning. For example, a chronically decerebrate cat has been conditioned to an eyelid response (Bard and Macht, 1958). There also has been research suggesting that classical conditioning occurs when the spinal cord is isolated from the brain, although the results are not entirely clear (Siegel, 1970).

Another strategy for investigating the neurophysiological bases of elementary learning principles has involved the use of organisms with primitive nervous systems. This is done, again, to get the simplicity necessary for experimental control so that elementary principles can be studied. Horridge (1962), for example, has utilized headless cockroaches, with all legs removed but one, to study the possibility that avoidance conditioning of a leg withdrawal could occur. The results were positive.

As another example, Kandel and Tauc (1964, 1965a, 1965b) have utilized another primitive organism to study the biological base of learning. The organism employed was the sea slug, a marine mollusk with relatively few, and very large, nerve cells. Procedures were used in which single nerve cells could be stimulated, and the electrical potential of the cells response could be recorded. Two cells having input into the same ganglion were employed in one experiment (Kandel and Tauc, 1965b). One cell was stimulated as the $C_S$ and the other cell was stimulated more strongly as the $UC_S$. The researchers found that the effect of $C_S$ stimulation could be enhanced by pairing it with the $UC_S$.

> Obviously, this approach to the analysis of conditioning at the cellular level can provide a tremendous amount of information about the biochemical and morphological changes that accompany simple learning. Research workers are currently attempting to use such a simplified preparation to study the physiological basis of learning and an initial attempt at a simple theoretical interpretation of the mechanisms of the conditioning analogue has been presented. . . . (Siegel, 1970, p. 402).

This summary is not meant to characterize the richness or significance of this type of research. The main point, however, is to illustrate the closeness of the behavioral and biological realms of study at this juncture point. The behavioral language, principles, and research methods move with no disjunction into the biological realm. Moreover, the interacting disciplines produce a vigorous research orientation with significant problems to investigate. The continuity again is smooth and productive.

## BIOLOGICAL THEORIES OF LEARNING

The processes by which the organism's experience results in a change in behavior that is retained is still not known, however. The mechanisms of learning and retention have been the subject of theoretical consideration nevertheless (Pfaff, 1969). One theory of learning and memory has been that greater efficiency of conduction from one neuron to another occurs through changes that take place in the synapse connecting the neurons. The theory suggests that the experience that provides conduction of the impulse from one neuron to another produces conditions that make it easier for such conduction to occur again in the future. Another hypothesis is that biochemical changes occur in the neuron in the amounts of differently coded RNA molecules formed. Stored learning (information) is represented through increase in some types of RNA at the expense of others. Another theory of learning is that change in the probability of firing of a neuron is caused by some metabolic change in the cell (John, 1960).

> The more a nerve cell is forced to fire at rates higher than its spontaneous rate of activity, the more ready it will become, metabolically, to respond to future excitatory input with long bursts of high frequency firing. Likewise, the more a cell is prevented from firing at rates as high as its spontaneous rate, the more it will become susceptible, metabolically, to inhibition. (Pfaff, 1969, p. 71)

Ultimately, of course, such biologically oriented theories of the basic mechanisms of learning must be evaluated on the basis of direct biological evidence. Theorists may hypothesize different possible biological mechanisms based upon the facts of learning (from evidence on the behavioral level) and these may serve as guides for direct search for the mechanisms. The aim, however, must be to find the mechanisms themselves. In the present context, nevertheless, it is important to see again the easy continuity by which science goes from the behavioral level to the biological level.

## Conclusions

It was mentioned in the introduction to the present chapter that there has not been an adequate philosophy within the traditional learning theories for indicating the relationship of behavioral and biological study. The philosophy of Skinner's learning theory has been that physiology and behavior theory should be independent realms of study. This approach suggested that psychology should be wary of theory, namely physiological theorization with respect to behavioral data.

> We are all familiar with the changes that are supposed to take place in the nervous system when an organism learns. Synaptic connections are made or broken, electrical fields are disrupted or reorganized, concentrations of ions are built up or allowed to diffuse away, and so on. In the science of neurophysiology statements of this sort are not necessarily theories in the present sense. But in a science of behavior, where we are concerned with whether or not an organism secretes saliva when a bell rings, or jumps toward a gray triangle, or says *bik* when a card reads *tuz*, or loves someone who resembles his mother, all statements about the nervous system are theories in the sense that they are not expressed in the same terms and could not be confirmed with the same methods of observation as the facts for which they are said to account. (Skinner, 1950, p. 193)

Skinner goes on to criticize theories that refer to another dimensional system, that is, another system of concepts and method of observation. This, of course, also has implications for separating social science, and various areas of psychology, from behavioral study—in addition to the biological-behavioral separatism that the above suggested. To continue, however, Skinner treated justified theory as a "formal representation of the data reduced to a minimal number of terms" (Skinner, 1950, p. 216). "Such a theoretical construction may yield greater generality than any assemblage of facts. But such a construction will not refer to another dimensional system and will not, therefore, fall within our present definition of theory" (Skinner, 1950, p. 216).

Modern behavioral approaches, in line with this rationale, have come to be considered to be atheoretical in nature, opposed to or isolated from the biological level of study on the one hand and the social level of study on the other. It is thus important to stress that a general, comprehensive behavioral approach needs to indicate its continuity in substance and method to the biological sciences and to the social sciences.

The purpose of the present chapter has been to illustrate the fact that the study of behavior, using basic learning-behavior theory, links very easily, well, and productively with the study of biological processes underlying behavior. There are no discontinuities with the philosophy of science in terms of the empiricist basis of science. The learning-behavior approach does not suggest any characteristics of scientific enquiry that would be unacceptable to the biological sciences. The traditions of the learning-behavior theory, as a matter of fact, and biological science have the same origins. The elaborations in the philosophy of science that are needed to seek the establishment of continuity between the study of behavior and the biology of behavior (and to the study of social phenomena as well) append to the basic philosophy of science with no difficulty.

It is equally important to indicate that there is great commonality in method between study within the learning-behavior theory orientation and biological study. That is, both have the tradition of laboratory experimental methods.

Evidence of the continuity of the biological and learning-behavioral realms may also be seen in the fact that investigations can begin in one realm and later progress to the other. Examples have been given of movement from the biological level to the behavioral level, and the reverse. This is an important element in characterizing levels of science that are intimately and directly related.

Thus, it is important to recognize the continuity between major and even minor subdivisions in the behavioral and biological sciences. As has been indicated, productive hypotheses may be derived from the behavioral level for study on the biological level, and vice versa. Moreover, it is in recognizing such continuities, and in exploiting the continuities, that support for the development of a classical, comprehensive, hierarchical theory is gained. Such a structure, once adopted, would be expected to have great productivity in stimulating further scientific development.

In addition, realization of the continuity between the learning-behavioral and the biological orientations would contribute towards avoidance of uselessly competitive conceptions and methods that retard progress. The fact is, it is not sufficient only to point out that there is potential continuity between the behavioral and biological levels. The manner in which the continuity is established is centrally important. There have been attempts to perform the biological-behavioral wedding by denying the importance of the behavioral. When one makes the assumption that complex human behavior, including complex social events, will be understood and dealt with by biological discoveries, this assumption constitutes the rejection of the vast and vastly important facts, principles, and methods of the science of learning, as well as those areas of biological science concerned with relevant matters. This approach may negate the real continuity between the biological and behavioral. Moreover, as will be further indicated, attempts to extrapolate biological principles directly to the social behaviors of man may not involve the same general method that has been described herein.

It should be noted that it is not suggested that all of the concerns of biological science have relevance for behavioral study, or vice versa. As the next chapter will indicate, continuity is only relevant where interests overlap—and much of each realm of science involves its own special concerns.

# Part five

# Epilogue

# 16

## *Unity of science in the study of man*

Each of the social sciences has a strong tradition of resistance to consideration of its subject matter in terms of the principles of another science. It is quite possible that this very strong concern stems from the historical development of the sciences. At one time there were no distinguishable fields of scientific study—one was simply a philosopher, and the study of philosophy included all of the events of the world (and many out of it). Breaking an area of study away from the parent body has always required an act of independence and a struggle for recognition.

Once separate science areas have been developed, however, for whatever reason, there are various factors that promulgate the separatism. Psychology, for example, is organized in a separatistic manner. There are divisions of psychology into experimental, developmental, social, personality, clinical, educational, and so on. There are social organizations formed around these divisions, with scientific journals devoted to publishing the activities of the divisions. Different audiences read the different journals and thereby become familiar with different facts, methods, concepts and principles.

Thus, there are organizational factors that work to continue scientific separatism. In addition, however, conceptual acceptance and rationalization of the separatism have also been influential. In psychology, for example, it has been said that general theory cutting across the different areas of study is by necessity a very distant possibility (M. H. Marx, 1970). In the 1930s, 1940s, and 1950s there was interest in general theory construction. However, when the issues between the classic learning theories of Tolman, Hull, Guthrie, and Skinner were not resolved into a general theory there was a retreat to the study of less comprehensive questions. Even in more specialized areas the diversity in theory has led to the conclusion that unified theory is an impossible goal.

> A truism we have come to in our review of personality theory is that the days of hoping to find a single, all-encompassing theory—about anything—are surely past. This plausible assumption of yesteryear was based upon an inaccurate interpretation. . . . In the future, the student of personality must come to know *all* theorists equally well. (Rychlak, 1968, p. 454)

There are also influential statements in opposition to unification in the social sciences. The emphasis has been on the need for the separation of the social sciences from a psychological foundation, as will be indicated in more detail. Similarly, the relationship between the biological sciences and psychology (as well as the biological sciences and the social sciences) has not been clear. Again, there have been powerful statements in favor of a philosophy of science that would keep these major science divisions separate. Several points in this area are relevant here.

## Reductionism

The term *reductionism*, as will be seen, is controversial. It refers to the fact that science has had the characteristic of seeking knowledge of the more elementary events that underlie the occurrences in which it is interested. Philosophically, the concept of reductionism has suggested that all phenomena are reducible to physical events and describable in a physicalistic language. It would be congruent with a simple reductionistic philosophy to assert that sociology is based upon the behavior of individual men, the behavior of individual men is based upon physiological laws (and other elementary laws), and these laws are based upon more elementary sets of laws such as those of chemistry and physics. The philosophy of reductionism has been contested in the various areas of study of concern, however.

### REDUCTIONISM AND THE SOCIAL SCIENCES

A very straightforward argument for isolation of scientific areas of study and for uniqueness of the terms and principles of a science area has been made by M. G. White (1943) in discussing a philosophy of history. He suggests that the status of a science depends in large part on its ability to discover terms that are unique to that area of study, and he sees an unfortunate difficulty in establishing this independence in the field of history. "If we proceed empirically, and examine history books in an effort to determine which terms are specific to history we find ourselves in a morass, chiefly because of the number of terms which come from other sciences" (p. 8). White proceeds to extricate the field of history from this "morass" by excluding the types of terms that are not specifically historical in nature. "It seems clear that one part of these terms are not specific to history, namely, those which come from what is called 'individual' psychology" (p. 12). He concludes that history is distinguished by the fact that, like sociology, it deals with social concepts.

In the social sciences, in general, most of the antipathy toward consideration of the science's events in terms of another science area involves psychological explanation. Gellner (1956), for example, has applied to history and sociology the Gestalt argument against analytic conceptualization of human behavior, stating that the events in these areas cannot be reduced to the principles of individual psychology. Many other historians and philosophers of history share the sentiment against psychological explanation which has been pejoratively termed "psychologism" (Popper, 1945).

In sociology it has been suggested that there are emergent social phenomena that are basic in and of themselves. The view is that the basic principles to account for social phenomena cannot come from other disciplines (such as a psychology), but they emerge in the more complex area itself. The functionalists in sociology have held this view. In promulgating this view, Durkheim suggested the following concerning social phenomena: "Since their essential characteristic consists in the power they possess of exerting, from outside, a pressure on individual consciousness, they do not derive from individual consciousness, and in consequence sociology is not a corollary of psychology" (Durkheim, 1927, pp. 124–25.) The theoretical approach of the functionalists in sociology has been that sociology should be independent of psychology, and it should concern concepts of societies, institutions, and social groups of various kinds.

The same type of functionalism has also pertained in anthropology, where Radcliffe-Brown and his successors have rejected the need for any type of underlying psychological theory in that area. A. F. C. Wallace, a social anthropologist who was trained in that tradition, relates the following:

> . . . I was informed that there was *no* future for any student who elected to specialize in culture-and-personality. It was a dying fad, I was told; it would be wiser for me to concentrate on a hard, reliable discipline like archaeology. In archaeology, I was told, one deals with tangible objects, like pieces of broken pottery and glass and flint and bone, and not shapeless, intangible chimeras like national character and the Oedipus complex. (p. 2)

Mandelbaum (1955), in stating a nonreductionist position, argues vigorously against psychologizing and for the uniqueness of emergent social concepts and principles, and he warns of an attempt to unify the social sciences within one set of concepts and basic principles, as well as one philosophy of science.

> . . . [W]hat some who plead for "integration" in social science seem to demand is that the various disciplines should merge into one larger whole. On such a view the goal of integration would be the achievement of a state in which all persons who work in the field of social science would operate with the same set of concepts and would utilize the same methods of inquiry. If I am not mistaken, it is sometimes assumed that the social sciences will have made their greatest advance when the individual social sciences which now exist will have lost their separate identities. In so far as this paper has a practical purpose, its purpose is to indicate that "integration," taken in this sense, is a mistaken goal for sociologists and psychologists and the other social scientists to pursue.
> . . . [I]t is clear that this paper has what might be termed an injunctive character. I am attempting to rule in advance that certain modes of procedure should or should not be adopted by practising social scientists.
> . . . My aim is to show that one cannot understand the actions of human beings as members of a society unless one assumes that there is a group of facts which I shall term "societal facts" which are as ultimate as are those facts which are "psychological" in character. In speaking of "societal facts" I refer to any facts concerning the forms of organization present in a society. In speaking of "psychological facts" I refer to any facts concerning the thoughts and the actions of specific human beings (Mandelbaum, 1955, pp. 306–7).

It should be noted that these philosophies are influential in the various social sciences. That is, scientists in the various social science fields strive for uniqueness, for separatism, for isolation from other social science fields. Even within one field there is little relating from subarea to subarea. Experiments which utilize the same principles and procedures but deal with different areas usually are not read by investigators in both areas. Scholars in one area of study of human behavior for the most part have little contact with the principles, procedures, findings, and knowledge of human behavior that exist in another area of study. The separatistic philosophy of science thus reflects the general practices.

## REDUCTIONISM AND PSYCHOLOGY

In the beginning the laboratory principles of conditioning were studied in a context that assumed a physiological basis for learning. For example, Ivan Pavlov was a physiologist, and his theoretical formulation was heavily oriented in this direction. "It appears that the cells predominantly excited at a given time become foci attracting to themselves nervous impulses aroused by new stimuli—impulses which on repetition tend to follow the same path and so to establish a conditioned reflex" (Pavlov, 1927, p. 38).

E. L. Thorndike's early analysis of learning also suggested a physiological basis for the process:

> When any neurone or neurone group is stimulated and transmits to or discharges into or connects with a second neurone or neurone group, it will, when later stimulated again in the same way, have an increased tendency to transmit to the same second neurone group as before, provided the act that resulted in the first instance brought a pleasant or at least indifferent mental state. If, on the contrary, the result in the first case was discomfort, the tendency to such transmission will be lessened. (Thorndike, 1905, p. 165)

Clark Hull has also been recognized for his speculations concerning the physiological basis of learning. As one example, a central concept of his theory, reinforcement, was given a physiological meaning when Hull posited that reinforcement depended upon drive reduction—the removal of a drive-instigated internal process (Hull, 1943). However, although Hull was free in physiological speculation—which B. F. Skinner criticized (1950)—and in the physiological overtones he gave to some of his concepts, he did not take a reductionistic position with respect to the explanation of learning principles. For Hull the higher-order (more basic) postulates of his theory were not physiological, nor was his methodology aimed at establishing the biological foundations for learning laws. He expected to obtain the higher-order principles of his learning theory on a logical basic, through more abstract, deductive principles and concepts. Hull did not consider in his methodological statements the empirical principles of learning themselves to be the higher-order laws of the learning theory, nor did he consider the higher-order laws to be theoretical or experimental principles of neurophysiology.

As was indicated in the last chapter, Skinner's approach has been opposed to physiological reductionism. Verplanck (1954), in analyzing the operant conditioning approach, indicates that Skinner does not accept physiological reasoning to the extent that Hull and Pavlov did:

> Skinner's approach . . . , bears no more than a terminological resemblance to Hull's or to Pavlov's, but it is at least first cousin to Kantor's system [Kantor, 1924], which explicitly rather than implicitly accepts a metaphysical position, naive realism, and rejects even the logical possibility of a reductionism. (pp. 308–309)

As indicated in the preceding chapter, Skinner's philosophy of science appears to reject reductionism between any of the realms of investigation. That is, his approach is against the use of methods of observation or concepts from one "dimensional system" (level of theory) to another, whether upward or downward. Thus, this philosophy does not provide the basis for extending behavioral psychology upward to a rapprochement with the social sciences, and in this sense it is in agreement with the statements of the social scientists referred to in the preceding section.

The present view is that a general paradigm for human behavior must indicate the manner in which the biological, behavioral, and social realms of study can productively contribute to one another. A goal of the preceding chapter was to indicate some of the rationale involved, in the context of relating the behavioral and biological levels of study. The same rationale applies to the behavioral and social levels, however.

## Reductionism and hierarchical theory

In the concept of hierarchical theory described in the first chapter elementary principles are elaborated to deal with more complex phenomena. Thus, as an example, the elementary principles of learning are derived from the basic animal laboratory. In addition, however, in a hierarchical extension, the elementary learning principles can be elaborated at the human level to deal with interactions between people, the development of personality repertoires, and so on. Hierarchical theoretical formulations such as this are reductionist in character. Theoretical level by theoretical level, the elementary principles are elaborated to deal with human behavior. Conversely, the behaviors of humans are reduced to the more elementary principles of animal learning.

This does not mean, it should be noted, that the analyses on the human behavior level are of less validity or value in science than those on the elementary level. In a true hierarchical theory, which includes direct derivation from the elementary level to the more complex level, there is actually mutual influence and relevance from one level to the other. Verification and disconfirmation of elementary principles with more complex phenomena is relevant to and influences the elementary theory level. Moreover, the source of creative developments can arise in either level. A number of examples have been included herein in which elaborations of the elementary theory level are suggested by the investigation at the more complex level. Thus, knowledge of behavior principles can suggest research in the physiology of learning. As another example, social facts of human motivation provided a source of knowledge for developing the principles of the A-R-D theory, which is a more elementary-level statement. Still another example is the productive relationship which was demonstrated between some of the principles in economics and A-R-D principles (Staats, 1963), including the use of the token-reinforcer system in treating hospitalized patients (Kagel and Winkler, 1972). The interdisciplinary research suggested would have significance both for economics and for behavior modification work.

The general point is that in a real hierarchical theory there is intimate interaction between adjacent theory levels. This interaction involves influence in both directions, not a dominance or preeminence of one over the other. Science is full of such interactions of levels.

## PSEUDOHIERARCHICAL THEORY

It is important to note that there are theories that suggest they are reductionistic or hierarchical in character but they really are not. There are numerous cases, for example, in which human behavior has been linked to the biological level of study, but not in the direct, reductionistic way described in Chapter 15. This requires some elaboration, because it has been the source of much misunderstanding. A number of theories of human behavior suggest that they are biological in nature, of a reductionistic sort. Part of the status of such theories derives from the status they receive from their relationship and harmony with biological science.

Examples may be taken from the field of personality. Actually, the term *personality* itself has been employed to suggest an internal personal process of a biological nature that accounts for the way the individual behaves. In addition, there are other terms such as traits, id, ego, the self, self-actualization forces, and so on that include the same implicit assumption of biological causation. However, unless there is translation of such terms to biological terms and continuity between research in these areas and the biological sciences, the relationship may not be of a direct variety.

As another example, developmental psychologists have long had a strong biological orientation. They have observed the changing behavior of the child with age and have inferred that internal biological processes account for the changes. However, the terms employed in the theories of the developmental psychologists usually do not display the type of continuity with biological areas of study described in the last chapter. The foremost theoretical orientation in the field of developmental psychology is presently that of Jean Piaget. His approach is nominally more closely related to biology than traditional learning theories, and it is thus relevant to examine the concepts that in Piaget's theory play the same role as do the principles of learning (Hunt, 1969) in the social behavioristic theory. Piaget employs the biological conception of adaptation; not everything the organism does is adaptive, but all adaptations have two components (Flavell, 1963). One component of adaptation is the incorporation of nutrients into the body; this can be used to illustrate the character of the approach. The food is first transformed into a form that coincides with the structures of the organism. This process is called *assimilation*. In this process the organism is also adjusting to the incorporated elements; this is called *accommodation*.

> What is the nature of cognitive as opposed to physiological assimilation and accommodation? Assimilation here refers to the fact that every cognitive encounter with an environmental object necessarily involves some kind of cognitive structuring (or restructuring) of that object in accord with the nature of the organism's existing intellectual organization. As Piaget says: "Assimilation is hence the very functioning of the system of which organization is the structural aspect" (Piaget, 1952, p. 410). Every act of intelligence, however rudimentary and concrete, presupposes an interpretation of something in external reality, that is, an assimilation of that something to some kind of meaning system in the subject's cognitive organization. . . . And it is Piaget's argument that intellectual assimilation is not different in principle from a more primary biological assimilation: in both cases the essential process is that of bending a reality event to the templet of one's ongoing structure.
>
> If intellectual adaptation is always and essentially an assimilatory act, it is no less an accommodatory one. . . . Reality can never be infinitely malleable, even for the most autistic of cognizers, and certainly no intellectual development can occur unless the organism in some sense adjusts his intellectual receptors to the shapes reality presents him. The essence of accommodation is precisely this process of adapting oneself to the variegated requirements or demands which the world of objects imposes upon one. And once again, Piaget underscores the essential continuity between biological accommodation, on the one hand and cognitive accommodation, on the other: a receptive and accommodating mouth and digestive system are not really different in principle from a receptive and accommodating cognitive system. (Flavell, 1963, p. 48)

As this indicates, Piaget has described cognitive activities in terms that have significance in biological science. However, it is suggested that this usage does not place the two levels of study into direct contact. The method involved, rather, is to use the biological concepts as *analogies* at the human behavioral level. The use of terms such as adaptation, assimilation, and accommodation does not establish an actual point of contact between the two fields in types of observations made, in philosophy of science, in principles and concepts employed, and so on. There is no continuity from the behavioral term *assimilation* to the biological processes involved, of the type described in Chapter 15. With the learning-behavior theory term *stimulus*, for example, there is direct continuity from the term to the physiology of the sense organs, to the functioning of certain areas of the brain, and so on. This is not the case in using such a term as accommodation. The concept of accommodation as used in the child development theory has no direct implications for biological study.

One other example will be given of theory that relates to the biological realm, but not in a direct, reductionistic manner. The concept is that of cultural evolution. Since antiquity there has been a conception that man's history is one of progressive development that occurs because of the internal nature of man, according to his imminent characteristics (Nisbet, 1969). This concept was joined with the concept of biological evolution to yield conceptions of social evolution and other "biologically" based interpretations of history. The analogy is drawn from the history of a culture to the biology of man. Various social philosophers, historians, and social scientists have, for example, employed the description of human biological growth as a model, or an explanation, of historical cultural changes. Toynbee (1947), for example, interprets the history of Western civilization in terms of growth stages like the ages of man:

> The germ of Western society first developed in the body of Greek society, like a child in the womb. The Roman Empire was the period of pregnancy during which the new life was sheltered and nurtured by the old. The "Dark Age" was the crisis of birth, in which the child broke away from its parent and emerged as a separate, though naked and helpless, individual. The Middle Ages were the period of childhood. . . . The fourteenth and fifteenth centuries . . . stand for puberty, and the centuries since the year 1500 for our prime. (p. 290)

Others have utilized more technical conceptions of biological development as their models for interpreting societal characteristics. The concept of cultural evolution has already been mentioned. The quotation of Sahlins in Chapter 14 indicates clearly the attempt to apply the specific principles of biological evolution to cultural change. Other social theorists have treated additional aspects of evolutionary principles in application to societal phenomena. As an example, Talcott Parsons, in a book entitled *Societies: Evolutionary and Comparative Perspectives* (1966), suggests that sociocultural evolution, in the same manner as biological evolution, proceeds "by variation and differentiation, from simple to progressively more complex forms" (p. 17).

Nisbet (1969) has indicated that the use of biological evolution principles in the consideration of societal change has been a metaphorical (in the present terms, analogical) extension. It is suggested that this extension does not represent a true hierarchical (or reductionistic) theory. Some of the principles of evolution theory, such as variation, may serve to describe certain social observations. But the general account of biological evolution, including its foundation in genetics, has no counterpart in theories of social evolution.

The lack of direct, reductionistic, interaction can be seen in the fact that the connection between biological evolution and social evolution does not entail continuity of empirical data. Challenges to social evolution theory, even its complete rejection by negative observations, would not in any way weaken the acceptance of biological evolution theory. Furthermore, one could expect no reciprocal extension of social evolution principles to the biological evolution field. Principles observed in the social realm do not appear to have significance in the biological realm. A situation in which one science area has productive suggestions for another area, but the relationship is not reciprocal, would suggest a dependence of the one science, and thus an inferiority. It is suggested that the onesidedness of pseudoreductionism also leads to rejection of actual reductionism—because of the implications of inferiority involved.

In the preceding chapter it was shown how concepts at the behavioral level could be translated readily and *directly* to concepts at the biological level. It is thus possible to derive hypotheses in either direction. Where this reciprocity is not present one must suspect that the relationship is analogical, not a direct and mutual (reductionistic) interrelationship. The analogical extension of the concepts of one to another, above or below, can be valuable—for example, the analogy of electrons moving about a nucleus in the atom as applied to the manner in which the planets move around the sun. But analogical extension is different from reductionistic extension. This should be recognized; lack of such recognition leads to various misunderstandings.

The relationships between levels of study must be examined in this light to establish the nature of the relationship. Moreover, it is important for the philosophy of a science to indicate what its mode of relationship is to be to other areas of study—whether the relationship will be isolation, analogical extension, or mutual reductionism.

A primary aspect of the present approach is that *there is a major continuity from biological events to the principles of behavior, and there is a major continuity from the elementary principles of behavior to complex human behavior and on to social events.* There is, it is suggested, relatively little continuity directly from the biological realm to complex human behavior events and to social events. The extensions of this type, it is suggested, have been largely of an analogical type—and the merits of such extensions must rest upon their value in the one area of study involved. No status should accrue to a theory because it uses basic science terms. Misunderstanding in this area has been the source of problems in the integration of knowledge.

## Hierarchical theory and the biological-psychological-social relationship

Chapter 1 described briefly some of the characteristics of the classic hierarchical theories of physical science, where higher-order (elementary and general) principles are employed to account for a realm of lower-order observations. It was suggested that such theory construction characteristics apply to the study of man. This was first suggested in an abstract sense. The succeeding chapters, however, have included content for such a possibility. It is relevant to schematize the hierarchical structure described, as shown in Figure 16.1.

It is suggested that social behaviorism exhibits hierarchical theory

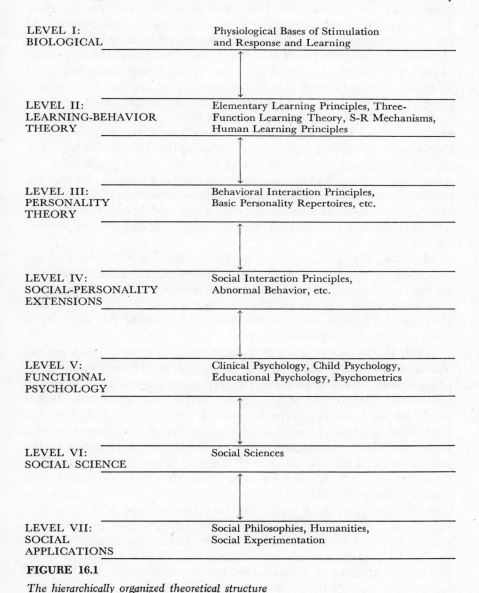

| LEVEL I:<br>BIOLOGICAL | Physiological Bases of Stimulation and Response and Learning |
| LEVEL II:<br>LEARNING-BEHAVIOR THEORY | Elementary Learning Principles, Three-Function Learning Theory, S-R Mechanisms, Human Learning Principles |
| LEVEL III:<br>PERSONALITY THEORY | Behavioral Interaction Principles, Basic Personality Repertoires, etc. |
| LEVEL IV:<br>SOCIAL-PERSONALITY EXTENSIONS | Social Interaction Principles, Abnormal Behavior, etc. |
| LEVEL V:<br>FUNCTIONAL PSYCHOLOGY | Clinical Psychology, Child Psychology, Educational Psychology, Psychometrics |
| LEVEL VI:<br>SOCIAL SCIENCE | Social Sciences |
| LEVEL VII:<br>SOCIAL APPLICATIONS | Social Philosophies, Humanities, Social Experimentation |

**FIGURE 16.1**

*The hierarchically organized theoretical structure*

characteristics. The paradigm does not employ biological concepts and principles as an analogy, for example. The elementary principles in the theory are quite different and independent of those of biology. However, as has been indicated, there is a close, direct relationship between these principles and relevant biological study. There is no variance in scientific method or in philosophy of science between the biological level and the learning-behavior theory level. Productive derivations in both directions have occurred and can be considered usual rather than extraordinary. As shown in the figure, the elementary principles of learning-behavior constitute the higher-order principles of the theory for dealing with the events of human behavior. These basic principles are to be explained themselves by biological principles concerning the neurological bases of learning, and these in turn will be related to biochemistry and biophysics.

Traditionally there has been a schism between the learning-behavior theory level and the personality level. This is one of the primary schisms that divides psychology and produces competitive theories. It has been suggested that with the introduction of human learning principles, behavior interaction principles, and the concept (and content) of the personality repertoires, a close and direct relationship between these two levels can be established. The theoretical statements and observations at each level may serve as the basis for deriving significant empirical hypotheses for the other.

Moreover, the elementary three-function learning principles and the personality level concepts can be elaborated to provide a theoretical basis for the consideration of the types of social interaction that have been important to social psychology. The elements of the higher-levels of the structure also provide the basis for the consideration of the observations and concepts of abnormal psychology.

Levels II, III, and IV provide the constituents for dealing in a general manner with much of the content of the various specialized areas of study of psychology. There is presently the very active accumulation of experimental and clinical data of a behavioral nature that verifies the higher levels and has functional properties as well. Studies in clinical, child, and educational psychology show the relevance of learning-behavior principles for understanding and dealing with human behavior. Elaboration of the constituents of Levels II, III, and IV suggests various ways for extending these developments. It should also be added that there are studies conducted in these areas that have implications at the more basic levels. For example, studies in child psychology have been central in showing that basic conditioning principles apply to man. Other studies in these areas are relevant to the development of the personality theory level.

This body of theoretical, experimental, and clinical concepts has important implications as a psychology for the social sciences. The possibilities have not yet begun to be recognized. The examples provided in Chapter 14 were limited and involved only the A-R-D principles. The manner in which social science problems can be productively treated within the social behaviorism paradigm will only be revealed when the various aspects of the paradigm are systematically extended to the social science levels. This type of extension is appropriate for the scientists in these areas. It should also be noted that an understanding of human behavior basic to clinical psychology, abnormal psychology, child psychology, and educational psychology should draw upon the knowledge of the social sciences. Again, the reciprocal continuity of reductionism is to be expected, if not presently developed.

Finally, although it has not been possible to treat these topics herein, it is suggested that the paradigm, at least when it has been suitably elaborated and supported, can serve as the basis for a social philosophy. Social philosophies are founded upon a very general psychology of man. The social behaviorism paradigm has that type of potential generality. In many cases, a social philosophy is the basis for social acts—establishing a welfare or penological system, social legislation of various types, and so on. It should be recognized that there are different social philosophies, and thus different injunctions for social action. *The empirical philosophy that is part of the social behaviorism paradigm also suggests that social experimentation could be employed to test the validity of competing social philosophies and their related social actions.*

## METHODOLOGICAL CONTINUITY AND DISCONTINUITY

Table 16.1 was drawn to depict the relationships between the science levels. These relationships, it is suggested, apply to methodological as well as to theoretical matters. There are continuities in science methods, especially between adjacent areas.

If there is one continuity that characterizes science in general, it is that the basis for scientific knowledge lies in observation. Scientific statements spring from observations of events, and the ultimate evaluation of the statements resides in their relevance for the observed events of nature. Moreover, much of the progress that occurs in a particular science area will involve the progressive improvement in observational methods. This progress can be seen in any branch of science. In the recognized sciences this progress, at least in later stages, has usually involved developments in laboratory experimentation and sophisticated apparatus.

This is not to say, however, that there is homogeneity in the methods of observation that the specific areas of study will employ. We must be prepared to find different observational methods that depend upon the different conditions that exist in those areas. This is often not well understood, especially in the sciences that have developed relatively precise (or small-grain) observational procedures. For example, physical scientists sometimes consider areas that depend upon observations which involve less implemented technologies not to be sciences at all. However, dogmatism with respect to science methods frequently gives impetus to the separatism that is presently evident.

It is suggested that what distinguishes a science at a basic level is its very basic philosophy. At the center of the philosophy is a dependence upon observations. The form which the observations will take will depend upon the nature of the events studied. It is straightforward in the physical, biological, and psychological sciences to observe not only single events but also two or more events that are causally related. One can manipulate the event of genetic structure, for example, and then observe the differences in the phenotypic characteristics of a biological specimen. One can manipulate the number of reinforced occurrences of a response in an animal and see the effect upon the frequency of occurrence of the response. Thus, the relationships between cause and effect conditions can be observed clearly in some areas of research—and in fact the effect can be produced because one has control of the cause and can manipulate it.

We could on this basis lay down the dogma that the only true sciences are those in which this type of observation and contact with causal conditions is available. This principle could then be employed to set up a hierarchy of areas of study, only some of which would be called scientific. Following such a practice, until recently astronomy would have to have been exluded, because although astronomy had cause and effect principles, it did not have the ability to direct (manipulate) the movement of bodies in space. The general point, at any rate, is that attributing the characteristics of specific areas of science to science in general may lead to a parochial view.

It is thus suggested here that the trick is to see continuities in science, through the obfuscations of variations in observational methods and technologies. The observational methods of different science areas can be expected to differ widely in precision, reliability, the use of instrumentation, the extent of direct contact with the events versus the necessity of interposition of other human or instrumented observers (as in historical data), the observation of single events versus the cause and effect relations of events, the ability to manipulate causes, the ability to simplify the events studied (as in the laboratory), and so on. This is not to disregard the need for reliability and objectivity of observations, or the differing degrees to which such standards are obtained. But these are matters of practice, not principle. In each area there must be striving to improve observational methods.

## Unity and independence

As Stevens (1939) has astutely noted in the context of psychological interests, there have been strivings for unified science for a long time.

Leibnitz in 1966 proposed such an endeavor based upon a "universal language or script" which would "direct the reason" (Bell, 1937). Stevens has suggested also that the Physicalism of Neurath (1931), which suggests that all scientific language should be a physical language, as contrasted to a language that includes substantive terms not reducible to physical observations, provides a universal language for science, including psychology.

> If every sentence can be translated into the physical language, then this language is an all-inclusive language—a universal language of science. And if the esoteric jargons of all the separate sciences can, upon demand, be reduced to a single coherent language, then all science possesses a fundamental logical unity (Stevens, 1939, p. 249).

In discussing the natural science ideal in the social sciences, Beck (1949), a philosopher, suggests that the social scientist should seek reductionistic explanations:

> . . . Admitting irreducible categories would not in the least exempt the social scientist from reducing all that he can in order to increase the likelihood that the remaining ones *are* irreducible and not simply nonreduced. He would still do everything in his power to diminish the scope and importance of the not-yet-reduced concepts. Following the principle of parsimony, he would and should try to account for as much as possible by means of reductive explanation. (p. 392)

It should be noted that the "law" of parsimony in science many times seems to be held almost as an aesthetic ideal. The less the unnecessary complexity of a theory, the more the theory conforms to the classic model, and thus the greater its value. There is, of course, greater significance to parsimony than as an artistic value. Parsimony in science involves several items of critical functional value that should be mentioned in this context.

One aspect of parsimony involves the number of different theories or concepts that attempt to deal with the same realm of events. First this aspect may be discussed in terms of *simplicity*, as simplicity has functional effects upon the behavior of scientists. That is, for one thing, the more complex the area of study is—the more duplication, overlap, superfluity, and outright error—the greater difficulty the student will have in mastering the area. Science progresses not only by the positive discovery of the true, *but also by the progressive deletion of the false and unproductive.*

As an example, the traditional study of general psychology involves a potpourri of different concepts, principles, logical and empirical methodologies, and research techniques. Many of these will not only be completely isolated from each other, they will be in conflict. The student of the several areas of attitudes, verbal learning, communication, behavior therapy, cognitive dissonance, word meaning, animal learning, physiological psychology, perception, conformity, language development, child development, social attraction, and so on must learn a number of different languages (theories). And what he learns is taught in separate, unrelated compartments, which makes the learning task more difficult. When this circumstance is multiplied across the major divisions of the social sciences, the lack of parsimony is truly monumental. This makes the task of becoming skilled in the various areas of human behavior correspondingly monumental for the scholar and scientist. The overlap and inconsistency, the need to learn a vast number of different languages (theories and concepts), the need to read large numbers of studies in areas that are superfluous because of overlap—all these make the possibility of attaining a general, integrated understanding an exceedingly difficult, perhaps impossible, task.

In this description the handicap the lack of parsimony creates can be seen, as can a glimmer of the productive simplicity that the parsimony of a common theory would yield. The simplicity engendered in having the student study the various areas of human behavior within a common theoretical framework would make the researcher's and the scholar's task much, much easier. The rapidity of training in the simplified situation would be great—and there would be a resulting increase in the depth and comprehensiveness of the scholar's grasp of the various subject matters that relate to man's behavior. It is this type of simplicity, through general acceptance of a common unifying basic theory or language, along with a generally accepted philosophy of science and general research method skills, that is characteristic of the recognized sciences.

A philosophy of science that leads us to seek uniqueness in and of itself would seem to be disadvantageous. Actually, the uniqueness (separatism) in areas of the social sciences helps create an isolation that prevents communication between areas that study the same events. That is, it is suggested that scholars and researchers presently study in different areas of the social sciences the same events concerned with man. However, in these areas, although the events can be the same, different languages (terms and principles) are employed, and so on. As a consequence, the findings in one area are not directly or quickly (and perhaps never) conveyed to the scholar in another field who is interested in the same realm of events. This state of affairs is a very great handicap. It means that each area has to make its own progress—there is no cumulation from large numbers of individuals working on the same problem and contributing to each other's knowledge. In fact, additional waste may occur in the competition between theories. It is a disadvantage, by any criteria, to deal with similar events, with similar types of observations, to be attempting to attain the same explanatory goals, and yet to have different systems with which to do this, with little communication across theories. It would seem that this state of affairs would provide an impetus to strive for unified theory.

The case has thus been given of the advantages of unification. Does the need and possibility for unification mean, however, that the goal is one vast, homogeneous science in the study of man, with no independence and diversity among the present sciences? Have those who have promulgated separatism between the sciences been entirely wrong? Are all statements within an area to be reduced to the next most elementary area?

This is not suggested. Again, there appears to be the need for rapprochement between a philosophy of unity and a philosophy of separatism. Not all, or even the major part, of all of the products of the various areas of study derive from or appear to be reducible to a set of common principles. Most of the biological sciences are concerned with methods, concepts, and observations that have little relevance for learning-behavior theory, for example. The same is true of the various areas, even the areas within psychology. Figure 16.2 can be employed to illustrate the relationship that is suggested.

As the figure indicates, not *all* of any of the science areas dealt with

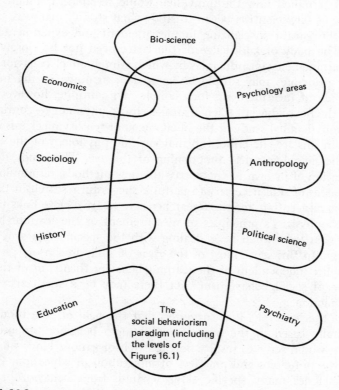

**FIGURE 16.2**

*Schematization of unity and independence in the study of man. Other areas of study should be included that are not schematized. The relative size of the different areas has no significance.*

can, or should, be reduced to the elements of the paradigm. This needs explication. It is being stated that the sciences concerned with human behavior can be unified and advanced by *basing* themselves in the paradigm or relating themselves to the paradigm. The areas dealt with have ranged from aspects of the biological sciences to some of the humanities. This does not mean that all of the strivings in the area of study—or, in some cases, even a major part—will stem from or be directly related to the learning-behavior theory. For example, it is characteristic of science to get more and more detailed in its observations. When this occurs a field may be opened to new methods of study, new terminology, new adjunct skill development, and so on. This may constitute a body of knowledge that has an autonomous being apart from the science that gave rise to these developments.

Take the study of man's sensory characteristics as an example. Let us say that the area of first interest was with human behavior, how external stimuli determined how the individual would respond. Through this interest, let us say, scientists began to study what stimuli man was sensitive to, and this meant specializing in measurement and experimental techniques. The body of knowledge of the latter need not be closely linked to the learning-behavior principles by which stimuli affect behavior. Those behavior principles may simply be ignored or assumed in the more detailed study of measurement, for example. To continue, however, let us say that the next adjacent science area—sensory physiology—continues on in the more detailed study of the phenomena of sensory mechanisms. The scientist in this area describes the anatomy and physiology of the eye, and finally the neural-chemical functioning of the eye. This study demands specialization of its own to encompass the new methods, terminology, and adjunct areas of skill necessary to perform successful research in this area. It may be unnecessary for the researcher here to have close links to either the psychophysicist's principles of measurement or the learning-behavior principles, at least when he is dealing with the specific topics in sensory physiology. In this sense most of the areas of study in sensory physiology will be quite independent of the learning-behavior theory, or of the larger paradigm of social behaviorism. Yet there may be a small area of mutual interaction.

As another example, it is suggested that the field of education should be generally based in the paradigm of human behavior proposed. The paradigm would serve to inspire new research of various kinds—for example, on the principles and practices of motivation in education. But education includes many specific skills separate from behavior principles. There are the subject matter disciplines of education, administrative methods of teacher training, and so on that may be removed from elements of the paradigm. The same, of course, is true of the other science areas. While certain parts of sociology or anthropology, for example, may be concerned with a conception of men, with principles of human behavior, with a behavioral philosophy of science, with research methods on aspects of human behavior, and so on, there are other areas where these matters are not of immediate concern.

This is to say that there are places of overlap in the various areas of science treated. This overlap or continuity could provide the basis for an intensely productive unification, without stifling the specializations in a science area that are relatively independent of the features that are common to the various areas. Thus, the general philosophy of the approach is that there is independence within the paradigm along with a commonality in basic principles, methods, philosophy, and the general conception of man. Thus, as Figure 16.1 depicts, the learning-behavior theory (and its associated philosophy, methods, and so on) would constitute a core structure. Upon this core is built the conception of human behavior, or the theory of personality, which treats the personality repertoires and areas such as communication, motivation, imitation, interaction, and so on. The principles of the personality theory come from the basic learning theory, but the content comes from observations that turn the basic theory into a conception of man's behavior. The central point here is that the resulting conception of human behavior can serve as a unifying structure for other areas of study, in a way that still provides for independence and identity.

As Figure 16.2 thus shows, the various areas in the social and behavioral sciences and professions are to a basic extent related to the social behaviorism paradigm. The *kind* of relationship is central in any proposed unification. Thus, the areas of education, sociology, economics, political science, history, psychiatry, and psychology are depicted as being *in part* anchored in social behaviorism's principles, concepts, analyses, and findings. Nevertheless, a large part of each area does not overlap with the elements of the paradigm. It is to be expected that much of each area will concern matters that are independent of a psychological paradigm's general principles and methods. These will not derive from the paradigm nor necessarily refer back to the paradigm. Neither will they be in conflict with the paradigm—only independent. Perhaps, as a good example, there are principles of organization in sociology that can be treated better considering organizations as units, rather than as groups of individuals in a psychological reductionism. As another example, the principles of learning by themselves are cause-and-effect principles. Their importance and justification as scientific principles does not rest upon demonstration of the biological mechanisms involved in learning. The learning principles were discovered and have been systematically studied without knowledge of the underlying biological events. These principles can be extended and employed in significant affairs without that biological knowledge, and in fact the biological knowledge may be of little use in such extensions. Thus, in any particular case, the extent of analysis, or reduction to more elementary principles, is a pragmatic consideration. It should be emphasized, however, that it does not detract from the independent worth of a set of empirical principles or observations to be related to a more elementary set of principles. Reductionistic search can lead to increased generality of the science area without loss of independence.

As Figure 16.1 indicates, the learning-behavior theory connects productively to the biological sciences. In this connection the social science areas may have little relevance. The biological science areas concerned with the principles of behavior do not derive productive avenues of research, concepts, or investigatory methods from the social sciences. And, as has been stressed, the biological sciences contribute only minimally concepts or principles that can be directly applied to the problems of the social sciences.

All of the other science areas should be shown as overlapping. There is overlap (in addition to that which occurs in the common learning-behavior theory and the common personality theory) between the various social and behavioral sciences. Topics in education are of concern to psychology, sociology, anthropology, economics, political science, ethics, and so on. The same is true in each case. There is overlap between the subject matters, concepts, and problems of the several areas, although this cannot be depicted in the relationship shown in Figure 16.2.

It may be added that when there is need or relevance for a psychological theory in the conduct of the activities of the specific science area, the area should have as part of its foundation the unified theory. Without this availability the special interests of the science area are likely to generate attempts to extend its specialized concepts in a manner that is limited and idiosyncratic. This is said to indicate that it is to be expected that a unified theory or paradigm would have general significance for the areas discussed. But it need not have confining characteristics that prevent the development of specializations in a science area that are unrelated to the unified theory. Nevertheless, it is important for the science area—even if the majority of its interests are unrelated to the unified theory—to have a paradigm in its background for the scientists in the area to consult when their interests take them in that direction. This is necessary to prevent the isolation, the separatism, the duplication, and the competition that exist now in understanding and dealing with human behavior.

## FUNCTIONAL BEHAVIORISM AND BASIC-APPLIED SCIENCE ISSUES

In psychology, as well as in some of the other sciences, there has been a traditional separatism between "basic" and "applied" science. This schism actually began to emerge early in the history of psychology. General psychology arose in the context of philosophy and physiology. However, impressed by the concepts of evolution and adaptation of the organism to the environment, individuals like Francis Galton began to study traits in terms of their functional properties. Later, John Dewey enunciated a pragmatic philosophy—that philosophy should deal with the problems of men. This served as a basis for a functional psychology. Dewey was opposed to analysis in terms of elementary constituents and urged, rather, the study of functional wholes. James Rowland Angell, another functional psychologist in the early 1900s, thought the field should be concerned with the utilities of the psychological act.

On the other hand, Edward Titchener considered functional psychology more commonplace and gave preeminence to the study of basic events. He thought that such events, because of their basic nature, should be studied first. As another example, Hugo Munsterberg asserted that scientific psychology could never deal with real-life events (Boring, 1950).

The same stands are taken in contemporary times. There are basically oriented investigators who see the primary needs of psychology to be development as a science. Frequently, the characteristics of science are identified with laboratory research, sophisticated apparatus and observational methods, and the use of mathematics and statistics. Positive values for these elements of study may also be teamed with denigration of the study of functional behaviors that are considered mundane in contrast to more abstract and pure psychological events. Negative attitudes toward applied work also frequently stem from the fact that applied work may not be conducted with the same scientific finesse that the experimental psychologist holds in high value.

> On the other hand, the applied worker saw the development of experimental psychology as less than of world shaking importance. While it might be a justifiable area of scientific study, it was obvious that it treated events that in the measure of worldly concerns were actually quite trivial. Furthermore, the basic worker seemed anxious to indicate that his work had little of practical significance, and was unlikely to develop any soon. . . . Without the necessary theoretical and experimental basis for the use of basic psychology in the study of functional human behaviors, the schism between basic and applied psychology has continued to exist to a large extent. (Staats, 1968c, pp. 196–97)

Perhaps the greatest impetus to separatism (and antipathy) resides in the separatism itself. When the activities of the applied worker do not derive from the basic science, the basic science is seen to have little relevance. On the other side, when the applied science is not based on the basic science, the basic scientist sees little relevance in applications. Neither adds to the other, and in the senses summarized above they can be seen to detract.

It is important that a general paradigm attempt to deal with this separatism, for there are important and related aspects of the study of man that fall into the basic category and others that fall into the applied category. To begin, it is suggested that there is no rationale in the present paradigm to give precedence to one emphasis at the expense of the other. The elementary principles have continuity from the basic study through the concern with functional behavior. There are changes in the principles in some cases, or elaborations, but with continuity. More advanced concepts and principles build upon the more basic. The philosophy of science extends from the basic through the applied areas. Moreover, although observational methods may differ—ranging from the controlled laboratory study through naturalistic and clinical case observations—there is no contradiction in principle.

Most importantly, however, there is relevance between the basic and applied activities. At the basis of the effort are the elementary principles. As these are extended to the other levels, however, there is mutual influence. The elementary principles are frequently employed in an analysis of a functional human behavior. However, the elementary principles may need elaboration in making such an analysis, and the observations at the human level may suggest changes and elaborations at the basic level. In any event, successful use of basic principles at the functional level serves to provide corroboration for the principles and indication of their generality. The same is true of methods of study, which can be and should be extended and elaborated in moving from the basic to applied levels of study. In support of these conclusions are the chapters in this book, which have moved from the "basic" and "pure" levels of the science to the applied concerns with functional human behavior—all within the same context of principles, methods, and philosophy.

Thus, the present paradigm does not draw a sharp demarcation between applied and pure science levels, certainly not in terms of status. All levels contribute to the classic theory development. It is true that the change of an elementary principle has the general significance of requiring changes throughout the hierarchical structure. This is of great importance to the theoretical structure and makes the elementary principle important. But at the other ends of the hierarchical theory there is great importance in showing the comprehensive nature of the human behavior theory, in providing understanding of human behavior, and in the great social significance of providing solution to human problems and forming the context for social decisions.

Several additional concepts that are involved in this philosophy will be mentioned in the next section.

THE BEHAVIORAL UNIVERSE, REPRESENTATIVE SAMPLING, AND THE PROGRESSION PRINCIPLE. It has been suggested in the hierarchical theory that there are a number of levels, ranging from the elementary principles into the concerns of the various areas of study of human behavior. This may be elaborated to suggest that there is a universe of events that constitutes the realm of study at each level. Thus, at the elementary principle level there are a number of principles and subprinciples. At the next level there are various S-R mechanisms, some of which were summarized in Chapter 2, and so on. Further on, the principles of the basic behavioral (personality) repertoires were elaborated—and it would be expected that there would be a universe of such principles. At another level several personality repertoires were described, and each of these constitutes a universe of events.

It is thus suggested that each level of study involves a universe of events. Described in this manner, the task of the scientists in each area, it is suggested, is to "sample" the universe in his realm of study. It is not usually possible to study each and every one of the events in any universe, the complexity ordinarily being too great. Science only makes contact with a portion of events in any area of study. The trick, then, is to sample well. Not every elementary learning principle has the same generality and importance as another. The principle of classical conditioning has greater significance, it is suggested, than that of partial reinforcement. General theories of learning can thus differ in the sample they make.

The same is true in the other levels. The manner of sampling is important. Moreover, although the various levels of study are related to one another, the interests of the one level may suggest a different sampling rationale for another level than that level would suggest for itself. Thus, for example, in constituting a basic learning theory with which to deal with human behavior, not all of the elementary principles that are studied in the basic area may be important. The sample taken for this purpose may be smaller and may include only general (heavyweight) principles that have an important impact on human behavior. As another example, in the realm of personality there are a myriad of behaviors. However, if one is interested in the study of the social sciences the sample of personality repertoires used as basic theory may be different from the sample made by a personality theorist whose focus of work is in that area. Thus, what will be considered a representative, and good, sample of the events in a universe may vary, depending upon one's purpose. Seeming differences between levels of study may involve this type of superficial variation.

Finally, however, general scientific advancement suggests a principle of progression which has implications for the basic-applied schism. That is, it is suggested that where the science involves hierarchically related sets of events, progression in dealing with the various levels *is required*. In the present case it has been suggested that the elementary principles of learning extend to a very wide number of human behavioral events and areas of events. If this is the case, the science is truncated if it is restricted only to the study of the samples of elementary principles. The progression principle, in the context of a hierarchical organization of levels of study, suggests that it is as important to the development of the science area to make the extensions as it is to study the elementary principles. The movement must be in the direction of continually extending the elementary principles, in a progressive way, to the more and more complex events to which the principles pertain. It is suggested that the philosophy of psychology must include this principle. Moreover, the principle indicates that separatism between areas so related is not justified.

## Conclusion

Because of the acceptance of separatism, it has been accepted that theorists work in restricted areas in constructing theories. Most frequently the theories are narrow, and it is not demanded that they apply widely to the various concerns in the study of man. The value on generality has not been widely held, and theories have not had to meet such a criterion.[1]

It is suggested, at any rate, that a central criterion for the evaluation of conceptions or theories of man should be the extent to which they display comprehensiveness. We must strive for unification and principles that have general applicability to various concerns. The philosophy of science involved, and the philosophy of research and research methodology, must have the same breadth—the capability of including within the structure the various studies of man, cutting across superficial differences. It is because of the contemporary acceptance of compartmentalization that theories are not commonly subjected to the criterion of generality—are never called upon to demonstrate relevance outside of a particular area of concern—and we therefore do not become aware of their limitations.

The foregoing materials have proposed that there is a unity of science extending through the biological, behavioral, and social sciences and the humanities. The suggestion is of a biological-behavioral-social theory or paradigm which can incorporate the various strivings of a wide number of areas, providing a productive framework with which to guide many of

---

[1] This is not to say that there has been no interest in general theory in psychology. For example, Freud attempted to generally apply his psychoanalytic theory. Another example, there has been an effort to link the concept of cognitive dissonance to physiological phenomena, in what may be seen as an attempt at hierarchical theory development. Cognitive dissonance refers to the case where the individual experiences two inconsistent or incompatible things, the result being considered as a state of drive. As an example of the attempt to relate the concept of cognitive dissonance as a drive state to an underlying physiology, Brehm (1962) reports a study by Back and Bogdanoff in which high dissonance subjects were compared to low dissonance subjects. Blood samples from the subjects revealed that the high-dissonance subjects had less mobilization of free fatty acids in the blood than did low-dissonance subjects. Later results were less clear, however (Brehm, Back, and Bogdanoff, 1964). Another relevant finding has been that high-dissonance subjects have a different (higher) threshold for pain than do low-dissonance subjects (Zimbardo, 1966). It would be important to see how far these and other theoretical systems can be extended on a very general level to various fields of psychology and the social sciences, as well as to the realm of biological study. In any event, the incipient interest in achieving comprehensiveness and reductive relationships may be seen in these endeavors.

the activities in the several areas. The paradigm has been named social behaviorism. It is suggested that such features as the learning-behavior theory, the theory of personality, the philosophy of science, applied activities of the clinic and of education, observations and problems of the social sciences, and related aspects of biological science can all be productively integrated within the same framework.

There is much potential power in such integration. Presently there are many scientists and practitioners working in the various areas, but they do so in isolation. Their work is not complementary; each does not build upon the others. What happens in one area has no effect upon the other areas. Thus, there is no incremental accrual of knowledge of the type found in some of the "more mature" sciences. Moreover, the fractionated, micro theories come into competition with one another. T. S. Kuhn (1962) has referred to similar circumstances in the history of areas of the physical sciences: "We shall note . . . that the early developmental stages of most sciences have been characterized by continual competition between a number of distinct views of nature, each partially derived from, and all roughly compatible with, the dictates of scientific observation and method" (p. 4).

It is suggested, however, that the possibility exists for general theory in the areas of study that have been treated. Such a theory can help guide relevant behavioral, social, and biological studies of man. Successful integration of these various areas—so that scientists, scholars, and professionals of the various areas could contribute to each other's knowledge— would release a vast store of energy. The potential for general development, it is suggested, is that of the scientific revolution of which Kuhn speaks.

> The success of a paradigm . . . is at the start largely a promise of success discoverable in selected and still incomplete examples. Normal science consists in the actualization of that promise, an actualization achieved by extending the knowledge of those facts that the paradigm displays as particularly revealing, by increasing the extent of the match between those facts and the paradigm's predictions, and by further articulation of the paradigm itself.
>
> Few people who are not actually practitioners of a mature science realize how much . . . work of this sort a paradigm leaves to be done or quite how fascinating such work can prove in the execution. (Kuhn, 1962, pp. 23–24)

The present book can only characterize the potentialities. It is clear, however, that as with other paradigms in their early stages, full development can only come when the scholars and researchers in the various fields become involved and provide the theoretical, empirical, and methodological developments and findings that convert promise to substance.

# Bibliography

Adams, J. A. Psychomotor performance as a function of intertrial rest interval. *Journal of Experimental Psychology*, 1954, 48, 131–33.

Adkins, D. C., Payne, F. D., and Ballif, B. L. Motivation factor scores and response set scores for ten ethnic-cultural groups of preschool children. *American Educational Research Journal*, 1972, 9, 557–72.

Alker, H. A. Is personality situationally specific or intrapsychically consistent? *Journal of Personality*, 1972, 40, 1–16.

Allen, K. D., and Harris, F. R. Elimination of a child's excessive scratching by training the mother in reinforcement procedures. *Behaviour Research and Therapy*, 1966, 4, 79–84.

Allen, K. E., Hart, B., Buell, J. S., Harris, F. R., and Wolf, M. M. Effects of social reinforcement on isolate behavior of a nursery school child. *Child Development*, 1964, 35, 511–18.

Allport, F. H. *Social psychology*. Cambridge, Mass.: Houghton Mifflin, 1924.

Allport, G. W. Attitudes. In C. A. Murchison (Ed.), *A handbook of social psychology*. Worcester, Mass.: Clark University Press, 1935.

Allport, G. W. *Personality: A psychological interpretation*. New York, Holt, 1937.

Allport, G. W. Traits revisited. *American Psychologist*, 1966, 21, 1–10.

Allport, G. W., Vernon, P. E., and Lindzey, G. *Study of Values*. (Rev. ed.) Boston: Houghton Mifflin, 1951.

American Psychiatric Association. *Diagnostic and statistical manual: Mental disorders* (DSM–I). Washington: American Psychiatric Association, 1952.

Anderson, N. H. Application of an additive model to impression formation. *Science*, 1962, 138, 817–18.

Anderson, N. H. Averaging versus adding as a stimulus-combination rule in impression formation. *Journal of Experimental Psychology*, 1965, 70, 394–400.

Anderson, N. H. Integration theory and attitude change. *Psychological Review*, 1971, 78, 171–206.

Andronico, M. P., and Guerney, B., Jr. The potential application of a filial therapy to the school situation. *Journal of School Psychology,* 1967, 6, 2–7.

Anisfield, M., and Gordon, M. Phonological features in morphological rules of English. Paper presented at the meetings of the Eastern Psychological Association, Washington, D. C., April, 1968.

Annon, J. S. The extension of learning principles to the analysis and treatment of sexual problems. Unpublished doctoral dissertation, University of Hawaii, 1971.

Aquinas, St. Thomas. Summa contra gentiles. In W. T. Jones, F. Sontag, M. V. Beckner, and R. J. Fogelin (Eds.), *Approaches to ethics.* New York: McGraw-Hill, 1969.

Arieti, S. *Interpretation of schizophrenia.* New York: Robert Brunner, 1955.

Arieti, S. Schizophrenia. In S. Arieti (Ed.), *American handbook of psychiatry.* New York: Basic Books, 1959.

Arnold, M. B. Perennial problems in the field of emotion. In M. B. Arnold (Ed.), *Feelings and emotions.* New York: Academic Press, 1970.

Aronfreed, J. *Conduct and conscience: The socialization of internalized control over behavior.* New York: Academic Press, 1968.

Aronson, E., and Linder, D. Gain and loss of esteem as determinants of interpersonal attractiveness. *Journal of Experimental Social Psychology,* 1965, 1, 156–71.

Asch, S. E. Forming impressions of personality. *Journal of Abnormal Sociology,* 1946, 41, 258–90.

Ashem, B. A., and Poser, E. G. (Eds.). *Adaptive Learning: Behavior modification with children.* London: Pergamon, 1973.

Atthowe, J. M., Jr., and Krasner, L. A preliminary report on the application of contingent reinforcement procedures (token economy on a "chronic" psychiatric ward). *Journal of Abnormal Psychology,* 1968, 73, 37–43.

Ax, A. F. The physiological differentiation between fear and anger in humans. *Psychosomatic Medicine,* 1953, 15, 433–42.

Ayllon, T., and Azrin, N. H. The measurement and reinforcement of behavior of psychotics. *Journal of the Experimental Analysis of Behavior,* 1965, 8, 356–83.

Ayllon, T., and Azrin, N. H. *The token economy: A motivational system for therapy and rehabilitation.* New York: Appleton-Century-Crofts, 1968.

Ayllon, T., and Haughton, E. Control of the behavior of schizophrenic patients by food. *Journal of Experimental Analysis of Behavior,* 1962, 5, 343–52.

Ayllon, T., and Michael, J. L. The psychiatric nurse as a behavioral engineer. *Journal of the Experimental Analysis of Behavior,* 1959, 2, 323–34.

Ayllon, T., and Roberts, M. D. Eliminating discipine problems by strengthening academic performance. *Journal of Applied Behavior Analysis,* 1974, 7, 71–76.

Azrin, N. H. Time-out from positive reinforcement. *Science,* 1961, 133, 382–83.

Azrin, N. H., and Lindsley, O. R. The reinforcement of cooperation between children. *Journal of Abnormal and Social Psychology,* 1956, 52, 100–102.

Bachman, J. A. Self-injurious behavior: A behavioral analysis. *Journal of Abnormal Psychology,* 1972, 80, 211–24.

Baer, D. M. Laboratory control of thumbsucking by withdrawal and re-presentation of reinforcement. *Journal of the Experimental Analysis of Behavior,* 1962, 5, 525–28.

Baer, D. M., Peterson, R., and Sherman, J. S. The development of imitation by reinforcing behavioral similarity to a model. *Journal of the Experimental Analysis of Behavior,* 1967, 10, 405–16.

Baer, D. M., and Sherman, J. A. Reinforcement control of generalized imitation in young children. *Jour-*

nal of Experimental Child Psychology, 1964, 1, 589–95.

Baker, S. J. A linguistic law of constancy. II. *Journal of General Psychology*, 1951, 44, 113–20.

Baldwin, A. L. *Behavior and development in childhood.* New York: Holt, 1955.

Bandura, A. Social learning through imitation. In M. R. Jones (Ed.), *Nebraska symposium on motivation.* Lincoln: University of Nebraska, 1962.

Bandura, A. Influence of models' reinforcement contingencies on the acquisition of imitative responses. *Journal of Personality and Social Psychology*, 1965, 1, 589–95.

Bandura, A. A social learning interpretation of psychological dysfunctions. In P. London and D. Rosenhan (Eds.), *Foundations of abnormal psychology.* New York: Holt, Rinehart and Winston, 1968. (a)

Bandura, A. Social-learning theory of identificatory processes. In D. A. Goslin (Ed.), *Handbook of socialization theory and research.* Chicago: Rand McNally, 1968. (b)

Bandura, A. *Principles of behavior modification.* New York: Holt, Rinehart and Winston, 1969.

Bandura, A. *Psychological modeling.* Chicago: Aldine-Atherton, 1971.

Bandura, A., Blanchard, E. B., and Ritter, B. The relative efficacy of desensitization and modeling approaches for inducing behavioral affective, and attitudinal changes. *Journal of Personality and Social Psychology*, 1969, 13, 173–99.

Bandura, A., Grusec, J. E., and Menlove, F. L. Vicarious extinction of avoidance behavior. *Journal of Personality and Social Psychology*, 1967, 5, 16–23.

Bandura, A., and Harris, M. B. Modification of syntactic style. *Journal of Experimental Child Psychology*, 1966, 4, 341–52.

Bandura, A., and Huston, A. C. Identification as a process of incidental learning. *Journal of Abnormal and Social Psychology*, 1961, 63, 311–18.

Bandura, A., and McDonald, F. J. Influence of social reinforcement and the behavior of models in shaping children's moral judgments. *Journal of Abnormal and Social Psychology*, 1963, 67, 274–81.

Bandura, A., Ross, D., and Ross, S. A comparative test of the status envy, social power, and the secondary reinforcement theories of identification learning. *Journal of Abnormal and Social Psychology*, 1963, 67, 527–34. (a)

Bandura, A., Ross, D., and Ross, S. A. Imitation of film-mediated aggressive models. *Journal of Abnormal and Social Psychology*, 1963, 66, 3–11. (b)

Bandura, A., and Walters, R. *Adolescent aggression.* New York: Ronald, 1959.

Bandura, A., and Walters, R. *Social learning and personality.* New York: Holt, Rinehart and Winston, 1963.

Bandura, A., and Whalen, C. K. The influence of antecedent reinforcement and divergent modeling cues on patterns of self-reward. *Journal of Personality and Social Psychology*, 1966, 3, 373–82.

Bard, P., and Macht, M. B. The behavior of chronically decerebrate cats. In *Neurological Basis of Behavior.* London: Churchill, 1958.

Barlow, D. H., Leitenberg, H., and Agras, W. S. The experimental control of sexual deviation through manipulation of the noxious scene in covert sensitization. *Journal of Abnormal Psychology*, 1969, 74, 596–601.

Barnett, P. E., and Benedetti, D. T. Vicarious conditioning of a GSR to sound. Paper read at the meeting of the Rocky Mountain Psychological Association, Glenwood Springs, Colo., 1960.

Baron, R. A., Byrne, D., and Griffitt, W. *Social psychology*, Boston: Allyn and Bacon, 1974.

Barrett, B. H. Reduction in rate of multiple tics by free operant conditioning methods. *Journal of Nervous and Mental Diseases*, 1962, 135, 187–95.

Beach, F. A. Characteristics of masculine sex drive. In M. R. Jones (Ed.), *Nebraska symposium on motivation*. Lincoln, Neb.: University of Nebraska Press, 1956.

Beard, C. A. *An economic interpretation of the Constitution of the United States*. New York: Macmillan, 1913.

Beck, A. T. Cognitive therapy: Nature and relation to behavior therapy. *Behavior Therapy*, 1970, 1, 184–200.

Beck, L. W. The "natural science ideal" in the social sciences. *Scientific Monthly*, 1949, 68, 386–94.

Becker, W. C., Madsen, C. H., Arnold, C. R., and Thomas, D. R. The contingent use of teacher attention and praise in reducing classroom behavior problems. *Journal of Special Education*, 1967, 1, 287–307.

Bell, E. T. *Men of mathematics*. New York: Simon and Schuster, 1937.

Bem, D. J. Self-perception: An alternative in interpretation of cognitive dissonance phenomena. *Psychological Review*, 1967, 74, 183–200.

Bem, D. J. Attitudes as self-descriptions: Another look at the attitude-behavior link. In A. G. Greenwald, T. C. Brock, T. M. Ostrom (Eds.), *Psychological foundations of attitudes*. New York: Academic Press, 1968.

Bem, D. J., and McConnell, H. K. Testing the self-perception explanation of dissonance phenomena: On the salience of premanipulation attitudes. *Journal of Personality and Social Psychology*, 1970, 14, 23–31.

Bem, S. L. Verbal self-control: The establishment of effective self-instruction. *Journal of Experimental Psychology*, 1967, 74, 485–91.

Bentham, J. In C. K. Ogden (Ed.), *The theory of legislation*. New York: Harcourt, Brace and Co., 1931.

Berger, S. M. Conditioning through vicarious instigation. *Psychological Review*, 1962, 69, 450–66.

Bergin, A. E. Some implications of psychotherapy for research for therapeutic practice. *Journal of Abnormal Psychology*, 1966, 71, 235–46.

Berko, J. The child's learning of English morphology. *Word*, 1958, 14, 150–77.

Berkowitz, L. The contagion of violence: An S-R mediational analysis of some effects of observed aggression. In W. J. Arnold and M. Page (Eds.), *Nebraska symposium on motivation*. Lincoln: University of Nebraska Press, 1970.

Berkowitz, L. Reactance and the unwillingness to help others. *Psychological Bulletin*, 1973, 79, 310–17.

Berkowitz, L., and Knurek, D. A. Label-mediated hostility generalization. *Journal of Personality and Social Psychology*, 1969, 13, 200–206.

Berkowitz, L., and Le Page, A. Weapons as aggression-eliciting stimuli. *Journal of Personality and Social Psychology*, 1967, 7, 202–7.

Berkowitz, S., Sherry, P. J., and Davis, B. A. Teaching self-feeding skills to profound retardates using reinforcement and fading procedures. *Behavior Therapy*, 1971, 2, 62–67.

Berlyne, D. E. Arousal and reinforcement. In D. Levine (Ed.), *Nebraska symposium on motivation*. Lincoln, Neb.: University of Nebraska Press, 1967.

Bernal, M. E. Behavioral feedback in the modification of brat behavior. *Journal of Nervous Mental Disorders*, 1969, 148, 375–86.

Bernal, M. E. The use of videotape feedback and operant learning principles in training parents in management of deviant children. In R. D. Rubin (Ed.), *Advances in behavior therapy*. Vol. VIII. New York: Academic Press, 1971.

Bernstein, B. Language and social class. *British Journal of Sociology*, 1960, 11, 271–76.

Bierstedt, R. *The social order: An introduction to sociology*. New York: McGraw-Hill, 1957.

Bijou, S. W. A systematic approach to an experimental analysis of young children. *Child Development*, 1955, 26, 161–68.

Bijou, S. W. Methodology for an experimental analysis of child behavior. *Psychological Reports*, 1957, 3, 243–50.

Bijou, S. W., and Baer, D. M. *Child development: Vol. 2, Universal stage of infancy.* New York: Appleton-Century-Crofts, 1965.

Bijou, S. W., Birnbrauer, J. S., Kidder, J. D., and Tague, C. Programmed instruction as an approach to teaching of reading, writing, and arithmetic to retarded children. In S. W. Bijou and D. M. Baer (Eds.), *Child development: Readings in experimental analysis.* New York: Appleton-Century-Crofts, 1967.

Bindra, D. Neuropsychological interpretation of the effects of drive and incentive-motivation on general activity and instrumental behavior. *Psychological Review,* 1968, 75, 1–22.

Birch, D. Verbal control of nonverbal behavior. *Journal of Experimental Child Psychology,* 1966, 4, 266–75.

Birkimer, J. C. Sensory preconditioning and higher-order conditioning with discriminative stimuli in instrumental reward learning. Unpublished doctoral dissertation, Ohio State University, 1966.

Bitterman, M. E., and Schoel, W. M. Instrumental learning in animals: Parameters of reinforcement. In P. H. Mussen and M. R. Rosenzweig (Eds.), *Annual Review of psychology,* 1970.

Blank, M. Cognitive functions of language in the preschool years. *Developmental Psychology,* 1974, 10, 229–45.

Blau, P. M. *Exchange power in social life.* New York: Wiley, 1964.

Bloom, L. Talking, understanding and thinking. In R. L. Schiefelbusch and L. L. Lloyd (Eds.), *Language perspectives: Acquisition, retardation, and intervention.* Baltimore, Md.: University Park Press, 1974.

Bloomfield, L., and Barnhart, C. L. *Let's read: A linguistic approach.* Detroit: Wayne State University Press, 1961.

Blum, G. S. *The Blacky Pictures, manual of instructions.* New York: Psychological Corporation, 1950.

Blumer, H. *Critiques of research in the social sciences: I. An appraisal of* Thomas and Znaniecki's "The Polish peasant in Europe and America." (Bull. 44) New York: Social Science Research Council, 1939.

Blumer, H. Sociological implications of the thought of George Herbert Mead. *American Journal of Sociology,* 1966, 72, 535–48.

Bogardus, E. S. Measuring social distance. *Journal of Applied Sociology,* 1925, 9, 299–308.

Bolles, R. C. *Theory of motivation.* New York: Harper and Row, 1967.

Bolles, R. C., and Grossen, N. E. Effects of an informational stimulus on the acquisition of avoidance behavior in rats. *Journal of Comparative and Physiological Psychology,* 1969, 68, 90–99.

Bolstad, O. D., and Johnson, S. M. Self-regulation in the modification of disruptive classroom behavior. Unpublished manuscript, University of Oregon, 1971.

Bond, G. L., and Dykstra, R. The cooperative research program in first grade reading instruction. *Reading Research Quarterly,* 1967, 6, 5–11.

Bookbinder, L J. Simple conditioning vs. the dynamic approach to symptoms and symptom substitution: A reply to Yates. *Psychological Reports,* 1962, 10, 71–77.

Boring, E. G. *A history of experimental psychology.* (2nd ed.) New York: Appleton-Century-Crofts, 1950.

Bortner, D. M. Pupil motivation and its relationship to the activity and social drives. *Progressive Education,* 1953, 31, 5–11.

Bostow, D. E., and Bailey, J. B. Modification of severe disruptive and aggressive behavior using brief time-out and reinforcement procedures. *Journal of Applied Behavior Analysis,* 1969, 2, 31–39.

Bowers, K. Situationism in psychology: On making reality disappear. *Research reports in psychology,* No. 137, University of Waterloo, 1972.

Brackbill, Y., and Koltsova, M. M. Conditioning and learning. In Y. Brackbill (Ed.), *Infancy and early childhood.* New York: Free Press, 1967.

Brady, J. V., Boren, J. J., Conrad, D., and Sidman, M. The effect of food and water deprivation upon intracranial self-stimulation. *Journal of Comparative and Physiological Psychology*, 1957, 50, 134–37.

Brainerd, C. J., and Allen, T. W. Training and generalization of density conservation: Effects of feedback and consecutive similar stimuli. *Child Development*, 1971, 42, 693–704.

Breger, L., and McGaugh, J. L. Critique and reformulation of "learning theory" approaches to psychotherapy and neurosis. *Psychological Bulletin*, 1965, 63, 338–58.

Brehm, J. W. Motivational effects of cognitive dissonance. In M. R. Jones (Ed.), *Nebraska symposium on motivation*. Lincoln, Nebr.: University of Nebraska Press, 1962.

Brehm, J. W. *A theory of psychological reactance*. New York: Academic Press, 1966.

Brehm, M. L., Back, K. W., and Bogdanoff, M. D. A physiological effect of cognitive dissonance under stress and deprivation. *Journal of Abnormal and Social Psychology*, 1964, 69, 303–10.

Bricker, W., and Bricker, D. Early language intervention. In R. L. Schiefelbusch and L. L. Lloyd (Eds.), *Language perspectives: Acquisition, retardation, and intervention*. Baltimore, Md.: University Park Press, 1974.

Bridgeman, P. W. *The logic of modern physics*. New York: Macmillan, 1928.

Broden, M., Hall, R. V., Dunlap, A., and Clark, R. Effects of teacher attention and a token reinforcement system, on study behavior in a junior high school special education class. *Exceptional Children*, 1970, 36, 341–49.

Broden, M., Hall, R. V., and Mitts, B. The effect of self-recording on the classroom behavior of two eighth-grade students. *Journal of Applied Behavior Analysis*, 1971, 4, 191–99.

Brogden, W. J. Higher order conditioning. *American Journal of Psychology*, 1939, 52, 579–91. (a)

Brogden, W. J. Sensory pre-conditioning. *Journal of Experimental Psychology*, 1939, 25, 323–32. (b)

Brogden, W. J. Sensory preconditioning of human subjects. *Journal of Experimental Psychology*, 1947, 37, 527–40.

Broom, L., and Selznick, P. *Sociology*. New York: Harper and Row, 1963.

Brown, J. S. Gradients of approach and avoidance responses and their relation to level of motivation. *Journal of Comparative and Physiological Psychology*, 1948, 41, 450–65.

Brown, P., and Elliott, R. Control of aggression in a nursery school class. *Journal of Experimental Child Psychology*, 1965, 2, 103–7.

Brown, R. *A first language*. Cambridge, Mass.: Harvard University Press, 1973.

Brown, R., and Bellugi, U. Three processes in the child's acquisition of syntax. *Harvard Educational Review*, 1964, 34, 133–51.

Brown, R., Cazden, C., and Bellugi, U. The child's grammar from I to III. In J. P. Hill (Ed.), *Minnesota symposium on child psychology*. Vol. II. Minneapolis: University of Minnesota Press, 1968.

Brown, R., and Fraser, C. The acquisition of syntax. In C. N. Cofer and B. S. Musgrave (Eds.), *Verbal behavior and learning*. New York: McGraw-Hill, 1963.

Brown, R., and Lenneberg, E. H. A study in language and cognition. *Journal of Abnormal and Social Psychology*, 1954, 49, 454–62.

Browning, R. M., and Stover, D. O. *Behavior modification in child treatment*. Chicago: Aldine-Atherton, 1971.

Buell, J., Stoddard, P., Harris, F. R., and Baer, D. M. Collateral social development accompanying reinforcement of outdoor play in a preschool child. *Journal of Applied Behavior Analysis*, 1968, 1, 167–74.

Bugelski, B. R., and Hersen, M. Conditioning acceptance or rejection of information. *Journal of Experimental Psychology*, 1966, 71, 619–23.

Bugental, J. F. T. The challenge that is man. In J. F. T. Bugental (Ed.),

*Challenges of humanistic psychology*. New York: McGraw-Hill, 1967.

Burdick, H. A., and Burnes, A. J. A test of "strain toward symmetry" theories. *Journal of Abnormal and Social Psychology*, 1958, 57, 367–70.

Bushell, D., Wrobel, P. A., and Michaelis, M. L. Applying "group" contingencies to the classroom study behavior of preschool children. *Journal of Applied Behavior Analysis*, 1968, 1, 55–61.

Butler, D. C., and Miller, N. "Power to reinforce" as a determinant of communication. *Psychological Reports*, 1965, 16, 705–9.

Butterfield, E. C., and Cairns, G. F. Discussion summary. In R. L. Schiefelbusch and L. L. Lloyd (Eds.), *Language perspectives: Acquisition, retardation, and intervention*. Baltimore, Md.: University Park Press, 1974.

Byrne, D. Interpersonal attraction and attitude similarity. *Journal of Abnormal and Social Psychology*, 1961, 62, 713–15.

Byrne, D. Response to attitude similarity-dissimilarity as a function of affiliation need. *Journal of Personality*, 1962, 30, 164–77.

Byrne, D. Attitudes and attraction. In L. Berkowitz (Ed.), *Advances in experimental social psychology*. New York: Academic Press, 1969.

Byrne, D. A reinforcement model of evaluative responses. *Personality*, 1970, 1, 103–28.

Byrne, D., Bond, M. H., and Diamond, M. J. Response to political candidates as a function of attitude similarity-dissimilarity. *Human Relations*, 1969, 22, 251–62.

Byrne, D., and Clore, G. L. Predicting interpersonal attraction toward strangers presented in three different stimulus modes. *Psychonomic Science*, 1966, 4, 239–40.

Byrne, D., and Clore, G. L. A reinforcement model of evaluative responses. *Personality: An International Journal*, 1970, 1, 103–28.

Byrne, D., Ervin, C. R., and Lamberth, J. Continuity between the experimental study of attraction and real-life computer dating. *Journal of Personality and Social Psychology*, 1970, 16, 157–65.

Byrne, D., London, O., and Reeves, K. The effects of physical attractiveness, sex, and attitude similarity on interpersonal attraction. *Journal of Personality*, 1968, 36, 259–71.

Byrne, D., and Nelson, D. Attraction as a linear function of proportion of positive reinforcements. *Journal of Personality and Social Psychology*, 1965, 1, 659–63.

Cameron, N. Paranoid conditions and paranoia. In S. Arieti (Ed.), *American handbook of psychiatry*. New York: Basic Books, 1959.

Cameron, N. *Personality development and psychopathology: A dynamic approach*. New York: Houghton Mifflin, 1963.

Cameron, N., and Magaret, A. *Behavior pathology*. Boston: Houghton Mifflin, 1951.

Camp, B. W. Remedial reading in a pediatric clinic. *Clinical Pediatrics*, 1971, 10, 36–42.

Camp, B. W., and van Doorninck, W. J. Assessment of 'motivated' reading therapy with elementary school children. *Behavior Therapy*, 1971, 2, 214–22.

Camp, B. W., and Staats, A. W. Manual for the Staats motivation attivation reading treatment. Copyrighted by the authors, 1972.

Campbell, D. P. *Handbook for the Strong Vocational Interest Blank*. Stanford, Calif.: Stanford University Press, 1971.

Campbell, D. P. The generality of a social attitude. Unpublished doctoral dissertation, University of California, Berkeley, 1947.

Campbell, D. P., and Fishe, D. W. Convergent and discriminant validation. *Psychological Bulletin*, 1959, 56, 81–105.

Cannon, W. B. *Bodily changes in pain, hunger, fear and rage*. (2nd ed.) New York: Appleton-Century-Crofts, 1929.

Cannon, W. B. *The wisdom of the body*. New York: Norton, 1932.

Caplow, T. *Principles of organization.* New York: Harcourt, Brace and World, 1964.

Carmichael, L. The development of behavior in vertebrates experimentally removed from the influence of external stimulation. *Psychology Review,* 1926, 33, 51–58.

Carnap, R. *Philosophy and logical syntax.* London: Kegan Paul, 1935.

Carriero, N. J. The conditioning of negative attitudes to unfamiliar items of information. *Journal of Verbal Learning and Verbal Behavior,* 1967, 6, 128–35.

Carlson, C. G. Extinction of conditioned meaning. Unpublished doctoral dissertation, University of Hawaii, 1970.

Carter, H. D. The development of vocational attitudes. *Journal of Consulting Psychology,* 1940, 4, 185–91.

Castro, B., and Weingarten, K. Toward experimental economics. *Journal of Political Economy,* 1970, 78, 598–607.

Cattell, R. B. *Description and measurement of personality.* London: Harrap, 1946.

Cattell, R. B. *Personality: A systematic theoretical and factual study.* New York: McGraw-Hill, 1950.

Cautela, J. R. Covert sensitization. *Psychological Reports,* 1967, 20, 459–68.

Cautela, J. R. Behavior therapy and self-control: Techniques and implications. In C. M. Franks (Ed.), *Behavior therapy: Appraisal and status.* New York: McGraw-Hill, 1969.

Cautela, J. R. Covert reinforcement. *Behavior Therapy,* 1970, 1, 33–50.

Cautela, J. R., and Kastenbaum, R. A Reinforcement Survey Schedule for use in therapy, training and research. *Psychological Reports,* 1967, 20, 1115–30.

Cava, E. L., and Rausch, H. L. Identification and the adolescent boy's perception of his father. *Journal of Abnormal and Social Psychology,* 1952, 47, 855–56.

Chall, J. *Learning to read.* New York: McGraw-Hill. 1967.

Chalmers, D. K. Meanings, impressions, and attitudes: A model of the evaluation process. *Psychological Review,* 1969, 76, 450–60.

Chapman, R., and Miller, J. Word order in early two and three word utterances: Does production precede comprehension? Paper presented to Stanford Child Language Research Forum, Stanford University, 1973.

Chen, H. P., and Irwin, O. C. Infant speech vowel and consonant types. *Journal of Speech Disorders,* 1946, 11, 27–29.

Chomsky, N. *Aspects of the theory of syntax.* Cambridge, Mass.: MIT Press, 1965.

Chomsky, N. *Linguistic contributions to the study of mind: II.* Berkeley, Calif.: Academic Publishing Co., 1967.

Chomsky, N. Language and the mind. *Psychology Today,* 1968, 1, 48 ff. (a)

Chomsky, N. Linguistic contributions to the study of mind. In N. Chomsky (Ed.), *Language and mind.* New York: Harcourt, Brace and World, 1968. (b)

Clark, E. V. Some aspects of the conceptual basis for first language acquisition. In R. L. Schiefelbusch and L. L. Lloyd (Eds.), *Language perspectives: Acquisition, retardation, and intervention.* Baltimore, Md.: University Park Press, 1974.

Clark, K. E. Problems of method in interest measurement. In W. L. Layton (Ed.), *The Strong Vocational Interest Blank: Research and uses.* Minneapolis: University of Minnesota Press, 1960.

Clark, M., Lackowicz, I., and Wolf, M. M. A pilot basic education program for school dropouts incorporating a token reinforcement system. *Behaviour Research and Therapy,* 1968, 6, 183–88.

Clarke-Stewart, K. A. Interactions between mothers and their young children: Characteristics and consequences. *Monographs of the Society for Research in Child Development,* 1973, 38, Nos. 6–7, Serial No. 153.

Clements, C. B., and McKee, J. M. Programmed instruction for institutionalized offenders: Contingency management and performance contracts. *Psychological Reports*, 1968, 22, 957–64.

Cofer, C. N. Comment on the paper by Brown and Fraser. In C. N. Cofer and B. S. Musgrave (Eds.), *Verbal behavior and learning*. New York: McGraw-Hill, 1963.

Cofer, C. N., and Musgrave, B. S. (Eds.) *Verbal behavior and learning*. New York: McGraw-Hill, 1963.

Cohen, B. D., Brown, G. W., and Brown, M. L. Avoidance learning motivated by hypothalamic stimulation. *Journal of Experimental Psychology*, 1957, 53, 228–33.

Cohen, B. D., Nachmani, G., and Rosenberg, S. Referent communication disturbances in acute schizophrenia. *Journal of Abnormal Psychology*, 1974, 83, 1–13.

Cohen, H. L. Educational therapy: The design of learning environments. In J. M. Shlien (Ed.), *Research in psychotherapy*. Vol. 3. Washington, D. C.: American Psychological Association, 1968.

Cohen, H. L., and Filipczak, J. A. *A new learning environment*. San Francisco: Jossey-Boss, 1971.

Coleman, D. E. The classical conditioning of attitudes toward selected educational concepts. *Dissertation Abstracts*, 1967, 27, 4125.

Coleman, F. J. *Contemporary studies in aesthetics*. New York: McGraw-Hill, 1968.

Coleman, J. C. *Abnormal psychology and modern life*. (2nd ed.) New York: Appleton, 1950.

Colman, A. D., and Baker, S. L., Jr. Utilization of an operant conditioning model for the treatment of character and behavior disorders in a military setting. Unpublished manuscript, Walter Reed Army Institute of Research, Washington, D. C., 1968.

Conant, M. B. Conditioned visual hallucinations. Unpublished manuscript, Stanford University, 1964.

Conger, J. C. The treatment of encopresis by the management of social consequences. *Behavior Therapy*, 1970, 1, 386–90.

Conger, J. J. The effects of alcohol on conflict behavior in the albino rat. *Quarterly Journal of Studies of Alcohol*, 1951, 12, 1–29.

Cooper, J. B. Emotion in prejudice. *Science*, 1959, 130, 314–18.

Cooper, J. B. Emotional response to statements congruent with prejudicial attitudes. *Journal of Social Psychology*, 1969, 79, 189–93.

Costanzo, P. R., Coie, J. D., Grumet, J. F., and Farnill, D. A reexamination of the effects of intent and consequence on children's moral judgements. *Child Development*, 1973, 44, 154–61.

Cowan, P. A., Langer, J., Heavenrich, J., and Nathanson, M. Social learning and Piaget's cognitive theory of moral development. *Journal of Personality and Social Psychology*, 1969, 11, 261–64.

Creer, T. L. The use of a time-out from positive reinforcement procedure with asthmatic children. *Journal of Psychosomatic Medicine*, 1969, 14, 117–20.

Critchley, M. W. *The dyslexic child*. London: Redwood Press, 1970.

Cromer, R. F. Receptive language in the mentally retarded: Processes and dragnostic distinctions. In R. L. Schiefelbusch and L. L. Lloyd (Eds.), *Language perspectives; acquisition, retardation, and intervention*. Baltimore. Md.: University Parke Press, 1974.

Crook, W. G. Child care. *Honolulu Star-Bulletin*, October 31, 1969, p. C-3.

Crowne, D. P., and Liverant, S. Conformity under varying conditions of personal commitment. *Journal of Abnormal and Social Psychology*, 1963, 66, 547–55.

Cuber, J. F. *Sociology: A synopsis of principles*. New York: Appleton, 1955.

Cullen, T. A. Invents alphabet of motion for dancers. *The Capital Times*, Madison, Wisconsin, January 19, 1966, p. B-1.

Dahlstrom, W. G., and Welsh, G. S. *An MMPI handbook: A guide for use in clinical practice and research.* Minneapolis: University of Minnesota Press, 1960.

Darley, J. G. The theoretical basis of interests. In W. L. Layton (Ed.), *The Strong Vocational Interest Blank: Research and uses.* Minneapolis: University of Minnesota Press, 1960.

Davison, G. C. Some problems of logic and conceptualization in behavior therapy research and theory. Paper presented at the first annual meeting of the Association for the Advancement of the Behavioral Therapies. American Psychological Association, Washington, D. C., 1967.

Davison, G. C. Elimination of a sadistic fantasy by a client-controlled counterconditioning technique: A case study. *Journal of Abnormal and Social Psychology,* 1968, 73, 84–90. (a)

Davison, G. C. Systematic desensitization as a counterconditioning process. *Journal of Abnormal Psychology,* 1968, 73, 91–99. (b)

Dawe, H. C. A study of the effect of an educational program upon language development and related mental functions in young children. *Journal of Experimental Education,* 1942, 11, 200–209.

Dechant, E. V. *Improving the teaching of reading.* Englewood Cliffs, N. J.: Prentice-Hall, 1964.

Delgado, J. M. R., Roberts, W. W., and Miller, N. E. Learning motivated by electrical stimulation of the brain. *American Journal of Physiology,* 1954, 179, 587–93.

Dember, W. N., and Jenkins, J. J. *General psychology.* Englewood Cliffs, N. J.: Prentice-Hall, 1970.

Deutsch, J. A., and Deutsch, D. *Physiological psychology.* (Rev. ed.) Homewood, Ill.: Dorsey Press, 1973.

Deutsch, M. The role of social class in language development and cognition. *American Journal of Orthopsychiatry,* 1965, 35, 73–88.

Dewey, J. The construction of good. In W. T. Jones, F. Sontag, M. O. Beckner, and R. J. Fogelin (Eds.), *Approaches to ethics.* New York: McGraw-Hill, 1969.

Diamond, I. C., and Hall, W. C. Evolution of Neocortex. *Science,* 1969, 164, 251–62.

DiCara, L. V., and Miller, N. E. Instrumental learning of systolic blood pressure responses by curarized rats: Dissociation of cardiac and vascular changes. *Psychosomatic Medicine,* 1968, 30, 487–94. (a)

DiCara, L. V., and Miller, N. E. Instrumental learning of vasomotor responses by rats: Learning to respond differentially in the two ears. *Science,* 1968, 159, 1485–86. (b)

DiVesta, F. J., and Stover, D. O. The semantic mediation of evaluative meaning. *Journal of Experimental Psychology,* 1962, 64, 467–75.

Dollard, J., Doob, L. W., Miller, N. E., Mowrer, O. H., Sears, R. R., Ford, C. S., Hovland, C. I., and Sollenberger, R. I. *Frustration and aggression.* New Haven, Conn.: Yale, 1939.

Dollard, J., and Miller, N. *Personality and psychotherapy.* New York: McGraw-Hill, 1950.

Doob, L. W. The behavior of attitudes. *Psychological Review,* 1947, 54, 135–56.

Doty, R. W. Discussion of Gastault's paper. In H. H. Jasper (Ed.), *The reticular formation of the brain.* Boston: Little, Brown, 1958.

Downing, J. A. *The i.t.a. reading experiment.* London: Evans Brothers, 1964.

Drach, K. The language of the parent: A pilot study. In Working Paper No. 14; The structure of linguistic input to children. Language-Behavior Research Laboratory, University of California, Berkeley, 1969.

Dulany, D. E. Awareness, rules, and propositional control: A confrontation with S-R behavior theory. In T. R. Dixon and D. L. Horton (Eds.), *Verbal behavior and general behavior theory.* Englewood Cliffs, N. J.: Prentice-Hall, 1968.

Durkheim, E. *Les regles de la metode sociologique* (8th ed.). Paris: Alcan, 1927.

Durup, G., and Fessard, A. L'electroencephalogramme de l'homme. *Annee Psycholgie*, 1935, 36, 1–32.

Dweck, C. S., and Reppucci, N. D. Learned helplessness and reinforcement responsibility in children. *Journal of Personality and Social Psychology*, 1973, 25, 109–16.

Edelman, R. I. Operant conditioning treatment of encopresis. *Journal of Behavior Therapy and Experimental Psychiatry*, 1971, 2, 71–74.

Early, J. C. Attitude learning in children. *Journal of Educational Psychology*, 1968, 59, 176–80.

Edwards, A. *Edwards Personal Preference Schedule*. New York: Psychological Corporation, 1953.

Eimas, P. D. Linguistic processing of speech by young infants. In R. L. Schiefelbusch and L. L. Lloyd (Eds.), *Language perspectives: Acquisition, retardation, and intervention*. Baltimore, Md.: University Park Press, 1974.

Eisenberger, R. Is there a deprivation-satiation function for social approval? *Psychological Bulletin*, 1970, 74, 255–75.

Eisman, B. S. Attitude formation: The development of a color-preference response through mediated generalization. *Journal of Abnormal and Social Psychology*, 1955, 50, 321–26.

Elder, G. H., Jr. Racial conflict and learning. *Sociometry*, 1971, 34, 151–73.

Ellis, A. *Reason and emotion in psychotherapy*. New York: Lyle Stuart, 1967.

Ellson, D. Hallucinations produced by sensory conditioning. *Journal of Experimental Psychology*, 1941, 28, 1–20.

Ellsworth, R. B. The regression of schizophrenic language. *Journal of Consulting Psychology*, 1951, 15, 387–91.

Emery, J. R., and Krumboltz, J. D. Standard versus individualized hierarchies in desensitization to re-

duce test anxiety. *Journal of Counseling Psychology*, 1967, 14, 204–9.

Engel, B. T., and Hansen, S. P. Operant conditioning of heart rate slowing. *Psychophysiology*, 1966, 3, 176–87.

Engler, R. *The politics of oil*. Chicago: University of Chicago Press, 1961.

Epstein, R. Aggression toward ontgroupe as a function of anthoritorianism and imitation of aggressive models. *Journal of Personality and Social Psychology*, 1966, 3, 574–79.

Ervin, S. M. Imitation and structural change in children's language. In E. H. Lenneberg (Ed.), *New directions in the study of language*. Cambridge, Mass.: MIT Press, 1964.

Ervin, S. M., and Miller, W. Language development. Child psychology: 62nd *Yearbook of the National Society for the Study of Education*, Chicago: University of Chicago Press, 1963.

Ervin-Tripp, S. An overview of theories of grammatical development. In D. I. Slibon (Ed.), *The ontogenesis of grammar*. New York: Academic Press, 1971.

Estes, W. K. Discriminative conditioning. II. Effects of a Pavlovian conditioned stimulus upon a subsequently established operant response. *Journal of Experimental Psychology*, 1948, 38, 173–77.

Etzel, B. C., and Gewirtz, J. L. Experimental modification of caretaker-maintained high-rate operant crying in a 6- and 20-week-old infant (Infans tyrannotearus): Extinction of crying with reinforcement of eye contact and smiling. *Journal of Experimental Child Psychology*, 1967, 5, 303–17.

Evans, I. M. A conditioning model of a common neurotic pattern—fear of fear. *Psychotherapy: Theory, Research and Practice*, 1972, 9, 238–41.

Evans, G. W., and Oswalt, G. L. Acceleration of academic progress through the manipulation of peer influence. *Behavior Research and Therapy*, 1968, 6, 189–95.

Eysenck, H. J. *Dimensions of personality.* London: Routledge, 1947.

Eysenck, H. J. The effects of psychotherapy: An evaluation. *Journal of Consulting Psychology,* 1952, 16, 319–24.

Eysenck, H. J. (Ed.) *Behavior therapy and the neuroses.* London: Pergamon, 1960. (a)

Eysenck, H. J. *The structure of human personality.* New York: Macmillan, 1960. (b)

Eysenck, H. J., and Rachman, S. *The causes and cures of neurosis.* London: Routledge and Kegan Paul, 1965.

Fales, E. Genesis of level of aspiration in children from one and one-half to three years of age. Reported in K. Lewin, T. Dembo, L. Festinger, and P. S. Sears, Level of aspiration. In J. McV. Hunt (Ed.), *Personality and the behavior disorders.* Vol. 1. New York: Ronald, 1944.

Farber, I. E. The things people say to themselves. *American Psychology,* 1963, 18, 185–97.

Fargo, G. A., Behrns, C., and Nolen, P. *Behavior modification in the classroom.* Belmont, Calif.: Wadsworth, 1970.

Feleky, A. *Feelings and emotions.* New York: Pioneer Press, 1922.

Ferster, C. B. Intermittant reinforcement of matching to sample in the pidgeon. *Journal of the Experimental Analysis of Behavior,* 1960, 3, 259–72.

Ferster, C. B., and Hammer, C. E., Jr. Synthesizing the components of arithmetic behavior. In Honig, W. K. (Ed.), *Operant behavior.* New York: Appleton-Century-Crofts, 1966.

Ferster, C. B., Nurnberger, J. I., and Levitt, E. B. The control of eating. *Journal of Mathematics,* 1962, 1, 87–109.

Ferster, C. B., and Skinner, B. F. *Schedules of reinforcement.* New York: Appleton-Century-Crofts, 1957.

Feshback, S., Stiles, W. B., and Bitter, E. The reinforcing effect of witnessing aggression. *Journal of*

*Experimental Research in Personality,* 1967, 2, 133–39.

Festinger, L. *A theory of cognitive dissonance.* Evanston, Ill.: Row, Peterson, 1957.

Fiedler, F. E. *A theory of leadership effectiveness.* New York: McGraw-Hill, 1967.

Finch, G. Hunger as a determinant of conditional and unconditional salivary response magnitude. *American Journal of Physiology,* 1938, 123, 379–82.

Fine, B. J. Conclusion-drawing, communicator credibility, and anxiety as factors in opinion change. *Journal of Abnormal and Social Psychology,* 1957, 54, 369–74.

Finley, J. R., and Staats, A. W. Evaluative meaning words as reinforcing stimuli. *Journal of Verbal Learning and Verbal Behavior,* 1967, 6, 193–97.

Flavell, J. H. *The developmental psychology of Jean Piaget.* New York: Van Nostrand, 1963.

Fodor, J. A. How to learn to talk: Some simple ways. In F. Smith and G. A. Miller (Eds.), *The genesis of language: A psycholinguistic approach.* Cambridge, Mass.: MIT Press, 1966.

Fowler, R. L., and Kimmel, H. D. Operant conditioning of the GSR. *Journal of Experimental Psychology,* 1962, 63, 563–67.

Fowler, W. Cognitive learning in infancy and early childhood. *Psychological Bulletin,* 1962, 59, 116–52.

Franks, C. M. Alcohol, alcoholism and conditioning. In H. J. Eysenck (Ed.), *Handbook of abnormal psychology.* New York: Basic Books, 1961.

Franks, C. M. (Ed.) *Behavior therapy: Appraisal and status.* New York: McGraw-Hill, 1969.

Fraser, C., Bellugi, U., and Brown, R. Control of grammar in imitation, comprehension and production. *Journal of Verbal Learning and Verbal Behavior,* 1963, 2, 121–35.

Freeman, T., Cameron, J. L., and McGhie, A. *Studies on psychosis.* New York: International Universities Press, 1966.

French, J. R. P., and Raven, B. The bases of power. In D. Cartwright (Ed.), *Studies in social power*. Ann Arbor: University of Michigan Press, 1959.

Frenkel-Brunswick, E. Motivation and behavior. *Genetic Psychology Monographs*, 1942, 26, 121–265.

Freud, S. *A general introduction to psycho-analysis*. New York: Liveright, 1935.

Freud, S. *An outline of psychoanalysis*. New York: Norton, 1949.

Freund, K. A laboratory method for diagnosing predominance of homo- or hetero-erotic interest in the male. *Behavior Reserach and Therapy*, 1963, 1, 85–93.

Friedlander, B. Z. The effect of speaker identity, voice inflation, vocabulary, and message redundancy on inputs' selections of vocal reinforcement. *Journal of Experimental Child Psychology*, 1968, 6, 443–59.

Friedman, G., and Carlson, J. G. Effects of a stimulus correlated with positive reinforcement upon discrimination learning. *Journal of Experimental Psychology*, 1973, 97, 281–86.

Froman, L. A., and Ripley, R. B. Conditions for party leadership: The case of the House Democrats. *American Political Science Review*, 1965, 59, 52–63.

Fryer, D. *The measurement of interests*. New York: Holt, 1931.

Fuster, J. M. Effect of stimulation of brain stem on tachistoscopic perception. *Science*, 1958, 127, 150.

Gagné, R. M., and Fleishman, E. A. *Psychology and human performance*. New York: Holt, 1959.

Gagné, R. M., and Foster, H. Transfer to a motor skill from practice on a pictured representation. *Journal of Experimental Psychology*, 1949, 39, 342–55.

Gallimore, R., Tharp, R. G., and Kemp, B. Positive reinforcing function of "negative attention." *Journal of Experimental Child Psychology*, 1969, 8, 140–46.

Gantt, W. H., and Dykman, R. A. Experimental psychogenic tachycardia. In P. H. Hock and J. Zubin (Eds.), *Experimental psychopathology*. New York: Grune and Stratton, 1957.

Geen, R. G., and Berkowitz, L. Some conditions facilitating the occurrence of aggression after the observation of violence. *Journal of Personality*, 1967, 35, 666–76.

Geer, J. H. The development of a scale to measure fear. *Behaviour Research and Therapy*, 1965, 3, 45–53.

Geer, J. H. A test of the classical conditioning model of emotion: The use of nonpainful aversive stimuli as unconditioned stimuli in a conditioning procedure. *Journal of Personality and Social Psychology*, 1968, 2, 148–56.

Gelber, H., and Meyer, V. Behaviour therapy and encopresis: The complexities involved in treatment. *Behaviour Research and Therapy*, 1965, 2, 227–31.

Gelfand, D. M. The influence of self-esteem on rate of verbal conditioning and social matching behavior. *Journal of Abnormal and Social Psychology*, 1962, 65, 259–65.

Gellner, E. Explanation in history. *Proceedings of the Aristotelian Society*, 1956.

Gelman, R. Conservation acquisition: A problem of learning to attend to relevant attributes. *Journal of Experimental Child Psychology*, 1969, 7, 167–87.

Gerst, M. D. Symbolic coding processes in observational learning. *Journal of Personality and Social Psychology*, 1971, 19, 7–17.

Gesell, A., Halverson, H. M., Thompson, H., Ilg, F. L., Castner, B. M., Ames, L. B., and Amatruda, C. S. *The first five years of life: A guide to the study of the preschool child*. New York: Harper, 1940.

Gesell, A., and Thompson, H. Learning and growth in identical infant twins: An experimental study by the method of co-twin control. *Genetic Psychological Monographs*, 1929, 6, 1–124.

Gesell, A., and Thompson, H. (assisted by C. Strunk). *The psychology of early growth.* New York: Macmillan, 1938.

Gewirtz, J. L., and Baer, D. M. Deprivation and satiation of social reinforcers as drive conditions. *Journal of Abnormal and Social Psychology,* 1958, 57, 165–72.

Gewirtz, J. L., and Stingle, K. G. The learning of generalized imitation as the basis for identification. *Psychological Review,* 1968, 75, 374–97.

Gibbs, J. P., and Martin, W. T. Toward a theoretical system of human ecology. *Pacific Sociological Review,* 1959, 2, 29–40.

Goldfried, M. R., and D'Zurilla, T. J. A behavioral-analytic model for assessing competence. In C. D. Spielberger (Ed.), *Current topics in clinical and community psychology.* Vol. I. New York: Academic Press, 1969.

Goldfried, M. R., and Kent, R. N. Traditional versus behavioral personality assessment: A comparison of methodological and theoretical assumptions. *Psychological Bulletin,* 1972, 77, 409–20.

Goldfried, M. R., and Sprafkin, J. *Behavioral personality assessment.* Morristown, N. J.: General Learning Press, 1974.

Goldiamond, I. The maintenance of ongoing fluent verbal behavior and stuttering. *Journal of Mathematics,* 1962, 1, 57–95.

Goldiamond, I. Self-control procedures in personal behavior problems. *Psychological Reports,* 1965, 17, 851–68. (a)

Goldiamond, I. Stuttering and fluency as manipulatable operant response classes. In L. Krasner and L. P. Ullmann (Eds.), *Research in behavior modification.* New York: Holt, Rinehart and Winston, 1965. (b)

Goldman, I. The Kwakiutl Indians of Vancouver Island. In M. Mead (Ed.), *Cooperation and competition among primitive peoples.* New York: McGraw-Hill, 1937. (a)

Goldman, I. The Zuni Indians of New Mexico. In M. Mead (Ed.), *Cooperation and competition among*

*primitive peoples.* New York: McGraw-Hill, 1937. (b)

Golightly, C., and Byrne, D. Attitude statements as positive and negative reinforcements. *Science,* 1964, 146, 798–99.

Goodenough, F. L. *Mental testing.* New York: Rinehart, 1949.

Gordon, J. E. *Personality and behavior.* New York: Macmillan, 1963.

Gore, P. M., and Rotter, J. B. A personality correlate of social action. *Journal of Personality,* 1963, 31, 58–64.

Gouldner, A. W. The norm of reciprocity: A preliminary statement. *American Sociological Review,* 1960, 25, 161–79.

Graham, J. T., and Graham, L. W. Language behavior of the mentally retarded: Syntactic characteristics. *American Journal of Mental Deficiency,* 1971, 75, 623–29.

Granoff, S. Classical conditioning of attitudes and agreement factors in persuasion. Unpublished manuscript based upon doctoral dissertation, University of Hawaii, 1973. (a)

Granoff, S. The influence of attitude conditioning upon "free choice" attitudinal advocacy. Unpublished manuscript based upon doctoral dissertation, University of Hawaii, 1973. (b)

Greenspoon, J. The effect of verbal and nonverbal stimuli on the frequency of members of two verbal response classes. Unpublished doctoral dissertation, Indiana University, 1950.

Greenwald, A. G. On defining attitude and attitude theory. In A. G. Greenwald, T. C. Brock, and T. M. Ostrom (Eds.), *Psychological foundations of attitudes.* New York: Academic Press, 1964.

Griffitt, W. Environmental effects on interpersonal affective behavior: Ambient effective temperature and attraction. *Journal of Personality and Social Psychology,* 1970, 15, 240–244.

Griffitt, W., and Jackson, T. The influence of information about ability and non-ability on personnel selection decisions. *Psychological Reports*, 1970, 27, 259–262.

Grinder, R. E. Relations between behavior and cognitive dimensions of conscience in middle childhood. *Child Development*, 1964, 35, 881–89.

Gross, M. C., and Staats, A. W. Interest inventory items as attitude eliciting stimuli in classical conditioning: A test of the A-R-D theory. Technical Report No. 4, 1969, University of Hawaii, Contract N00014–67–C–0387–0007, Office of Naval Research.

Grossberg, J. M. Behavior therapy: A review. *Psychological Bulletin*, 1964, 62, 73–88.

Grossman, S. P. *A textbook of physiological psychology*. New York: Wiley, 1967.

Guay, P. F. The effect of the attitudinal value of a social stimulus on imitation and sociometric choice: A test of the A-R-D theory. Technical Report No. 11, 1971, University of Hawaii, Contract N00014–67–C–0387–0007, Office of Naval Research.

Guess, D. A functional analysis of receptive language and productive speech: Acquisition of the plural morpheme. *Journal of Applied Behavior Analysis*, 1969, 2, 55–64.

Guess, D., Sailor, W., and Baer, D. M. To teach language to retarded children. In R. L. Schiefelbusch and L. L. Lloyd (Ed.), *Language perspectives: Acquisition, retardation, and intervention*. Baltimore, Md.: University Park Press, 1974.

Guess, D., Sailor, W., Rutherford, G., and Baer, D. M. An experimental analysis of linguistic development: The productive use of the plural morpheme. *Journal of Applied Behavior Analysis*, 1968, 1, 225–35.

Guilford, J. P. *Personality*. New York: McGraw-Hill, 1959.

Guilford, J. P., and Zimmerman, W. S. *Guilford-Zimmerman Temperament Survey*. Sheridan Supply Co., Beverly Hills, Calif., 1949.

Guthrie, E. R. *The psychology of learning*. New York: Harper, 1935.

Guthrie, G. The cultural origin of personality differences. Paper presented at the meeting of the Hawaii Psychological Association, Honolulu, December 1966.

Guttman, N., and Kalish, H. I. Experiments in discrimination. *Scientific American*, 1958, 198 (1), 77–82.

Gynther, M. D. Differential eyelid conditioning as a function of stimulus similarity and strength of response to the CS. *Journal of Experimental Psychology*, 1957, 53, 408–16.

Halford, G. S. A theory of the acquisition of conservation. *Psychological Review*, 1970, 77, 302–16.

Hall, C. S., and Lindzey, G. *Theories of personality*. New York: Wiley, 1957.

Hall, G. F. Association of neutral objects with rewards: Persistence of effect upon verbal evaluation. *Journal of Verbal Learning and Verbal Behavior*, 1967, 6, 291–94.

Hall, R. V., Willard, D., Fox, D. W., Goldsmith, Emerson, M., Owen, M., Davis, F., and Garcia E. The teacher as observer and experimenter in the modification of disputing and talking-out behaviors. *Journal of Applied Behavior Analysis*, 1971, 4, 141–49.

Hall, R. V., Lund, D., and Jackson, D. Effects of teacher attention on study behavior. *Journal of Applied Behavior Analysis*, 1968, 1, 1–12.

Hamblin, R. L., Buckholdt, D., Ferritor, D., Kozloff, M., and Blackwell, L. *The humanization processes: A social, behavioral analysis of children's problems*. New York: Wiley, 1971.

Hanley, E. M. Review of research involving applied behavior analysis in the classroom. *Review of Educational Research*, 1971, 40, 597–625.

Harbin, S. P., and Williams, J. E. Conditioning of color connotations. *Perceptual and Motor Skills*, 1966, 22, 217–18.

Harlow, H. F. The formation of learning sets. *Psychological Review*, 1949, 56, 51–65.

Harms, J. Y. Y. Deprivation conditions and the reinforcing value of attitude eliciting stimuli. Unpublished doctoral dissertation, University of Hawaii, 1973.

Harris, A. H., Gilliam, W. J., Findley, J. D., Brady, J. V. Instrumental conditioning of large-magnitude, daily, 12-hour blood pressure elevations in the baboon. *Science*, 1973, 182 (4108), 185–77.

Harris, F. R., Johnston, M. K., Kelley, C. S., and Wolf, M. M. Effects of positive social reinforcement on regressed crawling of a nursery school child. *Journal of Educational Psychology*, 1964, 55, 35–41.

Harris, M. C. A behavioral experimental analysis of dyslexia. Unpublished master's thesis, University of Hawaii, 1974.

Hart, B. M., Allen, K. E., Buell, J. S., Harris, F. R., and Wolf, M. M. Effects of social reinforcement on operant crying. *Journal of Experimental Child Psychology*, 1964, 1, 145–53.

Hart, B. M., Reynolds, N. J., Baer, D. M., Brawley, E. R., and Harris, F. R. Effect of contingent and noncontingent social reinforcement on the cooperative play of a preschool child. *Journal of Applied Behavior Analysis*, 1968, 1, 73–76.

Hartshorne, H., and May, M. A. *Studies in the nature of character*. Vol. I. *Studies in deceit*. New York: Macmillan, 1928.

Hartshorne, H., May, M. A., and Shuttleworth, F. K. *Studies in the nature of character*. Vol. 3. *Studies in the organization of character*. New York: Macmillan, 1930.

Haughton, E., and Ayllon, T. Production and elimination of symptomatic behavior. In L. P. Ullmann and L. Krasner (Eds.), *Case studies in behavior modification*. New York: Holt, Rinehart and Winston, 1965.

Hawkins, R. P. Stimulus/response: It's time we taught the young how to be good parents (and don't you wish we'd started a long time ago?). *Psychology Today*, November 1972, 28–40.

Hawkins, R. P., Peterson, R. F. Schweid, E., and Bijou, S. W. Behavior therapy in the home: Amelioration of problem parent-child relations with parents in a therapeutic role. *Journal of Experimental Child Psychology*, 1966, 4, 99–107.

Hayes, C. *The ape in our house*. New York: Harper and Row, 1951.

Hebb, D. O. Drives and the C.N.S. (conceptual nervous system). *Psychological Review*, 1955, 62, 243–54.

Hefferline, R. F. Proprioceptive discrimination of a covert operant without its observation by the subject. *Science*, 1963, 139, 834–35.

Heider, F. *The psychology of interpersonal relations*. New York: Wiley, 1958.

Hekmat, H. The role of imagination in semantic desensitization. *Behavior Therapy*, 1972, 3–223–31.

Hekmat, H. Systematic versus semantic desensitization and implosive therapy: A comparative study. *Journal of Consulting and Clinical Psychology*, 1973, 40, 202–9.

Hekmat, H. Three techniques of reinforcement modification: A comparison. *Behavior Therapy*, 1974, 5, 541–48.

Hekmat, H., and Vanian, D. Behavior modification through covert semantic desensitization. *Journal of Consulting and Clinical Psychology*, 1971, 36, 248–51.

Hempel, W. E., and Fleishman, E. A. A factor analysis of physical proficiency and manipulative skill. *Journal of Applied Psychology*, 1955, 39, 12–16.

Henderson, J. D. The use of dual reinforcement in an intensive treatment system. In R. D. Rubin and C. M. Franks (Eds.), *Advances in behavior therapy, 1968*. New York: Academic Press, 1969.

Herrick, R. M., Myers, J. L., and Korotkin, A. L. Changes in SD and S rates during development of an operant discrimination. *Journal of Comparative and Physiological Psychology*, 1959, 52, 359–64.

Hewett, F. M., Taylor, A., and Artuso, A. A. The Santa Monica project: Evaluation of an engineered classroom design with emotionally disturbed children. *Exceptional Children*, 1969, 35, 523–29.

Hilgard, E. R., and Marquis, D. G. Acquisiton, extinction, and retention of conditioned eyelid responses to light in dogs. *Journal of Comparative Psychology*, 1935, 19, 29–58.

Hillman, B., Hunter, W. S., and Kimble, G. A. The effect of drive level on the maze performance of the white rat. *Journal of Comparative and Physiological Psychology*, 1953, 46, 87–89.

Himmelfarb, S. Integration and attribution theories in personality impression formation. *Journal of Personality and Social Psychology*, 1972, 23, 309–13.

Hinshelwood, J. *Letter, word and mind blindness*. London: Lewis, 1900.

Hinshelwood, J. *Congenital word-blindness*. London: Lewis, 1917.

Hobbes, T. Of other laws of nature. In W. T. Jones, F. Sontag, M. O. Beckner, and R. Fogelin (Eds.), *Approaches to ethics*. New York: McGraw-Hill, 1969.

Hoebel, E. A. Anthropological perspectives on national character. *Annals of the American Academy of Political and Social Science*, 1967, 70, 1–7.

Hoffman, L. R., and Maier, N. R. F. Valence in the adoption of solutions by problem-solving groups: Concept, method, and results. *Journal of Abnormal Sociology*, 1964, 69, 264–71.

Holland, J. G. Teaching machines: An application of principles from the laboratory. *Journal of the Experimental Analysis of Behavior*, 1960. 3, 275–87.

Holland, J. G., and Skinner, B. F. *The analysis of behavior: A program for self-instruction*. New York: McGraw-Hill, 1961.

Homans, G. C. *Social behavior*. New York: Harcourt, Brace and World, 1961.

Homans, G. C. Bringing men back in. *American Sociological Review*, 1964, 29, 809–18.

Homans, G. C. *The nature of social science*. New York: Harbinger, 1967.

Hopkins, B. L., Schutte, B. C., and Garton, K. L. The effects of access to a playroom on the rate and quality of printing and writing of first- and second-grade students. *Journal of Applied Behavior Analysis*, 1971, 4, 77–87.

Horridge, G. A. Learning of leg position by the ventral nerve cord in headless insects. *Processes of the Royal Society* (London), Series B, 1962, 157, 33–52.

Horton, L. E. Generalization of aggressive behavior in adolescent delinquent boys. *Journal of Applied Behavior Analysis*, 1970, 3, 199–203.

Hosford, R. E. Behavioral counseling: A contemporary overview. *Counseling Psychologist*, 1969, 1, 1–33.

Hovland, C. I. The generalization of conditioned responses: I. The sensory generalization of conditioned responses with varying frequencies of tone. *Journal of General Psychology*, 1937, 17, 125–48. (a)

Hovland, C. I. The generalization of conditioned responses: IV. The effects of varying amounts of reinforcement upon the degree of generalization of conditioned responses. *Journal of Experimental Psychology*, 1937, 21, 261–76. (b)

Hovland, C. I., Janis, I. L., and Kelley, H. H. *Communication and persuasion*. New Haven, Conn.: Yale University Press, 1953.

Howes, D., and Solomon, R. L. A note on McGinnies' "Emotionality and perceptual defense." *Psychological Review*, 1950, 57, 229–34.

Hull, C. L. Quantitative aspects of the evolution of concepts. *Psychological Monographs*, 1920, No. 123.

Hull, C. L. Knowledge and purpose as habit mechanisms. *Psychological Review*, 1930, 37, 511–25.

Hull, C. L. The mechanism of the assembly of behavior segments in novel combinations suitable for problem solution. *Psychological Review*, 1935, 42, 219–45.

Hull, C. L. Simple trial-and-error learning—an empirical investigation. *Journal of Comparative Psychology*, 1939, 27, 233–58.

Hull, C. L. *Principles of behavior*. New York: Appleton Century, 1943.

Hull, C. L. *A behavior system.* New Haven, Conn.: Yale University Press, 1952.

Humphreys, L. G. The effect of random alternation of reinforcement on the acquisition and extinction of conditioned eyelid reactions. *Journal of Experimental Psychology,* 1939, 25, 141–58.

Hunt, J. McV. The impact and limitations of the giant of developmental psychology. In D. Elkind and J. H. Flavell (Eds.), *Studies in cognitive development.* New York: Oxford University Press, 1969.

Hunt, W. A., and Arnhoff, F. The repeat reliability of clinical judgment of test responses. *Journal of Clinical Psychology,* 1956, 12, 289–90.

Hunt, W. A., and Jones, N. F. Clinical judgment of some aspects of schizophrenic thinking. *Journal of Clinical Psychology,* 1958, 14, 235–39.

Ilg, F. L., and Ames, L. B. *School readiness: Behavior tests used at the Gesell Institute.* New York: Harper and Row, 1964.

Ingham, R. J., and Andrews, G. An analysis of a token economy in stuttering therapy. *Journal of Applied Behavior Analysis,* 1973, 6, 219–30.

Ingham, R. J., and Gavin, A. An analysis of a token economy in stuttering therapy. *Journal of Applied Behavior Analysis,* 1973, 6, 219–30.

Ingram, D. The relationship between comprehension and production. In R. L. Schiefelbusch and L. L. Lloyd (Eds.), *Language perspectives: Acquisition, retardation, and intervention.* Baltimore, Md.: University Park Press, 1974.

Insko, C. A., and Butzine, K. W. Rapport, awareness, and vrebal reinforcement of attitude. *Journal of Personality and Social Psychology,* 1967, 6, 225–28.

Insko, C. A., and Cialdini, R. B. A test of three interpretations of attitudinal-verbal reinforcement. *Journal of Personality and Social Psychology,* 1969, 12, 333–41.

Irwin, O. C. Infant speech: Consonantal sounds according to place of articulation. *Journal of Speech Disorders,* 1947, 12, 397–401.

Irwin, O. C. Development of vowel sounds. *Journal of Speech and Hearing Disorders,* 1948, 13, 31–34.

Irwin, O. C. Infant speech: Consonantal position. *Journal of Speech and Hearing Disorders,* 1951, 16, 159–61.

Irwin, O. C., and Chen, H. P. A reliability study of speech sounds observed in the crying of newborn infants. *Child Development,* 1941, 12, 351–68.

"Is intelligence racial?" *Newsweek,* May 10, 1971, 69–70.

Isaacs, W., Thomas, J., and Goldiamond, I. Application of operant conditioning to reinstate verbal behavior in psychotics. *Journal Speech and Hearing Disorders,* 1960, 25, 8–12.

Isen, A. M., and Levin, P. F. Effect of feeling good on helping: Cookies and kindness. *Journal of Personality and Social Psychology,* 1972, 21, 384–88.

Jackson, J. S. The impact of a favor and dependency of the reinforcing agent on special reinforcer effectiveness. Unpublished doctoral dissertation, Wayne State University, 1972.

Jakobson, R. *Child language aphasia and phonological universals.* Trans. A. R. Keiler. The Hague: Mouton, 1968.

James, W. What is emotion? *Mind,* 1884, 9, 188–204.

Janos, O. Age and individual differences in higher nervous activity in infants. *Halek's Collection of Studies in Pediatrics,* 1965, No. 8.

Jastak, J. F., Bijou, S. W., and Jastak, S. R. *Wide Range Achievement Test.* Wilmington, Del.: Guidance Associates, 1965.

Jeffrey, W. E. Variables in early discrimination learning: I. Motor responses in the training of a left-right discrimination. *Child Development,* 1958, 29, 269–75.

Jensen, A. R. How much can we boost IQ and scholastic achievement? *Harvard Educational Review,* 1969, 39 (1), 1–123.

Jervis, G. A. The mental deficiencies. In S. Arieti (Ed.), *American handbook of psychiatry*. New York: Basic Books, 1959.

Jespersen, J. O. H. *Language: Its nature, development, and origin*. London: Allen and Unwin, 1922.

John, E. R. Neural processes during learning. In R. Russell (Ed.), *Frontiers in physiological psychology*. New York: Academic Press, 1960.

Johnson, R. C. A study of children's moral judgments. *Child Development*, 1962, 33, 327–54.

Jones, E. E., and Harris, V. A. The attribution of attitudes. *Journal of Experimental Social Psychology*, 1967, 3, 1–24.

Jones, E. E., and Nisbett, R. E. *The actor and the observer: Divergent perceptions of the causes of behavior*. Morristown, N. J.: General Learning Press, 1971.

Jones, E. E., Rock, L., Shaver, K. G., Goethals, G. R., and Ward, L. M. Pattern of performance and ability attribution: An unexpected primacy effect. *Journal of Personality and Social Psychology*, 1968, 5, 317–40.

Jones, H. E. The galvanic skin reflex in infancy. *Child Development*, 1930, 1, 106–10.

Jones, M. C. The elimination of children's fears. *Journal of Experimental Psychology*, 1924, 7, 382–90.

Jones, M. C., and Mussen, P. H. Self-conceptions, motivations, and interpersonal attitudes of early- and late-maturing girls. *Child Development*, 1958, 29, 491–501.

Jones, W. T., Sontag, F., Beckner, M. O., and Fogelin, R. J. (Eds.) *Approaches to ethics*. New York: McGraw-Hill, 1969.

Judson, A. J., Cofer, C. N., and Gelfand, S. Reasoning as an associative process: II. "Direction" in problem solving as a function of prior reinforcement of relevant responses. *Psychological Reports*, 1956, 2, 501–7.

Kagan, J. Personality and the learning process. *Daedalus*, 1965, 94, 558–59.

Kagan, J., and Moss, H. A. *Birth to maturity*. New York: Wiley, 1962.

Kagel, J. H. Token economies and experimental economics. *Journal of Political Economy*, in press.

Kagel, J. H., and Winkler, R. C. Behavioral economics: Areas of cooperative research between economics and applied behavioral analysis. *Journal of Applied Behavior Analysis*, 1972, 5, 335–41.

Kandel, E. R., and Tauc, L. Mechanism of prolonged heterosynaptic facilitation. *Nature*, 1964, 145–47.

Kandel, E. R., and Tauc, L. Heterosynaptic facilitation in neurones of the abdominal ganglion of *Aplysia depilans*. *Journal of Physiology*, 1965, 181, 1–27. (a)

Kandel, E. R., and Tauc, L. Mechanism of heterosynaptic facilitation in the giant cell of the abdominal ganglion of *Aplysia depilans*. *Journal of Physiology*, 1965, 181, 28–47. (b)

Kanfer, F. H. Verbal conditioning: Reinforcement schedules and experimenter influence. *Psychological Reports*, 1958, 4, 443–52.

Kanfer, F. H. Influence of age and incentive conditions on children's self-rewards. *Psychological Reports*, 1966, 19, 263–74.

Kanfer, F. H. Verbal conditioning: A review of its current status. In T. R. Dixon and D. L. Horton (Eds.), *Verbal behavior and general behavior theory*. Englewood Cliffs, N. J.: Prentice-Hall, 1968.

Kanfer, F. H., and Duerfeldt, P. H. Effects of pretraining on self-evaluation and self-reinforcement. *Journal of Pesronality and Social Psychology*, 1967, 7, 164–68.

Kanfer, F. H., and Marston, A. R. Conditioning of self-reinforcing responses: An analogue to self-confidence training. *Psychological Reports*, 1963, 13, 63–70.

Kanfer, F. H., and Phillips, J. S. *Learning foundations of behavior therapy*. New York: Wiley, 1970.

Kanfer, F. H., and Saslow, G. Behavioral analyses. *Archives of General Psychiatry*, 1965, 12, 529–38.

Kantor, J. R. *Principles of psychology.* New York: Knopf, 1924.

Kaplan, D. The super-organic: Science or metaphysics? In R. A. Manners and D. Kaplan (Eds.), Theory in anthropology. Chicago: Aldine, 1968.

Kaplan, M. F., and Anderson, N. H. Information integration theory and reinforcement theory as approaches to interpersonal attraction. *Journal of Personality and Social Psychology,* 1973, 28, 301–12.

Kappenberg, R. P. Interrelatedness of stimulus function in Staats' A-R-D system. Unpublished doctoral dissertation, University of Hawaii, 1973.

Karolchuck, P., and Worrell, L. Achievement motivation and learning. *Journal of Abnormal and Social Psychology,* 1956, 53, 255–57.

Katkin, E. S. *Instrumental autonomic conditioning.* Morristown, N. J.: General Learning Press, 1971.

Katz, P. A., and Seavey, C. Labels and children's perception of faces. *Child Development,* 1973, 44, 770–75.

Kaye, H. Infant sucking behavior and its modification. In L. P. Lipsitt and C. C. Spiker (Eds.), *Advances in child development and behavior.* Vol. 3. New York: Academic Press, 1967.

Keller, F. S., and Schoenfeld, W. N. *Principles of psychology.* New York: Appleton, 1950.

Kelley, J. Attribution theory in social psychology. In D. Levine (Ed.), *Nebraska symposium on motivation.* Lincoln, Neb.: University of Nebraska Press, 1967.

Kelman, H. C., and Eagly, A. H. Attitude toward the communicator, perception of communication content, and attitude change. *Journal of Personality and Social Psychology,* 1965, 1, 63–78.

Kemp, J. A study of young children's spontaneous verbal imitations. Unpublished doctoral dissertation, University of Washington, 1972.

Kemp, J., and Dale P. Spontaneous imitations and free speech: A developmental comparison. Paper presented at the biennial meetings of the Society for Research in Child Development, Philadelphia, 1973.

Kendler, H. H. *Basic psychology.* (2nd ed.) New York: Appleton-Century-Crofts, 1968.

Kendler, H. H., and Kendler, T. S. Vertical and horizontal processes in problem solving. *Psychological Review,* 1962, 69, 1–16.

Kennedy, W. A., and Foreyt, J. P. Control of eating behavior in an obese patient by avoidance conditioning. *Psychological Reports,* 1968, 22, 571–76.

Kian, M., Rosen, S., and Tesser, A. Reinforcement effects of attitude similarity and source evaluation on discrimination learning. *Journal of Personality and Social Psychology,* 1973, 27, 366–71.

Kidd, J. S., and Campbell, D. T. Conformity to groups as a function of group success. *Journal of Abnormal and Social Psychology,* 1955, 51, 390–93.

Kiesler, C. A., Collins, B. E., and Miller, N. *Attitude change.* New York: Wiley, 1969.

Killackey, H., Diamond, I. T., Hall, W. C., and Hudgins, G. *Federation processes,* 1968, 27, 517.

Kimble, G. A. *Hilgard and Marquis' Conditioning and Learning.* New York: Appleton-Century-Crofts, 1961.

Kimmel, H. D., and Kimmel, E. An instrumental conditioning method for the treatment of enuresis. *Behavior Therapy and Experimental Psychiatry,* 1970, 1, 121–24.

Kingsley, R. C., and Hall, V. C. Training conservation through the use of learning sets. *Child Development,* 1967, 38, 1111–26.

Kirby, F. D., and Toler, H. C. Modification of preschool isolate behavior: A case study. *Journal of Applied Behavior Analysis,* 1970, 3, 309–14.

Klans, D. J. and Glaser, R. Reinforcement determinants of team proficiency. *Organizational Behavior and Human Performance,* 1970, 5, 33–67.

Kleinfeld, J. S. Intellectual strengths in culturally different groups: An Eskimo illustration. *Review of Educational Research*, 1973, 43, 341–60.

Klineberg, O. *Social psychology.* New York: Holt, 1954.

Kobashigawa, B. Repetitions in a mother's speech to her child. In The structure of linguistic input to children. Working Paper No. 14, Language-Behavior Research Laboratory, University of California, Berkeley, 1969.

Koegel, R. L., and Rincover, A. Treatment of psychotic children in the classroom environment: I. Learning in a large group. *Journal of Applied Behavior Analysis*, 1974, 7, 45–59.

Kohlenberg, R. J. Operant conditioning of human anal sphincter pressure. *Journal of Applied Behavior Analysis*, 1973, 6, 201–8.

Kolb, D. A., Winter, S. K., and Berlew, D. E. Self-directed change: Two studies. *Journal of Applied Behavioral Science*, 1968, 4, 353–71.

Konorski, J., and Miller, S. On two types of conditioned reflex. *Journal of Genetic Psychology*, 1937, 16, 264–72.

Kraepelin, E. *Textbook of psychiatry.* (8th ed.) New York: Macmillan, 1923.

Kramer, H. C. Effects of conditioning several responses in a group setting. *Journal of Counseling Psychology*, 1968, 15, 58–62.

Krasner, L. Studies of the conditioning of verbal behavior. *Psychological Bulletin*, 1958, 55, 148–70.

Krasner, L. Verbal operant conditioning and awareness. In K. Salzinger and S. Salzinger (Eds.), *Research in verbal behavior and some neurophysiological implications.* New York: Academic Press, 1967.

Krasner, L., and Ullmann, L. P. *Behavior influence and personality.* New York: Holt, Rinehart and Winston, 1973.

Krasnogorski, N. I. The formation of artificial conditioned reflexes in young children, *Russkii Vrach*, 1907, 36, 1245–46. (Translated and republished: In Y. Brackbill and G. G. Thompson (Eds.), *Behavior in infancy and early childhood: A book of readings.* New York: Free Press, 1967).

Krech, D., Crutchfield, R. S., and Ballachey, E. L. *Individual in society.* New York: McGraw-Hill, 1962.

Krumboltz, J. D. Behavioral counseling: Rationale and research. *Personnel and Guidance Journal*, 1965, 44, 383–87.

Krumboltz, J. D. Behavioral goals for counseling. *Journal of Counseling Psychology*, 1966, 13, 153–59.

Krumboltz, J. D., and Thoresen, C. E. *Behavioral counseling: Cases and techniques.* New York: Holt, Rinehart and Winston, 1969.

Kuhn, A. *The study of society.* Homewood, Ill.: Dorsey Press, 1963.

Kuhn, T. S. *The structure of scientific revolutions.* Chicago: University of Chicago Press, 1962.

Kunkel, J. H. *Society and economic growth: A behavioral perspective of social change.* New York: Oxford University Press, 1970.

Kuypers, D. S., Becker, W. C., and O'Leary, K. D. How to make a token system fail. *Exceptional Children*, 1968, 35, 101–9.

Lacey, J. I. Somatic response patterning and stress: Some revisions of activation theory. In M. H. Appley and R. Trumball (Eds.), *Psychological stress.* New York: Appleton-Century-Crofts, 1967.

Lacey, J. I., Bateman, D. E., and Van Lehn, R. Autonomic response specificity. *Psychosomatic Medicine*, 1953, 15, 433–42.

Lackner, J. R. A developmental study of language behavior in retarded children. *Neuropsychologic*, 1967, 6, 301–20.

Laffal, J., Lenkoski, L. D., and Ameen, L. "Opposite speech" in schizophrenic patient. *Journal of Abnormal and Social Psychology*, 1956, 52, 409–13.

Lahaderne, H. M. Attitudinal and intellectual correlates of attention: A study of four sixth-grade classrooms. *Journal of Educational Psychology,* 1968, 59, 320–24.

Lahey, B. B., McNees, M. P., and McNees, M. C. Control of an obscene "verbal tic" through timeout in an elementary school classroom. *Journal of Applied Behavior Analysis,* 1973, 6, 101–4.

Lal, H., and Lindsley, O. R. Therapy of chronic constipation in a young child by rearranging social contingencies. *Behaviour Research and Therapy,* 1968, 6, 484–85.

Landy, D., and Sigall, H. Beauty is talent: Task evaluation as a function of the performer's physical attractiveness. *Journal of Personality and Social Psychology,* 1974, 29, 299–304.

Lang, P. J., and Lazovik, A. D. Experimental desensitization of a phobia. *Journal of Abnormal and Social Psychology,* 1963, 66, 519–25.

Lang, P. J., Lazovik, A. D., and Reynolds, D. J. Desensitization, suggestibility and pseudotherapy. *Journal of Abnormal and Social Psychology,* 1968, 59, 320–24.

Langer, J. *Theories of development.* New York: Holt, Rinehart and Winston, 1969.

Lanzetta, J. T., and Hannah, T. E. Reinforcing behavior of "naive" trainers. *Journal of Personality and Social Psychology,* 1969, 11, 245–52.

LaPiere, R. T. Attitudes vs. actions. *Social Forces,* 1934, 13, 230–37.

Lashley, K. S. *Brain mechanisms and intelligence.* Chicago: University of Chicago Press, 1929.

Layzer, D. Heritability analyses of I Q scores: Science or numerology? *Science,* 1974, 183, 1259–266.

Lazarus, A. A. The elimination of children's phobias by deconditioning. In H. J. Eysenck (Ed.), *Behaviour therapy and the neuroses.* London: Pergamon, 1960.

Lazarus, R. S., and McCleary, R. A. Autonomic discrimination without awareness: A study of subception. *Psychological Review,* 1951, 58, 113–22.

Leeper, R. W. The motivational and perceptual properties of emotions as indicating their fundamental character and role. In M. Arnold (Eds.), *Feelings and emotions.* New York: Academic Press, 1970.

Lefcourt, H. M. Internal versus external control of reinforcement: A review. *Psychological Bulletin,* 1966, 65, 206–20.

Lefcourt, H. M., and Ladwig, G. W. The effect of reference group upon Negroes' task persistence in a biracial competitive game. *Journal of Personality and Social Psychology,* 1965, 1, 668–71.

Lefkowitz, M. M., Blake, R. R., and Mouton, J. S. Status factors in pedestrian violation of traffic signals. *Journal of Abnormal and Social Psychology,* 1955, 51, 704–6.

Lenneberg, E. H. *Biological foundations of language.* New York: Wiley, 1967.

Lerner, E. *Constraint areas and the moral judgement of children.* Menasha, Wis.: Banta, 1937.

Leuba, C. Images as conditioned sensations. *Journal of Experimental Psychology,* 1940, 26, 345–51.

Levenstein, P. Cognitive growth in preschoolers through verbal interaction with mothers. *American Journal of Orthopsychiatry,* 1970, 40, 426–32.

Leventhal, G. S., and Michaels, J. W. Locus of cause and equity motivation as determinants of reward allocation. *Journal of Personality and Social Psychology,* 1971, 17, 229–35.

Levy, D. M. Experiments on the sucking reflex and social behavior in dogs. *American Journal of Orthopsychiatry,* 1934, 4, 203–24.

Lewin, K. A *dynamic theory of personality.* New York: McGraw-Hill, 1935.

Lewin, K. Defining the "field at a given time." *Psychological Review,* 1943, 50, 292–310.

Lewis, D., and Shepard, A. H. Devices for studying associative interference in psychomotor performance: IV. The turret pursuit apparatus. *Journal of Psychology,* 1950, 29, 173–82.

Lewis, H. B. Studies in the principles of judgments and attitudes: IV.

The operation of "prestige suggestion." *Journal of Social Psychology*, 1941, 14, 229–56.

Liberman, R. A behavioral approach to group dynamics: Reinforcement and prompting of cohesiveness in group therapy. *Behavior Therapy*, 1970, 1, 141–75.

Limber, J. The genesis of complex sentences. Paper presented at the NSF Conference on Developmental Psycholinguistics, Buffalo, N. Y., August 1971.

Lindsley, D. B. Emotion. In S. S. Stevens (Eds.), *Handbook of experimental psychology*. New York, Wiley, 1951.

Lindsley, O. R. Operant conditioning methods applied to research in chronic schizophrenia. *Psychiatric Research Reports*, 1956, 5, 118–53.

Lipsitt, L. P., and Kaye, H. Conditioned sucking in the human newborn. *Psychonomic Science*, 1964, 1, 29–30.

Lipsitt, L. P., Kaye, H., and Bosack, T. N. Enhancement of neonatal sucking through reinforcement. *Journal of Experimental Child Psychology*, 1966, 4, 163–68.

Littman, R. A. Conditioned generalization of the galvanic skin reaction to tones. *Journal of Experimental Psychology*, 1949, 39, 868–82.

Littman, R. J. *The Greek experiment*. London: Thomas and Hudson, 1974.

Locke, E. A. What is job satisfaction? *Organizational Behavior and Human Performance*, 1969, 4, 309–36.

Locke, E. A. Is "behavior therapy" behavioristic? An analysis of Wolpe's psychotherapeutic methods. *Psychological Bulletin*, 1971, 76, 308–327.

Locke, E. A., and Bryan, J. F. The directing function of goals in task performance. *Organizational Behavior and Human Performance*, 1969, 4, 35–42.

Loevinger, J. Objective tests as instruments of psychological theory. *Psychological Reports, Monograph Supplement*, 1957, 3, 635–94.

Logan, F. A., and Wagner, A. R. *Reward and punishment*. Boston: Allyn and Bacon, 1965.

LoLordo, V. M. Similarity of conditioned fear responses based upon different aversive events. *Journal of Comparative and Physiological Psychology*, 1967, 64, 154–58.

Lombardo, J. P., Weiss, R. F., and Buchanan, W. Reinforcing and attracting functions of yielding. *Journal of Personality and Social Psychology*, 1972, 21, 359–68.

Lombardo, J. P., Weiss, R. F., and Stich, M. H. Effectance reduction through speaking in reply and its relation to attraction. *Journal of Personality and Social Psychology*, 1973, 28, 325–32.

Long, E. R., Hammack, J. T., May, F., and Campbell, B. J. Intermittent reinforcement of operant behavior on children. *Journal of the Experimental Analysis of Behavior*, 1958, 1, 315–40.

Longstreth, L. E. *Psychological development of the child*. New York: Ronald, 1968.

Lott, A. J., and Lott, B. E. Liked and disliked persons as reinforcing stimuli. *Journal of Personality and Social Psychology*, 1969, 11, 129–37.

Lott, B. E., and Lott, A. J. The formation of positive attitudes toward group members. *Journal of Abnormal and Social Psychology*, 1960, 61, 297–300.

Lovaas, O. I. Cue properties of words: The control of operant responding by rate and content of verbal operants. *Child Development*, 1964, 35, 245–56.

Lovaas, O. I. A behavior therapy approach to the treatment of childhood schizophrenia. In J. P. Hill (Ed.), *Minnesota symposium on child psychology*. Vol. I. Minneapolis: University of Minnesota Press, 1966.

Lovaas, O. I., Berberich, J. P., Perloff, B. F., and Schaeffer, B. Acquisition of imitative speech by schizophrenic children. *Science*, 1966, 151, 705–7.

Lovaas, O. I., Freitag, G., Gold, V. J., and Kassorla, I. C. Recording apparatus for observation of behaviors of children in free play settings. *Journal of Experimental Child Psychology*, 1965, 2, 108–30.

Lovaas, O. I., Freitag, G., Kinder, M., Rubenstein, B., Schaeffer, B., and Simmons, J. Experimental studies in childhood schizophrenia: Establishment of social reinforcers. Paper presented at the annual meeting of the Western Psychological Association, Portland, Ore., 1964.

Lovaas, O. I., Freitag, G., Kinder, M. L., Rubenstein, B. D., Schaeffer, B., and Simmons, J. Q. Establishment of social reinforcers in two schizophrenic children on the basis of food. *Journal of Experimental Child Psychology*, 1966, 4, 109–25.

Lovaas, O. I., Freitag, G., Nelson, K., and Whalen, C. The establishment of imitation and its use for the development of complex behavior in schizophrenic children. *Behavior Research and Therapy*, 1967, 5, 171–81.

Lovaas, O. I., and Koegel, R. L. Behavior therapy with autistic children. In C. E. Thoresen (Ed.), *Behavior modification in education: 72nd yearbook of the National Society for the Study of Education.* Chicago: University of Chicago Press, 1973.

Lovaas, O. I., Schaeffer, B., and Simmons, J. Q. Experimental studies in childhood schizophrenia: Building social behavior in children by use of electric shock. *Journal of Experimental Research in Personality*, 1965, 1, 99–109.

Lovitt, T. C., Guppy, T. E., and Blattner, J. E. The use of a free-time contingency with fourth graders to increase spelling accuracy. *Behaviour Research and Therapy*, 1969, 7, 151–56.

Loynd, J., and Barclay, A. A. A case study in developing ambulation in a profoundly retarded child. *Behaviour Research and Therapy*, 1970, 8, 207–8.

Luce, R. D., and Rogow, A. A. A game theoretic analysis of congressional power distributions for a stable two-party system. *Behavioral Science*, 1956, 1, 85–97.

Lundin, R. W. *Personality: An experimental approach.* New York: Macmillan, 1961.

Luria, A. R. Speech development and the formation of mental processes. In M. Cole and I. Maltzman (Eds.), *A handbook of contemporary Soviet psychology.* New York: Basic Books, 1969.

Maccoby, E. E., and Gibbs, P. K. Methods of child-rearing in two social classes. In W. E. Martin and C. B. Stendler (Eds.), *Readings in child development.* New York: Harcourt, 1954.

Mackie, J. D. *The earlier Tudors.* Oxford: Clarendon Press, 1952.

MacRae, D. A test of Piaget's theories of moral development. *Journal of Abnormal and Social Psychology*, 1954, 49, 14–18.

Macridis, R. C., and Brown, B. E. *Comparative politics.* Homewood, Ill.: Dorsey Press, 1964.

Maher, B. A. *Principles of psychopathology.* New York: McGraw-Hill, 1966.

Maher, B. A., McKean, K. O., and McLaughlin, B. Studies in psychotic language. In P. Stone (Ed.), *The general inquirer: A computer approach to content analysis.* Cambrdige, Mass.: M.I.T. Press, 1966.

Mahoney, M. J. Research issues in self-management. *Behavior Therapy*, 1972, 3, 45–63.

Maier, N. R. F. Reasoning humans: I. On direction. *Journal of Comparative Psychology*, 1930, 10, 115–43.

Maier, S. F., Seligman, M. E. P., and Solomon, R. L. Pavlovian fear conditioning and learned helplessness: Effects on escape and avoidance behavior of (a) the CS-US contingency and (b) the independence of the US and voluntary responding. In B. Campbell and R. M. Church (Eds.), *Punishment and aversive behavior.* New York: Appleton-Century-Crofts, 1969.

Malmo, R. B. Activation: A neuropsychological dimension. *Psychological Review*, 1959, 66, 367–86.

Malmo, R. B., and Belanger, D. Related physiological and behavioral changes: What are their determinants? In Association for Research in Nervous and Mental

Disease, *Sleep and altered states of consciousness.* Baltimore, Md.: Williams and Wilkins, 1967.

Maltzman, I., and Raskin, D. C. Effects of individual differences in the orienting reflex on conditioning and complex processes. *Journal of Experimental Research in Personality,* 1965, 1, 1–16.

Maltzman, I., Raskin, D. C., Gould, J., and Johnson, O. Individual differences in the orienting reflex and semantic conditioning and generalization under different UCS intensities. Paper delivered at the Western Psychological Association meetings, Honolulu, 1965.

Mandelbaum, M. Societal facts. *British Journal of Sociology,* 1955, 6, 305–17.

Mann, M. B. The quantitative differentiation of samples of spoken language. *Psychological Monographs,* 1944, 56 (255), 41–74.

Marak, G. E. The evolution of leadership structure. *Sociometry,* 1964, 27, 174–82.

March, J. G. The power of power. In D. Easton. (Ed.), *Varieties of political theory,* Englewood Cliffs, N. J.: Prentice-Hall, 1966.

Maretzki, T. W., and Maretski, H. Taira: An Okinawan village. In B. Whiting (Ed.), *Six cultures: Studies of child rearing.* New York: Wiley, 1963.

Marks, I. M., and Gelder, M. G. Transvestism and fetishism: Clinical and psychological changes during faradic aversion. *British Journal of Psychiatry,* 1967, 119, 711–30.

Marquis, D. P. Learning in the neonate: The modification of behavior under three feeding schedules. *Journal of Experimental Psychology,* 1941, 29, 263–82.

Marquis, D. G., and Hilgard, E. R. Conditioned lid responses to light in dogs after removal of the visual cortex. *Journal of Comparative Psychology,* 1936, 22, 157–78.

Marshall, G. R. Toilet training of an autistic eight-year-old through conditioning therapy: A case report. *Behavior Research and Therapy,* 1966, 4, 242–45.

Martin, C. H., and Staats, A. W. Attitude conditioning and imitation. Unpublished manuscript, University of Hawaii 1973.

Martin, M., Burkholder, R., Rosenthal, T. L., Tharp, R. L., and Thorne, G. L. Programming behavior change and reintegration into school milieu of extreme adolescent deviates. *Behaviour Research and Therapy,* 1968, 6, 371–83.

Marx, K. The materialist conception of history. In P. Gardiner (Ed.), *Theories of history.* Glencoe, Ill.: Free Press, 1959.

Marx, M. H. Observation, discovery, confirmation, and theory building. In A. R. Gilgen (Ed.), *Contemporary scientific psychology.* New York: Academic Press, 1970.

Maslow, A. H. A dynamic theory of human motivation. *Psychological Review,* 1943, 50, 370–96.

Maslow, A. H. Higher and lower needs. *Journal of Psychology,* 1948, 25, 433–36.

Maslow, A. H. *Motivation and personality.* New York: Harper, 1954.

Maslow, A. H. *The psychology of language.* Chicago: Regnery, 1966.

Maslow, A. H., and Mittelmann, B. *Principles of abnormal psychology.* New York: Harper and Row, 1951.

Mason, D. J., and Bourne, L. E. Complexities oversimplified. *Contemporary Psychology,* 1964, 9, 466–68.

Mathis, B. C., Cotton, J. W., Sechrest, L. *Psychological foundations of education.* New York: Academic Press, 1970.

Mausner, B. The effect of one partner's success in a relevant task on the interaction of observer pairs. *Journal of Abnormal and Social Psychology,* 1954, 49, 557–60. (a)

Mausner, B. The effect of prior reinforcement on the interaction of observer pairs. *Journal of Abnormal and Social Psychology,* 1954, 49, 65–68. (b)

May, M. A. Experimentally acquired drives. *Journal of Experimental Psychology,* 1948, 38, 66–77.

McCandless, B. R. The effect of enriched educational experiences upon the growth of intelligence of very superior children. Unpublished master's thesis, University of Iowa, 1940.

McCarthy, D. Language development. In L. Carmichael (Ed.), *Manual of child psychology*. New York: Wiley, 1954.

McClelland, D. C., Atkinson, R. A., Clark, R., and Lowell, E. *The achievement motive*. New York: Appleton-Century-Crofts, 1955.

McConaghy, N. Penile volume change to moving pictures of male and female nudes in heterosexual and homosexual males. *Behaviour Research and Therapy*, 1967, 5, 43–48.

McCoy, N., and Zigler, E. Social reinforcer effectiveness as a function of the relationship between child and adult. *Journal of Personality and Social Psychology*, 1965, 1, 604–12.

McDonald, F. *We the people*. Chicago: University of Chicago Press, 1958.

McDonald, F. J. Behavior modification in teacher education. In C. E. Thoresen (Ed.), *Behavior modification in education: 72nd yearbook of the National Society for the Study of Education*. Chicago: University of Chicago Press, 1973.

McGeoch, J. A., and Irion, A. L. *The psychology of human learning*. (Rev. ed.) New York: Longmans, 1952.

McGinnies, E. Emotionality and perceptual defense. *Psychological Review*, 1949, 56, 244–51.

McGinnies, E., and Ferster, C. B. (Eds.). *The reinforcement of social behavior*. Boston: Houghton Mifflin, 1971.

McIntire, R. W. Stimulus/response: Parenthood training or mandatory birth control: Take your choice. *Psychology Today*, October 1973, 34–143.

McNamara, H. J., and Wike, E. L. The effects of irregular learning conditions upon the rate and permanence of learning. *Journal of Comparative and Physiological Psychology*, 1958, 51, 363–66.

McNeil, E. G. *The concept of human development*. Belmont, Calif.: Wadsworth, 1966.

McNeill, D. The capacity for the ontogenesis of grammar. In D. I. Slobin (Ed.), *The ontogenesis of grammar*. New York: Academic Press, 1971.

Mead, G. H. *Mind, self, and society: From the standpoint of a social behaviorist*. Chicago: University of Chicago Press, 1934.

Meichenbaum, D. H., Bowers, K. S., and Ross, R. R. Modification of classroom behavior of institutionalized female adolescent offenders. *Behaviour Research and Therapy*, 1968, 6, 343–53.

Meichenbaum, D. H., and Goodman, J. Training impulsive children to talk to themselves: A means of developing self-control. *Journal of Abnormal Psychology*, 1971, 77, 115–26.

Melzack, R., and Scott, T. H. The effects of early experience on the response to pain. *Journal of Comparative and Physiological Psychology*, 1957, 50, 155–61.

Merton, R. K. *Social theory and social structure*. Glencoe, Ill.: Free Press, 1957.

Messer, S. Implicit phonology in children. *Journal of Verbal Learning and Verbal Behavior*, 1967, 6, 609–13.

Metz, J. R. Conditioning generalized imitation in autistic children. *Journal of Experimental Child Psychology*, 1965, 2, 389–99.

Meyer, D. R., and Miles, R. C. Intralist inter-list relations in verbal learning. *Journal of Experimental Psychology*, 1953, 45, 109–15.

Meyerson, L., Kerr, N., and Michael, J. L. Behavior modification in rehabilitation. In S. W. Bijou and D. M. Baer (Eds.), *Child development: Readings in experimental analysis*. New York: Appleton-Century-Crofts, 1967.

Meyerson, L., Michael, J. L., Mowrer, O. H., Osgood, C. E., and Staats, A. W. Learning, behavior and rehabilitation. In L. Loftquist (Ed.), *Psychological research in rehabilita-*

*tion*. Washington: American Psychological Association, 1963.

Michael, J. *Laboratory studies in operant behavior*. New York: McGraw-Hill, 1963.

Michael, J. L. Consequences in the control of working behavior. Paper read at Joseph P. Kennedy, Jr. Foundation Fourth International Scientific Symposium on Mental Retardation, Chicago, 1968.

Michael, J. L. Rehabilitation. In C. Neuringer and J. L. Michael (Eds.), *Behavior modification in clinical psychology*. New York: Appleton-Century-Crofts, 1970.

Mill, J. S. In W. Ashley (Ed.), *Principles of political economy*. New York: Augustus M. Kelley, 1969. (a)

Mill, J. S. Utilitarianism. In W. T. Jones, F. Sontag, M. O. Beckner, and R. J. Fogelin (Eds.), *Approaches to ethics*. New York: McGraw-Hill, 1969. (b)

Miller, D. R. Motivation and affect. In P. H. Mussen (Ed.), *Handbook of research methods in child development*. New York: Wiley, 1960.

Miller, J. F., and Yoder, D. E. Teaching language to retardates: A format, not a receipt. In R. L. Schiefelbusch and L. L. Lloyd (Eds.), *Language perspectives: Acquisition, retardation, and intervention*. Baltimore, Md.: University Park Press, 1974.

Miller, N. E. The influence of past experience upon the transfer of subsequent training. Unpublished doctoral dissertation, Yale University, 1935.

Miller, N. E. Experimental studies of conflict. In J. McV. Hunt (Ed.), *Personality and the behavior disorders*. Vol. 1. New York: Ronald, 1944.

Miller, N. E. Studies of fear as an acquired drive: I. Fear as motivation and fear-reduction as reinforcement in the learning of new responses. *Journal of Experimental Psychology*, 1948, 38, 89–101.

Miller, N. E. Liberalization of basic S-R concepts: Extension to conflict

behavior, motivation and social learning. In S. Koch (Ed.), *Psychology: A study of a science*. Vol. 2. New York: McGraw-Hill, 1959.

Miller, N. E. Learning of visceral and glandular responses. *Science*, 1969, 163, 434–45.

Miller, N. E., and Banuazizi, A. Instrumental learning by curarized rats of a specific visceral response, intestinal or cardiac. *Journal of Comparative and Physiological Psychology*, 1968, 65, 1–7.

Miller, N. E., and Dollard, J. *Social learning and imitation*. New Haven, Conn.: Yale University Press, 1941.

Miller, S. A., Shelton, J., and Flavell, J. H. A test of Luria's hypothesis concerning the development of verbal self-regulation. *Child Development*, 1970, 41, 651–65.

Miller, W., and Ervin, S. The development of grammar in child language. In U. Bellugi and R. Brown (Eds.), *The acquisition of language*. *Monographs of the Society for Research in Child Development*, 1964, 29 (1, Whole No. 92).

Mills, C. W. *The power elite*. New York. Oxford University Press, 1957.

Mills, J., and Aronson, E. Opinion change as a function of the communicator's attractiveness and desire to influence. *Journal of Personality and Social Psychology*, 1965, 1, 173–77.

Minke, K. A., and Carlson, J. G. Fixed and ascending criteria for unit mastery learning. *Journal of Educational Psychology*, in press.

Mischel, W. *Personality and assessment*. New York: Wiley, 1968.

Mischel, W. Continuity and change in personality. *American Psychologist*, 1969, 24, 1012–18.

Mischel, W. *Introduction to personality*. New York: Holt, Rinehart and Winston, 1971.

Mischel, W. Direct versus indirect personality assessment: Evidence and implications. *Journal of Consulting and Clinical Psychology*, 1972, 38, 319–24.

Mischel, W. Toward a cognitive social learning reconceptualization of personality. *Psychological Review,* 1973, 80, 252–83.

Mitchell, H. E., and Byrne, D. The defendant's dilemma: Effects of jurors' attitudes and authoritarianism on judicial decisions. *Journal of Personality and Social Psychology,* 1973, 25, 123–29.

Molfese, D. L. Cerebral asymmetry in infants, children and adults: Auditory evoked responses to speech and noise stimuli. Unpublished doctoral dissertation, Pennsylvania State University, 1972.

Moore, N. Behaviour therapy in bronchial asthma: A controlled study. *Journal of Psychosomatic Research,* 1965, 9, 257–76.

Morgan, C. T., and Stellar, E. *Physiological psychology.* (2nd ed.) New York: McGraw-Hill, 1950.

Morrell, F. Electrophysiological contributions to the neural basis of learning. *Physiological Review,* 1961, 41, 443–94.

Morse, P. A. Infant speech perception: A preliminary model and review of the literature. In R. L. Schiefelbusch and L. L. Lloyd (Eds.), *Language perspectives: Acquisition, retardation, and intervention.* Baltimore, Md.: University Park Press, 1974.

Mowrer, O. H. On the dual nature of learning—a re-interpretation of "conditioning" and "problemsolving." *Harvard Educational Review,* 1947, 17, 102–48.

Mowrer, O. H. *Learning theory and personality dynamics.* New York: Ronald, 1950.

Mowrer, O. H. The autism theory of speech development and some clinical applications. *Journal of Speech and Hearing Disorders,* 1952, 17, 263–68.

Mowrer, O. H. The psychologist looks at language. *American Psychologist,* 1954, 9, 660–94.

Mowrer, O. H. *Learning theory and the symbolic processes.* New York: Wiley, 1960.

Mowrer, O. H., and Mowrer, W. M. Enuresis: A method for its study and treatment. *American Journal of Orthopsychiatry,* 1938, 8, 436–59.

Muehl, S. The effects of visual discrimination pretraining on learning to read a vocabulary list in kindergarten children. *Journal of Educational Psychology,* 1960, 51, 217–21.

Murphy, G., Murphy, L. B., and Newcomb, T. M. *Experimental social psychology.* New York: Harper, 1937.

Murphy, J. V., and Miller, R. E. Higher-order conditioning in the monkey. *Journal of General Psychology,* 1957, 56, 67–72.

Murray, H. A. *Explorations in personality.* New York: Oxford University Press, 1938.

Murstein, B. I. Physical attractiveness and marital choice. *Journal of Personality and Social Psychology,* 1972, 22, 8–12.

Mussen, P. H. (Ed.) *Handbook of research methods in child development.* New York: Wiley, 1960.

Mussen, P. H., and Jones, M. C. Self conceptions of early and late maturing boys. *Child Development,* 1957, 29, 61–67.

Mussen, P. H., and Parker, A. L. Mother nurturance and girls' incidental imitative learning. *Journal of Personality and Social Psychology,* 1965, 2, 94–97.

Neale, D. H. Behaviour therapy and encopresis in children. *Behaviour Research and Therapy,* 1963, 1, 139–49.

Neisworth, J. T., and Moore, F. Operant treatment of asthmatic responding with the parent as therapist. *Behavior Therapy,* 1972, 3, 95–99.

Nelson, K. Structure and strategy in learning to talk. *Monographs of the Society for Research in Child Development,* 1973, 38 (Whole No. 149).

Neurath, O. Physiacalism: The philosophy of the Viennese Circle. *Monist,* 1931, 41, 618–23.

Newcomb, T. M. *Social psychology.* New York: Holt, 1950.

Newman, H. H., Freeman, F. N., and Holzinger, K. J. *Twins: A study of heredity and environment.* Chicago: University of Chicago Press, 1937.

Nisbet, R. A. *Aspects of the Western theory of development.* New York: Oxford University Press, 1969.

Nisbett, R. E., Caputo, C., Legant, P., and Marecek, J. Behavior as seen by the actor and as seen by the observer. *Journal of Personality and Social Psychology,* 1973, 27, 154–64.

Nord, W. R. Exchange theory: An integrative approach to social conformity. *Psychological Bulletin,* 1969, 71, 174–208.

Nunnally, J. C., Duchnowski, A. J., and Parker, R. K. Association of neutral objects with rewards: Effect in verbal evaluation, reward expectancy, and selective attention. *Journal of Personality and Social Psychology,* 1965, 1, 270–74.

Nuthman, A. M. Conditioning of a response class on a personality test. *Journal of Abnormal and Social Psychology,* 1957, 54, 19–23.

Offenback, S. I. Bibliography of learning in children. Unpublished manuscript, Purdue University, 1966.

Olds, J. Physiological mechanisms of reward. In M. R. Jones (Ed.), *Nebraska symposium on motivation.* Lincoln: University of Nebraska Press, 1955.

Olds, J. Runway and maze behavior controlled by basomedial forebrain stimulation in the rat. *Journal of Comparative and Physiological Psychology,* 1956, 49, 507–12.

Olds, J. Satiation effects in self-stimulation of the brain. *Journal of Comparative and Physiological Psychology,* 1958, 51, 675–78.

Olds, J., and Milner, P. Positive reinforcement produced by electrical stimulation of the septal area and other regions of the rat brain. *Journal of Comparative and Physiological Psychology,* 1954, 47, 419–27.

O'Leary, K. D., and Becker, W. C. Behavior modification of an adjustment class: A token reinforcement program. *Exceptional Children,* 1967, 33, 637–42.

O'Leary, K. D., Becker, W. C., Evans, M. B., and Saudargas, R. A. A token reinforcement program in a public school: A replication and systematic analysis. *Journal of Applied Behavior Analysis,* 1969, 2, 3–13.

O'Leary, K. D., and Drabman, R. Token reinforcement programs in the classroom: A review. *Psychological Bulletin,* 1971, 75, 379–98.

O'Leary, K. D., and O'Leary, S. G. (Eds.) *Classroom management: The successful use of behavior modification.* London: Pergamon, 1972.

O'Leary, K. D., O'Leary, S., and Becker, W. C. Modification of a deviant sibling interaction pattern in the home. *Behaviour Research and Therapy,* 1967, 5, 113–20.

Olim, E. G. Maternal language styles and cognitive development of children. In F. Williams (Ed.), *Language and poverty: Perspectives on a theme.* Chicago: Markham, 1970.

Orlando, R., and Bijou, S. W. Single and multiple schedules of reinforcement in developmentally retarded children. *Journal of the Experimental Analysis of Behavior,* 1960, 4, 339–48.

Osborne, J. G. Free-time as a reinforcer in the management of classroom behavior. *Journal of Applied Behavior Analysis,* 1969, 2, 113–18.

Osgood, C. E. *Method and theory in experimental psychology.* New York: Oxford University Press, 1953.

Osgood, C. E., and Sebeok, T. A. Psycholinguistics. *Supplement to the Journal of Abnormal and Social Psychology,* 1954, 49, 1–203.

Osgood, C. E., Suci, G. J., and Tannenbaum, P. H. *The measurement of meaning.* Urbana: University of Illinois Press, 1957.

Otto, W. The acquisition and retention of paired associates by good, average, and poor readers. *Journal of Educational Psychology,* 1961, 52, 241–48.

Page, M. M. Social psychology of a classical conditioning of attitudes experiment. *Journal of Personality and Social Psychology*, 1969, 11, 177–86.

Paivio, A. *Imagery and verbal process.* New York: Holt, Rinehart and Winston, 1971.

Palermo, D. Language acquisition. In H. W. Reese and L. P. Lipsitt (Eds.), *Experimental child psychology.* New York: Academic Press, 1970.

Palermo, D. S., and Eberhart, V. L. On the learning of morphological rules: An experimental analogy. *Journal of Verbal Learning and Verbal Behavior*, 1968, 7, 337–44.

Palkes, H., Stewart, M., and Kahana, B. Porteus Maze performance of hyperactive boys after training in self-directed verbal commands. *Child Development*, 1968, 39, 817–25.

Pandey, J., and Griffitt, W. Attraction and helping. *Bulletin of Psychonomic Science*, 1974, 3, 123–24.

Papousek, H. A method of studying conditioned food reflexes in young children up to the age of six months. *Pavlov Journal of Higher Nervous Activity*, 1959, 9, 136–40.

Papousek, H. Conditioning during postnatal development. In Y. Brackbill and G. G. Thompson (Eds.), *Behavior in infancy and early childhood: A book of readings.* New York: Free Press, 1967. (a)

Papousek, H. Experimental studies of appetitional behavior in human newborns. In H. W. Stevenson, E. H. Hess, and H. L. Rheingold (Eds.), *Early behavior: Comparative and developmental approaches.* New York: Wiley, 1967. (b)

Parsons, T. *The social system.* Glencoe, Ill.: Free Press, 1951.

Parsons, T. *Societies: Evolutionary and comparative perspectives.* Englewood Cliffs, N. J.: Prentice-Hall, 1966.

Parton, D. A., and Fonts, G. T. Effects of stimulus-response similarity and dissimilarity on children's matching performance. *Journal of Experimental Child Psychology*, 1969, 8, 461–68.

Patterson, G. R. An application of conditioning techniques to the control of a hyperactive child. In L. P. Ullmann and L. Krasner (Eds.), *Case studies in behavior modification.* New York: Holt, Rinehart and Winston, 1965. (a)

Patterson, G. R. Responsiveness to social stimuli. In L. Krasner and L. P. Ullman (Eds.), *Research in behavior modification.* New York: Holt, Rinehart and Winston, 1965. (b)

Patterson, G. R. Prediction of victimization from an instrumental conditioning procedure. *Journal of Consulting Psychology*, 1967, 31, 147–52.

Patterson, G. R. Reprogramming the families of aggressive boys. In C. E. Thoresen (Ed.), *Behavior modification in education: 72nd Yearbook of the National Society for the Study of Education.* Chicago: University of Chicago Press, 1973.

Patterson, G. R., Jones, R., Whittier, J., and Wright, M. A. A behaviour modification technique for the hyperactive child. *Behaviour Research and Therapy*, 1965, 2, 217–26.

Patterson, G. R., Shaw, C. A., and Ebner, M. J. Teachers, peers and parents as agents of change in the classroom. In F. A. Benson (Ed.), *Modifying Deviant Social Behaviors in Various Classroom Settings.* Monograph No. 1, Department of Special Education, University of Oregon, 1969.

Paul, G. L. *Insight vs. desensitization in psychotherapy: An experiment in anxiety reduction.* Stanford, Calif.: Stanford University Press, 1966.

Paul, G. L. Behavior modification research: Design and tactics. In C. M. Franks (Ed.), *Behavior Therapy*: Appraisal and status. New York: McGraw-Hill, 1969.

Pavlov, I. P. *Conditioned reflexes.* Trans. G. V. Anrep. London: Oxford University Press, 1927.

Penfield, W., and Roberts, L. *Speech and brain mechanisms.* Princeton, N. J.: Princeton University Press, 1959.

Perin, C. T. Behavior potentiality as a joint function of the amount of training and degree of hunger at the time of extinction. *Journal of Experimental Psychology*, 1942, 30, 93–113.

Perin, C. T. A quantitative investigation of the delay-of-reinforcement gradient. *Journal of Experimental Psychology*, 1943, 32, 37–51.

Perkins, C. C., Jr. An analysis of the concept of reinforcement. *Psychological Review*, 1968, 75, 155–72.

Peters, C. C., and McElwee, A. R. Improving functioning intelligence by analytical training in a nursery school. *Elementary School Journal*, 1944, 45, 213–19.

Peterson, D. R. *The clinical study of social behavior*. New York: Appleton-Century-Crofts, 1968.

Pfaff, D. Parsimonious biological models of memory and reinforcement. *Psychological Review*, 1969, 76, 70–81.

Pfuderer, C. Some suggestions for a syntactic characterization of baby talk style. In The structure of linguistic input to children. Working Paper No. 14, Language-Behavior Research Laboratory, University of California, Berkeley, 1969.

Phares, E. J. Expectancy changes in skill and chance situations. *Journal of Abnormal and Social Psychology*, 1957, 54, 339–42.

Phillips, E. L., Phillips, E. A., Fixsen, D. S., and Wolf, M. M. Achievement place: Modification of the behaviors of pre-delinquent boys within a token economy. *Journal of Applied Behavioral Analysis*, 1971, 4, 45–59.

Phillips, L. W. Mediated verbal similarity as a determinant of the generalization of a conditioned GSR. *Journal of Experimental Psychology*, 1958, 55, 56–62.

Piaget, J. *The child's conception of number*. New York: Humanities, 1952.

Piaget, J. How children form mathematical concepts. *Scientific American*, 1953, 189, 74–79.

Piaget, J. Piaget's theory. In P. Mussen (Ed.), *Carmichael's manual of child psychology*. New York: Wiley, 1970.

Piaget, J., and Inhelder, B. *The early growth of logic in the child*. London: Routledge and Kegan Paul, 1964.

Pick, A. D. Improvement of visual and tactual form discrimination. *Journal of Experimental Psychology*, 1965, 69, 331–39.

Pihl, R. O., and Greenspoon, J. The effect of amount of reinforcement on the formation of the reinforcing value of a verbal stimulus. *Canadian Journal of Psychology*, 1969, 23, 219–26.

Pines, H. A. An attributional analysis of locus of control orientation and source of informational dependence. *Journal of Personality and Social Psychology*, 1973, 26, 262–72.

Popper, K. R. *Open-society and its enemies*. New York: Harper and Row, 1945.

Porier, G. W., and Lott, A. J. Galvanic skin responses and prejudice. *Journal of Personality and Social Psychology*, 1967, 5, 253–59.

Postman, L., Bruner, J. S., and McGinnies, E. Personal values as selective factors in perception. *Journal of Abnormal and Social Psychology*, 1948, 43, 142–54.

Postman, L., and Sassenrath, J. The automatic action of verbal rewards and punishments. *Journal of General Psychology*, 1961, 65, 109–36.

Premack, D., and Premack, A. J. Teaching language to apes and language-deficient persons. In R. L. Schiefelbusch and L. L. Lloyd (Eds.), *Language perspectives: Acquisition, retardation, and intervention*. Baltimore, Md.: University Park Press, 1974.

Premack, D., and Schwartz, A. Preparations for discussing behaviorism with chimpanzee. In F. Smith and G. A. Miller (Eds.), *The genesis of language*, Cambridge, Mass.: M.I.T. Press, 1966.

Pribram, K. H. The biologic of mind: Neurobehavioral foundations. In A. R. Gilgen (Ed.), *Contempo-*

*rary scientific psychology.* New York: Academic Press, 1970. (a)

Pribram, K. H. Feelings as monitors. In M. Arnold (Ed.), *Feelings and emotions.* New York: Academic Press, 1970. (b)

Pylyshyn, Z. W. What the mind's eye tells the mind's brain: A critique of mental imagery. *Psychological Bulletin,* 1973, 80, 1–24.

Quay, J. C., Sprague, R. L., Werry, J. S., and McQueen, N. M. Conditioning visual orientation of conduct problem children in the classroom. *Journal of Experimental Child Psychology,* 1967, 5, 512–17.

Rabinovitch, R. D. Dyslexia: Psychiatric considerations. In J. Money (Ed.), *Reading disability.* Baltimore, Md.: Johns Hopkins Press, 1962.

Rabinovitch, R. D., Drew, A. L., de Jong, R. N., Ingram, W., and Withey, L. A research approach to reading retardation. In R. McIntosh and C. Hare (Eds.), *Neurology and psychiatry in childhood: Proceedings of the Association for Research in Nervous and Mental Diseases.* Vol. 34. Baltimore, Md.: Williams and Wilkins, 1954.

Rachman, S., and Teasdale, J. D. Aversion therapy: An appraisal. In C. M. Franks (Ed.), *Behavior therapy: Appraisal and status.* New York: McGraw-Hill, 1969.

Rankin, R. E., and Campbell, D. T. Galvanic skin responses to Negro and white experimenters. *Journal of Abnormal and Social Psychology,* 1955, 51, 30–33.

Rasmussen, T., and Penfield, W. Further studies of the sensory and motor cerebral cortex in man. *Processes of the American Society of Experimental Biology,* 1947, 6, 452–560.

Raven, B. H., and French, J. R. P., Jr. Group support, legitimate power, and social influence. *Journal of Personality,* 1958, 26, 400–409.

Raymond, M. Case of fetishism treated by aversion therapy. *British Medical Journal,* 1956, 2, 854–57.

Razran, G. H. S. The observable unconscious and the inferable conscious in current Soviet psychophysiology: Interoceptive conditioning, semantic conditioning, and the orienting reflex. *Psychological Review,* 1961, 68, 81–147.

Razran, G. H. S. Studies in configural conditioning: VI. Comparative extinction and forgetting of pattern and of single-stimulus conditioning. *Journal of Experimental Psychology,* 1939, 24, 432–38.

Redfield, R. The primitive world view. *Proceedings of the American Philosophical Society,* 1952, 96, 30–36.

Reitz, W. E., and McDougall, L. Interest items as positive and negative reinforcements: Effects of social desirability and extremity of endorsement. *Psychonomic Science,* 1969, 17, 97–98.

Rescorla, R. A., and Lolordo, V. M. Inhibition of avoidance behavior. *Journal of Comparative and Physiological Psychology,* 1965, 59, 406–12.

Rescorla, R. A., and Solomon, R. L. Two-process learning theory: Relationships between Pavlovian conditioning and instrumental learning. *Psychological Review,* 1967, 74, 151–82.

Reynolds, G. S. An analysis of interactions in a multiple schedule. *Journal of the Experimental Analysis of Behavior,* 1961, 4, 107–17. (a)

Reynolds, G. S. Relativity of response rate and reinforcement frequency in a multiple schedule. *Journal of the Experimental Analysis of Behavior,* 1961, 4, 179–84. (b)

Rheingold, H. L., Gewirtz, J. L., and Ross, H. W. Social conditioning of vocalizations in the infant. *Journal of Comparative and Physiological Psychology,* 1959, 52, 68–73.

Rickard, H. C., Dignam, P. J., and Horner, R. F. Verbal manipulation in a psychotherapeutic relationship. *Journal of Clinical Psychology,* 1960.

Risley, T. R., and Wolf, M. M. Establishing functional speech in echolalic children. *Behaviour Research and Therapy,* 1961, 5, 73–88.

Rogers, C. R. Some observations on the organization of personality. *American Psychologist*, 1947, 2, 358–68.

Rogers, C. R. *Client-centered therapy*. Boston: Houghton Mifflin, 1951.

Rohrer, J. H. Experimental extinction as a function of the distribution of extinction trials and response strength. *Journal of Experimental Psychology*, 1947, 37, 473–93.

Rokeach, M. Attitude change and behavioral change. *Public Opinion Quarterly*, 1966, 30, 529–50.

Rokeach, M. and Klienjunas, P. Behavior as a function of attitude-toward-object and attitude-toward-situation. *Journal of Personality and Social Psychology*, 1972, 22, 194–201.

Rosen, B. C. The achievement syndrome: A psychocultural dimension of social stratification. *American Sociological Review*, 1956, 21, 203–11.

Rosen, M., and Wesner, C. A behavioral approach to Tourette's syndrome. *Journal of Consulting and Clinical Psychology*, 1973, 41, 308–12.

Rosen, M., Wesner, C., Richardson, P., and Clark, G. R. A preschool program for promoting language acquisition. *Hospital and Community Psychiatry*, 1971, 22, 280–82.

Rosen, M., Zisfein, L., and Hardy, M. The clinical application of behaviour modification techniques: Three case studies. *British Journal of Mental Subnormality*, 1972, 18, 1–8.

Rosen, S., Johnson, R. D., Johnson, M. J., and Tesser, A. Interactive effects of news valence and attraction on communicator behavior. *Journal of Personality and Social Psychology*, 1973, 28, 298–300.

Rosenthal, R., Archer, D., Koivumaki, J. H., DiMatteo, M. R., and Rogers, P. L. Assessing sensitivity to nonverbal communication: The PONS test. *Division 8 Newsletter*, January 1974.

Rotter, J. B. *Social learning and clinical psychology*. Englewood Cliffs, N. J.: Prentice-Hall, 1954.

Rotter, J. B., Liverant, S., and Crowne, D. P. The growth and extinction of expectancies in chance controlled and skilled tasks. *Journal of Psychology*, 1961, 52, 161–77.

Rouse, R. O. and Vernis, S. J. The effect of associative connections on the recognition thresholds for words. *Journal of Experimental Psychology*, 1952, 43, 195–206.

Routh, D. K. Conditioning of vocal response differentiation in infants. *Developmental Psychology*, 1969, 1, 219–26.

Ryback, D., and Staats, A. W. Parents as behavior therapy-technicians in treating reading deficits (dyslexia). *Journal of Behavior Therapy and Experimental Psychiatry*, 1970, 1, 109–19.

Rychlak, J. F. A *philosophy of science for personality theory*. Boston: Houghton Mifflin, 1968.

Sachs, D. H., and Byrne, D. Differential conditioning of evaluative responses to neutral stimuli through association with attitude statements. *Journal of Experimental Research in Personality*, 1970, 4, 181–85.

Sahlins, M. D. Evolution: Specific and general. In R. A. Manners and D. Kaplin (Eds.), *Theory in anthropology*. Chicago: Aldine, 1968.

Sailor, W. Reinforcement and generalization of productive plural allomorphs in two retarded children. *Journal of Applied Behavior Analysis*, 1971, 4, 305–10.

Sailor, W., and Tamar, T. Stimulus factors in the training of prepositional usage in three autistic children. *Journal of Applied Behavior Analysis*, 1972, 5, 183–90.

Salzinger, K. Experimental manipulation of verbal behavior: A review. *Journal of General Psychology*, 1959, 61, 65–94.

Salzinger, K. *Psychology: The Science of behavior*. New York: Springer, 1969.

Salzinger, K., Feldman, R. S., and Portnoy, S. Training parents of brain-injured children in the use of operant conditioning procedures. *Behavior Therapy*, 1970, 1, 4–32.

Samuels, S. J. Effect of experimentally learned word associations on the acquisition of reading responses.

*Journal of Educational Psychology,* 1966, 57, 159–63.

Samuels, S. J. An experimental program for teaching letter names of the alphabet. Project No. 9–F–009. Washington, D. C.: U. S. Office of Education, 1970.

Samuels, S. J. Attention and visual memory in reading acquisition. Research Report No. 26, 1971, University of Minnesota, Grant No. OE–09–332189–4533 (032), U. S. Office of Education. (a)

Samuels, S. J. Success and failure in learning to read: A critique of the research. Occasional Paper No. 9, 1971, University of Minnesota, Grant No. OE–09–332189–4533 (032), U. S. Office of Education. (b)

Samuels, S. J. Effect of distinctive feature training on paired-associate learning. *Journal of Educational Psychology,* 1973, 64, 164–70.

Samuelson, P. A. *Economics: An introductory analysis.* New York: McGraw-Hill, 1958.

Santayana, G. The sense of beauty. In K. Aschenbrenner and A. Isenberg (Eds.), *Aesthetic theories: Studies in the philosophy of art.* Englewood Cliffs, N. J.: Prentice-Hall, 1965.

Sarnoff, I. Psychoanalytic theory and social attitudes. *Public Opinion Quarterly,* 1960, 24, 251–79.

Schachter, S. The assumption of identity and peripheralist-centralist controversies in motivation and emotion. In M. B. Arnold (Ed.), *Feelings and emotions.* New York: Academic Press, 1970.

Schachter, S., and Singer, J. E. Cognitive, social, and physiological determinants of emotional state. *Psychological Review,* 1962, 69, 379–99.

Schaefer, H., and Martin, P. L. Behavioral therapy for "apathy" of hospitalized schizophrenics. *Psychological Reports,* 1966, 19, 1147–58.

Schimmel, S. Conditioned discrimination, number conception, and response inhibition in two- and three-year-old children. Unpublished doctoral dissertation, Wayne State University, 1971.

Schlosberg, H. The relationship between success and the laws of conditioned. *Psychological Review,* 1937, 44, 379–94.

Schlosberg, H. The description of facial expressions in terms of two dimensions. *Journal of Experimental Psychology,* 1952, 44, 229–37.

Schopenhauer, A. The world as will and idea. In W. T. Jones, F. Sontag, M. O. Beckner, and R. J. Fogelin (Eds.) *Approaches to ethics.* New York: McGraw-Hill, 1969.

Schopler, J., and Layton, B. Determinants of the self-attribution of having influenced another person. *Journal of Personality and Social Psychology,* 1972, 22, 326–32.

Scott, W. A. Attitude change through reward of verbal behavior. *Journal of Abnormal and Social Psychology,* 1957, 55, 72–75.

Sears, R. R., Hovland, C. I., and Miller, N. E. Minor studies of aggression: I. Measurement of aggressive behavior. *Journal of Psychology,* 1940, 9, 275–93.

Sears, R. R., and Wise, G. Relation of cup feeding in infancy to thumb-sucking and the oral drive. *American Journal of Orthopsychiatry,* 1950, 20, 123–38.

Secord, P. F., and Backman, C. W. Interpersonal congruency, perceived similarity and friendship. *Sociometry,* 1964, 27, 115–27.

Senf, G. M. Development of immediate memory for bisensory stimuli in normal children and children with learning disorders. *Developmental Psychology Monographs,* 1969, 1 (Whole No. 6).

Seward, J. P., and Seward, G. H. Studies on reproductive activities of the guinea pig: IV. A comparison of sex drive in males and females. *Journal of Genetic Psychology,* 1940, 57, 429–40.

Shapiro, M. M., Mugg, G. J., and Ewald, W. Instrumental preferences and conditioned preparatory responses in dogs. *Journal of Comparative and Physiological Psychology,* 1971, 74, 227–32.

Sheffield, F. D. Theoretical considerations in the learning of complex sequential tasks from demonstration and practice. In A. A. Lumsdaine (Ed.), *Student response in programmed instruction.* (Publication 943) Washington, D. C.: National Academy of Sciences—National Research Council, 1961.

Sherif, M. An experimental study of stereotypes. *Journal of Abnormal and Social Psychology,* 1935, 29, 371–75.

Sherman, J. A. Use of reinforcement and imitation to reinstate verbal behavior in mute psychotics. *Journal of Abnormal Psychology,* 1965, 70, 155–164.

Shipley, E., Smith, C., and Gleitman, L. A study in the acquisition of language: Free responses to commands. *Language,* 1969, 45, 322–42.

Shipley, W. C. An apparent transfer of conditioning. *Psychological Bulletin,* 1933, 30, 541.

Shoben, E. J. Psychotherapy as a problem in learning theory. *Psychological Bulletin,* 1949, 46, 366–92.

Shockley, W. Negro IQ deficit: Failure of a "malicious coincidence" model warrants new research proposals. *Review of Educational Research,* 1971, 41, 227–48.

Shontz, F. C. *Research methods in personality.* New York: Appleton-Century-Crofts, 1965.

Siegel, S. The physiology of conditioning. In M. H. Marx (Ed.), *Learning: Interactions.* London: Macmillan, 1970.

Sigall, H., and Aronson, E. Liking for an evaluator as a function of her physical attractiveness and nature of the evaluations. *Representative Research in Social Psychology,* 1971, 2, 19–25.

Sigall, H., and Landy, D. Radiating beauty: Effects of having a physically attractive partner on person perception. *Journal of Personality and Social Psychology,* 1973, 28, 218–24.

Sigall, H., Page, R., and Brown, A. C. Effort expenditure as a function of evaluation and evaluator attractiveness. *Representative Research in*

*Social Psychology,* 1971, 2, 19–25.

Silverstein, A. Secondary reinforcement in infants. *Journal of Experimental Child Psychology,* 1972, 13, 138–44.

Silverstein, A. Acquired pleasantness and conditioned incentives in verbal learning. In D. E. Berlyne and K. B. Madsen (Eds.), *Pleasure, reward, preference.* New York: Academic Press, 1973.

Siqueland, E. R., and Lipsitt, L. P. Conditioned head turning in new borns. *Journal of Experimental Child Psychology,* 1966, 356–76.

Siqueland, E. R. Reinforcement patterns and extinction in human newborns. *Journal of Experimental Child Psychology,* 1968, 6, 431–42.

Siqueland, E. R. Basic learning processes: I. Classical conditioning. In H. W. Reese and L. P. Lipsitt (Eds.), *Experimental child psychology.* New York: Academic Press, 1970.

Skiba, E. A., Pettigrew, L. E., and Alden, S. E. A behavioral approach to the control of thumbsucking in the classroom. *Journal of Applied Behavior Analysis,* 1971, 4, 121–29.

Skinner, B. F. Two types of conditioned reflex and a pseudo type. *Journal of Genetic Psychology,* 1935, 12, 66–77.

Skinner, B. F. *The behavior of organisms.* New York: Appleton, 1938.

Skinner, B. F. Are theories of learning necessary? *Psychological Review,* 1950, 57, 193–96.

Skinner, B. F. *Science and human behavior.* New York: Macmillan, 1953.

Skinner, B. F. *Verbal behavior.* New York: Appleton-Century-Crofts, 1957.

Skinner, B. F. *Cumulative record.* New York: Appleton, 1959.

Skinner, B. F. Philogeny and ontogeny of behavior. *Science,* 1966, 153, 1205–13.

Skinner, B. F. *Contingencies of reinforcement.* New York: Appleton-Century-Crofts, 1969.

Slobin, D. I. Imitation and grammatical development in children. In N. S.

Endler, L. B. Boultier, and H. Osser (Eds.), *Contemporary issues in developmental psychology*. New York: Holt, Rinehart and Winston, 1968.

Slobin, D. I. (Ed.) *The ontogenesis of grammar*. New York: Academic Press, 1971.

Slobin, D. I. Cognitive prerequisites for the development of grammar. In C. A. Ferguson and D. I. Slobin (Eds.), *Studies of child language development*. New York: Holt, Rinehart and Winston, 1973.

Smelser, N. J., and Smelser, W. T. Personality and social systems. New York: Wiley, 1963.

Smith, I. D. The effects of training procedures upon the acquisition of conservation of weight. *Child Development*, 1968, 39, 515–26.

Smith, M. B., Bruner, J. S., and White, R. W. *Opinions and personality*. New York: Wiley, 1956.

Smith, M. E. An investigation of the development of the sentence and the extent of vocabulary in young children. *University of Iowa Studies in Child Welfare*, 1926, 3 (No. 5).

Snyder, A., Mischel, W., and Lott, A. J. Value, information, and conformity behavior. *Journal of Personality*, 1960, 28, 333–41.

Snygg, D., and Combs, A. W. *Individual behavior*. New York: Harper, 1949.

Soffieti, J. P. Why children fail to read: A linguistic analysis. *Harvard Educational Review*, 1955, 25, 63–84.

Solarz, A. K. Latency of instrumental responses as a function of compatibility with the meaning of eliciting verbal signs. *Journal of Experimental Psychology*, 1960, 59, 239–45.

Solomon, R. L., and Wynne, L. C. Traumatic avoidance learning: The principles of anxiety conservation and partial irreversibility. *Psychological Review*, 1954, 61, 353–85.

Sopolsky, A. Effect of interpersonal relationship upon verbal conditioning. *Journal of Abnormal and Social Psychology*, 1960, 60, 241–46.

Sorokin, P. *Sociological theories of today*. New York: Harper and Row, 1966.

Spelt, D. K. The conditioning of the human fetus in utero. *Journal of Experimental Psychology*, 1948, 38, 375–76.

Spence, K. W. The nature of theory construction in contemporary psychology. *Psychological Review*, 1944, 51, 47–68.

Spence, K. W. *Behavior theory and conditioning*. New Haven, Conn.: Yale University Press, 1956.

Spence, K. W. *Behavior theory and learning*. Englewood Cliffs, N. J.: Prentice-Hall, 1960.

Spires, A. M. Subject-experimenter interaction in verbal conditioning. Unpublished doctoral dissertation, New York University, 1960.

Staats, A. W. A behavioristic study of verbal and instrumental response hierarchies and their relationship to human problem solving. Unpublished doctoral dissertation, University of California, Los Angeles, 1956.

Staats, A. W. Learning theory and "opposite speech." *Journal of Abnormal and Social Psychology*, 1957, 55, 268–69. (a)

Staats, A. W. Verbal and instrumental response hierarchies and their relationship to problem-solving. *American Journal of Psychology*, 1957, 70, 442–446 (b)

Staats, A. W. Verbal habit families, concepts, and the operant conditioning of word classes. Technical Report No. 10, August 1959, Arizona State University Contract Nonr 2794 (02), Office of Naval Research. (Republished in revised form: *Psychological Review*, 1961, 68, 190–204.)

Staats, A. W. (with contributions by C. K. Staats). *Complex human behavior*. New York: Holt, Rinehart and Winston, 1963.

Staats, A. W. Conditioned stimuli, conditioned reinforcers, and word meaning. In A. W. Staats (Ed.), *Human learning*. New York: Holt, Rinehart and Winston, 1964. (a)

Staats, A. W. (Ed.) *Human learning.* New York: Holt, Rinehart, and Winston, 1964. (b)

Staats, A. W. Outline of an integrated learning theory of attitude formation and function. In M. Fishbein (Ed.), *Attitude theory and measurement.* New York: Wiley, 1967.

Staats, A. W. Categories and underlying mental processes, or representative behavior samples and S-R analyses: Opposing heuristic strategies. *Ontario Journal of Educational Research,* 1968, 10, 187–201. (a)

Staats, A. W. A general apparatus for the investigation of complex learning in children. *Behaviour Research and Therapy,* 1968, 10, 187–201 (b)

Staats, A. W. *Learning, language, and cognition.* New York: Holt, Rinehart and Winston, 1968 (c)

Staats, A. W. Social behaviorism and human motivation: Principles of the attitude-reinforcer-discriminative system. In A. G. Greenwald, T. C. Brock, and T. M. Ostrom (Eds.), *Psychological foundations of attitudes.* New York: Academic Press, 1968. (d)

Staats, A. W. Development, use and social extensions of reinforcer motivational) systems in the solution of human problems. In J. T. Nagoshi and R. T. Omura (Eds.), *Progress in behavior modification: Programs and results:* Proceedings of Youth Development Center Conference, University of Hawaii, 1969. (a)

Staats, A. W. Experimental demand characteristics and the classical conditioning of attitudes. *Journal of Personality and Social Psychology,* 1969, 11, 187–92. (b)

Staats, A. W. A learning-behavior theory: A basis for unity in behavioral-social science. In A. R. Gilgen (Ed.), *Contemporary scientific psychology.* New York: Academic Press, 1970. (a)

Staats, A. W. Reinforcer systems in the solution of human problems. In G. A. Fargo, C. Behrns, and P. Nolen (Eds.), *Behavior modification in the classroom.* Belmont, Calif.: Wadsworth Publishing Co., 1970. (b)

Staats, A. W. Social behaviorism, human motivation, and the conditioning therapies. In B. A. Maher (Ed.), *Progress in experimental personality research.* (New York: Academic Press, 1970. (c)

Staats, A. W. *Child learning, intelligence, and personality.* New York: Harper and Row, 1971. (a)

Staats, A. W. Linguistic-mentalistic theory versus an explanatory S-R learning theory of language development. In D. I. Slobin (Ed.), *The ontogenesis of grammar.* New York: Academic Press, 1971. (b)

Staats, A. W. Language behavior therapy: A derivative of social behaviorism. *Behavior Therapy,* 1972, 3, 165–92.

Staats, A. W. Social behaviorism: A science of man with freedom and dignity. Invited address presented at the meetings of the American Psychological Association, September 1972.

Staats, A. W. Behavior analysis and token reinforcement in educational behavior modification and curriculum research. In C. E. Thoresen (Ed.), *Behavior modification in education: 72nd yearbook of the National Society for the Study of Education.* Chicago: University of Chicago Press, 1973.

Staats, A. W. Behaviorism and cognitive theory in the study of language: A neopsycholinguistics. In R. L. Schiefelbusch and L. L. Lloyd (Eds.), *Language perspectives: Acquisition, retardation, and intervention.* Baltimore, Md.: University Park Press, 1974.

Staats, A. W., Brewer, B. A., and Gross, M. C. Learning and cognitive development: Representative samples, cumulative-hierarchical learning, and experimental-longitudinal methods. *Monographs of the Society for Research in Child Development,* 1970, 35 (8, Whole No. 141).

Staats, A. W., and Butterfield, W. H. Treatment of nonreading in a culturally-deprived juvenile delinquent: An application of reinforcement principles. *Child Development*, 1965, 36, 925–42.

Staats, A. W., Finley, J. R., Minke, K. A., and Wolf, M. M. Reinforcement variables in the control of unit reading responses. *Journal of the Experimental Analysis of Behavior*, 1964, 7, 139–49.

Staats, A. W., Gross, M. C., Guay, P. F., and Carlson, C. C. Personality and social systems and attitude-reinforcer-discriminative theory: Interest (attitude) formation, function, and measurement. *Journal of Personality and Social Psychology*, 1973, 26, 251–61.

Staats, A. W., and Hammond, O. W. Natural words as physiological conditioned stimuli: Food-word-elicited salivation· and deprivation effects. *Journal of Experimental Psychology*, 1972, 96, 206–8.

Staats, A. W., and Higa, W. R. Effects of affect-loaded labels on interpersonal attitudes across ethnic groups. Unpublished manuscript, 1970.

Staats, A. W., Higa, W. R., and Reid, I. E. Names as reinforcers: The social value of verbal stimuli. Technical Report No. 9, 1970, University of Hawaii, Contract N00014–67–C–0387–0007, Office of Naval Research.

Staats, A. W., Minke, K. A., and Butts, P. A token-reinforcement remedial reading program administered by black instructional technicians to backward black children. *Behavior Therapy*, 1970, 1, 331–53.

Staats, A. W., Minke, K. A., Finley, J. R., Wolf, M. M., and Brooks, L. O. A reinforcer system and experimental procedure for the laboratory study of reading acquisition. *Child Development*, 1964, 35, 209–31.

Staats, A. W., Minke, K. A., Goodwin, W., and Landeen, J. Cognitive behavior modification: "Motivated learning" reading treatment with sub-professional therapy technicians. *Behaviour Research and Therapy*, 1967, 5, 283–99.

Staats, A. W., Minke, K. A., Martin, C. H., and Higa, W. R. Deprivation-satiation and strength of attitude conditioning: A test of attitude-reinforcer-discriminative theory. *Journal of Personality and Social Psychology*, 1972, 24, 178–85.

Staats, A. W., and Staats, C. K. Attitudes established by classical conditioning. *Journal of Abnormal and Social Psychology*, 1958, 57, 37–40.

Staats, A. W., and Staats, C. K. Effect of number of trials on the language conditioning of meaning. *Journal of General Psychology*, 1959, 61, 211–23.

Staats, A. W., and Staats, C. K. A comparison of the development of speech and reading behaviors with implications for research. *Child Development*, 1962, 33, 830–46.

Staats, A. W., Staats, C. K., and Biggs, D. A. Meaning of verbal stimuli changed by conditioning. *American Journal of Psychology*, 1958, 71, 429–31.

Staats, A. W., Staats, C. K., and Crawford, H. L. First-order conditioning of a GSR. *Journal of General Psychology*, 1962, 67, 159–67.

Staats, A. W., Staats, C. K., and Heard, W. G. Denotative meaning established by classical conditioning. *Journal of Experimental Psychology*, 1961, 61, 300–303.

Staats, A. W., Staats, C. K., Heard, W. G., and Finley, J. R. Operant conditioning of factor analystic personality traits. *Journal of General Psychology*, 1962, 66, 101–14.

Staats, A. W., Staats, C. K., Schutz, R. E., and Wolf, M. M. The conditioning of reading responses using "extrinsic" reinforcers. *Journal of the Experimental Analysis of Behavior*, 1962, 5, 33–40.

Staats, A. W., and Warren, D. R. Motivation and three-function learning: Deprivation-satiation and approach-avoidance to food words. *Journal of Experimental Psychology*, in press.

Staats, C. K., and Staats, A. W. Meaning established by classical conditioning. *Journal of Experimental Psychology*, 1957, 54, 74–80.

Staats, C. K., Staats, A. W., and Schutz, R. E. The effects of discrimination pretraining on textual behavior. *Journal of Educational Psychology*, 1962, 53, 32–37.

Stalling, R. B. Personality similarity and evaluative meaning as conditioners of attraction. *Journal of Personality and Social Psychology*, 1970, 14, 77–82.

Stanley-Jones, D. The biological origin of love and hate. In M. Arnold (Ed.), *Feelings and emotions*. New York: Academic Press, 1970.

Stang, D. J. Effect of interaction rate on ratings of leadership and liking. *Journal of Personality and Social Psychology*, 1973, 27, 405–8.

Stevens, S. S. Psychology and the science of science. *Psychological Bulletin*, 1939, 36, 221–63.

Stevens, S. S. Mathematics, measurement, and psychophysics. In Stevens, S. S. (Ed.), *Handbook of experimental psychology*. New York: Wiley, 1951.

Storms, M. D. Videotape and the attribution process: Reversing actors' and observers' points of view. *Journal of Personality and Social Psychology*, 1973, 27, 165–75.

Strodtbeck, F. L. Family interaction, values and achievement. In C. C. McClelland (Ed.), *Talent and society*. Princeton, N. J.: Van Nostrand, 1958.

Strong, E. K., Jr. *Vocational interests of men and women*. Stanford, Calif.: Stanford University Press, 1943.

Strong, E. K., Jr. *Vocational Interest Blank for men: Manual*. Stanford, Calif.: Stanford University Press, 1952.

Stuart, R. B. Behavioral control of overeating. *Behaviour Research and Therapy*, 1967, 5, 357–65.

Stuart, R. B. Operant-interpersonal treatment for marital discord. *Journal of Consulting and Clinical Psychology*, 1969, 33, 675–82.

Studdert-Kennedy, M., and Hadding-Koch, K. Auditory and linguistic processes in the perception of intonation contours. *Haskins Status Reports*, 1971, SR-27, 153–74.

Studdert-Kennedy, M., and Shankweiler, D. P. Hemispheric specialization for speech perception. *Journal of the Acoustical Society of America*, 1970, 48, 579–94.

Suinn, R. M. The desensitization of test-anxiety by group and individual treatment. *Behaviour Research and Therapy*, 1968, 6, 385–87.

Suinn, R. M. *Abnormal psychology*. New York: Wiley, 1970.

Sullivan, E. Acquisition of conservation through film modeling techniques. In D. W. Brision and E. Sullivan (Eds.), *Recent research on the acquisition of substance*. (Educational Research Series No. 27) Ontario: Ontario Institute for Studies in Education, 1967.

Super, D. E., and Crites, J. O. *Appraising vocational fitness*. New York: Harper and Brothers, 1962.

Szasz, T. S. *The myth of mental illness: Foundations of a theory of personal conduct*. New York: Hueber-Harper, 1961.

Talbot, E. The effect of note taking upon verbal responses and its implications for the interview. Unpublished doctor's dissertation, University of California, Los Angeles, 1954.

Taylor, P. W. *Problems of moral philosophy*. Belmont, Calif.: Dickenson, 1967.

Templin, M. C. *Certain language skills in children: Their development and interrelationships*. (Institute of Child Welfare Monograph Series, No. 26) Minneapolis: University of Minnesota Press, 1957.

Terman, L. M., and Merrill, M. A. *Measuring intelligence*. Boston: Boughton Mifflin, 1937.

Tharp, R. G., and Wetzel, F. J. *Behavior modification in the natural environment*. New York: Academic Press, 1969.

Thibaut, J. W., and Kelley, H. H. *The social psychology of groups*. New York: Wiley, 1959.

Thomas, W. I., and Znaniecki, F. *The Polish peasant in Europe and America*. (2nd ed.) New York: Knopf, 1927. 2 vols.

Thoresen, C. E. (Ed.) *Behavior modification in education: 72nd yearbook of the National Society for the Study of Education.* Chicago: University of Chicago Press, 1973. (a)

Thoresen, C. E. Behavioral humanism. In C. E. Thoresen (Ed.), *Behavior modification in education: 72nd yearbook of the National Society for the Study of Education.* Chicago: University of Chicago Press, 1973. (b)

Thoresen, C. E., and Hosford, R. E. Behavioral approaches to counseling. In C. E. Thoresen (Ed.), *Behavior modification in education: 72nd yearbook of the National Society for the Study of Education.* Chicago: University of Chicago Press, 1973.

Thoresen, C. E., and Mahoney, M. J. *Behavioral self-control.* New York: Holt, Rinehart and Winston, 1974.

Thorndike, E. L. *The elements of psychology.* New York: Seiler, 1905.

Thorndike, E. L. The fundamentals of learning. New York: Columbia University, 1932.

Thorndike, E. L. *Adult interests.* New York: Macmillan, 1935.

Thurstone, L. L. The measurement of social attitudes. *Journal of Abnormal and Social Psychology,* 1931, 26, 249–69.

Thurstone, L. L. *Multiple factor analysis: A development and expansion of the vectors of the mind.* Chicago: University of Chicago Press, 1947.

Tolman, E. C. *Purposive behavior in animals and men.* New York: Appleton-Century, 1932.

Tolman, E. C. An operational analysis of "demands." *Erkenntnis,* 1936, 6, 383–90.

Tolman, E. C. The determiners of behavior at a choice-point. *Psychological Review,* 1938, 45, 1–41.

Tooley, J. T., and Pratt, S. An experimental procedure for the extinction of smoking behavior. *Psychological Record,* 1967, 17, 209.

Toynbee, A. J. History. In R. W. Livingstone (Ed.), *The legacy of Greece.* Oxford: Clarendon Press, 1921.

Toynbee, A. J. *A study of history* (abridged D. C. Somervell). New York: Oxford University Press, 1947.

Trapold, M. A., and Overmier, J. B. The second learning process in instrumental learning. In A. H. Black and W. F. Prokasy (Eds.), *Classical conditioning, II.* New York: Appleton-Century-Crofts, 1972.

Trapold, M. A., and Winokur, S. W. Transfer from classical conditioning and extinction to acquisition, extinction, and stimulus generalization of a positively reinforced instrumental response. *Journal of Experimental Psychology,* 1967, 73, 517–25.

Trowill, J. A. Instrumental conditioning of the heartrate in the curarized rat. *Journal of Comparative and Physiological Psychology,* 1967, 63, 7–11.

Tryon, W. W. A reply to Staats' language behavior therapy: A derivative of social behaviorism. *Behavior Therapy,* 1974, 5, 273–76.

Turner, R. K., Young, G. C., and Rachman, S. Treatment of nocturnal enuresis by conditioning techniques. *Behavior Research and Therapy,* 1970, 8, 367–81.

Tuttle, H. S. Creating motives. *Journal of General Psychology,* 1940, 23, 17–29.

Tyler, V. O., Jr., and Brown, G. D. Token reinforcement of academic performance with institutionalized delinquent boys. *Journal of Educational Psychology,* 1968, 59, 164–68.

Ullmann, L. P., and Krasner, L. *A psychological approach to abnormal behavior.* Englewood Cliffs, N. J.: Prentice-Hall, 1969.

Ulmer, M. J. *Economics: Theory and practice.* Boston: Houghton Mifflin, 1959.

Ulrich, R. E., and Azrin, N. H. Reflexive fighting in response to aversive stimulation. *Journal of the Experimental Analysis of Behavior,* 1962, 5, 511–20.

Underwood, B. J., and Schultz, R. W. *Meaningfulness and verbal learning.* Chicago: Lippincott, 1960

Vaughan, G. M., and Mangan, G. L., Conformity to group pressure in relation to the value of the task material. *Journal of Abnormal and Social Psychology*, 1963, 66, 179–83.

Vernon, P. E. *Personality assessment: A critical survey.* New York: Wiley, 1964.

Verplanck, W. S. Burrhus F. Skinner. In W. K. Estes, S. Koch, K. Mac-Corquodale, P. E. Meehl, C. G. Mueller, W. N. Schoenfeld, and W. S. Verplanck, *Modern learning theory.* New York: Appleton-Century-Crofts, 1954.

Verplanck, W. S. The control of the content of conversation: Reinforcement of statements of opinion. *Journal of Abnormal and Social Psychology*, 1955, 51, 668–76.

Wahler, R. G. Behavior therapy with oppositional children: Attempts to increase their parents' reinforcement value. Paper presented at the meetings of the Southeastern Psychological Association, Atlanta, April 1967.

Wahler, R. G. Setting generality: Some specific and general effects of child behavior therapy. *Journal of Applied Behavior Analysis*, 1969, 2, 239–46.

Wahler, R. G., and Erickson, M. Child behavior therapy: A community program in Appalachia. *Behavior Research and Therapy*, 1969, 7, 71–78.

Wahler, R. G., Winkel, G. H., Peterson, R. F., and Morrison, D. C. Mothers as behavior therapists for their own children. *Behaviour Research and Therapy*, 1965, 3, 113–23.

Walder, L. O. Teaching parents to modify the behaviors of their autistic children. Paper presented at the meetings of the American Psychological Association, New York, September 1966.

Walker, H. M., Mattson, R. H., and Buckley, N. K. The functional analysis of behavior within an experimental class setting. In A. M. Benson (Ed.), *Modifying deviant social behaviors in various classroom*

settings. (Monograph 1) Eugene: University of Oregon, Department of Special Education, College of Education, 1969.

Wallace, A. F. C. The new culture-and-personality. In T. Gladwin and W. C. Sturtevant (Eds.), *Anthropology and human behavior.* Washington, D. C.: Anthropological Society of Washington, 1962.

Wallace, W. L. Overview of contemporary sociological theory. In W. L. Wallace (Ed.), *Sociological theory.* Chicago: Aldine, 1969. (a)

Wallace, W. L. *Sociological theory.* Chicago: Aldine, 1969. (b)

Walster, E., Aronson, V., Abrahams, D., and Rottman, L. The importance of physical attractiveness in dating behavior. *Journal of Personality and Social Psychology*, 1966, 4, 508–16.

Walters, R. H., and Brown, M. Studies of reinforcement of aggression: III. Transfer of responses to an interpersonal situation. *Child Development*, 1963, 34, 563–71.

Wang, M. C., Resnick, L. B., and Boozer, R. The sequence of development of some early mathematics behaviors. *Child Development*, 1971, 42, 1767–778.

Warner, R. W. Alienated students: Six months after receiving behavioral group counseling. *Journal of Counseling Psychology*, 1971, 18, 426–30.

Warner, R. W., and Hansen, J. C. Verbal-reinforcement and model-reinforcement group counseling with alienated students. *Journal of Counseling Psychology*, 1970, 17, 1968–72.

Warren, A. B., and Brown, R. H. Conditioned operant response phenomena in children. *Journal of General Psychology*, 1943, 38, 181–207.

Warren, H. C. (Ed.) *Dictionary of psychology.* Boston: Houghton Mifflin, 1934.

Watson, D. L., and Tharp, R. G. *Self-directed behavior: Self-modification for personal adjustment.* Monterey, Calif.: Brooks/Cole, 1972.

Watson, J. B. *Behaviorism*. (Rev. ed.) Chicago: University of Chicago Press, 1930.

Watson, J. B., and Rayner, R. Conditioned emotional reactions. *Journal of Experimental Psychology*, 1920, 3, 1–14.

Wechsler, D. *The measurement of adult intelligence*. Baltimore, Md.: Williams and Wilkins, 1944.

Weiner, B., Heckhausen, H., Meyer, W., and Cook, R. E. Causal ascriptions and achievement behavior: A conceptual analysis of effort and re-analysis of locus of control. *Journal of Personality and Social Psychology*, 1972, 21, 239–48.

Weiner, B., and Kukla, A. An attributional analysis of achievement motivation. *Journal of Personality and Social ·Psychology*, 1970, 15, 1–20.

Weiss, R. F. Persuasion and the acquisition of attitudes: Models from conditioning and selective learning. *Psychological Reports*, 1962, 11, 709–32.

Weiss, R. F. An extension of Hullian learning theory to persuasive communication. In A. G. Greenwald, T. C. Brock, and T. M. Ostrom (Eds.), *Psychological foundations of attitudes*. New York: Academic Press, 1968.

Weiss, R. F. The drive theory of social facilitation. *Psychological Review*, 1971, 78, 44–57.

Weiss, R. F., Boyer, J. L., Lombardo, J. P., and Stich, M. H. Altruistic drive and altruistic reinforcement. *Journal of Personality and Social Psychology*, 1973, 25, 390–400.

Weiss, R. F., Buchanan, W., Alstatt, L., and Lombardo, J. P. Altruism is rewarding. *Science*, 1971, 171, 1262–63.

Weiss, R. L. Operant conditioning techniques in psychological assessment. In P. McReynolds (Ed.), *Advances in psychological assessment*. Palo Alto, Calif.: Science and Behavior Books, 1968.

Weiss, S. J. Stimulus compounding in free-operant and classical conditioning: A review and analysis. *Psychological Bulletin*, 1972, 78, 189–208.

Wenzel, B. M. Changes in heart rate associated with responses based on positive and negative reinforcement. *Journal of Comparative and Physiological Psychology*, 1961, 54, 638–44.

Westie, F. R., and DeFleur, M. L. Antonomic responses and their relationship to race attitudes. *Journal of Abnormal and Social Psychology*, 1959, 58, 340–47.

Wetzel, R. J., Baker, J., Roney, M., and Martin, M. Out-patient treatment of autistic behavior. *Behavior Research and Therapy*, 1966, 4, 169–77.

White, C. T., and Schlosberg, H. Degree of conditioning of the GSR as a function of the period of delay. *Journal of Experimental Psychology*, 1952, 43, 357–62.

White, M. G. Historical explanation. *Mind*, 1943, 52, 212–29.

White, R. (in collaboration with Fallaci, C.). The dead body and the living brain. *Look*, November 28, 1967.

White, R. W. Motivation reconsidered: The concept of competence. *Psychological Review*, 1954, 66, 297–333.

White, S. H. Bibliography: Psychological studies of learning in children. Unpublished manuscript, Harvard University, 1962.

Whitehurst, G. J. Production of novel and grammatical utterances by young children. *Journal of Experimental Child Psychology*, 1972, 13, 502–15.

Whitehurst, G. J., Novak, G., and Zorn, G. A. Delayed speech studied in the home. *Developmental Psychology*, 1972, 7, 169–77.

Whitlock, C. Note on reading acquisition: An extension of laboratory principles. *Journal of Experimental Child Psychology*, 1966, 3, 83–85.

Whitlock, C., and Bushell, D. Some effects of "back-up" reinforcers on reading behavior. *Journal of Experimental Child Psychology*, 1967, 5, 50–57.

Whorf, B. L. Science and linguistics. In J. B. Carroll (Ed.), *Language*, *thought and reality.* Cambridge, Mass.: Wiley, 1956.

Wicker, A. W. Attitudes versus actions: The relationship of verbal and overt behavioral responses to attitude objects. *Journal of Social Issues,* 1969, 25, 41–78.

Wilkins, W. Desensitization: Social and cognitive factors underlying the effectiveness of Wolpe's procedure. Psychological Bulletin, 1971, 76, 311–17.

Williams, C. D. The elimination of tantrum behavior by extinction procedures. *Journal of Abnormal and Social Psychology,* 1959, 59, 269.

Wilson, F. S., and Walters, R. H. Modification of speech output of near-mute schizophrenics through social-learning procedures. *Behaviour Research and Therapy,* 1966, 4, 59–67.

Winkler, R. C. The relevance of economic theory and technology to token reinforcement systems. *Behavior Research and Therapy,* 1971, 9, 81–88.

Wolf, M. M., Giles, D. K., and Hall, V. R. Experiments with token reinforcement in a remedial classroom. *Behaviour Research and Therapy,* 1968, 6, 51–64.

Wolf, M. M., Risley, T., and Mees, H. Application of operant conditioning procedures to the behaviour problems of an autistic child. *Behaviour Research and Therapy,* 1964, 1, 305–12.

Wolfe, J. B. The effect of delayed reward upon learning in the white rat. *Journal of Comparative Psychology,* 1934, 17, 1–21.

Wolpe, J. *Psychotherapy by reciprocal inhibition.* Stanford, Calif.: Stanford University Press, 1958.

Wolpe, J., and Lang, P. J. A Fear Survey Schedule for use in behaviour therapy. *Behaviour Research and Therapy,* 1964, 2, 27–30.

Wolpe, J., and Lazarus, A. A. *Behavior therapy techniques: A guide to the treatment of neuroses.* London: Pergamon, 1966.

Wood, C. C., Goff, W. P., and Day, R. S. Auditory evoked potentials during speech perception. Science, 1971, 173, 1248–251.

Woodworth, R. S. Heredity and environment: A critical survey of recently published material on twins and foster children. *Social Science Research Council Bulletin,* 1941 (Whole No. 47).

Woodworth, R. S., and Schlossberg, H. *Experimental psychology.* (Rev. ed.) New York: Holt, Rinehart and Winston, 1954.

Wulfeck, W. W. Learning the two-hand coordination test. *Journal of Applied Psychology,* 1942, 26, 41–49.

Wynne, L. C., and Solomon, R. L. Traumatic avoidance learning: Acquisition and extinction in dogs deprived of normal peripheral autonomic function. *Genetic and Psychological Monographs,* 1955, 52, 240–84.

Yakovleva, S. V. Certain features of the process of formation of voluntary behavior in preschool-age children. In *Problems of higher nervous activity in the normal and abnormal child.* Vol. 2. Moscow: Izd. Adad. Pedag. Nauk RSFSR, 1958.

Yarrow, L. J. Interviewing children. In P. H. Mussen (Ed.), *Handbook of research methods in child development.* New York: Wiley, 1960.

Yates, A. J. Symptoms and symptom substitution. *Psychological Review,* 1958, 65, 371–74.

Young, P. T. *Motivation of behavior.* New York: Wiley, 1936.

Young, P. T. Affective arousal: Some implications. *American Psychologist,* 1967, 22, 32–40.

Zander, A. F. A study of experimental frustration. *Psychological Monographs,* 1944, 56 (Whole No. 256).

Zanna, M. P. On inferring one's beliefs from one's behavior in a low-choice setting. *Journal of Personality and Social Psychology,* 1973, 26, 386–94.

Zanna, M. P., Kiesler, C. A., and Pilkonis, P. A. Positive and negative attitudinal affect established by classical conditioning. *Journal of*

*Personality and Social Psychology,* 1970, 14, 321–28.

Zeaman, D., and House, B. J. The role of attention in retardate discrimination learning. In N. R. Ellis (Ed.), *Handbook of mental deficiency.* New York: McGraw-Hill, 1963.

Zelazo, P. R., Zelazo, N. A., and Kolb, S. "Walking" in the newborn. *Science,* 1972, 176, 314–15.

Zeiberger, J., Sampen, S., and Sloane, H. Modification of a child's problem behaviors in the home with the mother as therapist. *Journal of Ap-*plied Behavior Analysis,* 1968, 1, 47–53.

Zimbardo, P. G. The cognitive control of motivation. Glenview, Ill.: Scott-Foresman, 1969.

Zimmerman, D. W. Durable secondary reinforcement: Method and theory. *Psychological Review,* 1957, 64, 373–83.

Zimmerman, E. H., and Zimmerman, J. The alteration of behavior in a special classroom situation. *Journal of Experimental Analysis of Behavior,* 1962, 5, 59–60.

# Indexes

# Name index

# Subject index

*This book has been set in 10 and 9 point
Electra, leaded 2 points. Part numbers and
titles are in 24 point Scotch Roman; chapter
numbers are in 30 point Bernhardt italic, and
chapter titles are in 18 point Scotch Roman
italic. The size of the type page is 27 x 45½
picas.*